T0406109

A Geoinformatics Approach to Water Erosion

Tal Svoray

A Geoinformatics Approach to Water Erosion

Soil Loss and Beyond

Tal Svoray
Geography and Environmental
Development
Ben-Gurion University of the Negev
Beer-Sheva, Israel

ISBN 978-3-030-91535-3 ISBN 978-3-030-91536-0 (eBook)
https://doi.org/10.1007/978-3-030-91536-0

This Springer imprint is published by the registered company Springer Nature Switzerland AG
The registered company address is: Gewerbestrasse 11, 6330 Cham, Switzerland

For my family: Yona, Yuval, Yotam, Lior & Nevo
"mo chuisle mo chroi"

Preface

Degradation of agricultural catchments due to water erosion is a major environmental threat at the global scale, with long-lasting destructive consequences valued at tens of billions of USD per annum. Eroded soils lead to reduced crop yields and deprived agroecosystem's functioning through, for example, decreased water holding capacity, poor aeration, scarce microbial activity, and loose soil structure. These types of malfunctions can result in reduced carbon sequestration, limited nutrient cycling, contamination of water bodies due to eutrophication, and low protection from floods. Furthermore, eroded fields have low recreational values, and they provide weak therapeutic experience (Ulrich 1984).

The destruction to societies caused by soil erosion was documented in inspirational books such as *Collapse* by Jared Diamond (2005), Dirt by David Montgomery (2007), and *Earth Matters: How Soil Underlies Civilization*, by Richard D. Bardgett (2016). These authors show how misuse of soils causes reduced food production that ultimately may lead to starvation and community collapse. The trap is simple: Poor soil management leads to extreme soil degradation which depletes agricultural production, and causes scarcity to persist for many decades. One key action that is needed to shift society away from this adverse cycle onto a fruitful path is to promote an educated soil management policy across wide regions (Banwart 2011).

The crucial role of spatial information technology in improving soil management—through monitoring, data processing, and quantifying water erosion—was already recognized by previous authors such as Guertin and Goodrich (2010). Geoinformatics is a modern scientific discipline that integrates all spatial information technologies—including remote sensing, spatial data science modeling, geostatistics, and spatial decision-support systems. As such, it offers promising paradigms, theories, and techniques for analyzing water erosion processes and forms and it plays a crucial role in the earth sciences, in general, and in the soil sciences and water erosion studies, in particular.

With the rapid rise in the use of geoinformatics, the time is ripe to design and implement new tools for quantifying water erosion processes, and for analyzing climatic, human-induced and environmental conditions, in a spatially and temporally explicit manner. Remotely sensed data from drones, airplanes, and satellites may be used to extract two and three dimensional information on catchment characteristics, and channeled and sheet erosion. Big-data analysis and machine-learning methodologies provide the means to identify areas under threat, to map soil health over wide regions, to prioritize expensive soil conservation solutions, to quantify soil movement, and to make educated decisions for agricultural soils management.

The Challenges

The recent progress in geoinformatics has made it possible to tackle four weighty challenges in the extraction of supportive spatial information for studying water erosion. First: *Data collection and storage*. Data collection on environmental factors, human activities, and soil and vegetation properties, at the hillslope and catchment scale, by means of field visits, is complex, expensive, and labor-intensive. There is a burning need, therefore, to use rapid and extensive methods, that exploit various parts of the electromagnetic spectrum, in conjunction with GIS capabilities, to collect, store, retrieve, and analyze the information necessary for studying and preventing water erosion. Second: *Analysis of interacting real-world states or input variables*. Water erosion processes are affected by surface heterogeneity that is governed largely by intrinsic and extrinsic factors. It is difficult to unravel the inter-related effects of all these factors through traditional field-mapping and classical statistics alone. Thus, to estimate water erosion risks at the hillslope and catchment scale, we must be able to disentangle these effects in a spatially and temporally explicit manner, and to predict their combined effect on water erosion. Third: *Quantifying soil and agroecosystem consequences after erosion events*. Mapping soil health over large areas requires expanding information from point measurements to areal maps. Hence, soil health studies at the regional scale are still very limited (Lehmann et al. 2020), and geoinformatics methods are needed to allow interpolation of the status of the remaining soil. Fourth: *The difficulty in allocating a specific soil conservation solution to a given soil degradation risk*. If soil conservationists wish to mitigate the consequences of water erosion, they must be able to suggest the most suitable solutions to the problem at hand, among the many available, while taking into account surface variance.

The Questions

In tackling these comprehensive and complex four challenges, the framework suggested in this book attempts to discuss a whole range of questions, starting with the very basic questions on the extent of the phenomena studied: What is the actual effect of soil erosion on our lives? What locations on earth suffer from water erosion consequences more than other locations? How much does water erosion cost us every year? Is soil erosion a recent problem due to aggressive modern agrotechnology, or is it a long-standing problem since the days of our ancestors? How can we counteract the process, and encourage farmers to collaborate with national efforts to prevent soil erosion? What are the main causes of accelerated water erosion from agricultural catchments? What geoinformatics data models can be used to quantify spatial variation at the hillslope and catchment scales? What is the appropriate sample size and spatial distribution of soil samples for estimating soil properties and the health of the remaining soil after an erosion event? What are the soil and vegetation characteristics relevant to water erosion in agricultural catchments, and how can they be mapped through remote sensing techniques? Do we need more data, or more expert knowledge, to analyze water erosion risks? How can we map soil health at the hillslope and catchment scales? Are soil properties continuous in space? How can we prioritize soil conservation policy to reduce water erosion?

These are but a few of the questions that are discussed in this book. Broadly speaking, by attempting to answer them, this book demonstrates how cutting edge geoinformatics paradigms can help reduce soil degradation due to water erosion.

Book Structure

The aim of this book is to demonstrate how geoinformatics and data science algorithms may be used to monitor and regulate consequences of accelerated water erosion, by drawing on sources of remotely sensed data. Accordingly, it does not set out to review all existing geoinformatics methods for water erosion studies, but rather to offer simple and robust methods for tackling water erosion challenges with a relatively high degree of certainty.

The book is divided into four parts (Fig. 1): (1) Background: The "story so far" with regard to water erosion in different parts of the world, and the modeling approaches used to study the governing processes [Chaps. 1–3]; (2) Data: Spatial data models used in geoinformatics to store and model measurements of water erosion and surface characteristics [Chap. 4]; (3) Indicators of soil erosion risks and consequences, and those of soil health [Chaps. 5–7]; and (4) Solutions: The spatial decision-support systems used to allocate resources and suggest solutions to the consequences of water erosion [Chap. 8]. Finally, Chap. 9 summarizes the book's findings.

Chapter 1 reviews and discusses soil variation, soil terminology, water erosion processes, and on-site/off-site consequences of water erosion and their impact on agroecosystems functioning. Chapter 2 is all about the uniqueness of agricultural catchments, the water erosion processes and the environmental conditions they operate in, and the possible contributions farmers can make to soil conservation, through pro-environmental behavior. Chapter 3 reviews principles and models of water erosion and discusses the mutual use of models, geoinformatics, and data science techniques to study water erosion. The use of spatially explicit erosion models is tightly coupled with geoinformatics, and has a bearing on all topics examined in this book including erosion risks, soil health, and soil conservation actions. Although soil erosion models use geoinformatics data for explicit representations, their aim is to simulate the erosion process, and to estimate water and sediment yields. Conversely, geoinformatics tools are aimed at representing operational aspects of curbing soil erosion. In that regard, geoinformatics methods make it possible to extend the approach of quantifying water erosion from the

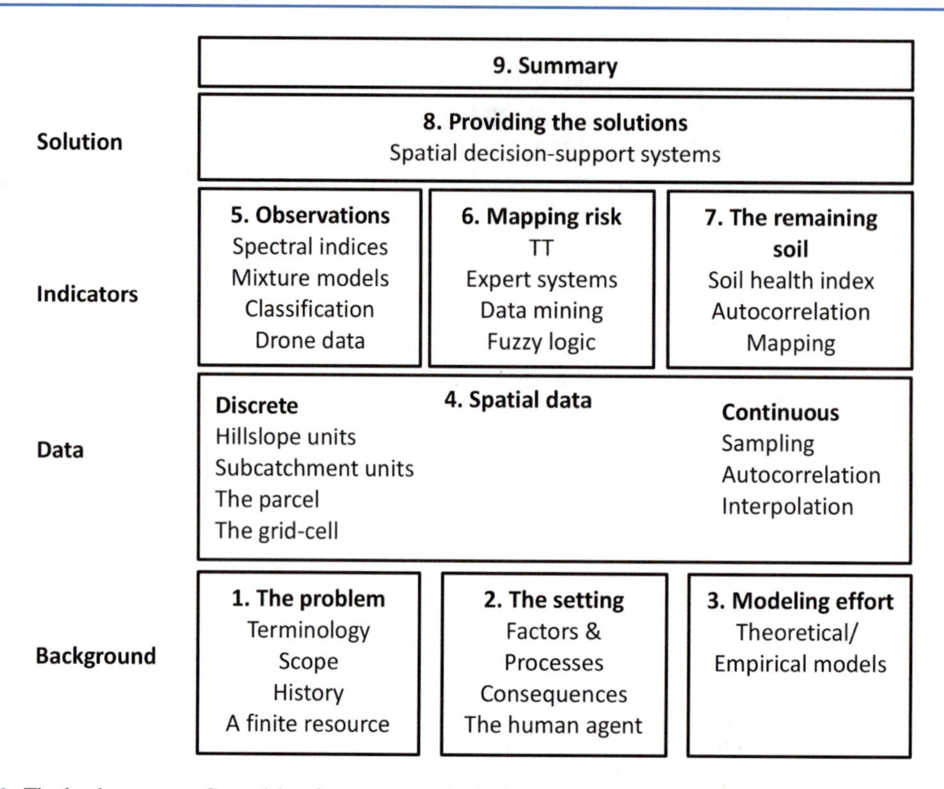

Fig. 1 The book structure. Comprising four parts, the book describes the workflow of using geoinformatics to provide information on water erosion from the scientific background, to the decision-making stage

limited scope of soil budgets (loss/deposition) to the more comprehensive examination of soil health. The multifaceted use of erosion models and geoinformatics including data modeling, parameterization, validation, and heuristic simulation is described in Fig. 2.

Chapter 4 describes both the discrete and continuous spatial data models used in geoinformatics for analyzing water erosion systems at the hillslope and catchment scales. Chapter 5 revisits earth observation techniques with particular focus on spectral indices, soft and hard classification, and change-detection techniques. Chapter 6 is perhaps the most comprehensive one, as it discusses a highly contentious topic in the data-science realm: Data-driven methodologies versus expert-based systems. To this end, it provides a comparison between topographic threshold, analytical hierarchal processes, data mining approaches, and space-time dynamic fuzzy logic as estimators of water erosion risks. Chapter 7 charts a new direction for the notorious problem of mapping soil health over large areas including sampling, autocorrelation analyses, and interpolation techniques. It also suggests the soil health index as an indicator of deterioration/rehabilitation of an agrosystem, beyond mere loss of soil. In Chap. 8, the use of spatial decision-support tools is reviewed, to provide the reader with tools for choosing the most suitable course of action in the treatment of agricultural catchments, with a view to providing a range of ecosystem services.

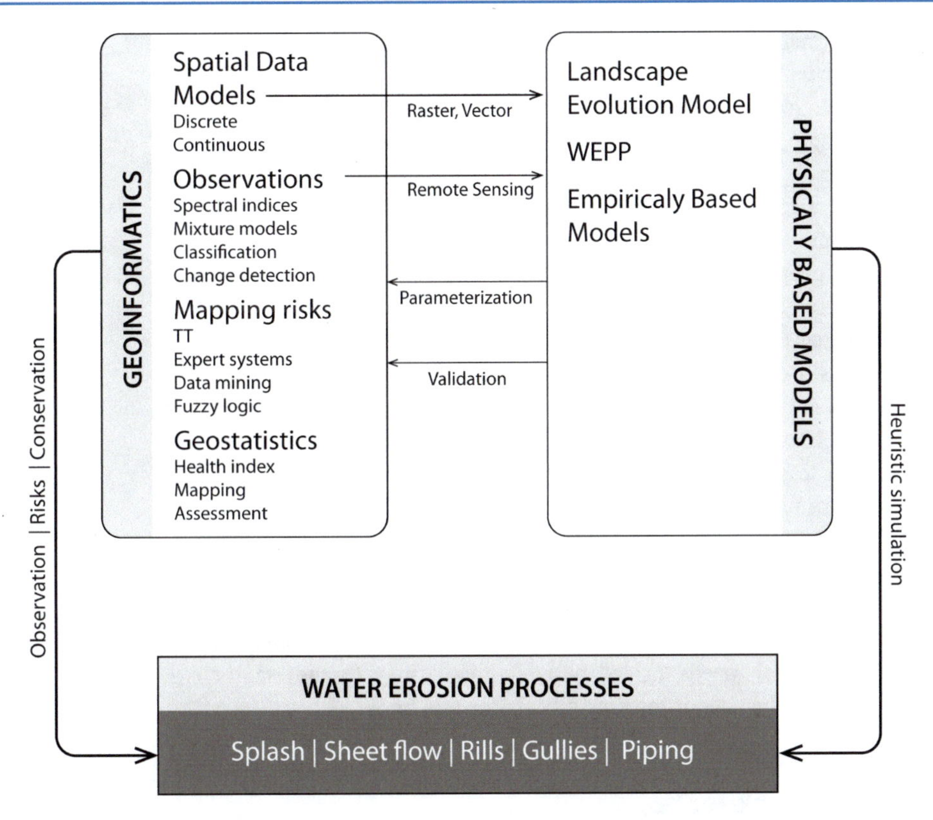

Fig. 2 The integrated use of geoinformatics and models in the study of water erosion. The arrows in the center part signify uses of geoinformatics sources and data models as the spatial data infrastructure of erosion modeling, and the use of models for parameterizing and validating geoinformatics tools. The upper boxes represent the methods reviewed in this book

Such a spatially and temporally explicit study of water erosion and soil degradation is fundamental to the earth sciences (especially the soil sciences), as it seeks to explore in climatic, biological, topographic, parent material characteristics, as well as those of human activity, small-scale and historical processes. As such, this book may contribute to the ongoing quest of addressing three major gaps in water erosion studies (Poesen 2018):

1. Understanding natural and anthropogenic soil erosion processes, and their interactions;
2. Providing tools for upscaling soil erosion processes and rates, in space and time;
3. Providing innovative techniques and strategies for mitigating erosion rates.

At a more practical level, the book includes numerous examples of published datasets from the Harod catchment in northern Israel, as a supersite (see Sect. 1.1.3.4 for details). The Harod supersite is indeed used for standardization purposes but the algorithms described in this book can be applied also to other semiarid agricultural catchments, and practically to any other agroecosystem that suffers water erosion. Well-known examples are the Bradano catchment in southern Italy (Aiello et al. 2015), the Loess Plateau of China (Jin et al. 2021), the Burdekin catchment in Australia (Bartley et al. 2007) and the agroecosystems of Walnut Gulch Experimental Watershed in Arizona (Goodrich et al. 2021). However, any other agricultural catchment that is mapped with spatial and temporal data can fit with necessary adaptations to climatic and landuse conditions.

In addition, every chapter ends with review questions for providing the reader with exercises related to the algorithms discussed. The book also provides strategies to answer these questions, but, it does not provide the full answers. This is done to encourage the reader to make herself the full quest for the answers in order to reach the insights from the book by exploring the possible research avenues inside the different chapters.

Last but not least, while this book highlights methods developed at the Geographic Information Laboratory of Ben-Gurion University, many other methods described here were developed by other authors in various other places around the world and were published in the scientific literature in other contexts. Those author's names are cited alongside their algorithms as is customary but I wish to stress here that I am indebted to all those who identified the research needs in the past, so that I may "stand on their shoulders" and use their algorithms as a basis for this book.

Whom is this Book for?

Because water erosion is a major environmental threat, it attracts the attention of many researchers who study the nature of the problem, the mechanisms and processes behind it, its on-site and off-site consequences, and possible solutions. The geoinformatics framework described in this book may be of interest to academics and students in the soil sciences, agricultural engineering, earth sciences, environmental and agrosystem management, and geoinformatics. The book may also be useful to practitioners including soil conservationists and drainage engineers, nature conservation agents, range managers, farmers, policymakers and decision makers in the public service, and anyone else who wishes to use geoinformation to minimize soil degradation in agricultural catchments.

To make use of this book, the reader will need GIS and remote sensing skills, and to be able to work with simple algebraic expressions and basic statistics. In preparing the examples provided in this book, in most cases we used *R* but the algorithms presented can be also executed with other spatial tools or GIS software.

Looking Ahead

This book illustrates ideas that can be used for implementing geoinformatics algorithms to curb water erosion processes. Exemplification is done through case studies from the Harod catchment in northern Israel and by providing synthetic explanations involving toy models. Despite occasional pitfalls, possible uncertainties and various generalizations and simplifications, these algorithms can be suitable for real world problems. The increased availability and the plummeting costs of spatial data, coupled with growing computing capabilities, may allow to apply these algorithms to different catchments of interest. With the expanding use of geoinformatics worldwide, a book such as this one can be used to provide the essential geoinformatics toolbox for exploring and solving water erosion problems in agricultural catchments.

Beer-Sheva, Israel Tal Svoray

References

Aiello A, Adamo M, Canora F (2015) Remote sensing and GIS to assess soil erosion with RUSLE3D and USPED at river basin scale in southern Italy. Catena 131:174–185. https://doi.org/10.1016/j.catena.2015.04.003

Banwart S (2011) Save our soils. Nature 474(7350):151–152. 10.1038/474151a

Bardgett R (2016) Earth Matters. Oup Oxford, GB

Bartley R, Hawdon A, Post DA, Roth CH (2007) A sediment budget for a grazed semi-arid catchment in the Burdekin basin, Australia. Geomorphology 87(4): 302–321. https://doi.org/10.1016/j.geomorph.2006.10.001

Diamond JM (2005) Collapse: How Societies Choose to Fail or Survive. Centro de Investigaciones Sociológicas, london

Goodrich DC, Heilman P, Nearing M, Nichols M, Scott RL, Williams C J, Biederman J (2021) The USDA-Agricultural Research Service's Long Term Agro-ecosystems Walnut Gulch Experimental Watershed (WGEW), Arizona, USA. Hydrological Processes. https://doi.org/10.1002/hyp.14349

Guertin DP, Goodrich DC (2010) The future role of information technology in erosion modelling. In: Morgan R, Nearing MA (eds) Handbook of Erosion Modelling John Wiley & Sons, Ltd, Chichester, UK, p 1–4

Jin F, Yang W, Fu J, Li Z (2021) Effects of vegetation and climate on the changes of soil erosion in the Loess Plateau of China. Science of The Total Environment 773: 145514. https://doi.org/10.1016/j.scitotenv.2021.145514

Lehmann J, Bossio DA, Kögel- Knabner I et al (2020) The concept and future prospectsof soil health. Nature Reviews Earth & Environment 1:554–563

Montgomery DR (2007) Soil erosion and agricultural sustainability. Proceedings of the National Academy of Sciences 104(33). 10.1073/pnas.0611508104

Poesen J (2018) Soil erosion in the Anthropocene: Research needs. Earth Surface Processes and Landforms 43(1):64–84. 10.1002/esp.4250

Ulrich RS (1984) View through a window may influence recovery from surgery. Science 224(4647):420–421

Acknowledgments

The water erosion geoinformatics approach described in this book evolved over fifteen years of research that I conducted with graduate students and colleagues at the Geographic Information Laboratory at Ben-Gurion University of the Negev. Without this ongoing collaboration, I would have never been able to complete this book. Among others, who are acknowledged below, particular appreciation is reserved for my friend and colleague Rami Zaidenberg, who first introduced me to the consequences of water erosion in agriculturally important areas.

Special thanks are extended to the excellent team that I worked with closely on the publication of this book. Roni Blushtein-Livnon, Head of the Geoinformatics Unit of our Department, read every chapter of the manuscript, prepared many of the figures, and provided valuable comments on the geoinformatics computations and algorithms presented in the book. Michael Dorman—my former Ph.D. student, and now a colleague—provided significant comments on Chaps. 4 and 6, and programmed many of the R codes used in this book. Finally, my research assistant, Gili Mizrachi, edited the manuscript with care and punctuality. I enjoyed every moment working with these three extremely talented individuals.

I wish to express gratitude to my graduate students who worked with me throughout the years on the accumulated research presented in this book. First and foremost, Sagy Cohen developed FuDSEM, and subsequently worked with me on various SEM projects. Sagy also provided valuable comments on Chap. 3, and is now an associate professor working on global erosion models at the University of Alabama in Tuscaloosa. Other former students who worked with me on water erosion geoinformatics include: Shlomo Dabora, Yael Storz-Peretz, Hila Markovitch, Orr Gevili, Shimon Ben-Said, Roei Levi, Inbar Hassid, David Hoober, Hodaya Bithan-Guedj, Dori Katz, Galia Barshad, Arthur Khozin, Ariel Nahalieli, Vladislav Dubinin, and Din Danino. All now hold various positions in the geoinformatics world.

I also wish to acknowledge several others who helped in various other ways. Yair Neuman, Golan Shahar, and Raz Jelinek of BGU and Dov Rupp of Israel Aerospace Industries were very instrumental in encouraging me to write this book during this endeavor. Jonathan Orr-Stav, the English Editor of this book, provided valuable proofreading, and my son, Yotam Svoray, provided important interpretations of some of the more complex equations. Oren Ackerman is thanked for his comments on Chap. 1 and for our collaborations along the years. I thank my student Eran Grossman for the photo of the modern plow and his help in fieldwork, and Eng. Benny Yaacobi—Deputy Director of the Soil Conservation and Drainage Division of the Israeli Ministry of Agriculture and Rural Development—for discussions in the field and the provision of several photos that I used in this book. I am indebted to Harold van Es, who opened up the world of soil health to me while I was on sabbatical at his lab at Cornell University in 2009, and to Peter Atkinson of Lancaster University, who first acquainted me with geostatistics. Shmuel Assouline inspired me during our many research collaborations on runoff and infiltration modeling, and Gil Eshel is thanked for leading the huge research project on soil health in Israel. The expertise of all these people is gratefully acknowledged.

Funding for the research that this book is based-on came from the Ben-Gurion University of the Negev, the Harry Levy Chair in Geography, the Soil Conservation and Drainage Division of the Israeli Ministry of Agriculture and Rural Development, the Chief Scientist of the Israeli Ministry of Agriculture and Rural Development (as part of the Centre of Excellence's study project of Water erosion as a soil-loss determinant: Dynamic models and avenues to minimize damage), and the Britain-Israel Research and Academic Exchange Partnership (BIRAX) Fund.

The drone data in Chap. 5 was provided by Eli Argaman—then the Academic Manager of the Soil Erosion Research Station at the Israeli Ministry of Agriculture. The meteorologic radar data was processed by Efrat Morin's Lab at the Hebrew University in Jerusalem during a previous collaboration of ours. I also thank the Venμs project and the Israeli Ministry of Science and Technology for the various Venμs images used in this book, and the USGS for: The Shuttle Radar Topography Mission (SRTM) elevation data, the LANDSAT 8 images and The Advanced Spaceborne Thermal Emission and Reflection Radiometer (ASTER) data. Thanks are also extended to the farmers of the Harod catchment and to the Soil Conservation and Drainage Division, who provided me with valuable data and information that helped executing many of the procedures described in this book. The support of my colleagues at the Ben-Gurion University of the Negev—my professional home over the past twenty years—and especially that of my Department Chair Izhak Katra, and colleagues Pua Bar and Itai Kloog, is much appreciated. The help of the staff at SPRINGER-NATURE—particularly that of Doris Bleier—was invaluable.

The first draft of this book was written in 2019, during an excellent sabbatical year at Bob Gifford's laboratory at the Department of Psychology of the University of Victoria, B.C. This is also how the section on how the Dragons of Inactions relate to the pro-environmental behavior of farmers was born.

Finally, and mostly, I wish to thank my best friend and wife, Yona Ore-Svoray, for her constant and unstinting support, and for laying the groundwork for the bibliographic database of the literature cited in this book.

It would be great if this book was error-free—but no book is. If there are any errors in this book, they are, of course, solely my responsibility. If you spot any, I do hope that you will write to me about it, as well as about any comment, observation or suggestion related to any idea presented in this book.

Beer-Sheva, Israel
September 2021

Tal Svoray
tsvoray@bgu.ac.il

Contents

1 Soil Erosion: The General Problem 1
1.1 The Soil and Erosion 1
1.1.1 The Soil Layer 1
1.1.2 Extrinsic and Intrinsic Factors 4
1.1.3 Basic Terms in Soil Erosion Studies 9
1.2 Scope of Soil Erosion 15
1.2.1 The Monetary Cost 15
1.2.2 Geographical Extent 16
1.2.3 Implications for Food Supply 18
1.3 A Brief History of Soil Loss 19
1.3.1 Hunters-Gatherers 21
1.3.2 Agricultural Societies 22
1.3.3 Modern Farmers 25
1.4 Soil as a Finite Resource 27
1.5 Summary 30
References 34

2 The Case of Agricultural Catchments 39
2.1 Erosion Factors in a Distinct Landform 41
2.1.1 Human Factors 41
2.1.2 Environmental Factors 48
2.2 Processes and Forms of Water Erosion in Agricultural Catchments 51
2.2.1 Splash Erosion 52
2.2.2 Sheet Erosion 53
2.2.3 Rill Erosion 53
2.2.4 Gully Erosion 54
2.2.5 Piping Erosion 55
2.3 The Damage: On-Site and Off-Site Consequences 56
2.3.1 On-Site Consequences 57
2.3.2 Off-Site Consequences 59
2.4 The Human Agent 61
2.4.1 Agricultural Soils as Social Traps 61
2.4.2 The Psychological Barriers 62
2.4.3 Evidence for Farmer's Conservation Actions 65
2.5 Summary 67
References 70

3 Modeling the Erosion Process . . . 75
3.1 Basics of a Hillslope—The Regolith Profile . . . 77
3.2 Landscape Evolution Models . . . 79
3.2.1 Background . . . 79
3.2.2 The CAESAR-Lisflood . . . 81
3.2.3 The Model Operation . . . 82
3.2.4 The Model Output . . . 85
3.3 Water Erosion Prediction Project (WEPP) . . . 86
3.3.1 Background . . . 86
3.3.2 Model Operation . . . 87
3.3.3 The Model Output . . . 90
3.4 Morgan–Morgan–Finney (MMF) . . . 93
3.4.1 Background . . . 93
3.4.2 Model Operation . . . 94
3.4.3 Further Development and Model Application . . . 97
3.5 Summary . . . 100
References . . . 103

4 Spatial Variation in Soils . . . 107
4.1 Discrete Spatial Units . . . 108
4.1.1 The Topographic Approach . . . 109
4.1.2 The Field-Based Approach . . . 121
4.1.3 The Image Pixel Approach . . . 123
4.2 A Suite of Continuous Variables . . . 126
4.2.1 Spatial Sampling . . . 127
4.2.2 Autocorrelation in Soil Properties . . . 131
4.2.3 Interpolation of Soil Properties . . . 137
4.3 Summary . . . 144
References . . . 147

5 Earth Observations . . . 151
5.1 Spectral Indices: Spectral Signatures, and Algebraic Expressions . . . 153
5.1.1 Vegetation Indices . . . 157
5.1.2 Soil Indices . . . 159
5.2 Image Classification Techniques . . . 161
5.2.1 Hard Classification . . . 162
5.2.2 Soft Classification—Spectral Mixture Modeling . . . 172
5.3 Synergy of RS Data in Catchment Models . . . 177
5.3.1 Background . . . 177
5.3.2 The Procedure . . . 178
5.3.3 Results . . . 189
5.4 Drone Remote Sensing . . . 191
5.4.1 DEM Extraction . . . 191
5.5 Summary . . . 196
References . . . 200

6 Assessments of Erosion Risk . . . 205
6.1 Topographic Threshold . . . 207
6.1.1 The Approach . . . 207
6.1.2 Procedure . . . 208
6.1.3 Results and Discussion . . . 210
6.2 Expert-Based Systems . . . 216
6.2.1 The Approach . . . 216
6.2.2 Weighting . . . 217
6.2.3 Decision Rules . . . 220
6.2.4 Procedure for Estimating Risk Levels . . . 223
6.2.5 Simulations . . . 225
6.2.6 Results and Discussion . . . 226
6.3 Data Mining (DM) . . . 233
6.3.1 Theoretical Basis . . . 236
6.3.2 Procedures . . . 241
6.3.3 Results and Discussion . . . 242
6.3.4 Summary . . . 244
6.4 Fuzzy Logic . . . 246
6.4.1 Theoretical Background . . . 247
6.4.2 Procedure . . . 250
6.4.3 Results and Discussion . . . 255
6.4.4 Summary . . . 258
6.5 Summary . . . 258
References . . . 260

7 The Health of the Remaining Soil . . . 265
7.1 Soil Health Indicators . . . 268
7.1.1 Selection of Soil Properties . . . 269
7.1.2 Scoring Functions . . . 272
7.1.3 Composite Soil Health Index . . . 274
7.1.4 Summary . . . 276
7.2 Autocorrelation in Space . . . 276
7.2.1 Spatial Variation in Soil Properties . . . 277
7.2.2 Autocorrelation of Soil Properties in the Harod Catchment—Procedure . . . 279
7.2.3 Autocorrelation of Soil Properties of the Harod—Results . . . 280
7.2.4 Summary . . . 283
7.3 Soil Health Maps . . . 283
7.3.1 Procedure . . . 284
7.3.2 Results and Discussion . . . 284
7.3.3 Summary . . . 295
7.4 Summary . . . 297
References . . . 300

8 Spatial Decision Support Systems . . . 305
8.1 Introduction . . . 306
8.1.1 The Decision-Making Process . . . 306
8.1.2 Basic Terms in SDSS . . . 307
8.1.3 The Single-Criterion System . . . 308
8.2 SDSS in Soil Conservation . . . 310
8.2.1 Applications of SDSS in Water Erosion Studies . . . 310
8.2.2 Beyond Soil Loss . . . 311
8.3 MCDM . . . 313
8.3.1 Background . . . 313
8.3.2 Procedure . . . 314
8.3.3 Decision Rules . . . 315
8.4 GISCAME . . . 320
8.4.1 Background . . . 320
8.4.2 Application of GISCAME to the Harod Catchment . . . 324
8.5 Summary . . . 328
References . . . 331

9 Final Thoughts . . . 335
9.1 The General Framework . . . 335
9.2 Usage of Geoinformatics to Study Water Erosion . . . 337
9.2.1 Known Knowns . . . 338
9.2.2 Known Unknowns . . . 339
9.2.3 Unknown Knowns . . . 340
9.2.4 Unknown Unknowns . . . 341
9.3 Summary . . . 342
References . . . 342

Index . . . 345

1 Soil Erosion: The General Problem

Abstract

Soils are thin heterogenous layers covering the Earth's surface. Space-time variation in soil properties depends on environmental, climatic, and human-induced factors, as well as on intrinsic factors/pedogenic processes that occur at the local scale and past events that are "etched" in the soil's memory. Water erosion processes can degrade soil health to the point of no return—which may result in reduced yields and threaten food supply. The monetary cost of soil degradation globally is on a scale of tens of billions of USD per year—with some continents suffering more physical and economic consequences than others. Soil degradation due to water erosion is not a new or modern phenomenon—humans have degraded soil health since the days of the hunters-gatherers—but it has increased dramatically since the dawn of the Agricultural Revolution, and has only been mildly kept in check by modern farmers. A major problem with water erosion is that the rate of soil loss is an order of magnitude larger than the rate of soil production. Under such conditions, once the soil is lost from a field, it will almost certainly never be restored.

Keywords

CLORPT · Economic threat · Extrinsic factors · History of water erosion · Intrinsic factors · Irreversible process · Soil

1.1 The Soil and Erosion

1.1.1 The Soil Layer

Earth is the largest of the rocky planets orbiting the Sun and is composed of layers. The innermost core is a dense spheroid covered by a viscous and hot outer core. The outer core is wrapped by a thick mantle, which in turn is covered by the viscous asthenosphere layer sealed by a solid crust layer (Anderson 1989). The continental crust—comprising molten magmatic rocks, metamorphic rocks shaped by heat or pressure, and sedimentary rocks that have accreted in water bodies or on the land due to the weathering of ingenious rocks—is wrapped by an irregular, granular, and much thinner layer—the soil. The soil is usually structured in layers parallel to the Earth crust, denoted as soil horizons O, A, E, B, and C, which differ in their physical, chemical, and biological properties (Fig. 1.1).

T. Svoray, *A Geoinformatics Approach to Water Erosion*,
https://doi.org/10.1007/978-3-030-91536-0_1

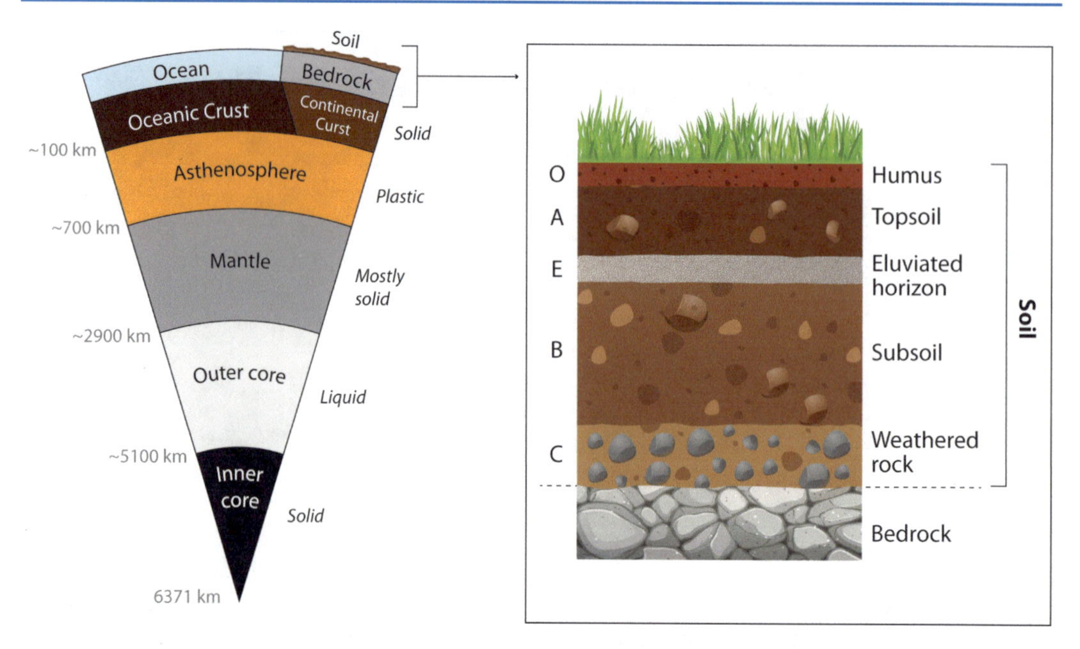

Fig. 1.1 The soil on Earth. The left panel (adapted from USGS 1999) illustrates the spherical layers of the Earth including the very thin soil layer (not to scale). The right panel shows a conceptual sketch of the soil horizons. The soil horizons—composing the soil profile—usually retain two general contrasting trends: (1) Concentration of organic matter decreases from the topsoil to the bedrock and (2) Rock fractions proportions increase in the same direction

The uppermost soil layer below the vegetation strata is the organic residue of humus (O horizon)—which is a collection of organic compounds, usually formed by animals, and mainly plants necromass. The soil becomes less fertile in the lower horizons: Horizon A (the topsoil) is where most of the biological processes in the soil occur and Horizon E is a mineral horizon with the main feature of eluvial loss of heavy metals, clay, and silicate while the remaining layer is composed of sand and silt particles (Hartemink et al. 2020). Horizon B (the subsoil) is marked by a decrease in biological activity, and comprises iron oxides and clay minerals that were accumulated by weathering. Horizon C—the weathered bedrock below—consists mainly of rock fractions. Thus, in the upper layers, the soil is a mix of organic matter and minerals, and it is the only part of the terrestrial crust that is biologically active. The thickness of the entire soil profile, from the plant roots to bedrock, varies between $\sim$0 m, in the most eroded deserts, and $\sim$6 m, in the most stable high plateaus of the world (Richter and Markewitz 1995). However, the boundary between the bedrock and the soil layer is not a clear line: The interface between the two is indeterminate, and often it is difficult to say where one ends and the other begins.

The soil layer has attracted the attention of humans for thousands of years—particularly since the agricultural revolution approximately 12,000 years ago. But, as we learn more about soils, science poses new questions about their significance, function, and characterization. One example of the adjustment of soil science paradigms to the new knowledge acquired over generations of studies is the recent debate over the meaning of the fundamental term soil, between the members of the Soil Science Society of America (SSSA).

For many years, until recently, according to the Glossary of Soil Science Terms by the SSSA, soil was defined:

> Either using the short version, (1) "The unconsolidated mineral or organic material on the immediate surface of the Earth that serves as a

natural medium for the growth of land plants."; or by using the more comprehensive definition, (2) "The unconsolidated mineral or organic matter on the surface of the Earth that has been subjected to—and shows effects of—genetic and *environmental factors* of: *Climate* (including water and temperature effects) and macro- and microorganisms, conditioned by relief, acting on parent material over a period of time. A product-soil differs from the material from which it is derived in many physical, chemical, biological, and morphological properties and characteristics.

Although these two definitions appear to be very logical and have served the soil community for many years, they have recently been found to be insufficient for modern soil science requirements, in light of advancements in the field. Accordingly, the definition of soil has been reassessed by the SSSA members in a bid to establish a more accurate description. In particular, the members found the old two definitions to be deficient on four major counts (van Es 2017): (1) they refer to the soils on Planet Earth only; (2) some soils have been found to be consolidated; (3) the definitions pertain to the solid component of the soil layer, when, in fact, soils are multiphase systems that include liquid and vaporous phases as well; and (4) the comprehensive definition is primarily a summary of soil formation processes. Consequently, the SSSA members agreed on the following alternative definition (van Es 2017):

> Soil is "The layer(s) of generally loose mineral and/or organic material that are affected by physical, chemical, and/or biological processes at or near the planetary surface and usually hold liquids, gases, and biota and support plants."

Figure 1.2 illustrates the soil structure and soil processes as set out in the new definition. The new definition (or, shall we say, approach) may have operational ramifications for soil studies—especially for the modeling of soil processes. For example—and most relevant to the geoinformatics approach suggested in this book—the new definition underlines the notion that the soil is a multiphase system, and is affected by physical, biological, and chemical processes. This view of the soil may also lead to consideration of the chemical and biological consequences of *soil erosion*, in addition to those arising from horizontal and vertical water movements or soil mineral movements, which were the focus of past water erosion research. Modeling the chemical and biological effects is specifically prescribed in this book when it comes to computing and modeling spatial *soil health* attributes of the remaining soil after a given erosion event.

The soil definition also highlights the fact that the soil is a complex system that is affected by several environmental, atmospheric, and human factors, as well as various small-scale processes. It also emphasizes the role of the soil in ecosystem functioning, and the importance of the

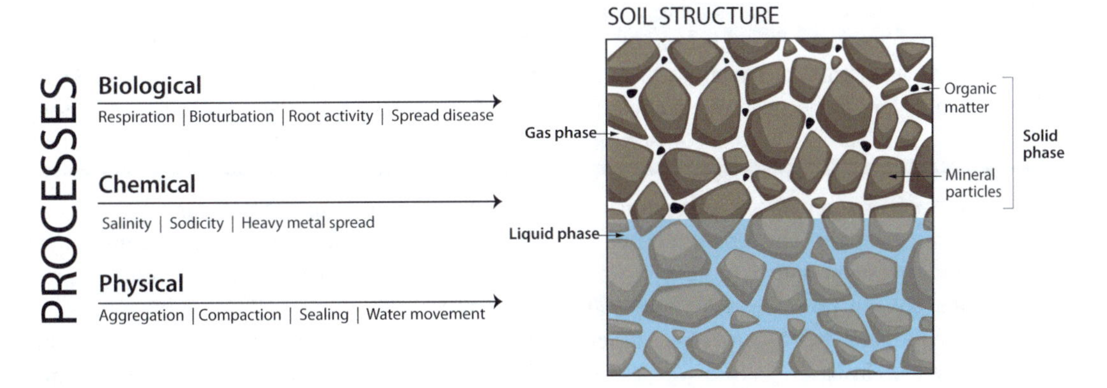

Fig. 1.2 Soil processes and structure. The left panel illustrates a few examples of the biological, chemical, and physical processes that operate within the soil layer. The right panel shows the soil structure and the three soil phases: gaseous, liquid, and solid. The small elements are the cement material—the organic matter

soil layer in enabling biological activities and plant growth—including agriculture as the prime source of food supply. In other words, soil is the infrastructure for nutrients—primarily but not only, nitrogen, phosphorus, and potassium (NPK)—whose accumulation and availability are the prerequisite for biomass production. Soil is also an important factor in processes of plant growth, as the water reservoir for the plant's own consumption. The importance of the soil as a water container is due to the fact that plants can use water stored in the soil under dry spell conditions during dry seasons with no rainfall supply. This, of course, requires a sufficiently thick soil profile to store the necessary moisture. As the infrastructure for plant growth, soils must also have sufficient pore size to allow aeration of plant roots, to provide the plants' vascular system with oxygen for their healthy function, and to allow the plants to use the oxygen to absorb water through their roots. Improved soil properties are key to enabling better quality crop production in every agricultural catchment.

However, the soil on Earth is not a homogenous system; soil properties may vary greatly with location, and some soil properties may also vary with time (Hartemink et al. 2013; Scull et al. 2016). Variation in soil properties has been detected in all scales of observation—from particle and pore scale to variation in soil properties between catchments (McBratney et al. 2003). Homogeneity in soil properties can be found in specific and very rare areas, but in most catchments (especially agricultural ones), large variation has been observed. A prediction of this variation using mathematical equations has been highly sought after for decades (Heuvelink and Webster 2001). That is because space–time variation greatly affects soil processes, soil formation, and water erosion processes in particular. Indeed, to some extent, the entire arsenal of geoinformatics tools provided in this book to combat water erosion processes is required, due to pronounced spatial heterogeneity in agricultural catchments. However, before we describe these tools for quantifying soil variation (mainly in Chap. 4), in the following section we review the *extrinsic* and *intrinsic factors* that cause space–time variation in soil properties.

1.1.2 Extrinsic and Intrinsic Factors

Variation in soil properties depends on two main groups of factors: (1) extrinsic environmental, climatic, and human-induced factors and (2) intrinsic soil factors that refer to small-scale processes throughout the soil history and have a notable impact on soil heterogeneity at fine resolutions. This section describes the two groups, how they vary with location and their use as powerful tools in predicting variation in soil properties.

1.1.2.1 Extrinsic Factors

Extrinsic factors—namely, the environmental conditions that affect the spatial variability in soil properties in a hillslope, a catchment, or a region—are fundamental to soil formation. They were first formalized as the controls of soil development by the geographer Dokuchaev (1893), who worked in Russia in the 1880s, on the typology of the various types of world soils. His work was so comprehensive that it subsequently led to the establishment of the scientific discipline known today as pedology, which is aimed at the study of soils in their natural environments. In his work, Dokuchaev provides a qualitative structure of the principles of soil formation and the consequent heterogeneity of the soil. The following factors were defined by Dokuchaev as the most important in determining a given soil formation: parent material (bedrock) and topographic inclination, climate, living organism's activity, and the time in which the soil formation processes take place. Dokuchaev's work, and that of his successors in the Russian soil studies group, seeped into other environmental sciences, such as ecology. In a short paper published in *Ecology*, Shaw (1930) proposed a semi-quantitative model based on the Russian group's earlier work, but interestingly stressed the importance of the parent material as a multiplier factor, and erosion and deposition processes as mechanisms that modify the Earth's surface, and consequently the properties of the future soil layer. Furthermore, Shaw saw time as having an exponential effect on the role of soil maturity in determining soil properties, as set out in Eq. (1.1):

$$s = m \cdot (c + v)^t + d, \quad (1.1)$$

where s is the character of the soil formed from the parent material of the bedrock m, based on specific conditions of c (climate) and v (vegetation) during a given time progress toward maturity of the soil profile t, and altered by the balance of erosion and deposition processes d. But, although Shaw's equation looks formal with some sort of predictive ability, and he defined Dokuchaev's factors as mathematical terms, Shaw's model was not developed with a view to producing quantitative predictions, but rather as a list of possible factors affecting soil character, and their possible roles in the process (Minasny et al. 2008).

The extrinsic factors in the model of soil formation and their effects on soil properties were taken a step further, and given a more quantitative expression, by Hans Jenny in his pivotal book: Factors of Soil Formation: A System of Quantitative Pedology (Jenny 1941), as the basic components of what is now the very familiar and commonly used *CLORPT* model (Eq. 1.2):

$$S = f(cl, o, r, p, t, \ldots). \quad (1.2)$$

CLORPT is an acronym of the external factors affecting soil formation S. The climatic conditions *cl*—include, but not exclusively, rainfall characteristics, such as rainfall amount, duration, and intensity; *o*—organisms or biotic factors in general, including bioturbation, the effects of plants, and even necromass and root activities; *r*—relief includes the *contributing area*, *hillslope gradient*, orientation, and curvature, which are of paramount importance; *p*—parent material including the effect of the chemistry and physical state of the underpinning rocks (rock fractures, for example, play a major role in the lower boundary of vertical soil water flow); and *t*—time. As will be shown later, CLORPT models have been used for soil mapping and their variation across a given region, as illustrated in Fig. 1.3, which shows four clorpt variables mapped in the Yehezkel catchment in northern Israel.

In a later work, Jenny (1961) reformulated the CLORPT variables—more holistically and, in particular, more dynamically—by reflecting the soil system in a way that pedogenetic processes take place as an open system (Eq. 1.3):

$$I_{s,v,a} = f(L_0, P_x, t), \quad (1.3)$$

where $l_{s,v,a}$ are the ecosystem, soil, vegetation, and animal properties, respectively—all determined as a function of L_0—the initial state of the ecosystem, based on mineral and organic fraction of soil materials; P_x being the external flux potential, based on climate and biotic factors; and t being the time elapsed for the ecosystem formation. Equation (1.3) represents a much broader approach, which includes the soil as part of an ecosystem, and has since been adopted by other researchers (Rinot et al. 2019).

However, subsequent research revealed that the clorpt is essentially a spatial generalization, and that the clorpt variables have an effect at broader spatio-temporal scales—including an organizational scale of soil types and geographical scale of a region. At a much finer scale, the effect of small-scale pedogenetic processes was acknowledged to operate as well (Paton 1978). Thus, while the clorpt model features are widely used in the literature, the model lacks the necessary means to quantify the small-scale pedogenetic processes. Birkeland (1984) and Hoosbeek and Bryant (1992), for example, suggested mathematical simulations as quantitative studies of pedogenesis and they hypothesized that pedogenetic processes at the small scale can change soil properties within the spatial boundaries of permutations of the environmental state defined by clorpt factors. In other words, the clorpt factors are not enough, in and of themselves, to model, quantify, and predict variation in soil properties at

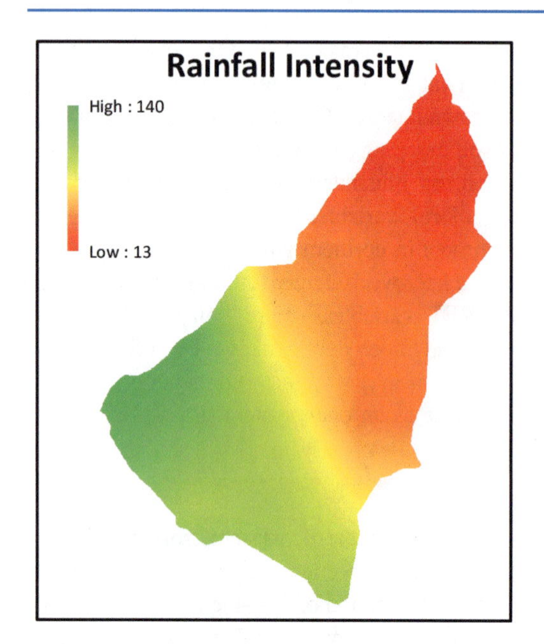

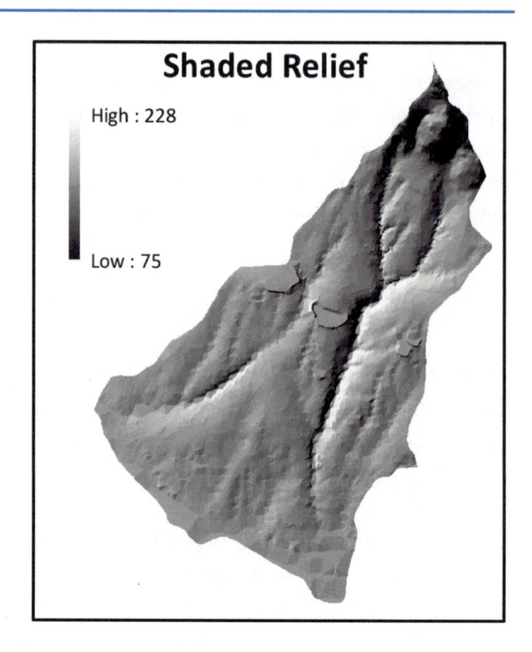

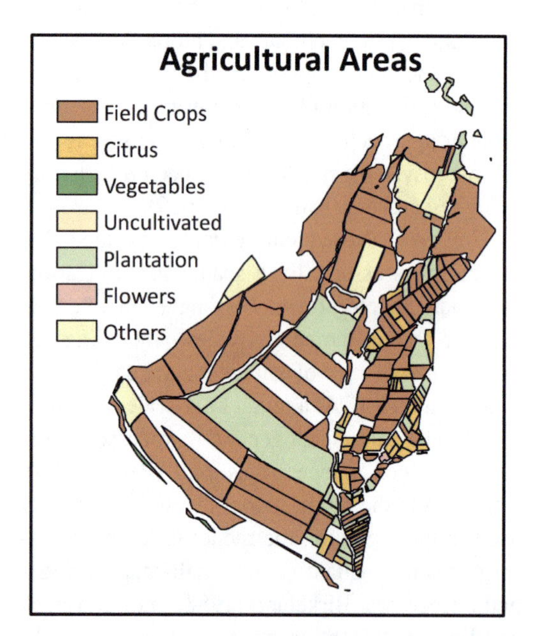

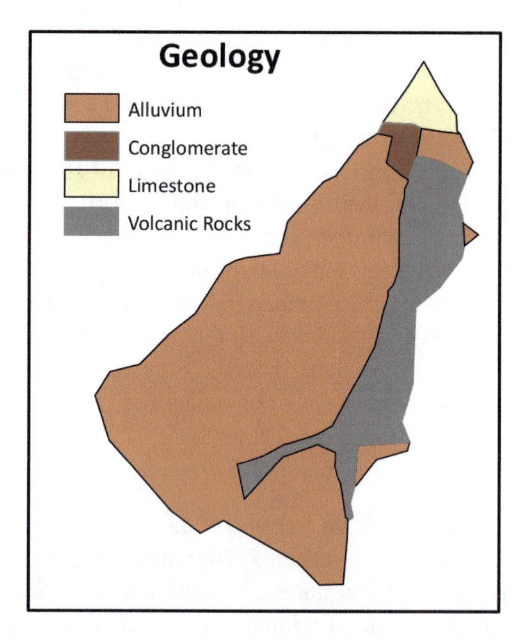

Fig. 1.3 Four GIS layers of the Yehezkel catchment in northern Israel, featuring the spatial variables of the clorpt model. Rainfall intensity (from a meteorological radar) varies from southwest to northeast; relief is expressed in the shades of gray with very clear channels dissecting the catchment, and parent material in the geological patches. Both the organisms and the human activity of CLORPT are expressed in the various crops and orchards of the Yehezkel catchment in the fourth layer

high spatio-temporal resolution, and are especially insufficient to simulate soil dynamics at the local scale of between a few centimeters and meters.

Nonetheless, despite its limitations in mapping soil variation, several studies have used clorpt factors to explore soil dynamics and their effect on plant productivity (Burrough et al.

1992; Zhu et al. 2001). This has been done using knowledge extracted from field studies and remote sensing data analysis that was merged in a geoinformatics framework using advanced mathematical modeling—such as fuzzy logic, physically based equations, and, more recently, machine-learning and data-mining tools (Cohen et al. 2008; Svoray et al. 2012).

A notable step in formalizing the effect of spatial location on soil prediction—including the effect of small-scale processes in soil properties—is the SCORPAN model put forward by McBratney et al. (2003). In it, the authors use a Jenny-like formulation to quantify the association between a given soil property and the aforementioned environmental factors, with predictive ability. However, a key addition of the SCORPAN model is the consideration of other soil properties in the predictions, based on the assumed association between the various soil properties. In this way, the authors linked between soil properties and environmental factors using an empirical framework (Eq. 1.4):

$$S = f(s, c, o, r, p, a, n), \tag{1.4}$$

where S is the soil property that one tries to predict and s is the local properties of the soil at a given location. Data on s is extracted from a prior map, or from remotely sensed data or expert knowledge, which most importantly reflects local small-scale processes; c refers to climate, and in water erosion usually referred to rainfall characteristics; o refers to the biotic component of the system including vegetation species, bioturbation, and the effect of domesticated animals and plants; r refers to the relief of topographic and morphometric characteristics; p is the bedrock effect; a refers to the time elapsed and the effect of age on the soil; and n is the location in space. According to McBratney et al. (2003), spatially and temporally explicit predictions are applied by using the known geographical coordinates longitude and latitude, and to characterize the age, time coordinate $\sim t$ is added to the equation which follows (Eq. 1.5) while each factor is represented by several continuous or categorical variables:

$$\begin{aligned} S[x, y, \sim t] = f(&s[x, y, \sim t], c[x, y, \sim t], o[x, y, \sim t], r[x, y, \sim t], \\ &p[x, y, \sim t], a[x, y], [x.y]). \end{aligned} \tag{1.5}$$

In essence, Eq. (1.5) renders SCORPAN a framework for soil properties prediction in space and time. As such, it also considers spatially and temporally auto-correlated errors that will be further discussed in Chaps. 4 and 7. This approach is particularly relevant to the current book, as it provides a mathematical framework that encompasses traditional theoretical soil formation and variation as well as empirically based and applied geoinformatics.

1.1.2.2 Intrinsic Factors

As was noted in Eqs. (1.4) and (1.5), the s (for soil) factor is added to the clorpt list in SCORPAN to express the effect of soil processes on soil variation within the framework of the small scale (centimeters to meters). Another framework for quantifying spatial variation in soil properties at the finer scale is the concept of intrinsic factors (Phillips 2017) that was developed in the ecological and soil sciences (Ibáñez et al. 1998; Webster 2000). These studies and others have led to the emergence of the *system complexity approach* that stresses the impact of soil history and the local processes that occur within the clorpt model permutations on current soil variation (Ibàñez 2013; Petersen et al. 2010).

Thus, intrinsic factors may be treated as additional source of information about soil variation which is usually related to the effect of local soil processes (Montagne et al. 2013). Simulation of the effect of small-scale processes may allow to improve quantification of the soil system dynamics and to extend the ability to map soil properties by clorpt (Seydel 1988). Intrinsic factors may also affect current patterns of soil properties by intensifying historical pedogenetic processes—such as the chemical consequences of ancient wildfires, or exceptionally long periods of heavy rainfall (Phillips 2001). The effect of intrinsic factors is not limited to mathematical theory, and observed patterns of soil variation in physical, chemical, and biological soil properties were linked to intrinsic factors in the 1990s

(Skidmore and Layton 1992). Since the traditional clorpt model cannot hold the dynamic information provided by the intrinsic factors, the study of extrinsic and intrinsic factors is complementary. The simplest equation that can be used to quantify the effect of intrinsic soil factors on soil variation in a multiple scale approach is adopted from the species-distribution curve that is widely used in the ecological sciences to quantify the arrangement of a given biological taxon in space. According to the *species-distribution theory*, the number of species increases with the habitat area in line with the well-known power law curve. The power law describes a relationship between two variables, where one variable varies as a power of another. Such a non-linear relationship leads naturally to an augmentation of the model output.

Phillips (2001) proposed exploiting this non-linear relationship as a means of quantifying variation in soil intrinsic factors, through the following formula (Eq. 1.6):

$$S_i = c_i \cdot A_i^{bi}, \qquad (1.6)$$

where S denotes the number of unique homogenous subset areas with a particular level of soil property (such as soil moisture content—or, in the ecological context, the number of species) in a spatial field i. The model coefficient c (the multiplier) represents the innate variation of the spatial field i and A is the area [m^2] covered by the soil group—i.e., some sort of clorpt permutation or, in the ecological context, the habitat. In accordance, the model coefficient b is the exponential effect of the inclination of soil variation to increase with area size independently of the variation in clorpt factors. The two coefficients, multiplier and exponent, are extracted for every studied catchment using empirical measurements.

A valuable index to dissociate between the effect of intrinsic and extrinsic factors on the soil variation is the $\frac{b_i}{b_t}$ ratio (Phillips 2017), where b_i is the mean value of the exponent coefficients of the richness-area curve (Eq. 1.6) of all soil groups that participate in the analysis, and is not representing variation associated with the extrinsic factors. The variable b_t is the same exponent computed for the entire area and therefore represents the variation caused by the extrinsic factors. Therefore, if the $\frac{b_i}{b_t}$>1, then the intrinsic factors are the major factors to create the soil heterogeneity, otherwise the variation within the soil group is smaller than the variation between the soil groups and therefore the major determinants of the soil variability are the intrinsic factors.

This may be the most important distinction between intrinsic and extrinsic factors: Variation in soil properties increases within a given unit (field), which may be homogenous in clorpt terms. Figure 1.4 illustrates an example of variation in Organic Matter (OM) content [%] in a given hypothetical plot: The grid-cell values represent measured OM content, and the polygon boundaries represent unit area with uniform clorpt characteristics, namely, the same parent material, topography, rainfall characteristics, biological activity, and human cultivation of the soil. As a result, variation in OM content within the polygon boundaries is determined by local pedogenetic processes—such as the impacts of historical wildfires and heavy rainfall events, presence of fecal and urine of domestic animals, and other random processes/events.

The results of Fig. 1.4 illustrate how larger supposedly homogenous units yield larger variation within the unit. For example, the smallest units exhibit two different observations, the larger units show four different observations, and so on. From the largest units (72 m^2), we see no further increase in variation, and the curve enters the asymptotic plateau. This phenomenon—which has been observed in many soils around the world—can be quantified for each research field, using the coefficients c and b that can be extracted using measurements.

The ability to quantify the effect of intrinsic factors in real-world cases has already been demonstrated in the literature. For example, in an SAR-based remote sensing study in the Negev

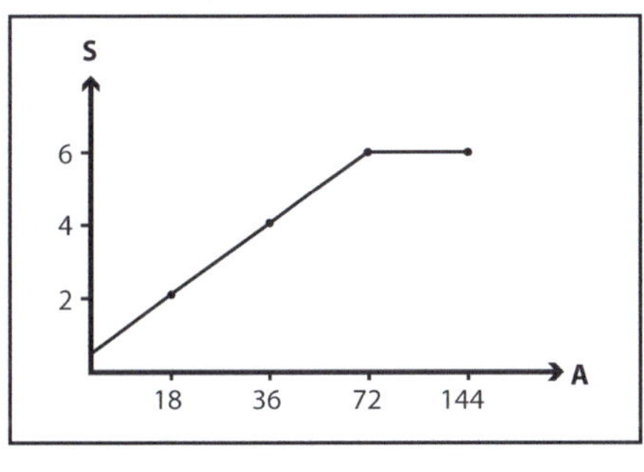

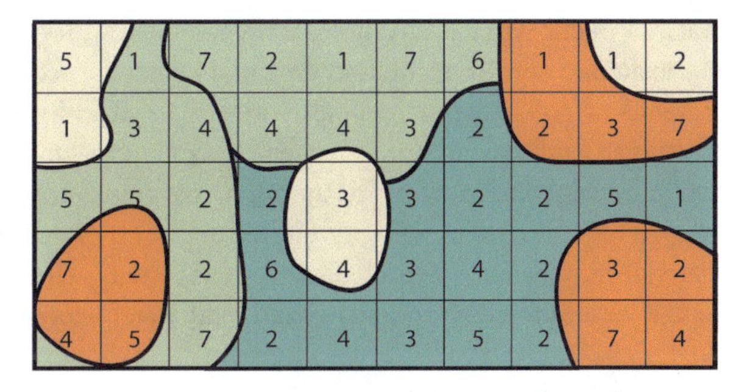

$$S_i = c_i \cdot A_i^{b_i}$$

Fig. 1.4 Intrinsic factors operating in a hypothetical agricultural plot. The right panel shows OM values [%] in the different polygons—which differ in area (at a cell resolution of 9 m^2); the left panel shows the relationship between area size of the field and the number of different OM values in the raster cell. Thus, for example, the upper left polygon has an area size of 18 m^2 (2 grid-cells), and two different OM groups, the lower right polygon covers an area of 36 m^2 (4 grid-cells) and four different groups of OM, etc. Thus, up to an area of 72 m^2, the greater the unit, the larger the variation. At 72 m^2, the curve reaches an asymptote and S does not increase anymore as a function of A. The computed function for the example in Fig. 1.4 is $S = 0.5 \cdot A^{0.54}$

Desert (southern Israel), Svoray and Shoshany (2004) quantified the effect of intrinsic soil factors on soil moisture space–time variation, using the synergy of optical and microwave data from Landsat TM and ERS-2. Their results show the variation in the drying rate and soil hydraulic conductivity of various soils and an increase in the value of these properties with the size of polygons of soil types.

To sum up, soil is a highly diverse system that evolves in response to extrinsic and intrinsic factors. The latter express the inherent diversity due to various physical, chemical, and highly localized biological processes. As we will see in the next chapters, the intrinsic factors affect soil processes in general and soil processes related to water erosion in particular.

Soil formation and the impact of spatio-temporal variation in soil properties on soil evolution are very extensive topics and therefore their full extent lies beyond the scope of this book. Their inclusion here is as a basis for soil erosion studies. Let us now focus, therefore, on four key terms that play a major role in water erosion studies of agricultural catchments and will serve us throughout the entire book: *Soil erosion*, *soil loss,* soil quality/health, and *soil degradation.*

1.1.3 Basic Terms in Soil Erosion Studies

Soil, soil formation, and heterogeneity, in general, and soil erosion in particular, are studied by researchers from various disciplines including, for example, soil scientists, pedologists, geologists, geographers, and other Earth scientists. Therefore, definitions and terms being used in these studies may vary between schools. Considering the differences in jargon, four basic terms require special attention and therefore will be described here as a basis for further discussions in this book.

1.1.3.1 Soil Erosion

Soil erosion may be loosely defined as the transformation of the soil into sediments. In other words, it is a process whereby soil mass is moved from one place to another on the planet's surface. This soil displacement may be instigated by wind or water, or by less extensive agents such as snow, bioturbation, or the labor of humans, their tools, or domestic animals. To put it another way, although humans accelerate erosion processes (as shall be clearly demonstrated in Sect. 1.3), soil erosion is a natural process on the Earth's surface that can occur without any human involvement. For example, organic matter—the natural cement material of soil mineral particles—may be disassembled and lost or simply absent in a given soil layer. In the resulting destruction, or prevention of formation, of soil aggregates, soil material may become dislodged and transported by the natural forces of gravity, wind, and water flow.

A soil erosion process is usually divided into three sub-processes: *Detachment*, the process of tearing loose of soil particles and the formation of non-cohesive sediments for subsequent transport; *transportation*, namely, the process of soil movement along a path, usually downslope; and *deposition*, namely, soil accumulation at another location—usually, but not exclusively, downslope (Pelletier 2008 and Fig. 1.5).

As previously noted, water is not the only agent to initiate and affect soil erosion. However, water has been found, in several studies (Trimble and Crosson 2000), to be the most destructive agent of soil displacement and the cause of most soil erosion damage in many countries around the world. The impact of water erosion on humanity is particularly evident in agricultural ecosystems due to the enormous harm that it inflicts on the food production industry. Thus, the ramifications of soil erosion extend well beyond the loss of soil minerals; water erosion has also been linked to nutrient cycling and carbon sequestration; to agricultural yields and food supply; and, in turn, to worldwide socio-economic, climatic, and environmental conditions (Borrelli et al. 2017). This is probably why studies of soil erosion are not limited to the physical sciences of Earth surface processes, but are also conducted in social science disciplines, such as public policy and environmental economics (see Chap. 2).

1.1.3.2 Soil Loss

Soil loss is defined as the mass of soil removed by the erosion process in a given time frame from a given plot (a well-defined unit area such as an agricultural plot, or a catenary hillslope unit, or a sub-catchment). The contributing factors to soil loss were formalized in the work of Wischmeier and Smith (1978), in the form of a simple mathematical expression known as the *Universal Soil Loss Equation* (USLE). The USLE was designed to compute mean annual rate of soil loss on a plot located on a hillslope that is cultivated for agricultural purposes (Eq. 1.7):

$$A = R \cdot K \cdot L \cdot S \cdot C \cdot P, \tag{1.7}$$

where A is the rate of soil loss, per unit area, per unit time [ton $\cdot$ Ha^{-1} $\cdot$ y^{-1}], and R is the factor of erosivity power by rainfall and runoff [MJ $\cdot$ mm $\cdot$ Ha^{-1} $\cdot$ h^{-1} $\cdot$ y^{-1}]. The greater the intensity, duration, and amount of rainfall and runoff, the larger the erosivity power. K is the soil erodibility [ton $\cdot$ Ha $\cdot$ y $\cdot$ Ha^{-1} $\cdot$ mm^{-1} $\cdot$ MJ^{-1}].

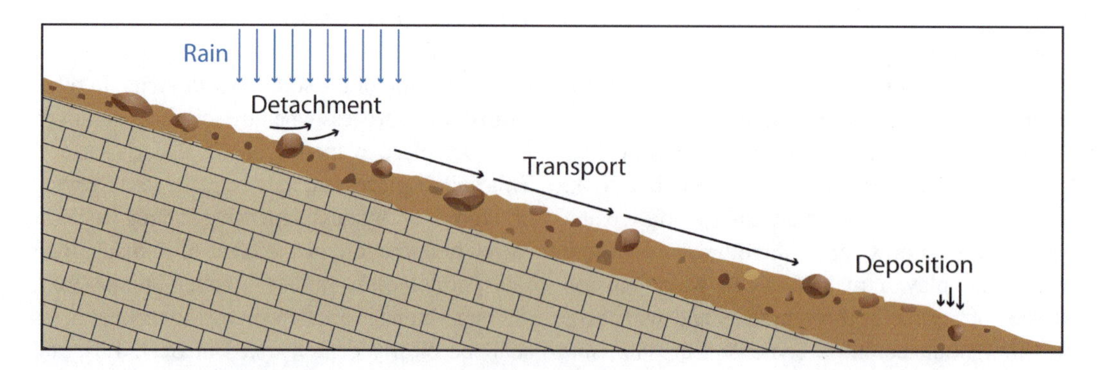

Fig. 1.5 A conceptual diagram of a simple model for the three sub-processes of soil erosion: Detachment, transport, and deposition. Under intensive rainfall event conditions, soil particles are detached from the surface and being transported and deposited downslope. Therefore, in many hillslopes, the soil profile is thick and wet in the footslope and thin and dry in the interfluve

Erodibility is an index that represents the level of exposure of the soil to detachment and then transport by the water force (initiated by runoff or rainfall). *K* is primarily affected by soil structure, organic matter content as a cement material, hydraulic conductivity, and pore permeability. *L* is the hillslope length [m] and *S* is the hillslope gradient [°]: The longer and steeper the field/plot, the higher the potential for soil loss. The *C* factor is strongly related to human intervention in the soil. It quantifies how much the soil and crop management systems are oriented toward conservation tillage and how much it helps to reduce soil loss. *C* is expressed as an algebraic ratio between actual soil loss amount from a soil under specific crop and cultivation system, and the soil loss from a similar uncultivated soil. Finally, *P* stands for the effect of practices to reduce water runoff and its consequent soil loss—calculated as the ratio of soil loss by a supporting practice to the soil loss from a field cultivated by straight-row farming along the hillslope orientation. Among the effective practices *P* for soil conservation are: *contour plowing* to delay overland flow by creating tillage mounds perpendicular to the hillslope aspect; *terrace cultivation* that are applied to slow down overland flow by reducing flow length; and *strip cropping,* namely, plowing the field in elongated strips of different crops to regulate the overland flow. Note that *C* and *P* are dimensionless (Foster et al. 1981).

Wischmeier and Smith (1978) have based their formulation of the USLE on empirical field data gathered from uniquely large number of overland flow and erosion plots across the continental US over a relatively long period of time: 1930s until the 1950s. The plots were initiated and managed by Hugh Hammond Bennet (1881–1960)—the father of soil conservation, and founder of what is known today as the Natural Resources Conservation Service (NRCS). The USLE has been extensively used by researchers since the 1970s, and has since been developed into a spatially explicit and more updated successor model, dubbed the RUSLE (Revised USLE, Hu et al. 2019).

Figure 1.6 illustrates the general framework of the USLE, and the five factors affecting soil loss. The USLE itself is still in use, and the controlling factors of soil loss were mapped very recently through the use of remote sensing and GIS data by Patil (2018).

A very detailed example of the computation of the USLE, using the necessary conversion tables available for the five factors, is provided in

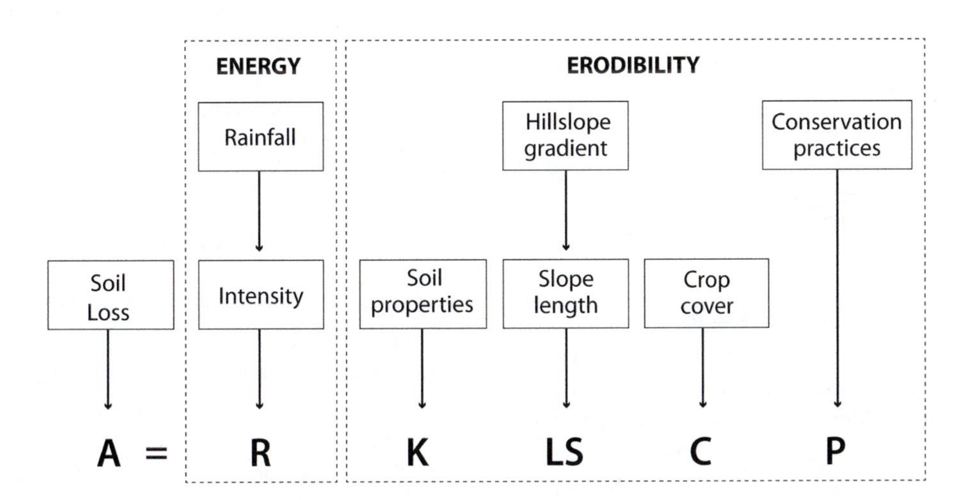

Fig. 1.6 Soil loss general framework according to the USLE (Universal Soil Loss Equation). The soil loss is the outcome of rainfall intensity, soil cohesive properties, the topography including plot length and orientation, cultivation methods including tillage method, *crop rotation*, and type and soil conservation methods including terracing, contouring, and strip cropping

the paper of Stone and Hilborn (2012). The full procedure will not be repeated here but in short, a general practice of USLE for a given field can be expressed as follows:

Step 1: the user determines the rainfall factor *R* based on measured rainfall data;
Step 2: the *K* factor is determined based on texture measurements and the division of the field into uniform polygons of hillslope gradient and length *LS*;
Step 3: cropping system and cultivation methods for the given field are determined and quantified as factors. These two factors are then multiplied to obtain the *C* factor;
Step 4: the user selects the *P* factor, based on the conservation practice applied in the field; and.
Step 5: the user multiplies the five factors to predict soil loss.

While the USLE provides a quantification of soil loss from a field, the question remains as to the function of the remaining soil on agricultural productivity and other ecosystem services. In other words, after a given period of water erosion, is the soil in the field still suitable for agricultural uses? This question is especially crucial, as the layer that is usually lost through the water erosion process is the more biologically active upper layer. For this reason, the definition and the ability to quantify soil health at such sites has attracted significant attention by scientists over the past three decades (Lehmann et al. 2020). The effect of water erosion on soil health and ecosystem services will be further discussed in Chaps. 7 and 8.

1.1.3.3 Soil Quality/Health

Soil quality is defined as "the capacity of a soil to sustain biological productivity, maintain environmental quality, and promote plant and animal health, within ecosystem boundaries" (Karlen et al. 1997; Doran and Parkin 1994). Soil quality and, consequently, agricultural productivity may diminish due to soil erosion processes and misuse of the land. In the past 20 years or so, the term soil quality has been superseded by the term soil health, which is defined as: "The continued capacity of soil to function as a vital living system, within ecosystem and land-use boundaries, to sustain biological productivity, maintain the quality of air and water environments, and promote plant, animal, and human health" (Doran et al. 1996). This definition highlights the integration of biological, physical, and chemical domains, enhancing the paradigm of the soil as a dynamic system, as reflected in the new definition of soil in van Es (2017). Soil quality therefore denotes the capacity for the intended use as determined by the bedrock chemical and physical characteristics, morphometric characteristics and drainage structure, and rainfall properties (Carter et al. 1997), while soil health refers mainly to the soil's instantaneous state, which is affected by, for example, cultivation method, within general and preconditioned soil quality. Chapter 7 will demonstrate an application of geostatistical measures to map soil health in agricultural catchments that have undergone water erosion processes. The concept includes mapping of soil properties to quantify the effect of soil loss not only on the regularly quantified soil profile, but on the entire soil as a system. Moreover, working in agricultural catchments with the specific aim of maintaining agricultural productivity and food security in a sustainable fashion leads to the quantification of soil health not only as a measure to preserve yields, but also as a measure of other ecosystem services, such as soil erosion regulation, gas exchange, and water conservation (Rinot et al. 2019).

1.1.3.4 Soil Degradation

Continuously growing soil loss rates and diminished soil quality can cause soil degradation. The literature provides several definitions for soil degradation, but for the purposes of this book, and in line with the recently established definition of the soil itself, the following definition will serve: "Soil degradation is a process that leads to a decrease in physical, biological and chemical qualities of the soil and ecosystem productivity for a long period" (Conacher 1995). Soil degradation therefore represents a long-term damage

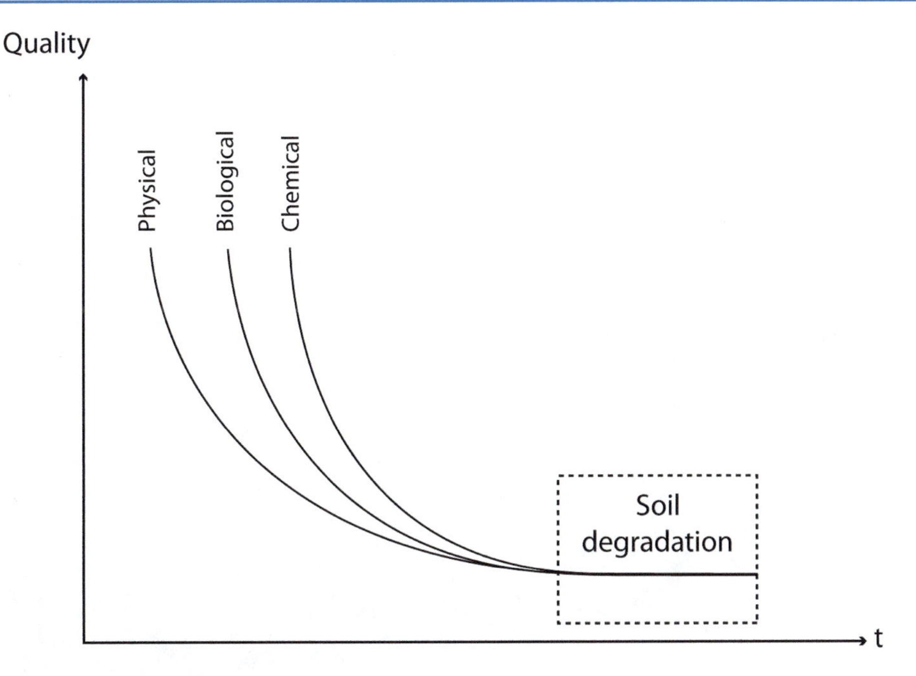

Fig. 1.7 Soil degradation is the outcome of decreased soil quality over time. Once physical, biological, and chemical processes diminish to a low level (to be determined in situ) and remain low over time, the soil is considered degraded (dashed rectangle)

to the soil that may be caused by erosion (Fig. 1.7). Mapping the extent of soil degradation remains a long sought-after challenge, with recent efforts using remotely sensed data (see Chap. 5), whereby vegetation status and change may be used as a proxy to soil health. The use of remotely sensed data to monitor changes in soil properties is more difficult. Mapping soil health and land degradation based on soil properties is done usually using field measurements—including close-range remote sensing at point measurements and kriging-type spatial interpolations to create maps (for more details see Chap. 7).

Based on Conacher's definition, soil degradation might be said to represent a continuous range that is measured in terms of decreased soil quality. Reduction in soil quality and soil degradation can indeed occur due to water erosion, but can be caused by natural processes that are not directly related to soil erosion—such as wildfires, reduced rainfall, or increased temperatures due to global warming and uncontrolled large flood events. The literature shows that: (1) soil degradation may occur in areas of different climatic conditions around the globe (Amundson et al. 2015; Oldeman et al. 1991) and (2) soil degradation may become not only a destructive process in the long term, but also, in the case of soil degradation due to soil loss, an irreversible one that can threaten national food security (Godfray et al. 2010).

One tangible example of accelerated soil erosion processes and consequent soil loss leading to a long-term reduction in soil quality to the point of soil degradation is the Harod catchment in the North of Israel (Fig. 1.8).

The Harod catchment is located in northern Israel (32°37′13″N,35°17′51″E) covering an area of 193 km^2. The climatic zone according to Koppen is warm Mediterranean climate (Csa) with cool winter (mean annual rainfall 300–600 mm decreasing eastward) and hot and dry summer. Elevation also diminishes eastward—from 138 m AMSL at the western part to (−100 m) in the eastern part near the Jordan River. Soils are mainly vertisols, with dark brown clayey soils, and organic matter content of $\sim 3\%$. The Harod is an agricultural catchment which is

Fig. 1.8 The Harod agricultural catchment in northern Israel—the supersite of this book. The blue lines represent the boundaries of the settlements in the catchment

being cultivated with increasing pressure for the last ~100 years. The main crop is wheat but other crops such as sunflowers, clover, olive, almonds, and orange orchards are also grown in the catchment. The Harod is also heterogeneous in parent material and topographic conditions and it is due to these heterogeneities that the catchment was selected as a supersite to this book. The catchment suffers the consequences of water erosion every year, especially after the first rainfall event, before plant growth. Following a particularly intensive rainfall event in October 2006, a large amount of soil was lost from the fields—estimated by the soil experts of the Israeli Ministry of Agriculture to be the equivalent to the load of 40,000 trucks (Rami Zaidenberg, Personal Communication). Apart from this soil loss, the surrounding environment suffered enormous damage. Every object in the path of this drifting soil mass was severely damaged: crops of neighboring fields were buried under an avalanche of fine material; drainage trenches along roadsides became clogged; and bridges, structures, and waterways were completely smothered in soil. The soil mass also accumulated on roads, causing road accidents and casualties. In addition to the costs of repairing this damage, the long-term and most expensive consequence was the diminished quality of the remaining soil at the site. Figure 1.9 illustrates water erosion damages that have occurred in northern Israel over the years, and not necessarily from the 2006 event.

The damage caused in the Harod catchment in a single event—and in other events over the years—is disturbing and demonstrates that water erosion can have dire consequences for agriculture and to the environment. But is this problem unique to the Harod catchment, or is it a worldwide phenomenon? In other words, what is the extent of the threat by soil erosion around the

Fig. 1.9 A collection of photos of soil loss phenomena in northern Israel (photos taken by Benny Yaacobi, the Israeli Ministry of Agriculture). A, B, and C are flooded orchard and fields, D and E are eroded agricultural roads, F is a wheat field with C-horizon exposed due to sheet erosion, and G and H are deep gullies developed in agricultural fields

world, in geographical and economic terms? Is it limited to the socio-economic state of a single country or specific climatic conditions? Can the economic damage be rectified? Are the decision-makers aware of the scale of the threat, and are they doing anything about it? Sect. 1.2 will try to provide some of the possible answers to these questions.

1.2 Scope of Soil Erosion

To the general public—which is usually more aware of sudden disasters caused by global warming, tsunamis, earthquakes, stock market collapse, or traffic jams—soil erosion appears to be an esoteric problem that is of concern only to farmers and Ministry of Agriculture officials. In reality, however, soil erosion poses a clear and present economic danger at the global scale that is recognized by governments and the UN as afflicting large parts of the world population, and casting a looming shadow on the future of agriculture, food supply, and food security. The following two sections will provide some evidence on the cost of soil erosion at the continental scale, and the geographical extents of the global regions currently subject to erosion processes and soil loss threats.

1.2.1 The Monetary Cost

UN researchers have estimated the cost of soil erosion for the entire world to be tens of billions of USD per annum (FAO 2014). These estimates represent only the consequences of water erosion—due to lack of data, wind erosion damage estimates were not included. Indeed, these estimates do not even cover all water erosion damage, as they only refer to the costs arising from yield losses, and only for specific crops: oil crops and legumes, vegetables, starchy roots, grains, fruits and meat growth, and the secondary damage to milk and egg production.

The reduction in yields, and their consequent costs, as a result of water erosion, occurs because, naturally, the uppermost layer is the one that is lost during the water erosion process and this layer is usually rich in nutrients and organic matter. These losses result in lower quality of chemical, biological, and physical soil components as a seedbed for agricultural plants, and resulting in diminished productivity, crop yields, and available planting area (Telles et al. 2011). These costs cannot be ignored, and must be made up for, because the agricultural products in question play a major role in food supply.

Closer inspection of the distribution of soil erosion costs across the continents reveals that the highest costs were recorded in the 2014 FAO report in North America and Oceania—at nearly 12 billion USD per year, for water erosion alone. Second in the rankings is the group of south Asia and southeast Asia regions and Latin America, with a cost of a little over 4 billion USD per year for each of these regions. The lowest costs have been recorded in Europe, sub-Saharan Africa, North Africa, and West/Central Asia, at between 2 and 4 billion USD per year in each region.

These costs add up to a global amount of approximately 33 billion USD per annum loss in yields in the aforementioned agricultural products, as a result of water erosion—which is higher than the annual national expenditure of two-thirds of the countries in the world (The World Bank 2018).

But these billions of USD are not the final costs of soil erosion to humanity. Another factor that is increasing costs for farmers—and is becoming even more expensive in the era of global warming—is the reduction in soil depth, which means decreased water-holding capacity. This affects farmers all over the world, but is particularly critical in dry environments, where water can be a limiting factor to agricultural growth.

These monetary costs of soil erosion impose a burden on humanity for centuries to come—yet they were already reflected in the literature in the 1990s, have been presented as a concern, and could have been avoided, or at least minimized, had they met with a proper response. Had the damage from soil erosion and its enormous consequences been treated in the 1990s or earlier, the state of soils around the world today would likely not have been so dire. Pimentel et al. (1995) identified the financial threat posed by soil erosion (wind and water) on humanity, and even proposed solutions (see Sect. 2.1 on soil conservation tools). In a paper published in *Science*, the authors estimated the cost of reducing (water and wind) erosion in US croplands to sustainable levels (1 tons $\cdot$ ha^{-1} $\cdot$ yr^{-1} instead of 17 tons $\cdot$ ha^{-1} $\cdot$ yr^{-1}) at 6.4 billion USD per year. They also computed the cost required to sustain uncultivated pasture lands at an additional 2 billion USD per year—making a total investment of 8.4 billion USD per annum to maintain US erosion at a sustainable level.

1.2.2 Geographical Extent

As we shall see below, soil erosion is a worldwide problem, affecting a large part of the global land area and is evident under diverse environmental, climatic, and human-induced conditions. Gibbs and Salmon (2015), for example, have provided a global mapping of soil degradation, based on four independent data sources and methodologies: experts assessments, satellite observations, biophysical models, and a database of abandoned agricultural field. The total global area of degraded soil that they estimated varied between ~1 and 6 billion ha, depending on the method and the data source used. These are large differences, but even the lowest estimate of 1 billion ha of degraded soil (~7.5% of the entire world terrestrial land area, without Antarctica) represents a global problem.

Another, rather comprehensive, attempt to map the world degraded soils based on detailed field campaigns was made in the early 1990s.

The Global Assessment of Soil Degradation (GLASOD) project, led by Oldeman et al. (1991), is the largest and most wide-ranging effort to map the geographical extent of soil degradation. This 3-year project sought to compile a report—including a map—of degraded soils due to various human activities at the scale of tens of meters through collaboration between the United Nations Environment Programme (UNEP) and the International Soil Reference and Information Centre (ISRIC). This effort, performed by dozens of professionals, is fully described in its final report, and while its methodology will not be repeated here, its findings were astounding: approximately, 15% of the total global soil surface was found to be degraded due to human intervention. Human-induced degraded soil areas were defined by Oldeman et al. (1991, p. 13) as:

> [...] Regions where the balance between the attacking forces of climate and the natural resistance of the terrain against these forces has been broken by human intervention, resulting in a decreased current and/or future capacity of the soil to support life.

In other words, the degraded soil areas are those where human intervention in the natural governing forces had inflicted severe harm to the soil. The project's findings revealed that the scope of degraded soil is much higher in some continents than in others. While in central Europe the total degraded area is over 20%, in Africa, Asia, South America, and Australia, it varies between 10 and 20%, and in North America it amounts to only 5% of the total continental area.

In addition to mapping soil degradation, Oldeman et al. (1991, p. 24) have also pointed at the main causes of soil degradation worldwide:

1. Systematic removal of the vegetation layer by humans, through deforestation and land clearings for infrastructure, agricultural, and urban purposes.
2. Overgrazing, which diminishes or even eliminates areas of vegetation.
3. A wide variety of agricultural cultivation practices with regard to soil plowing as well as irrigation with poor-quality water, excessive use of fertilizers, use of relatively heavy agricultural vehicles, etc.
4. Extensive planting of non-conservational vegetation types for domestic use. Namely, overuse of vegetation that does not provide protection against soil erosion, for fuel and fencing, for example, instead of planting species that may serve for soil conservation.
5. Soil contamination by industrial use, to the point where it can no longer resist water erosion processes.

In the course of their work, Oldeman et al. (1991, p. 34) have also found water erosion to be the most influential factor in soil degradation:

> Water erosion is by far the most important type of soil degradation occupying around 1094 million ha or 56% of the total area affected by human-induced soil degradation. On a world scale the area affected by wind erosion occupies 548 million ha (or 38% of the degraded terrain).

Their map (Fig. 1.10) provides a spatially explicit illustration of the impact of water erosion on the world soils—with water erosion severity ranging from low (light blue) to high (dark blue). It clearly shows that the entire continental US (with the exception of the large central plains, which were victim to wind erosion during the great "Dust Bowl" storms of the 1930s) has suffered various degrees of soil degradation—as well as large swathes of South America; sub-Saharan Africa; almost all of Europe; India; and the Far East, including large parts of China and large parts of Australia. This means that a large part of the world population lives under the threat of water erosion soil degradation of some degree of severity. Oldeman et al. (1991) also found soil degradation due to wind erosion to be widespread—albeit mainly in dry environments—afflicting 38% of arable soils in Africa, 30% in Asia, 17% in South America, 7% in Central America, 36% in North America, 19% in Europe, 16% in Oceania, and 28% of the world overall.

Among the world largest hotspots of water erosion, the Loess Plateau of Northern China is a prominent example for an area that experiences exceedingly severe soil erosion for many centuries. The Loess Plateau area is cultivated for

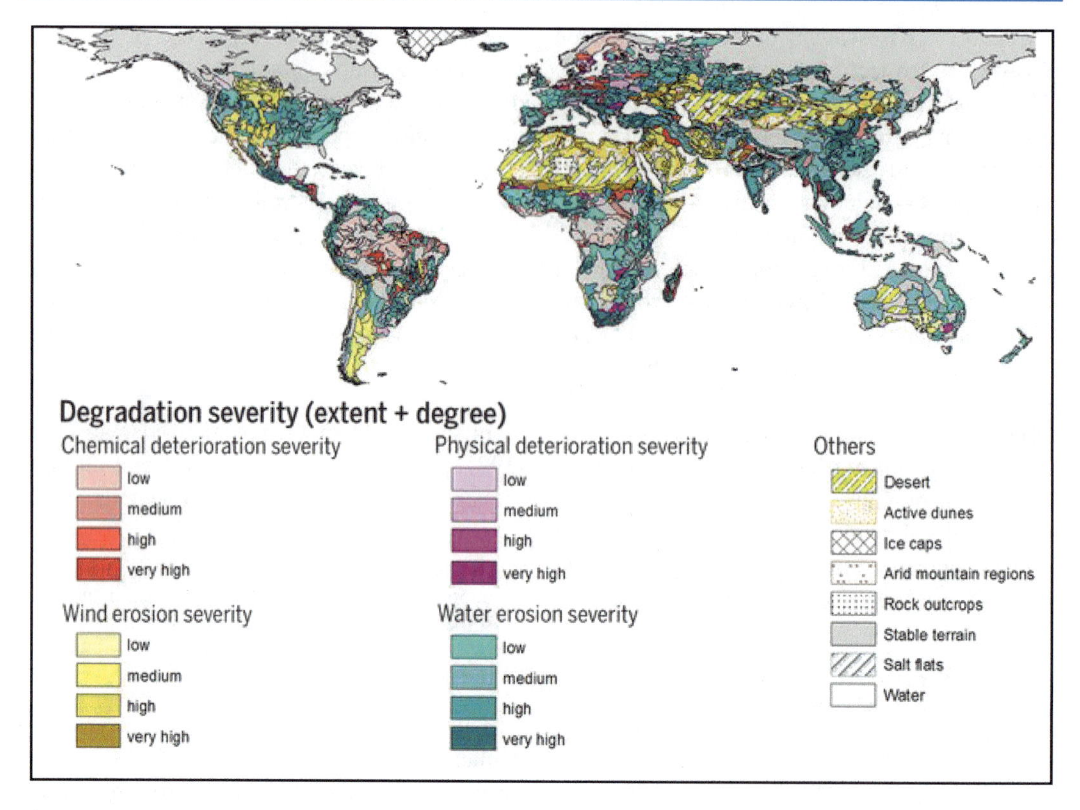

Fig. 1.10 Global mapping of degradation severity. Reprinted from Amundson et al., *Science* 08 May 2015:Vol. 348, Issue 6235, 1,261,071 (https://doi.org/10.1126/science.1261071)

grain production for approximately 2000 years, and with the population expansion, soil erosion damage became extremely high. Soil loss in most areas of the Loess Plateau reaches 50–100 tons $ha^{-1} \cdot yr^{-1}$ and up to 200 tons $\cdot$ $ha^{-1} \cdot yr^{-1}$ in the most degraded locations (Chen et al. 2007). Many studies indicate that soil erosion on the Loess Plateau—although its sources lie in the Holocene—is mainly induced by incautious agricultural cultivation and low vegetation coverage due to lack in water resources.

1.2.3 Implications for Food Supply

While the wide geographical extent of eroded soils under diverse environmental conditions is worrying, what makes soil degradation in agricultural catchments an even more acute threat to humanity is the relentless growth of world population, leading to ever growing need for food supply and, consequently, more agricultural activities. According to The World Population Prospect report by the United Nations Population Division (available online), world population growth rate is currently around 83 million individuals per annum—nearly an increase of 1% of world population every year. Based on UN data, the entire world population increased from around 1 billion people during the Industrial Revolution era to 7.6 billion people today, and is forecasted to reach 8.6 billion by 2030. This growth in population size has dramatically increased the number of mouths to feed, resulting in ever greater pressure on farmers, governments, and local authorities to increase the size of cultivated areas and yield production. Under these circumstances, and assuming a "business-as-usual scenario," the pressure on arable soils globally is expected to continue to increase substantially. Population growth, coupled with the threat to agricultural catchments, produces a vicious cycle that hinders society development due to restricted basic food supply (Kraay and McKenzie 2014). This cycle will be highly dependent on the feedback between soil degradation and food supply: increased population means greater demand on the soil, which causes less considered decision-making, which causes further soil degradation, etc. This conceptual outline of the problem unfortunately dates back to the 1990s. Kendall and Pimentel (1994) predicted that the availability of global arable land will place a severe constraint on food supply by 2050. They predicted that the business-as-usual strategy of the decision-makers (in 1994) will result in loss of arable lands to the point of food shortage for the world population. As a possible solution, they called on governments around the world to invest more money and effort in soil conservation practices. In their words:

> Attempts to markedly expand global food production would require massive programs to conserve land, much larger energy inputs than at present, and new sources as well as more efficient use of fresh water, all of which would demand large capital expenditures.

However, since 1994 the problem has not been solved, and the need to reduce soil degradation and to secure food production is even more urgent today than in the twentieth century.

In a more recent work, Ramankutty et al. (2018) reviewed the trends in global agricultural land uses and highlighted their implications for societal needs, especially agricultural products, yields, and food supply. They found, inter alia, that "agriculture is a major cause of global environmental degradation, and its impacts are expected to increase even further with rising future demand for agricultural products due to human population growth and increasing per capita consumption." This highlights the fact that soil degradation due to soil erosion, like any other environmental hazard in the global scale, has a social dimension. While the impact of soil degradation on developed countries and food consumption is bad enough (Thibier and Wagner 2002), in poorer countries, its consequences can be lethal, because in large populations diminished crop yields can result in malnutrition, hunger, or even death (Smith et al. 1998). For example, in parts of Africa soil erosion is so widespread that almost two-thirds of the arable lands in these areas have been diagnosed as degraded soils (Muchena et al. 2005). The main threat to the population in these degraded soils is that in many agricultural catchments, farmers keep increasing the pressure on the soil to the level that damage can be irreversible (Boardman 2006). Not surprisingly, this dramatically increases the threat that soil erosion poses to society.

In light of all of the above, soil erosion is clearly a major global problem, bringing about large swathes of degraded land, prompting very high expenditure on added fertilizers, and threatening the food supply to an ever-burgeoning world population. Is this a new phenomenon? In other words, has the soil erosion threat emerged only in the twenty-first century, or was it already fully apparent in the 1990s, when Pimentel et al. raised the alarm in their paper? Did it begin in the Industrial Revolution, following the introduction of mechanized agrotechnology—or it has originated even further in the past, with our ancient ancestors? To answer this question, we must go back in history and explore the roots of the misuse of the land by humans.

1.3 A Brief History of Soil Loss

We live in an industrial era, characterized by a massive use of agrotechnology, where farmers are equipped with heavy machinery and cultivate large areas in two and sometimes even three cycles a year (Lal et al. 2007). When farmers apply conventional tillage—as they still do in many places around the world—they penetrate the soil with plows featuring an elongated steel blade, or a subsoiler, cultivate with coil-tine harrows or disks, level the soil surface with the use of roller and leveling frames, followed by fertilization of the soil, and further leveling with a leveling frame. These intensive activities put enormous stress on agricultural soils, and, in many cases, the damage to the soil structure is irreversible (Godfray et al. 2010).

We tend to associate this kind of assembly-line style of cultivation, and the threats that it poses to the economy and the food supply, with large agricultural mechanization, and believe therefore that soil degradation is a modern problem arising from the "industrialization" of agriculture—whereas in the past, exploitation of the soil was more limited and more in tune with nature, and that our ancestors, with their more limited tools, caused less damage to the soil (Fig. 1.11).

However, in recent decades, research has discovered that soil erosion has been a continual problem with roots as far back as our hunter-gatherer days. Geo-archeologists now divide the history of the impact of humans on the environment into two periods: (1) the Paleo-Anthropocene era, from the earliest hominids (the emergence of the genus *Homo*, ~2.5 myr ago) to the Industrial Revolution (1870) (Foley et al. 2013) and (2) the Anthropocene, from 1870 to today (Crutzen and Stoermer 2000). Henceforward, these two terms will be used throughout this book. However, within the context of these two periods, and more specifically to the effect of humans on soil erosion, we may suggest to divide human history into the following three distinct categories:

1. The hunter-gatherer society, which existed for an exceptionally long period on Earth, with relatively low impact on the soil.
2. The advent of agriculture (~12,000 years ago), which was practiced by groups ranging from local tribes and chiefdoms to entire empires.
3. Farmers, but ones who have begun to adopt some form of conservational agriculture, dating from the 1800s onward (for more on soil conservation strategies, see Chap. 2).

The chronology of these periods may vary between different regions, depending on the development of the local society in each case. One key point about the latter category—the conservationalist farmers—is that they have operated under rapidly increasing pressures on the soil, and therefore notwithstanding their efforts to conserve the soil, they adversely affected the soil far more than their predecessors. According to the evidence of various studies (described below), all three categories had the capacity to affect the soil—and they did. Accordingly, soil erosion is not a new problem arising from powerful modern agrotechnology. Moreover, today, as a result of this history, we live in a human-induced world: The cumulative

Fig. 1.11 An old plow versus a new one. The old plow on the left is a Bissok plow that was pulled by two horses, as was used in the past at Kibbutz Gaash in Israel. In the right panel, there is a photo of a modern plow (photo courtesy of Eran Grossman), which is pulled by a tractor used in 2019 at the Cooperation of Merchavia, Kfar HaHoresh, and Yizre'el kibbutzes. Note the size and the number of the blades in each plow

impact of human activity on Earth is such that, as of 2018, 77% of the world land area (outside Antarctica) has been modified by direct actions of humans (Watson et al. 2018).

1.3.1 Hunters-Gatherers

We begin with the acknowledgment that understanding the influence of hunters-gatherers on the environment is still limited, due to lack of data (Burger and Trevor 2018). Despite this paucity of information, some record does exist, and it shows humans to be a rather successful species, capable of establishing communities—and subsequently settlements—in almost every environment on Earth. However, the preferred habitats for human inhabitation were usually characterized by high biomass production, high biodiversity with access to favored plant species, and reduced environmental pathogen stress (Tallavaara et al. 2018). The desirability of these habitats naturally influenced the spatial distribution pattern of hunter-gatherer populations, and Tallavaara et al. (2018) have found that productivity and biodiversity exerted the strongest influence on hunter-gatherer societies in ecosystems of the high and mid-latitudes, while in the tropics, the absence of pathogens was more significant in the selection for human settlement and its survival. Adaptation to the preferred habitats, however, was not without its limitations. Our ancestors had to maintain and defend the preferred territories, and in practical terms to interfere with the natural settings. Such modification of the environment was made possible by the human aptitude for ecosystem engineering (Jones et al. 1994)—including the use of various tools to change the soil, vegetation, and topographic properties to human benefit. This aptitude gradually grew over time, as hunter-gatherers learned to become sophisticated farmers (Lal et al. 2007).

More specifically, to live safely and more comfortably in their preferred habitats, groups of hunters-gatherers built shelters or dwellings, usually from debris, woody material, mud, or rocks that had been transported from one place to another. In due course, they began building dams and water channels to use the water more efficiently. The dwelling structures were used to defend their preferred territories from predators and from other tribes, and to provide shelter from harsh weather, such as rainfall events and snowstorms. Stone tools—particularly, flint—were used to hunt and process food; people became extensive users of stone resources, which required stone mining operations and transportation from one place to another (Leakey 1981).

Groups of hunter-gatherers also used wildfire management to control plant composition, to steer animals to specific (usually trap) locations and possibly to expel a rival tribe out of the clan's territory (Anderson 1994). Studies on wildfire show that even low-intensity wildfire events can affect the soil properties (Fig. 1.12). For example, exchangeable bases, potassium, phosphorus, and exchangeable acidity can change substantially after a wildfire event. Also, due to the removal of the vegetation strata by the wildfire event, water and sediment flow on the exposed surface can increase immediately after the event, contributing to soil degradation (Fonseca et al. 2017).

However, despite the relatively wide use of wildfire and stone—and rock-mining and movement by gatherers-hunters—most scholars are of the opinion that at the landscape scale, these societies did not have a very large footprint, and mainly lived in a world dominated by natural atmospheric, hydrological, and ecological processes (Messerli et al. 2000; Crombé et al. 2011). One possible reason for the relatively little impact that hunters-gatherers had on the soil may have been their high reliance on animal food. Cordain et al. (2000) showed that, wherever habitat conditions allowed, hunter-gatherers throughout the world consumed large amounts of meat, rather than vegetarian food. Although hunting can affect the environment by changing the food chain, and it is believed that hunters caused the decline in mammal species in six continents (Ceballos and Ehrlich 2002), their impact on the soil itself and on soil erosion is believed to have been relatively low.

Another reason for the low impact of hunter-gatherer groups on the environment is the slow

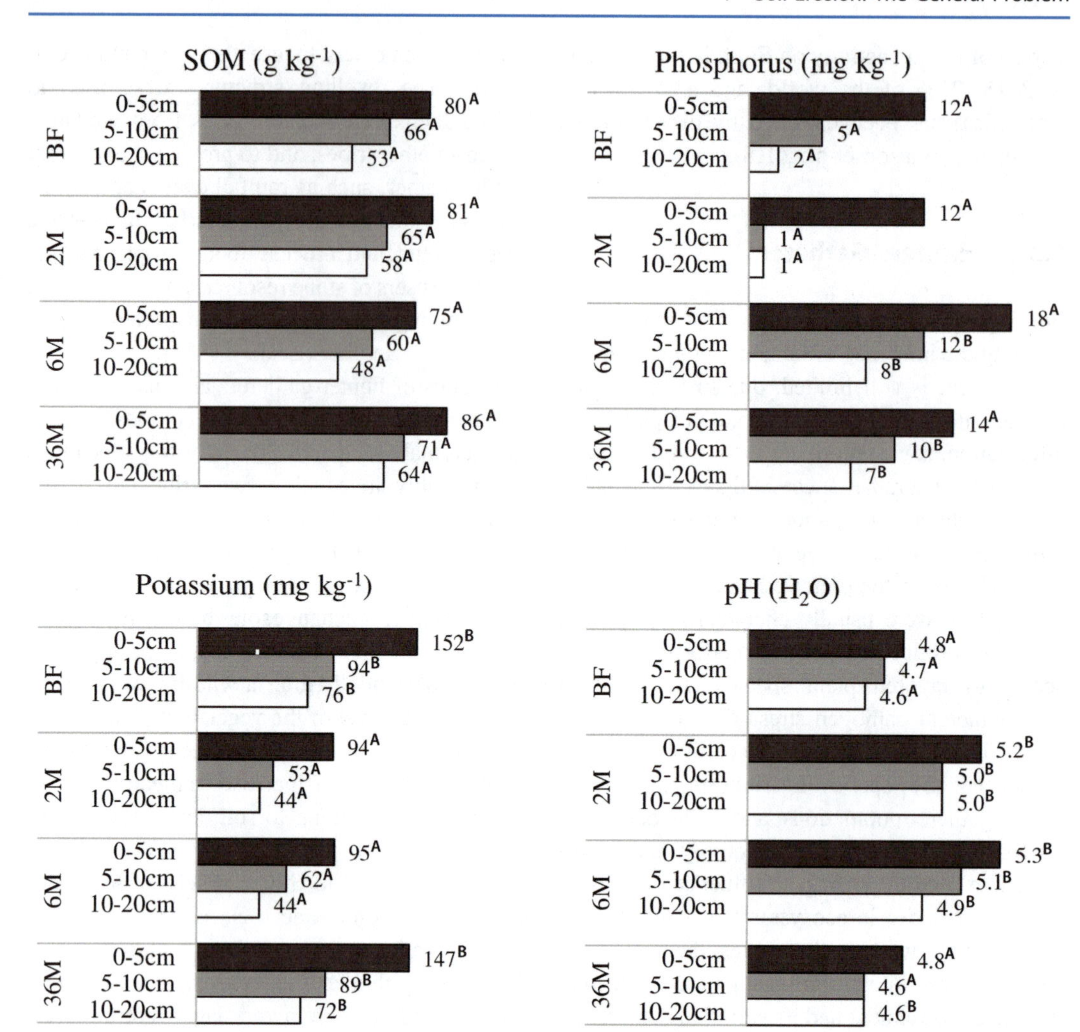

Fig. 1.12 The effect of wildfire on soil properties. Mean values of amount of four soil properties, in three different soil depths, for different sampling dates: control plot, before the fire (BF), and three more sampling dates: two (2 M), six (6 M), and thirty-six (36 M) months after the wildfire event. Bars with same color followed by different letters (a, b) are significantly different ($p < 0.05$). Reprinted from Geoderma, 307, Fonseca, de Figueiredo, Nogueira et al., Effect of prescribed fire on soil properties and soil erosion in a Mediterranean mountain area, 172–180, Copyright (2017), with permission from Elsevier

population growth rate during this period, and the fact that their exploitation of environmental resources was widely dispersed. At the dawn of the agricultural era, the entire human population on Earth, over hundreds of thousands of years, is estimated to have been approximately a few millions and land use was rather limited (Ellis et al. 2013)—so while hunters-gatherers did exert a pressure on soils, their total impact was relatively small.

1.3.2 Agricultural Societies

A much more extensive impact of humans on the soil was clearly apparent in the wake of the agricultural revolution, approximately 12,000 years ago (Braidwood 1960). Agriculture—in the form of extensive raising of domesticated plants through the use of cultivation tools—originated mostly along the world's large rivers that provided both drinking and irrigation water, a means of transportation, as well as fertile soils. As a result, soils were cultivated extensively in the Fertile Crescent, around the Tigris and Euphrates Rivers and the Nile in Africa, the Indus and Yangtze in Asia, the network of rivers between the Yukatan and the Guatemalan Islands in America and along other rivers around the world.

In all these agricultural regions, the plow was a central means in the agricultural revolution. The first farmers used a wooden plow, which eventually evolved into the iron-blade plow that came to be known in southern Europe and Near East as the "Roman plow" (Lal et al. 2007). Vestiges of these ancient farming practices in the soil are remarkable, and evident to this day. In Israel, for example, Ackermann et al. (2017) showed that the current landscape and ecosystem of the Judean Plateau was shaped in the Paleo-Anthropocene. Ackermann et al. and others (e.g., Smejda et al. 2017) have found that past farmers designed the soil physical infrastructure, and inhibited spatial distribution patterns of vegetation plants, and animal's diversity as well as species composition. They have also found that ancient farmers changed chemical composition of sediments and soils, biased plant species composition, destroyed weed populations, and maintained agricultural species. Similarly, in Mediterranean Europe, during the past 3000 years, erosion processes have been found to be largely influenced by human activities. This observation is specifically true for Greek and subsequently Roman areas along the Mediterranean coasts. Brückner (1986) who studied different parts of the Mediterranean Basin, including Spain, Italy, Greece, Turkey, and elsewhere, showed that human impact on the soil was occasionally catastrophic including the formation of badlands in hinterlands and on valley hillslopes, and enormous accumulation of sediments in valleys and coastal plains. In particular, he found that the establishment of settlements and their growth, the removal of protective vegetation cover, deforestation, and over-cultivation of the soil and grazing activities in the wake of increased population density, contributed to accelerated erosion processes that caused the filling of valley bottoms, and the disproportionate growth of deltas. In another study, van Andel et al. (1990) found that soil cultivation played a similar dominant role in destabilizing erosion–deposition budgets in ancient Greece. Although it is reasonable to assume that the Greek empire administration was aware of this damage, they did little to rectify the policy. Measurements of soil properties by Andel et al. (1986) at the cultivated fields reveal that Greek farmers were legally precluded from deciding which land to cultivate based on considerations of soil quality, not to mention sustainability. According to Andel et al., soil quality and soil conservation considerations played only a secondary role in decisions about land use allocation in the Aegean Region. Instead, water availability and access to irrigation were the main factors in determining land use and settlement patterns in Greece between the early Neolithic period and the nineteenth century. This has resulted in overuse of the soil and cultivation of marginal lands.

Under the Roman Republic and Empire rule (~509 BC–476 AD), soil conservation was no better than under Greek administration. Stock et al. (2016) have found that humans triggered significant soil erosion in the past three thousand years surrounding a Roman harbor and the harbor canal of Ephesus, in western Turkey. The authors found a high sedimentation rate since the Hellenistic period, indicated by rapid deposition of sediments in the Küçük Menderes River in conjunction with soil loss from the surrounding hillslopes, brought on primarily by deforestation, soil plowing agriculture, and goat grazing. This erosion–deposition process also caused water pollution of the entire harbor and its immediate environment. Human impact on soils continued to be strongly evident from Hellenistic/Roman period to the Byzantine times (~330 AD–1453 AD), as evident from high sedimentation rates, growing incidence of heavy metal contamination, and increased fruit tree pollen and intestinal parasites. Although evidence shows that some landscapes in the Roman Empire remained stable despite intensive agricultural activities (Dotterweich 2013), it is reasonable to assume, based on the aforementioned evidence and alike, that the

Greek and Roman regimes sought first and foremost to ensure a continual supply of agricultural products and timber, rather than conserve their soils in the long run.

Another empire that was highly active in the Mediterranean Region, and cultivated soil with minimum—if any—conservation considerations, is the Ottoman Empire. The oblivious attitude to the importance of soil conservation became all the more apparent in the face of the growing requirement for wood as a fuel for trains in the late eighteenth century. In the past, deforestation was brought on by the need for wood for cooking, heating, and construction—but in the Ottoman Empire deforestation for the sake of wood fuel in steam locomotive accelerated environmental deterioration. Although the full extent of this environmental deterioration is unclear, it certainly contributed to political and economic weakening of the Ottoman Empire—much as it did to several other civilizations in the past (Hughes and Thirgood 1982). Curebal et al. (2015) have studied the factors affecting the Turkish region in the early Anthropocene under the Ottoman rule. As expected, deforestation accelerated erosion and heightened water scarcity, and misuse of agricultural catchments led to loss of fertile fields and increased water erosion in Turkey in past centuries. In the final decades of the empire in the late nineteenth and early twentieth century, the Sultans lost the power to control agricultural activities and mainly forest clearing and could no longer stop degradation of natural resources—including soils and particularly agricultural soils—even if the consequences were apparent. Curebal et al. data shows that in the eighteenth and nineteenth centuries, illegal logging and land clearing fragmented ecosystems across wide swaths of Asia Minor and other parts of the Ottoman empire.

Poor conservation practices of early farmers and their administrations are not limited to European civilizations. In Latin America, evidence from the ancient Mayan empire between 1000 BC and 900 AD reveals severe soil degradation due to incautious cultivation (Beach et al. 2006). During that period, the last generations (550–830 AD) were found to be the most destructive, resulting in the highest erosion rates—in some instances, as much as the loss of 1–3 m in soil depth—in the wake of land use changes by the population.

Evidence from Asia is no better: even in the celebrated Chinese regime, soil was not treated with more care. According to Zhao et al. (2013), the Loess Plateau of China suffered dramatic climatic changes throughout the entire Holocene, resulting in large variation in soil erosion rates due to climatic changes and their consequent reduction in vegetation productivity. However, before the increase in population density in the area, levels of soil erosion and sediment yield from the Plateau were relatively low. The sudden significant increase in erosion rates coincided with the increase of settlements density and human use of the soil. According to Zhao et al., population growth and decline of vegetation cover in China were found to be responsible for the growing sediment loss for the past three millennia.

In summary, evidence from around the world since the dawn of agriculture shows that accelerated water erosion due to human use of the soil is neither new, nor limited to the post-Industrial Revolution era. This conclusion is evident not only from a range of local studies from selected areas, but from global scale analysis, as well. Citing historical documents and current global data, Dotterweich (2013) has shown that most agricultural catchments in the world were eroded as far back as ancient times. In another comprehensive study of the history of humans as soil-changing agents, and based on several data sources, Hooke (2000) estimated the amount of soil loss due to various human activities—intentionally or otherwise—from 3000 BC to the present to be equivalent of a huge 4 km × 40 km × 100 km dimensions ridge. However, the most disturbing finding is that the vast increase in population density and agrotechnology has dramatically increased the pressure on soils in the past century, especially during the period of great acceleration in technology since 1950. The small decrease in Earth moved per capita during the modern era may be due to the increased use of conservation tillage in current times. Hooke predicts that if current rates of population growth persist, the size of the estimated mountain range of lost soil will double in the next 100 yrs. As he points out, perhaps the most important question arising from this finding is how long can such growth rates remain sustainable and is it not our own responsibility to stop the impending catastrophe in its tracks?

1.3.3 Modern Farmers

The concluding questions of the previous section highlight the third category of users of the soil in human history, namely, farmers who do provide some kind of response to the soil erosion challenge. As will be shown below, humanity has taken note of the soil erosion threat, and consequently modern-day governments, scientists, professionals, and farmers are mobilizing in response to provide what may be the answer for the soil erosion threat: soil conservation policy.

Professional and structured soil conservation has been mooted or practiced in Europe since the nineteenth century. One example is a report published in Germany (Heusinger 1815) on the effects of vegetation on preventing erosion, and the negative impact to the soil if it is removed. Heusinger discussed the problems caused by siltation and by soil loss and their consequences to soil fertility, as well as the failings of ridge-and-furrow systems, which trigger runoff development and gully erosion. The report went on to describe and discuss the futile efforts to stop water erosion by means of control structures (Dotterweich 2013).

It is not surprising that German engineers identified the problem. Europe has suffered accelerated soil erosion processes throughout virtually the entire continent. But early efforts at soil conservation were sporadic, and failed to ease the pressure on the soil. Water erosion processes intensified even further in Europe in the twentieth century, which ultimately prompted an important milestone in soil conservation policy on the continent: The initiative of the Good Agricultural Environmental Conditions (GAEC) requirements in 2003, by the European Union's Common Agricultural Policy (CAP). GAEC is a collection of standards at the national and regional levels, aimed at achieving a sustainable agriculture, and keeping soil in good condition, in a bid to maintain rangelands, protect and manage water, and, in particular, prevent soil erosion and protect soil organic matter and soil structure. This initiative has made it possible to reduce soil erosion rates and preserve the vital soil organic matter in the topsoil (Panagos et al. 2016).

Soil conservation efforts have also been made elsewhere in the world. In the US, soil conservation initiatives began only in the 1930s, as part of the New Deal program. These conservation efforts were in response to a series of events collectively known as the "Dust Bowl"—a series of severe dust storms that had dramatic effects on the North American prairies. The Dust Bowl was, at least in part, the outcome of the invention and wide use, of the cast-iron moldboard plow, that resulted in widespread and severe breakdown of soil crust and soil structure in North America. However, the response of the authorities and the farmers was very slow, and a real transition from the moldboard plow to conservation tillage began only after World War II (Lal et al. 2007).

Soil conservation in various forms is currently being practiced in many countries, and has been found useful in reducing soil water losses to evaporation, sequestering carbon in soil, and reducing runoff generation and flow and the consequent soil loss (Maetens et al. 2012). Maetens et al. reviewed soil conservation studies in the first decade of the twenty-first century in a comprehensive meta-analysis of works that made actual measurements fields to test the efficiency of conservation tools in reducing overland flow and consequent soil transportation and deposition downslope. Their findings show that current soil conservation techniques are useful and can make a big difference in soil and water loss. Furthermore, they found that routine soil conservation methods that are based on crop management—such as buffer strips, mulching, cover crops, and mechanical methods such as geotextiles, terraces, and contour bunds—are even more effective in decreasing water erosion processes than conservational cultivation methods such as no tillage, reduced tillage, and contour tillage. The authors have also found that all the aforementioned conservation methods have more successfully reduced soil loss than they reduced runoff.

These findings can be important to combating water erosion, because enhancing the infiltration rate by reducing runoff can be just as important as the actual loss of soil, especially in drylands, where the loss of water downslope can be a major hazard to crop productivity and other biological processes needed for plant growth. Notably, however, Maetens et al. (2012) found variations in the efficiency of conservation methods in space and time suggesting that atmospheric (such as rainfall characteristics) and environmental (such as land use and topography) factors also figure prominently in reducing water erosion through soil conservation practices.

Such spatio-temporal variations invite the use of geoinformatics procedures in deploying soil conservation uses more accurately and efficiently, in a bid to minimize the damage to the soil by water erosion. In light of these findings, the main challenge in identifying the impact of environmental factors on soil conservation effectiveness in preventing soil loss and runoff development remains the lack of spatially and temporally explicit data. Current research shows that conservation techniques are indeed more effective under extreme conditions such as high-intensity rainfall events and steeper hillslopes (Svoray et al. 2015). However, these are still generalized conclusions that require much subtler analysis to understand the processes governing the conservation of soil and water in agricultural catchments. Therefore, current research needs to provide sufficient spatially explicit data on key environmental factors such as rainfall data, soil type and properties, vegetation cover and parent material and topography—to facilitate a more detailed analysis of the effect of soil conservation and runoff and soil loss.

Furthermore, the efficiency of soil conservation techniques may vary over time. Blanco-Canqui and Lal (2008) have shown that the rotation of row crops with legumes and perennial grass can be useful in restraining water erosion processes. That is because crop rotation may promote soil aggregation and macro-porosity, but as previously noted, our knowledge of the temporal effect is rather limited, due to shortage in historical data. This issue should be further explored, since temporal practices such as crop rotation can be more easily adopted by the farmer than, say, redesigning and modifying the topography. It is particularly needed in the study of long-term effects of conservation circles, since current studies in the literature are limited to 2 or 3 years only.

Several of these research gaps are tackled in this book, which attempts to provide the necessary algorithms and platforms to store and analyze spatially and temporally explicit data, to prevent or even decrease soil loss in agricultural catchments and hillslopes. In summary of this short review of studies on the history of human impact on soil erosion, we can state that soil erosion is a long-lived problem in human history (Fig. 1.13).

Humans have used the soil to construct tools that allow them to improve agricultural productivity, but usually without sufficient attention to preserving the land for future generations. We appear to have begun practicing care in soil cultivation only when the pressure on the soil reached crisis proportions, in the face of ever larger populations, and when we became educated enough to be capable of predicting the impending catastrophe. As Hooke (2000) has succinctly put it: "As hunter-gatherer cultures were replaced by agrarian societies to feed this expanding population, erosion from agricultural fields also, until recently, increased steadily." However, the more optimistic aspect of this state of affairs is that once actions are taken, they have proven effective in several respects, and may perhaps point the way to sustainable agriculture. Conservation activities have been found to reduce runoff and subsequent soil movement, and to decrease soil loss and diminished soil health. Because the first layers to be washed away are the topmost and most fertile layers of the soil, good conservation strategy is expected

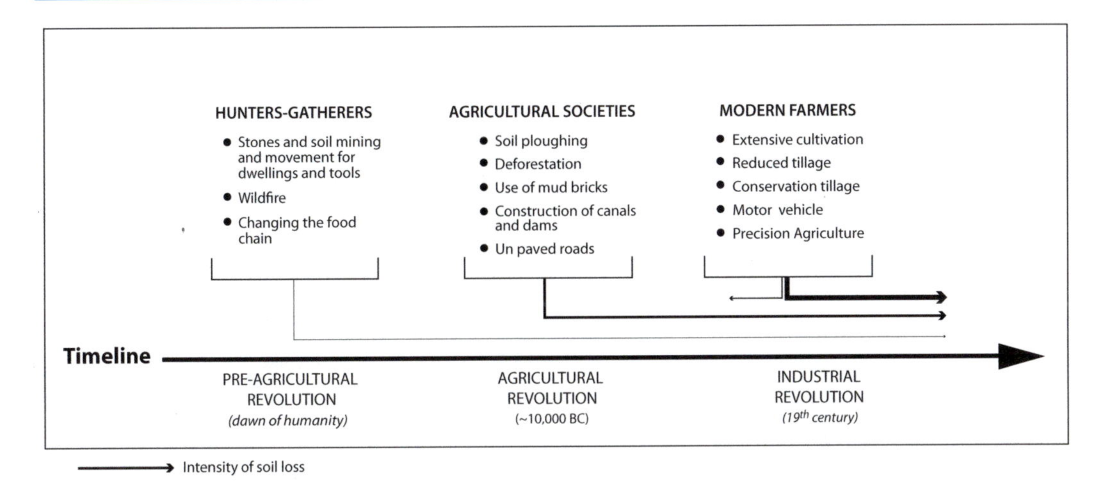

Fig. 1.13 A conceptual sketch of the brief history of the impact of humans on soil loss. The X-axis represents time, divided into the three eras. The solid arrowheads show the natural progression trends, in line with the evolution from one era and society to the next. The line width illustrates the intensity of the human effect on the soil. The small arrow under the category of the modern farmers implies that this group, as opposed to the other groups, uses conservation agriculture which reduces, to some extent, the soil loss damage it causes with modern agrotechnology

to foster better and more sustainable agriculture. With the geoinformatics procedures set out in this book, it is hoped that readers will be better equipped to study erosion risks and soil conservation strategies at the catchment scale.

1.4 Soil as a Finite Resource

The estimated annual monetary expenses due to losses in yields, water, and nutrients resulting by water erosion processes (as outlined in Sect. 1.2) have been found to inflict considerable short-term economic burden on farmers and governmental budgets. However, a striking long-term damage to agricultural cultivation through soil erosion is the permanent loss of the soil itself. Continuous reduction of soil thickness may cause, in some instances, the disappearance of the entire soil layer. The loss of a soil layer, or parts of it, represents an irreversible damage to the farmer, because rates of soil loss in agricultural fields involving traditional tillage are much higher than the rates of soil production through natural processes (Tables 1.1 and 1.2).

Studies that actually measured soil production in different parts of the world have found that it is a slow process—so much so, that it takes thousands of years to reform a soil layer that is lost in only a few decades through intensive water erosion processes.

A comparison between Tables 1.1 and 1.2 shows that the differences are stark: mean soil loss rates in some locations are in excess of 35 tons $\cdot$ ha^{-1} $\cdot$ yr^{-1}—up to 25 times more rapid than the mean rate of soil production. The ramifications of this huge discrepancy between soil loss and production rates are evident in the hotspots of degraded soils worldwide such as those in the US, Europe, northwest China, Inner Mongolia, north-central Africa, and many other regions (see Fig. 1.10). Needless to say, rebuilding the soil profile in these catchments cannot be done artificially, because we still lack the ability to develop artificial soils over large areas.

This rapid loss of soil from agricultural fields can be illustrated by simulations, based on the mean values in Tables 1.1 and 1.2. The conditions of the simulation in Fig. 1.14 include: soil loss rate (35 tons per annum), soil production rate (1.4 ton per annum), and a hypothetical field measuring 1 ha in size, with an initial soil mass of 1000 tons. In this illustration, the x-axis

Table 1.1 Rates of soil loss through tillage erosion in agricultural catchments of various parts of the world. Note the different methodologies applied for soil loss rate assessments and that the range between a few tons · ha^{-1} · yr^{-1} to hundreds of tons · ha^{-1} · yr^{-1} occurs in extreme cases. RS stands for Remote Sensing

Source	Site	Mean rainfall (mm)	Rate (tons · ha^{-1} · yr^{-1})	Method
Cerdan et al. (2010)	Europe	196–1304	4.4–15	Experimental plots
Verachtert et al. (2011)	Belgium	800	2.3–4.6	RS and interviews, field measurements
Prasannakumar et al. (2012)	India	3046	17.73	RUSLE
Liu et al. (2012)	China	535	36–90	Experimental plots
López-Vicente et al. (2013)	Spain	586	1.5–30	Revised MMF
Alexakis et al. (2013)	Cyprus	300–1000	20.95	RS and RUSLE
Zhu (2013)	China	500	150	Experimental plots
Napoli et al. (2016)	Italy	750	6.4–22	RUSLE
Ollobarren et al. (2016)	Italy	500	18–41	Field measurements
Rubio-Delgado et al. (2018)	Spain	500	27.2	Field measurements
Mean			**35.148**	

Table 1.2 Rates of soil production through natural processes in agricultural catchments in various parts of the world. Note the range between 0.3 tons · ha^{-1} · yr^{-1} and 5 tons · ha^{-1} · yr^{-1} in the very best case. The acronym kyr stands for a millennia

Source	Location	Mean rainfall (mm)	Rate (mm · kyr^{-1})	Rate (tons · ha^{-1} · yr^{-1})	Method
Heimsath et al. (2009)	Australia	1408	20	0.3	Geochemical mass balance
Ma et al. (2010)	Pennsylvania, USA	1070	27	0.405	Measured activity ratios of U-series isotopes in soils
Dosseto et al. (2012)	Puerto Rico	3000–4000	334	5.01	Measured activity ratios of U-series isotopes in soils
Suresh et al. (2013)	Australia	500	24	0.36	Weathering Index and measured activity ratios of U-series isotopes in soils
Huang et al. (2013)	China	1585	66	0.99	Geochemical mass balance
Mean				**1.413**	

represents the time elapsed from the initial timepoint, and the y-axis represents the soil mass in tons. The steeply sloped line represents the rate of soil loss, and the dashed flat line (level) represents the case of no erosion or production. Under these conditions, after 29 years, the new

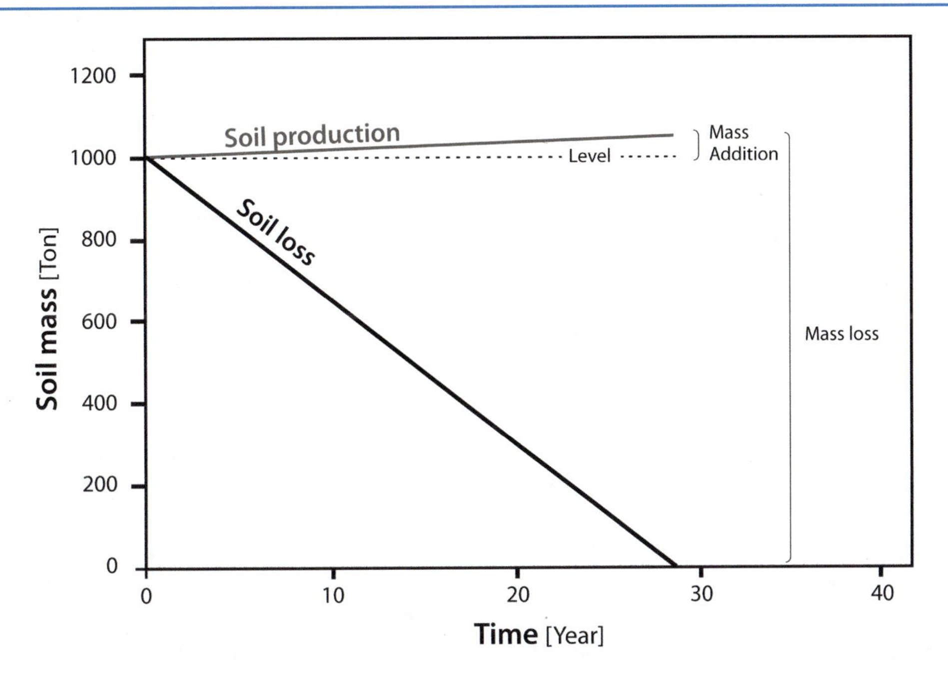

Fig. 1.14 Computed rates of soil loss versus soil production in a hypothetical field measuring 1 ha in size, with 1000 ton soil mass at year 0. Additional soil mass under the rate of 1.4 ton · yr^{-1} leads to ~1041 ton after 29 years. But, based on 35 ton · yr^{-1} rate of mass loss, it losses all 1041 tons during these 29 years

soil produced adds ~40 tons to the field's soil mass but the rapid soil erosion results in the loss of all ~1040 tons during that period.

In a seminal work on the subject of soil loss from different surfaces, Montgomery (2007) compiled hard data from studies in different parts of the world and determined with a high degree of certainty that soil loss rates from agricultural fields that applied conventional cultivation method—is, on average, 1–2 orders of magnitude greater than: (1) soil loss rates from areas covered with native vegetation, (2) geological erosion rate, and (3) soil production. Montgomery has found that the large discrepancy between rates of soil loss and production in agricultural catchments has led to inevitable reduction in soil profile of approximately half a centimeter a year in some places around the world. In other words, when viewed in terms of soil loss rates, there is a clear distinction between agricultural and non-agricultural catchments. While in natural catchments, soil mass can be relatively stable over a period of decades or centuries, with little or no threat of degraded soil profile, in agricultural catchments it is under considerable threat. However, despite these worrying figures, there is still hope: although conventionally plowed fields in agricultural catchments can lose up to 1 mm yr^{-1} through soil erosion over periods of a few generations, Montgomery found that with no-till cultivation, erosion rates are roughly on a par with soil production, thereby offering the foundation for sustainable agriculture. No-till cultivation involves leaving the crop straw from last season to remain on the soil. In this technique, the seeds are implanted into the soil layer using a dedicated drilling tool. The remaining straw also increases surface roughness and rainfall infiltration and in turn decreases substantially overland flow and subsequent soil loss.

Soil loss consequences are not merely hypothetical danger. Severe threat of soil loss has been observed in some of the most densely populated regions of the world. Europe is a notable example, with soil loss problems evident across the densely populated continent, from the southern Mediterranean countries to the more temperate northern regions. In their study,

Verheijen et al. (2009) compared *actual soil erosion* in Europe (defined as "the total amount of soil lost by all recognized erosion types") with *tolerable soil erosion* (defined as "any actual soil erosion rate at which a deterioration or loss of one or more soil functions does not occur"). To this end, they collected data on rates of soil production and loss by known erosion types, to estimate the current state of soil loss budget across the entire European continent. Even when dust deposition is taken into account, the highest rate of soil production (and, therefore, of sustainable soil loss) was found to be ~ 1.4 tons $\cdot$ $ha^{-1} \cdot yr^{-1}$ but in conditions prevalent in Europe, often as little as ~ 0.3 tons $\cdot$ $ha^{-1} \cdot yr^{-1}$. Rates of soil erosion showed much higher variability—up to 3–40 times greater, on average, than the upper limit of soil production in tilled, arable land across the continent. Similar number to those we have seen in Tables 1.1 and 1.2.

Another densely populated region suffering from severe soil loss problems is India. Therefore, soil degradation due to water erosion processes is widely recognized as a national environmental problem. As part of a project to combat soil degradation, Singh et al. (1992) prepared a map of soil erosion rates in India, using soil base maps, rainfall erosivity measures, hillslope gradient and land use, forests, degraded soil, sand dunes, and irrigation as well as measured soil loss data. Soil loss values were estimated using the aforementioned ULSE (see Sect. 1.1.3.2). Annual erosion rates due to water erosion were found <5 tons $\cdot$ $ha^{-1} \cdot yr^{-1}$ in dense forests with over 40% canopy cover, hot and cold arid regions. The Indo-Gangetic plains, including salt-affected lands, exhibited moderate rates (5–10 tons $\cdot$ $ha^{-1} \cdot yr^{-1}$). The areas with the severest erosion rates, >20 tons $\cdot$ $ha^{-1} \cdot yr^{-1}$, include the northwestern Himalayan regions and intensively cultivated agricultural catchments. In the rest of India, soil loss was found to vary between 10 and 20 tons $\cdot$ $ha^{-1} \cdot yr^{-1}$ while in several challenging regions, soil loss has even reached the level of >20 tons $\cdot$ $ha^{-1} \cdot yr^{-1}$.

In the US, soil erosion was recognized as a national threat by the Franklin Delano Roosevelt administration in the wake of the Dust Bowl storms of the 1930s. As part of a national effort to monitor the hazard, the National Resources Inventory (NRI) was established, to be responsible for assessments of soil erosion (wind and water) budgets in cultivated and non-cultivated croplands, and land enrolled in the Conservation Reserve Program (CRP). Citing data from the 2003 Annual NRI, Kertis and Iivari (2006) found that average soil erosion rates in these areas dropped in almost 40% between 1982 and 2003 (from 16.36 to 10.09 tons $\cdot$ $ha^{-1} \cdot yr^{-1}$). Similarly, total soil loss nationwide reduced from >3 billion tons $\cdot$ yr^{-1} to ~ 1.7 billion tons $\cdot$ yr^{-1}. According to the authors, this improvement may be due to: (1) new administration and legislation that changed incentives and caused pro-environmental behavior; (2) increased use of conservation practices; and (3) a reduction in cultivated cropland area. Notwithstanding these gains in curbing erosion, however, they noted that the soil loss rates are still extremely high (especially in relation to soil production), so further reductions may require new approaches.

Soil loss threat has also been observed in global scale analyses. In a recent work—using a global potential soil erosion model and a combination of spatial modeling, census data, and remotely sensed data at the pixel resolution of 250×250 m^2—Borrelli et al. (2017) estimated annual global soil erosion values to be 35.9×10^9 ton $\cdot$ yr^{-1} in 2012. They have also estimated an increase in global soil erosion due to the increase in agricultural land uses and they expect that the worst cases will be in sub-Saharan Africa, South America, and southeast Asia.

1.5 Summary

The role of the soil layer in agricultural cultivation is clearly the reason why it is—while extremely thin compared with the other Earth layers—so critical to humanity, and why the issue of soil erosion attracts so much attention by scientists, practitioners, and decision-makers. However, despite major efforts in recent years—such as the announcement of 2015 as the International Year of Soils by the United Nations

General Assembly, nomination of December 5 as World Soil Day, and recent popular-science books and documentaries about soil erosion—the importance of the soil layer has not yet sufficiently entered the public consciousness, nor reflected enough in public opinion, and consequently little public pressure has been brought to bear upon governments and local administrations to pursue and adequately fund soil conservation and rehabilitation. The present chapter has presented a rather grim picture of the threat that soil erosion—and water erosion in particular—poses to humanity. Under conventional agricultural cultivation practices, the soil is disaggregated, leaving it vulnerable to surface processes, soil erosion, and degradation. For thousands of years, since the hunter-gatherer days of humanity, countless generations have placed huge pressures on the world soils, and today conditions are so severe that many soils have been removed from cultivation, due to loss of soil and severely diminished soil quality and soil health. Moreover, the rate of soil production is extremely low compared with erosion rates, so lost soils cannot be rebuilt.

The financial costs arising from the drops in crop productivity as a result of water erosion are estimated to be tens of billions of USD every year. Yet decision-makers, scientists, and farmers are still insufficiently aware of the need for soil conservation and rehabilitation. Thanks to the digital revolution (Sanchez et al. 2009), advanced science, tools, and technologies have been developed by soil experts to better understand soil erosion processes, and to respond to the threat. The increased awareness and understanding, coupled with improved computerized tools from the worlds of geoinformatics and data science, allow us to make more reasonable and sustainable use of the land. This requires collaboration between the soil community and geoinformatics experts, which must take full advantage of modern tools and precise methods. In particular, the spatial resolution of input data must be able to offer storage of catchment topographic and coverage data, identification of areas under risk, decision-making tools, and soil loss modelling. Although wind erosion has also been implicated in harm to arable land, in this book, we focus on the methods to combat water erosion processes, because water erosion has been found to be the most destructive process of soil erosion in various parts of the world. Perhaps for this reason, more conservation tools are available in practice to combat water erosion.

Chapter 2 describes why agricultural catchments are a distinct landform with regard to soil erosion. Specifically, it tackles the factors and surface processes that affect soil erosion in agricultural catchments, and the consequences of soil erosion. Finally, it estimates the main psychological barriers that prevent farmers—the main agent in the process—from acting in an environmentally sustainable fashion.

Review Questions

1. Illustrate the three sub-processes of soil erosion, and explain the consequences of the erosion process on soil depth along the hillslope profile.
2. What are the new notions designated in the new soil definition that are crucial to studies of soil degradation in general, and water erosion in particular?
3. What are the physical, chemical, and biological processes that the soil is affected by? Give three examples to each process.
4. Why organic matter content is so important to the stability of the soil layer? What is the role of organic matter in the process of water erosion?
5. What are the differences between extrinsic and intrinsic soil factors ? What are the extrinsic factors, according to Jenny and others? Which processes and conditions can cause a species-area trend in soil characteristics?
6. What is the value of the parameter of innate variation of the spatial field associated with any spatial plot *i*; and the value of the parameter of the tendency of large soil plots to increase diversity, independently of environmental heterogeneity, in Carol's agricultural fields, represented by the number of soil types with $S_{1,2,3} = 4, 7, 9$ and, respectively, areas $A_{1,2,3} = 100, 200, 300$?

7. The agricultural fields in Bob's catchment were analyzed by Alice using species-area curve. The exponent of the entire catchment was equal to 4.4 and the mean exponent of all fields was equal to 8. Which are the factors that determine the diversity of soils in this catchment? Intrinsic or extrinsic factors? Explain why.
8. Calculate the soil loss [tons $\cdot$ ha^{-1} $\cdot$ yr^{-1}] of Eve's field with a rainfall erosivity factor of 30 MJ $\cdot$ mm $\cdot$ Ha^{-1} $\cdot$ h^{-1} $\cdot$ y^{-1}; soil erodibility of 35 ton $\cdot$ Ha $\cdot$ y $\cdot$ Ha^{-1} $\cdot$ mm^{-1} $\cdot$ MJ^{-1}; slope length of 100 [dimensionless]; slope steepness of 15 [dimensionless]; cover management of 20 [dimensionless]; and the supporting practices of 12 [dimensionless].
9. Provide a single-line definition for each of the four basic terms defined in the chapter: soil erosion; soil loss; soil quality; and soil degradation. Also explain the link between the four terms, and the difference between soil quality and soil health.
10. What are the three categories of human activities that have affected the soil over history? What is the conflict that the most recent group faces? In light of this, how long have humans affected soil erosion? Is it a new problem since the beginning of the Anthropocene? Explain.
11. What is the global annual cost of soil erosion due to reductions in crop yield?
12. List and describe the areas under threat of soil erosion around the world, in order of severity, from Low, Medium, High to Very high. What are the human and physical characteristics that these sites have in common?
13. A hypothetical wheat field of Oded, measuring 1 ha in size, with 1000 ton soil mass at year 0, is suffering from intensive water erosion: The soil production rate is 7 tons ha^{-1} $\cdot$ yr^{-1}, but soil loss has been measured at 70 tons $\cdot$ ha^{-1} $\cdot$ yr^{-1}. At what point in time will the field lose its entire soil depth? Assume linear soil loss and soil production processes.
14. According to the data in Tables 1.1 and 1.2, is there any association between rainfall amount and soil loss? Plot on a graph, and explain the result.

Strategies to Address the Review Questions

1. In hillslope erosion, soil is removed from a given part of the hillslope along the catena, and then transported and deposited in a lower part. Section 1.1.3.1 describes the three sub-processes of soil erosion including: detachment, transportation, and deposition. The consequences of hillslope water erosion processes on soil depth, along the hillslope catena, are illustrated in Fig. 1.5. In short, because soil particles are being transported downslope, in many hillslopes the soil profile is relatively thick and wet in the footslope and thin and dry in the interfluve.
2. Section 1.1.1 describes the new and the old soil definitions. The new definition underlines the notion that the soil is a multiphase system (including gas, liquid, and solid components), and is affected by physical, biological, and chemical processes. This view of the soil may also lead to the consideration of the chemical and biological consequences of soil erosion, in addition to those arising from horizontal and vertical water movements or soil mineral movements, which were the focus of past water erosion research.
3. Soil processes make a basis for indicating about the damage of soil erosion to soil health. Figure 1.2 describes the physical, chemical, and biological soil processes and soil structure. The following are just a few examples of the different three soil processes, but the literature suggests many more:
   - Physical processes: aggregation, compaction, sealing, and water movement.
   - Chemical processes: salinity, sodicity, and heavy metal spread.
   - Biological processes: respiration, bioturbation, root activity, and spread disease.
4. Organic matter is important for soil productivity because it is the source of carbon-

based compounds in the soil. Section 1.1.1 and Fig. 1.1 provide a description of the soil on Earth and illustrate the trend of decrease in organic matter from shallower to deeper soil layers. Organic matter content plays an important role on the stability of soil aggregates and the soil resistance to erosion. That is because organic matter is the cement material of the soil aggregates.

5. Extrinsic and intrinsic factors are those factors that affect variation in soil properties. Sections 1.1.2.1 and 1.1.2.2 describe the extrinsic and intrinsic factors in detail. In short:
   - Extrinsic factors are environmental conditions (including climate, biota, terrain, time, parent material, and human impact) that affect soil variation in a hillslope, a catchment, or a region. They stem from Dokuchaev's work about the development of soils in different environments.
   - Hans Jenny has given the extrinsic factors a quantitative expression in his book, with Eqs. (1.2)–(1.5) defining the evolution of the clorpt model.
   - Intrinsic factors are additional source of information about soil variation which is usually related to the effect of small-scale soil processes over time.
   - The difference: extrinsic factors represent the wider scale and present environmental conditions while intrinsic factors signify small-scale processes and the soil's history.
6. The c and b parameters can be computed using Eq. (1.6) in Sect. 1.1.2.2 on the intrinsic factors. We can regress the Si and Ai series using a power function. The result of the regression shows that the innate variation of the spatial field i is $c = 0.13$ and the exponential effect of the inclination of soil variation to increase with area size, $b = 0.75$ with $R^2 = 0.99$ between the two series.
7. We can use the $\frac{b_i}{b_t}$ ratio (Sect. 1.1.2.2) as an index to dissociate between the effect of intrinsic and extrinsic factors on the soil variation. If $\frac{b_i}{b_t}>1$, (which is the answer in our case of Bob's fields, $\frac{b_i}{b_t} = 1.81$) then the intrinsic factors are the major factors to create the soil heterogeneity.
8. Soil loss in Eve's field can be computed using the universal soil loss equation, USLE, (Eq. 1.7). The resulted soil loss, using the data in question 8, is $A = 378 \times 10^6$ tons $\cdot$ ha^{-1} $\cdot$ yr^{-1}.
9. The four single-line definitions can be formulated from Sect. 1.1.3.1–1.1.3.4:
   - Soil erosion is the transformation of soils into sediments.
   - Soil loss is defined as the mass of soil removed by the erosion process, in a given time frame, from a given tangible area.
   - Soil health measures the continued capacity of a soil to function as a vital living system, sustain biological productivity, and maintain the quality of air and water environments.
   - Soil degradation is a process of a long-term decrease in the physical, biological, and chemical qualities of the soil.
   - The four terms are connected because if soil erosion occurs, the soil is lost, and as a result soil health decreases. A continuous decrease in soil health leads to soil degradation.
   - The difference between soil quality and soil health is explained in Sect. 1.1.3.3. Soil quality denotes the capacity for the intended use as determined by the bedrock chemical and physical characteristics, morphometric characteristics and drainage structure, and rainfall properties, while soil health refers to the soil's state, which is affected by cultivation within general and preconditioned soil quality.

10. Section 1.3 suggests three categories (groups) of human activities that have affected soil degradation over history: hunters-gatherers, agricultural societies, and modern farmers. The conflict that the most recent group faces (Sect. 1.3.3) is that the rapid population increase causes large pressure on the soil, which in turn causes less educated decision-making and therefore increased soil degradation. The effect of humans on soils is not new because humans affected soil erosion even as hunter-gatherers since the ancient days.
11. Section 1.2.1 provides an overview of the monetary cost of reduction in crop yield due to soil erosion. The global annual cost sums up to 33 billion USD per annum.
12. Water erosion damage is widespread but not equal between countries. Figure 1.10 describes the areas under threat of soil erosion around the world. Particularly water erosion severity is divided into four levels symbolized in colors between dark and light blue. You can make the required list based on the bluish color plate. Also, Sect. 1.2.1 describes the cost of soil erosion in different parts of the world.
13. Figure 1.14 shows computed rates of soil loss vs. soil production in a hypothetical field (based on empirical data). We can compute the graph for Oded's field, on an annual basis, in the same manner. Oded's field is starting with 1000 ton soil mass in year 0, with soil production rate 7 tons $\cdot$ $ha^{-1} \cdot yr^{-1}$, and soil loss rate 70 tons $\cdot$ $ha^{-1} \cdot yr^{-1}$. Under these conditions, the field will lose the entire soil profile in 14 years.
14. You should draw a scatter plot of the variables according to the data in Tables 1.1 and 1.2, with mean annual rainfall [mm] in the X-axis and soil loss rate [$tons{\cdot}ha^{-1} \cdot yr^{-1}$] in the Y-axis. After you will do that, the results will show that there is no association between the two variables.

References

Ackermann O, Maeir AM, Frumin SS et al (2017) The Paleo-Anthropocene and the genesis of the current landscape of Israel. J Landsc Ecol 10(3):109–140. https://doi.org/10.1515/jlecol-2017-0029

Alexakis DD, Hadjimitsis DG, Agapiou A (2013) Integrated use of remote sensing, GIS and precipitation data for the assessment of soil erosion rate in the catchment area of "Yialias" in Cyprus. Atmos Res 131:108–124. https://doi.org/10.1016/j.atmosres.2013.02.013

Amundson R, Berhe AA, Hopmans JW et al (2015) Soil science. Soil and human security in the 21st century. Science (New York, NY) 348(6235):1261071

Andel THV, Runnels CN, Pope KO (1986) Five thousands years of land use and abuse in the southern Argolid, Greece. Hesperia: J Am Sch Class Stud Athens 55 (1):103–128. https://doi.org/10.2307/147733

Anderson DL (1989) Theory of the earth. Blackwell Scientific Publications, Boston, MA

Anderson MK (1994) Prehistoric anthropogenic wildland burning by hunter-gatherer societies in the temperate regions: a net source, sink, or neutral to the global carbon budget? Chemosphere 29(5):913–934. https://doi.org/10.1016/0045-6535(94)90160-0

Beach T, Dunning N, Luzzadder-Beach S et al (2006) Impacts of the ancient Maya on soils and soil erosion in the central Maya lowlands. Catena 65(2):166–178. https://doi.org/10.1016/j.catena.2005.11.007

Birkeland PW (1984) Soils and geomorphology. Oxford University Press, New York [u.a.]

Blanco-Canqui H, Lal R (2008) No-tillage and soil-profile carbon sequestration: an on-farm assessment. Soil Sci Soc Am J 72(3):693–701. https://doi.org/10.2136/sssaj2007.0233

Boardman J (2006) Soil erosion science: Reflections on the limitations of current approaches. Catena 68(2):73–86. https://doi.org/10.1016/j.catena.2006.03.007

Borrelli P, Robinson DA, Fleischer LR et al (2017) An assessment of the global impact of 21st century land use change on soil erosion. Nat Commun 8(1):1–13. https://doi.org/10.1038/s41467-017-02142-7

Braidwood R (1960) The agricultural revolution. Sci Am 203(3):130–152

Brückner H (1986) Man's impact on the evolution of the physical environment in the Mediterranean region in historical times. GeoJournal 13(1):7–17. https://doi.org/10.1007/BF00190684

Burger J, Trevor SF (2018) Hunter-gatherer populations inform modern ecology [Anthropology]. http://scholar.aci.info/view/1555a4c688800130002/1617ae3a1bf0001fc475326

Burrough PA, Macmillan RA, Deursen W (1992) Fuzzy classification methods for determining land suitability from soil profile observations and topography. J Soil Sci 43(2):193–210. https://doi.org/10.1111/j.1365-2389.1992.tb00129.x

Carter MR, Gregorich EG, Anderson DW et al (1997) Concepts of soil quality and their significance. In: Anonymous developments in soil science, vol 25. Elsevier Science & Technology, pp 1–19

Ceballos G, Ehrlich PR (2002) Mammal population losses and the extinction crisis. Science 296(5569):904–907. https://doi.org/10.1126/science.1069349

Cerdan O, Govers G, Le Bissonnais Y et al (2010) Rates and spatial variations of soil erosion in Europe: a study based on erosion plot data. Geomorphology 122(1):167–177. https://doi.org/10.1016/j.geomorph.2010.06.011

Chen L, Wei W, Fu B, Lü Y (2007) Soil and water conservation on the Loess Plateau in China: review and perspective. Prog Phys Geogr 31(4):389–403

Cohen S, Svoray T, Laronne JB et al (2008) Fuzzy-based dynamic soil erosion model (FuDSEM): modelling approach and preliminary evaluation. J Hydrol (Amsterdam) 356(1–2):185–198. https://doi.org/10.1016/j.jhydrol.2008.04.010

Conacher AJ (1995) Rural land degradation in Australia. Oxford University Press, Rural Land Degradation Australia Melbourne

Cordain L, Miller JB, Eaton SB et al (2000) Plant-animal subsistence ratios and macronutrient energy estimations in worldwide hunter-gatherer diets. Am J Clin Nutr 71(3):682–692. https://doi.org/10.1093/ajcn/71.3.682

Crombé P, Sergant J, Robinson E et al (2011) Hunter–gatherer responses to environmental change during the Pleistocene-Holocene transition in the southern North Sea basin: final Palaeolithic–final Mesolithic land use in northwest Belgium. J Anthropol Archaeol 30(3):454–471. https://doi.org/10.1016/j.jaa.2011.04.001

Crutzen PJ, Stoermer EF (2000) The anthropocene. IGBP Newsl 41:12

Curebal I, Efe R, Soykan A et al (2015) Impacts of anthropogenic factors on land degradation during the anthropocene in Turkey. J Environ Biol (36):51–58. PMID: 26591882.

Dokuchaev VV (1893) The Russian steppes: study of the soil in Russia, its past and present. Department of Agriculture Ministry of Crown Domains for the World's Columbian Exposition at Chicago, St. Petersburg

Doran JW, Parkin TB (1994) Defining and assessing soil quality. Defin Soil Qual Sustain Environ (35):3–21

Doran JW, Sarrantonio M, Liebig MA (1996) Soil health and sustainability. In: Advances in Agronomy, vol 56, 1–54. http://dx.doi.org/10.1016/S0065-2113(08)60178-9

Dosseto A, Buss HL, Suresh PO (2012) Rapid regolith formation over volcanic bedrock and implications for landscape evolution. Earth Planet Sci Lett 337–338:47–55. https://doi.org/10.1016/j.epsl.2012.05.008

Dotterweich M (2013) The history of human-induced soil erosion: geomorphic legacies, early descriptions and research, and the development of soil conservation—a global synopsis. Geomorphology 201:1–34. https://doi.org/10.1016/j.geomorph.2013.07.021

Ellis EC, Fuller DQ, Kaplan JO et al (2013) Dating the anthropocene: towards an empirical global history of human transformation of the terrestrial biosphere. Elementa (Washington, DC) 1:000018. https://doi.org/10.12952/journal.elementa.000018

FAO (2014) Food wastage footprint full-cost accounting. FAO

Foley SF, Gronenborn D, Andreae MO et al (2013) The Palaeoanthropocene—the beginnings of anthropogenic environmental change. Anthropocene 3:83–88. https://doi.org/10.1016/j.ancene.2013.11.002

Fonseca F, de Figueiredo T, Nogueira C et al (2017) Effect of prescribed fire on soil properties and soil erosion in a Mediterranean mountain area. Geoderma 307:172–180. https://doi.org/10.1016/j.geoderma.2017.06.018

Foster GR, McCool DK, Renard KG et al (1981) Conversion of the universal soil loss equation to SI metric units. J Soil Water Conserv 36(6):355

Gibbs HK, Salmon JM (2015) Mapping the world's degraded lands. Appl Geogr 57:12–21. https://doi.org/10.1016/j.apgeog.2014.11.024

Godfray HCJ, Beddington JR, Crute IR et al (2010) Food security: the challenge of feeding 9 billion people. Science 327(5967):812–818. https://doi.org/10.1126/science.1185383

Hartemink AE, Krasilnikov P, Bockheim JG (2013) Soil maps of the world. Geoderma 207–208:256–267. https://doi.org/10.1016/j.geoderma.2013.05.003

Hartemink AE, Zhang Y, Bockheim JC et al (2020) Soil horizon variation: a review. Adv Agron 160

Heimsath AM, Fink D, Hancock GR (2009) The 'humped' soil production function: Eroding Arnhem land Australia. Earth Surf Process Landf 34(12):1674–1684. https://doi.org/10.1002/esp.1859

Heusinger F (1815) About the drainage of the fields and the uprooting of the trench beds. Hannoversches Mag 83–94:1313–1510

Heuvelink GBM, Webster R (2001) Modelling soil variation: past, present, and future. Geoderma 100(3):269–301. https://doi.org/10.1016/S0016-7061(01)00025-8

Hooke RLB (2000) On the history of humans as geomorphic agents. Geology 28(9):843–846

Hoosbeek MR, Bryant RB (1992) Towards the quantitative modeling of pedogenesis—a review. Geoderma 55 (3):183–210. https://doi.org/10.1016/0016-7061(92)90083-J

Hu B, Shao S, Fu Z et al (2019) Identifying heavy metal pollution hot spots in soil-rice systems: a case study in south of Yangtze River Delta, China. Sci Total Environ 658:614–625. https://doi.org/10.1016/j.scitotenv.2018.12.150

Huang LM, Zhang GL, Yang JL (2013) Weathering and soil formation rates based on geochemical mass balances in a small forested watershed under acid precipitation in subtropical China. CATENA 105:11–20. https://doi.org/10.1016/j.catena.2013.01.002

Hughes JD, Thirgood JV (1982) Deforestation, erosion, and forest management in ancient Greece and Rome. J for Hist 26(2):60–75
Ibàñez JJ (2013) Pedodiversity. CRC Press, Boca Raton
Ibáñez JJ, De-Alba S, Lobo A et al (1998) Pedodiversity and global soil patterns at coarse scales (with discussion). Geoderma 83(3):171–192
Jenny H (1941) Factors of soil formation; a sytem of quantitative pedology. Dover Publications, New York
Jenny H (1961) Derivation of state factor equations of soils and ecosystems. Soil Sci Soc Am J 25(5):385
Jones CG, Lawton JH, Shachak M (1994) Organisms as ecosystem engineers. Oikos 69(3):373–386. https://doi.org/10.2307/3545850
Karlen DL, Mausbach MJ, Doran JW et al (1997) Soil quality: a concept, definition, and framework for evaluation (a guest editorial). Soil Sci Soc Am J 61(1):4
Kendall HW, Pimentel D (1994) Constraints on the expansion of the global food supply. Ambio 23(3):198–205. https://doi.org/10.2136/sssaj2001.6551463x
Kertis CA, Iivari TA (2006) Soil erosion on cropland in the United States: status and trends for 1982–2003. In: Anonymous eighth federal interagency sedimentation conference (8thFISC), Reno, NV, USA, April 2–6, 2006, pp 961–969
Kraay A, McKenzie D (2014) Do poverty traps exist? Assessing the evidence. J Econ Perspect 28(3):127–148. https://doi.org/10.1257/jep.28.3.127
Lal R, Reicosky DC, Hanson JD (2007) Evolution of the plow over 10,000 years and the rationale for no-till farming. Soil Tillage Res 93(1):1–12. https://doi.org/10.1016/j.still.2006.11.004
Leakey MD (1981) Tracks and tools. Philos Trans R Soc B-Biol Sci 292(1057):95. https://doi.org/10.1098/rstb.1981.0017
Lehmann J, Bossio DA, Kögel- Knabner I et al (2020) The concept and future prospects of soil health. Nat Rev Earth Environ 1:554–563
Liu Y, Fu B, Lü Y et al (2012) Hydrological responses and soil erosion potential of abandoned cropland in the Loess Plateau, China. Geomorphology 138(1):404–414. https://doi.org/10.1016/j.geomorph.2011.10.009
López-Vicente M, Poesen J, Navas A et al (2013) Predicting runoff and sediment connectivity and soil erosion by water for different land use scenarios in the Spanish Pre-Pyrenees. Catena 102:62–73. https://doi.org/10.1016/j.catena.2011.01.001
Ma L, Chabaux F, Pelt E et al (2010) Regolith production rates calculated with Uranium-series isotopes at Susquehanna/Shale hills critical zone observatory. Earth Planet Sci Lett 297(1):211–225. https://doi.org/10.1016/j.epsl.2010.06.022
Maetens W, Poesen J, Vanmaercke M (2012) How effective are soil conservation techniques in reducing plot runoff and soil loss in Europe and the Mediterranean? Earth Sci Rev 115(1–2):21–36. https://doi.org/10.1016/j.earscirev.2012.08.003
McBratney AB, Mendonça Santos ML, Minasny B (2003) On digital soil mapping. Geoderma 117(1):3–52. https://doi.org/10.1016/S0016-7061(03)00223-4
Messerli B, Grosjean M, Hofer T et al (2000) From nature-dominated to human-dominated environmental changes. Quatern Sci Rev 19(1):459–479. https://doi.org/10.1016/S0277-3791(99)00075-X
Minasny B, McBratney AB, Salvador-Blanes S (2008) Quantitative models for pedogenesis—a review. Geoderma 144(1):140–157. https://doi.org/10.1016/j.geoderma.2007.12.013
Montagne D, Cousin I, Josière O et al (2013) Agricultural drainage-induced Albeluvisol evolution: a source of deterministic chaos. Geoderma 193–194:109–116. https://doi.org/10.1016/j.geoderma.2012.10.019
Montgomery D (2007) Soil erosion and agricultural sustainability. Proc Natl Acad Sci 104(33). https://doi.org/10.1073/pnas.0611508104
Muchena FN, Onduru DD, Gachini GN et al (2005) Turning the tides of soil degradation in Africa: capturing the reality and exploring opportunities. Land Use Policy 22(1):23–31
Napoli M, Cecchi H, Orlandini S et al (2016) simulation of field-measured soil loss in Mediterranean hilly areas (Chianti, Italy) with RUSLE. Sci Lett 145:246–256
Oldeman LR, Hakkeling RTA, Sombroek WG (1991) World map of the status of human-induced soil degradation: an explanatory note. GLASOD
Ollobarren P, Capra A, Gelsomino A et al (2016) Effects of ephemeral gully erosion on soil degradation in a cultivated area in Sicily (Italy). Catena 145:334–345. https://doi.org/10.1016/j.catena.2016.06.031
Panagos P, Imeson A, Meusburger K et al (2016) Soil conservation in Europe: wish or reality? Land Degrad Dev 27(6):1547–1551. https://doi.org/10.1002/ldr.2538
Patil RJ (2018) Spatial techniques for soil erosion estimation. Springer International Publishing AG, Cham
Paton TR (1978) The formation of soil material. George Allen & Unwin
Pelletier JD (2008) Quantitative modeling of earth surface processes, illustrated edition edn. Cambridge University Press—M.U.A., GB
Petersen A, Gröngröft A, Miehlich G (2010) Methods to quantify the pedodiversity of 1 km 2 areas—results from southern African drylands. Geoderma 155(3):140–146. https://doi.org/10.1016/j.geoderma.2009.07.009
Phillips JD (2001) The relative importance of intrinsic and extrinsic factors in pedodiversity. Ann Assoc Am Geogr 91(4):609–621. https://doi.org/10.1111/0004-5608.00261
Phillips JD (2017) Soil complexity and pedogenesis. Soil Sci 182(4):117–127. https://doi.org/10.1097/SS.0000000000000204
Pimentel D, Harvey C, Resosudarmo P et al (1995) Environmental and economic costs of soil erosion and conservation benefits. Science 267(5201):1117–1123. https://doi.org/10.1126/science.267.5201.1117
Prasannakumar V, Vijith H, Abinod S et al (2012) Estimation of soil erosion risk within a small mountainous sub-watershed in Kerala, India, using revised universal soil loss equation (RUSLE) and geo-

information technology. Geosci Front 3(2):209–215. https://doi.org/10.1016/j.gsf.2011.11.003
Ramankutty N, Mehrabi Z, Waha K et al (2018) Trends in global agricultural land use: implications for environmental health and food security. Ann Rev Plant Biol 69 (1):789–815. https://doi.org/10.1146/annurev-arplant-042817-040256
Richter D, Markewitz D (1995) How deep is the soil? Bioscience 45(9):600–609. https://doi.org/10.2307/1312764
Rinot O, Levy GJ, Steinberger Y et al (2019) Soil health assessment: a critical review of current methodologies and a proposed new approach. Sci Total Environ 648:1484–1491. https://doi.org/10.1016/j.scitotenv.2018.08.259
Rubio-Delgado J, Schnabel S, Gómez-Gutiérrez Á et al (2018) Estimation of soil erosion rates in dehesas using the inflection point of holm oaks. Catena 166:56–67. https://doi.org/10.1016/j.catena.2018.03.017
Sanchez PA, Ahamed S, Carré F et al (2009) Environmental science. Digital soil map of the world. Science 325(5941):680
Scull P, Franklin J, Chadwick OA et al (2016) Predictive soil mapping: a review. Prog Phys Geogr 27(2):171–197. https://doi.org/10.1191/0309133303pp366ra
Seydel R (1988) From equilibrium to chaos. Elsevier, New York
Shaw JB (1930) A practical method of controlling the quartz content of fedspar. J Am Ceram Soc 13(7):470–474. https://doi.org/10.1111/j.1151-2916.1930.tb16304.x
Singh G, Babu R, Narain P et al (1992) Soil erosion rates in India. J Soil Water Conserv 47(1):97
Skidmore EL, Layton JB (1992) Dry-soil aggregate stability as influenced by selected soil properties. Soil Sci Soc Am J 56(2):557–561
Smejda L, Hejcman M, Horak J et al (2017) Ancient settlement activities as important sources of nutrients (P, K, S, Zn and Cu) in Eastern Mediterranean ecosystems—the case of biblical Tel Burna, Israel. Catena 156:62–73. https://doi.org/10.1016/j.catena.2017.03.024
Smith SR, Woods V, Evans TD (1998) Nitrate dynamics in biosolids-treated soils. III. Significance of the organic nitrogen, a twin-pool exponential model for nitrogen management and comparison with the nitrate production from animal wastes. Bioresourc Technol 66(2):161–174
Stock F, Knipping M, Pint A et al (2016) Human impact on Holocene sediment dynamics in the eastern Mediterranean—the example of the Roman harbour of Ephesus. Earth Surf Proc Land 41(7):980–996. https://doi.org/10.1002/esp.3914
Stone RP, Hilborn D (2012) Universal soil loss equation. Ministry of Agricutlure, Food and Rural Affairs. Queens Printer, Ontario
Suresh PO, Dosseto A, Hesse PP et al (2013) Soil formation rates determined from Uranium-series isotope disequilibria in soil profiles from the Southeastern Australian highlands. Earth Planet Sci Lett 379:26–37. https://doi.org/10.1016/j.epsl.2013.08.004
Svoray T, Shoshany M (2004) Multi-scale analysis of intrinsic soil factors from SAR-based mapping of drying rates. Remote Sens Environ 92(2):233–246. https://doi.org/10.1016/j.rse.2004.06.011
Svoray T, Michailov E, Cohen A et al (2012) Predicting gully initiation: comparing data mining techniques, analytical hierarchy processes and the topographic threshold. Earth Surf Proc Land 37(6):607–619. https://doi.org/10.1002/esp.2273
Svoray T, Hassid I, Atkinson PM et al (2015) Mapping soil health over large agriculturally important areas. Soil Sci Soc Am J 79(5):1420–1434. https://doi.org/10.2136/sssaj2014.09.0371
Tallavaara M, Eronen JT, Luoto M (2018) Productivity, biodiversity, and pathogens influence the global hunter-gatherer population density. Proc Natl Acad Sci USA 115(6):1232–1237. https://doi.org/10.1073/pnas.1715638115
Telles TS, Guimarães MD, Dechen SCF (2011) The costs of soil erosion. Rev Bras Ciênc Solo 35(2):287–298. https://doi.org/10.1590/S0100-06832011000200001
The World Bank (2018) The World Bank Data. https://data.worldbank.org/
Thibier M, Wagner HG (2002) World statistics for artificial insemination in cattle. Livest Prod Sci 74 (2):203–212. https://doi.org/10.1016/S0301-6226(01)00291-3
Trimble SW, Crosson P (2000) Land use: U.S. soil erosion rates—myth and reality. Science 289(5477):248–250. https://doi.org/10.1126/science.289.5477.248
USGS (1999) Cutaway views showing the internal structure of the Earth. https://www.usgs.gov/media/images/cutaway-views-showing-internal-structure-earth-left
van Es H (2017) A new definition of soil. CSA News 62 (10):20
van Andel TH, Zangger E, Demitrack A (1990) Land use and soil erosion in prehistoric and historical Greece. J Field Archaeol 17(4):379. https://doi.org/10.2307/530002
Verachtert E, Maetens W, Van Den Eeckhaut M et al (2011) Soil loss rates due to piping erosion. Earth Surf Proc Land 36(13):1715–1725. https://doi.org/10.1002/esp.2186
Verheijen FGA, Jones RJA, Rickson RJ et al (2009) Tolerable versus actual soil erosion rates in Europe. Earth-Sci Rev 94(1):23–38. https://doi.org/10.1016/j.earscirev.2009.02.003
Watson JEM, Venter O, Lee J et al (2018) Protect the last of the wild. Nature 563(7729):27–30. https://doi.org/10.1038/d41586-018-07183-6
Webster R (2000) Is soil variation random? Geoderma 97 (3):149–163. https://doi.org/10.1016/S0016-7061(00)00036-7
Wischmeier WH, Smith DD (1978) Predicting rainfall erosion losses, 1st edn. US Department of Agriculture, Washington DC

Zhao G, Mu X, Wen Z et al (2013) Soil erosion, conservation, and eco-environment changes in the Loess plateau of China. Land Degrad Dev 24(5):499–510. https://doi.org/10.1002/ldr.2246

Zhu T (2013) Spatial variation and interaction of runoff generation and erosion within a semi-arid, complex terrain catchment: a hierarchical approach. J Soils Sediments 13(10):1770–1783. https://doi.org/10.1007/s11368-013-0760-9

Zhu AX, Hudson B, Burt J et al (2001) Soil mapping using GIS, expert knowledge, and fuzzy logic. Soil Sci Soc Am J 65(5):1463–1472. https://doi.org/10.2136/sssaj2001.6551463x

2 The Case of Agricultural Catchments

Abstract

The on-site and off-site consequences of water erosion in agricultural catchments are caused by five types of hillslope processes: splash and sheet erosion, rilling, gulling, and piping. These processes are widely affected by the composition of clorpt factors in the catchment and on the hillslopes. First and foremost of the clorpt factors are human-induced factors, such as cultivation method, tillage direction, unpaved roads, and cropping systems. These interact with environmental factors such as rainfall characteristics, topography, vegetation cover, parent material, and bioturbation. This chapter reviews the processes, factors, and consequences of water erosion in agricultural catchments around the world. In its final section, it focuses on the human agent, proposes an explanation as to why agricultural soils are not always treated in a pro-environmental manner by farmers, and explores the psychological barriers that may prevent farmers from cooperating with conservational efforts initiated by the authorities. Finally, this chapter also reviews the actions that have been found to be effective in spurring farmers to adopt conservation practices.

Keywords

Human intervention · Off-site consequences · On-site consequences · Soil conservation · Soil processes

For anthropocentric reasons, when professionals refer to a soil as "degraded", they usually mean that it can no longer provide sufficient *ecosystem services* (Lal 2001). Such services may include, for example, drinking water and food provision, carbon sequestration, disease regulation, nutrient cycling, aesthetic inspiration, and cultural identity (for more details on ecosystem services see Chap. 7). Said otherwise, soil erosion attracts our attention mainly when it affects environments in the service of man. Sites under construction for human settlement and infrastructure, rangelands, planted forests, and cultivated fields are just a few examples for these environments. All these environments can suffer substantially from water erosion processes, soil degradation, and consequent reduction in the services they provide (Nosrati et al. 2018).

But, there are notable differences in the rate and extent of erosion processes, their consequences and the conservation and regulation actions between these human-induced environments. Sites under construction cover relatively

T. Svoray, *A Geoinformatics Approach to Water Erosion*,
https://doi.org/10.1007/978-3-030-91536-0_2

small areas compared with rangelands, forests, and cultivated fields, but possess very high economic value per square meter. In addition, erosion processes in built areas can be life threatening, so they are usually controlled through the use of expensive drainage devices and engineering solutions such as culverts, concrete drops, and terraces (Blanco-Canqui and Lal 2008c). Rangelands and forests make a more complex case because their vulnerability to water erosion processes may vary greatly with vegetation coverage. Sparsely covered rangelands and forests—especially deforested areas—are highly vulnerable to erosion processes and may suffer from a loosened vegetation shield. Denser forests are more protected due to over- and under-story vegetation shield and therefore they are usually less degraded (Borrelli et al. 2017). In agricultural catchments, soils have a high economic value due to their role in life supporting. However, owing to the nature of cultivation methods, agricultural soils are subjected to constant year-in, year-out, mechanical, chemical, and biological pressure which is much more intense than in the aforementioned human-induced environments.

Agricultural fields are cultivated every year using heavy machinery, they cover large areas, and every growing season they experience a very vulnerable phase after the soil is tilled, the seeds are sown, and the first rainfall event arrives and encounters a disturbed, loosened, and unshielded soil. At this time of the season, the soil is exposed, usually un-crusted, and easily transported by water and also wind (Katra 2020). This vulnerability of agricultural fields to soil erosion processes can have significant impact on the economy—even to the point of collapse of civilizations (Diamond 2005). Accelerated erosion in agricultural catchments can cause decline in soil productivity, which may lead to increased use of marginal soils, and consequently lead to reduced food supply and higher food prices (Pimentel et al. 1987). Thus, when erosion occurs in agriculturally important areas, practitioners are interested in the consequences not only as part of the soil evolution process, but they wish also to examine it in terms of yield growth and other ecosystem services.

The reduction in the capabilities of ecosystems to provide services can be quantified using the recently emerged soil health indices which usually encompass the following three soil traits (Lal 2001):

1. A decline in *physical* soil health of agricultural soils can be caused by weakening of soil structure and the formation of soil crust that seals the topmost layer and reduces water infiltration and aeration, leading to root anoxia.
2. The *chemical* quality of the soil may be reduced by standing runoff water, alkalization, and acidification, as well as leaching and illuviation.
3. *Biologically*, biodiversity may be reduced due to poor habitat conditions.

Degraded soils may lead to reduced biomass productivity, water pollution, poorer air quality, emissions of trace gases into the atmosphere, and many more other negative effects.

Water erosion models, both physically based and empirical, make excellent heuristic tools to study how water erosion processes may lead to soil loss and deposition in unwanted areas. The models may provide realistic simulations of earth surface processes in natural or semi-natural catchments (Nearing et al. 2017; Cohen et al. 2015); however, they are mainly focused on water/sediment budgets with little if any reference to chemical and biological consequences of the water erosion processes to the remaining soil onsite. Predictions of soil loss may be very useful to gaining more knowledge on water erosion in open areas, but may be limited for actual conservation planning in agricultural sites. Erosion models will be reviewed in Chap. 3 of this book in a bid to quantify water erosion processes for better understanding the phenomena, but this book suggests a more comprehensive approach to quantify soil health in a spatially and temporally explicit manner using geoinformatics and link it with ecosystem services. This framework will be presented in the next chapters. In the meantime, the present chapter will focus on the unique nature of agricultural catchments and on human activities in agricultural catchments in the context of water erosion. Specifically, this chapter shall describe the extrinsic factors that distinguish agricultural catchments from natural ones (Sect. 2.1); quantify processes and forms of water erosion in agricultural catchments (Sect. 2.2); classify the consequences of water erosion for agricultural catchments (Sect. 2.3); and highlight the leading agent in water erosion in agricultural catchments—namely, the farmer (Sect. 2.4).

2.1 Erosion Factors in a Distinct Landform

A distinct characteristic of water erosion processes in agricultural catchments is the large impact of farmers' activities, namely, the various actions that farmers engage in to cultivate the soil and to access the fields with designated vehicles. These are coupled, of course, with the environmental conditions in the catchment (Fig. 2.1). While other human activities may also affect water erosion in agricultural catchments, the ones cited in Fig. 2.1 and detailed below are among the key factors, and they will guide us throughout the case studies illustrated in this book.

The factors illustrated in Fig. 2.1 are drawn from the clorpt model (see Sect. 1.1.2), and they may interact with each other. Thus, for example, high-intensity rainfall events may be much more destructive to the soil in steep hillslope gradient conditions than in flat areas. Similarly, the effect of *conventional tillage* may be harsher on parts of the field with large contributing area than in parts of the field with small contributing area, or in areas of low vegetation cover conditions compared with those of high coverage. We shall describe the specific effect of each of these factors and their quantification below, but the following fundamental ideas can be summarized already at this point.

Soil cultivation creates major changes in the soil properties that increase soil erosivity and expose the soil to detachment, transport, and sedimentation. This occurs especially at the start of winter when the unshielded soil encounters the first rainfall events. The forces on the soil are determined by rainfall characteristics—in particular, rainfall intensity—with topography determining the velocity and connectivity of runoff flow. *Parent material* may have a smaller effect, and *bioturbation* is usually suppressed in agricultural catchments by the farmers. The comparative importance and contribution of these factors to the erosion process may be determined either by expert systems or based on statistical analysis of data measured in specific fields. These procedures of parameters weighting are illustrated and further discussed in Chap. 6. In the following section, we describe each of the factors and their function in water erosion processes.

2.1.1 Human Factors

2.1.1.1 Cultivation Method

Agricultural soils are reshaped by farmers to allow efficient seed sowing, seed germination, and biomass production. The soil is therefore cultivated to preserve a high concentration of nutrients, to support conditions for biological processes (Li et al. 2008), and to ensure that its structure supports vertical water movement and root aeration (Schwen et al. 2011). In addition, the farmer must curb weed species (Kettler et al. 2000), to provide the crops with an advantage in competition over resources. However, while cultivation processes provide certain benefits, they impose shear and compaction forces on the soil layer. These forces cause changes in the soil structure and properties that may increase erosivity (Shmulevich et al. 2007). A typical conventional cultivation process often consists of plowing, harrowing by disk, tillage with a coil tine, loosening, smoothing, rolling, and subsoiling (Svoray et al. 2015). These activities release direct several forces on the soil (Lal and Shukla 2004). In physical terms, shearing forces act upon the connecting forces between the soil particles, and once they overcome the connecting forces, structural failure in the soil occurs (Arvidsson and Keller 2011). In addition, the agricultural machines that operate in the fields

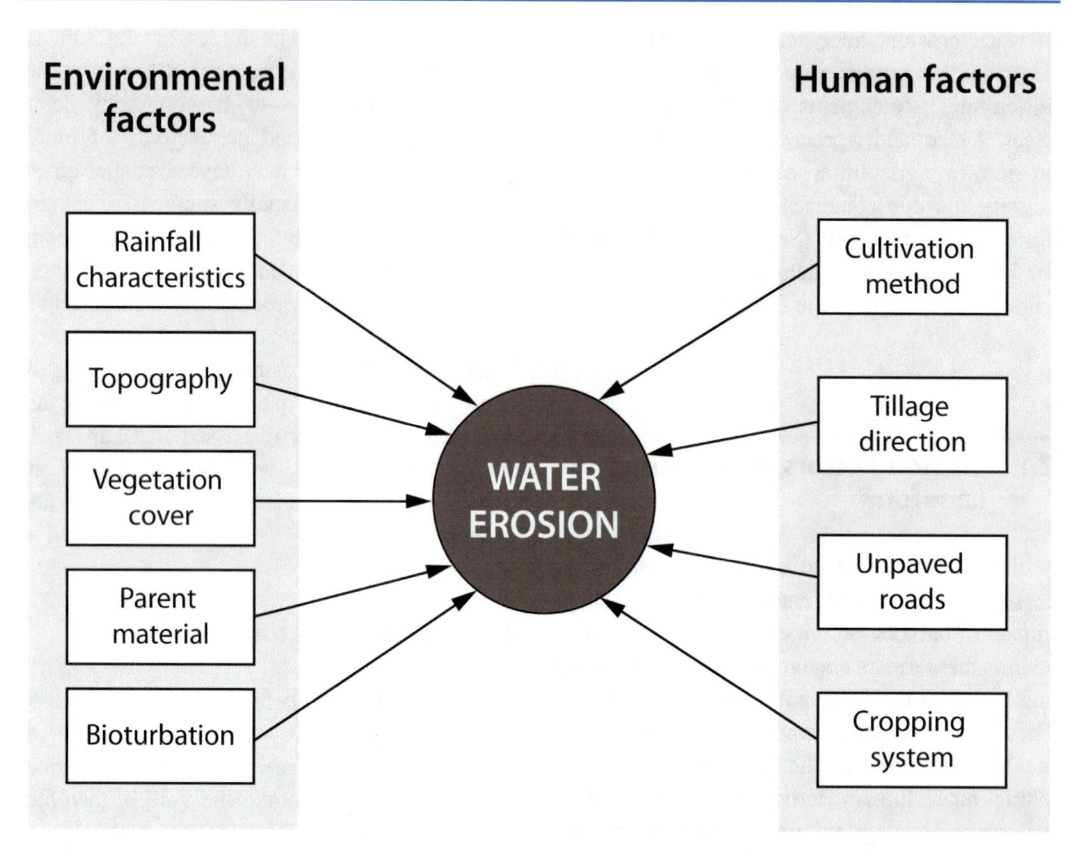

Fig. 2.1 Extrinsic water erosion factors in agricultural catchments, as derived from the clorpt model. The human factors in the focus relate to the various distinct activities of the farmer. Note that some of the factors—such as topography—include several measures such as contributing area, hillslope curvature, hillslope length, and hillslope gradient

are usually heavy, and their own weight can cause soil compaction and compression (Arvidsson et al. 2011)—which can change the size of the soil pores, and the connectivity between them. These changes in structure of the pores adversely affect soil productivity, as they may reduce soil porosity (Schaeffer et al. 2007) and infiltration rate (Martinez et al. 2008). Since soil porosity also controls hydraulic conductivity (Arvidsson and Keller 2011), this may lead to the development of—an artificial and impenetrable —soil horizon below the field, which may obstruct water drainage from the upper soil layer.

Despite this gloomy description, there are several other soil cultivation methods that may be more protective to the soil. Their actual application in the fields may vary in location and time, the farmer's knowledge, habits, budget, and, in particular, her willingness to apply more conservational and demanding cultivation methods (see Sect. 2.4). Blanco-Canqui and Lal (2008c) have summarized the characteristics of field crop cultivation methods into four categories: (1) *Tillage systems* (including no-till, chisel, mulch, and strip tillage, as well as residue removal and burning); (2) *Cropping systems* (including monoculture, fallow, strip, and multiple cropping, as well as cover crops and organic farming); (3) *Water and nutrient management* (including the use of manure, compost, N fixation, irrigation practices, and water harvesting); and (4) *Erosion control practices* (including buffers, windbreaks, terraces and engineering devices, and sedimentation basins). Between these combinations, farmers seek to increase crops biomass production, yields and soil health, reduce soil loss, and maintain soil structure, biological activity, and the cycle of nutrients.

Protective soil conservation actions can be very efficient. Erosion control on the Loess Plateau in China, for example, has been a primary goal for the Chinese government for several decades. Construction of dams and terraces, planting trees and grasses were applied and much was achieved. For example, in the most severe erosion area—the Coarse Sandy Hilly Catchments—that covers only $\sim 15\%$ of the whole Yellow River Basin but produces $\sim 80\%$ of the coarse sediment in the Yellow River—the average annual sediment yield was 75 $\text{tons}\cdot\text{ha}^{-1}\cdot\text{yr}^{-1}$ during 1955–1969 and was decreased to 9.5 $\text{tons}\cdot\text{ha}^{-1}\cdot\text{yr}^{-1}$ in 2000–2009—a reduction of 87% due to soil conservation practices (Fu et al. 2017).

A full review on cultivation methods is provided by Blanco-Canqui and Lal (2008c) but for the purposes of demonstrating the algorithms presented in this book, the next lines divide the myriad possible cultivation methods into three categories, based on their effect on soil erosivity: *conventional tillage, reduced tillage*, and *conservational tillage.* These cultivation methods have been applied in the supersite of this book—the Harod catchment, in field crops and orchards.

1. Conventional tillage is among the most common cultivation methods, used traditionally in many countries. In terms of water erosion, it is also the most destructive one. In practical terms, it consists of a series of actions that usually begins with soil plowing. The soil plowing (or, alternatively, disking) leads to the development of a compacted layer, just below the tillage depth, depending on the length of the plow blade. This sealing of the belowground soil layer reduces hydraulic conductivity with the layers below. After plowing (or disking), the farmer grinds the large clods of earth created by the plow. To do so, farmers usually use a leveling machine while this process may decrease the pore size even more and therefore increase soil compaction. The farmer then breaks the soil further still, using a disk, followed by a coil-tine harrow (USDA 2011). Finally, they conduct another disking operation to aerate the soil and spread out nearby soil particles (Dabney et al. 2011). The natural seal layer shield that is formed by raindrop impact that protects against erosion is, in most cases, destroyed as a result, because the cohesion of the soil bulks is disintegrated, leaving soil particles that are susceptible to detachment and transport downslope in the next rainfall event.
2. Reduced tillage is not a specific technique, but a combination of several different cultivation activities that vary in time and place, with a view to reducing the frequency of soil disturbance by the plow blade. As such, it is also referred to as minimum tillage, because farmers apply it to reduce the pressure on the soil layer to the minimum necessary to grow the sought-after crops (Blanco-Canqui and Lal 2008b). In its simplest form, reduced tillage can be merely a reduction in the number of tillage lines. Another relatively simple application of reduced tillage is the replacement of the regular plow with a subsoiler which facilitates aeration without mixing the soil particles (Fielke 1999)—followed by a supplementary tillage with a light disk, which unties the particles in the upper layer to some extent, but without grinding the topsoil seal layer. Other reduced cultivation activities may include: the use of cover crops, cutting crop residues instead of disking, and planting cover crops with deep roots. Being less intrusive, reduced tillage may allow the soil to recover from previous damage caused by conventional tillage, and to maintain healthy soils. Among the less intrusive practices of reduced tillage is a complete cultivation circle without soil penetration, where the seeds are planted in the previous year's residue known as conservation tillage that will be described in the next paragraph.
3. Conservation tillage was developed to cultivate the soil with minimal penetration or overturn to sow the crop seeds. As an alternative, last-year crop residue remains to protect the topsoil from adverse rain impact, thus

decreasing erosion risk (Leys et al. 2010). Conservation tillage was found in previous studies to decrease runoff initiation and flow (Schwen et al. 2011), and to contribute to stabilizing surface aggregates (Bertolino et al. 2010). Aggregate stability, as an example, was the subject of many studies of cultivation methods, and in fields that were using conservation tillage, aggregates were found more stable than in fields cultivated in reduced or conventional tillage (Hernández-Hernández and López-Hernández 2002). Conservation tillage has also been found to improve the soil's hydrological characteristics (Abid and Lal 2009) and decrease erosivity (Daraghmeh et al. 2008). Furthermore, when the field is sowed and fertilized in a single action, the frequency of movement of agricultural vehicles in the field is reduced and with that the compaction forces on the soil diminish (Friedrich 2000).

Additionally, in conservation tillage the soil is plowed only once in a few years' time while in reduced tillage, partial plowing (plow and roller, or subsoiler and cultivator) is applied every year. The lower frequency of field visits also reduces the pressure that agricultural machinery puts on the soil.

This description of cultivation methods pertains to the cultivation of field crops. In the case of orchards, the cultivation process is even less intrusive to the soil, because the soil is plowed only at the establishment phase, for planting purposes. On a seasonal basis, however, the farmers destroy only the surface seal layer between the tree lines using a coil-tine harrow. They also displace the weeds from the growing rows because they compete for water with the orchard trees. In addition, since tree canopy usually does not provide cover to the exposed separation rows, these slots are unprotected from erosion in intensive rainfall events (Hillel 2006). The conservation treatment in orchards is done by increasing herbaceous vegetation cover in the separation rows. Namely, to minimize the erosion rate in the separation rows, once the large weeds that may compete for water have been removed, herbaceous vegetation of small individual plants is sowed by the farmer to increase infiltration, protect the soil from *splash* effects, and reduce surface flow. This thin herbaceous vegetation layer also helps to stabilize the upper soil layer with an entangled root system (De Baets et al. 2011).

Figure 2.2 summarizes the typical activities involved in cultivation methods in both field crops and orchards.

Figure 2.3 illustrates some of the differences between the five groups of cultivation methods and their consequences for water erosion in orchards and field crops. The close-range photographs on the left panel clearly show the difference in aggregate size between the treatments and the graphs in the right panel show the destructive effect of the less conservational treatments on surface hardness in both orchards and field crops.

Cultivation method is among the most influential factors involved in water erosion, and as such will be treated carefully in the algorithms presented in this book. In Chap. 6, we shall demonstrate a procedure for spatial erosion risk assessment in the entire Harod catchment by simulating the three cultivation methods in field crops and the two cultivation methods in orchards.

2.1.1.2 Tillage Direction

Within the tillage process, the soils can be tilled in several geometrical directions with reference to the hillslope contour lines. The decision regarding the tillage direction may depend on the size and shape of the field, the local topography, and the farmer's decision considerations, be they historical (within the family, for example), cultural (e.g., convenience), or other (Takken et al. 2001a). Tillage direction has been found to substantially determine channeling water flow on the hillslopes of agricultural catchments (Ludwig et al. 1995). Since the mounds created by tilling are relatively high, they can easily direct the flow direction away from the expected topographic downslope in the direction of the till line. In this case, the mounds act as barriers and may reduce runoff and subsequent erosion. The complex

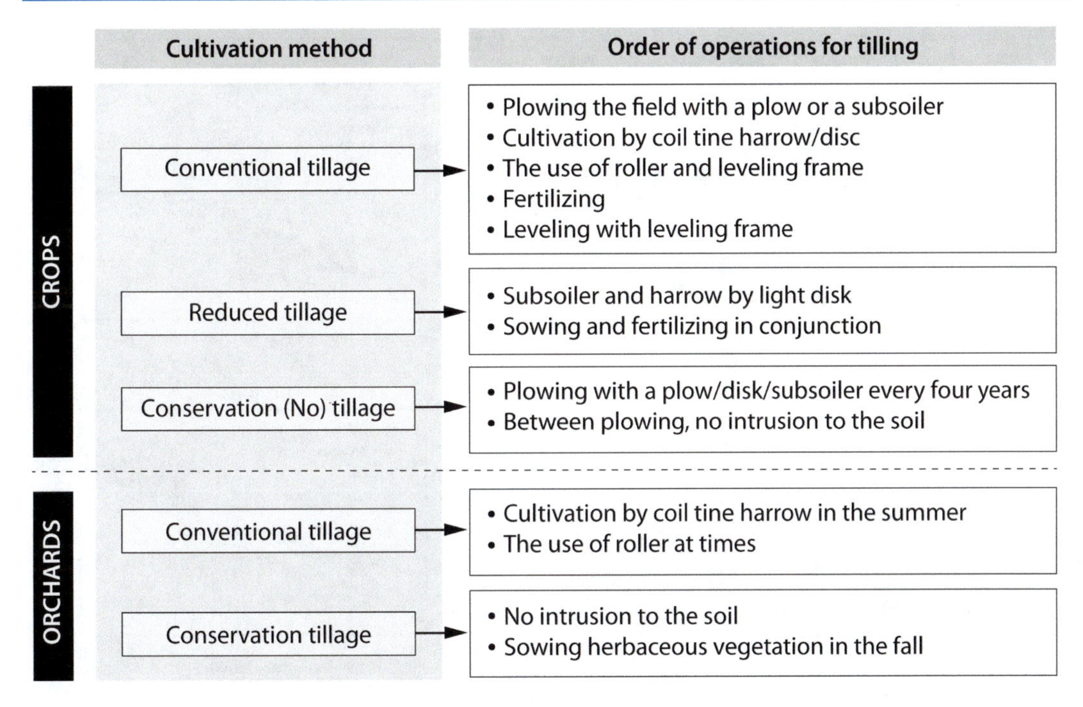

Fig. 2.2 The five cultivation methods applied in the Harod catchment and studied in this book (*Source* Svoray T, Levi R, Zaidenberg R et al. (2015). The effect of cultivation method on erosion in agricultural catchments: integrating AHP in GIS environments. Earth Surface Processes and Landforms 40(6):711–725. https://doi.org/10.1002/esp.3661, With permission from Wiley, Copyright © 2014 John Wiley & Sons, Ltd.)

interaction between hillslope orientation (aspect) and tillage direction may be approximated by means of a cosine function in Eq. 2.1 (Svoray and Markovitch 2009).

$$t = S^{\cos^2 \beta}, \tag{2.1}$$

where t denotes the intensity of the impact of tillage direction on runoff flow, S denotes the hillslope gradient [°], and β is the angle between hillslope aspect [°] and the azimuth of tillage direction [°].

The hillslope gradient, S, in the direction of any α [in degrees from the east and positive in counterclockwise direction] is computed as described in Takken et al. (2001a, b), assuming the gradients in the X- and Y-directions as $\frac{\partial z}{\partial X}$ and $\frac{\partial z}{\partial Y}$, respectively, (Eq. 2.2):

$$S(\alpha) = \frac{\partial z}{\partial x} \cdot \cos\alpha + \frac{\partial z}{\partial y} \cdot \sin\alpha. \tag{2.2}$$

For the hillslope orientation direction used for α, the steepest hillslope gradient S_{max} is computed using Eq. 2.3 with z as the elevation dimension (rise) whereas the hillslope orientation is determined as the aspect of the steepest hillslope gradient.

$$S_{\max} = \sqrt{\left(\frac{\partial x}{\partial z}\right)^2 + \left(\frac{\partial y}{\partial z}\right)^2}. \tag{2.3}$$

According to Eq. 2.1, the effect of tillage direction on runoff flow and consequent soil transportation increases with: (1) the increase in hillslope gradient (base) and (2) the square of the cosine of the angle between tillage direction and hillslope orientation (exponent). For a given field, the lowest effect of tillage direction on runoff flow t occurs when the earlier is perpendicular to the grid-cell hillslope orientation, and the lowest effect of tillage on runoff occurs when the tillage direction aligns with the grid-cell

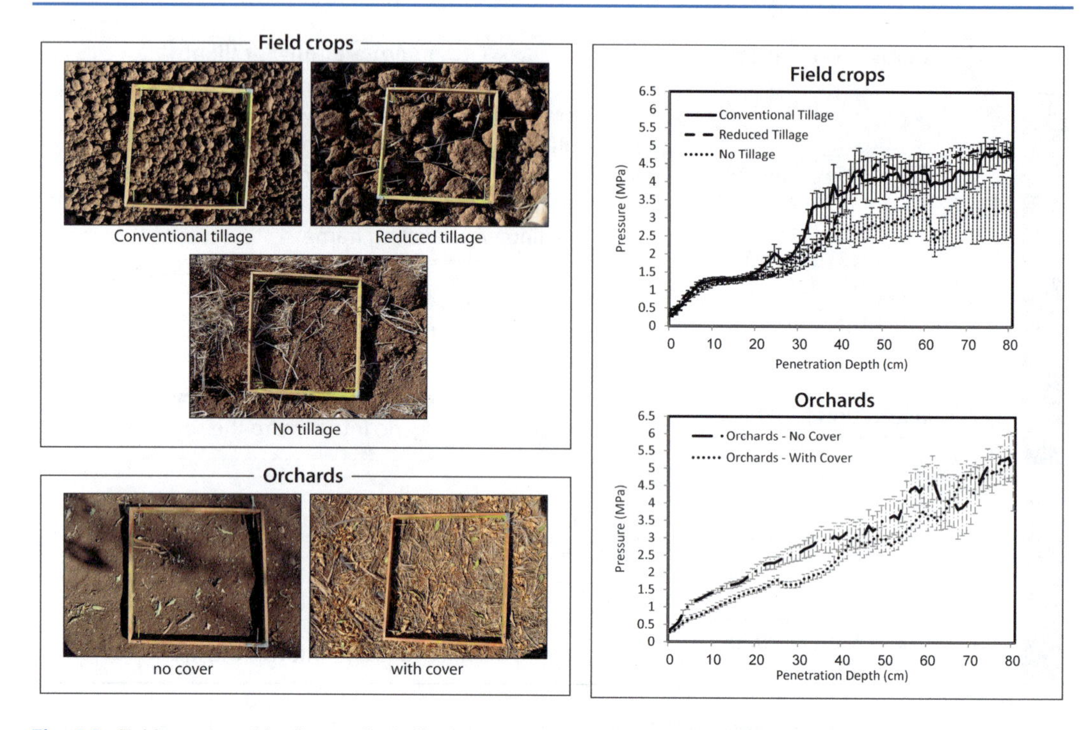

Fig. 2.3 Evidence on cultivation method effect on aggregate size (left panel) and soil penetration depth (right panel). The aggregate size increases from conventional tillage to no tillage while the herbaceous vegetation cover in orchards shields the topsoil. Regarding soil penetration, no tillage plot shows the lowest pressure and the conventional and reduced tillage plots show increased resistance at 30–40 cm depth—that is, the seal layer. The seal layer is not observed in orchards and in the covered orchard the pressure required for penetration is significantly lower (*Source* Svoray T, Levi R, Zaidenberg R et al. (2015). The effect of cultivation method on erosion in agricultural catchments: integrating AHP in GIS environments. Earth Surface Processes and Landforms 40(6):711–725. https://doi.org/10.1002/esp.3661, With permission from Wiley, Copyright © 2014 John Wiley & Sons, Ltd.)

hillslope orientation. Otherwise, t is adjusted by the cosine function (Fig. 2.4).

The map in the left panel of Fig. 2.4 shows per-field mean t values that were originally computed, using Eq. 2.1, on a grid-cell basis. A field with the highest mean t values (7), colored in red, means that the field is tilled perpendicular to the contour lines, while the lowest mean t values (0), colored in blue, mean that the field is tilled equivalent to the contour lines. Lower values are naturally observed along the Harod River, in the lower part of the valley with mild hillslope gradients. The t values increase northward with increasing hillslope gradient values.

As will be demonstrated in Chap. 6, Eq. 2.1 may be applied using the procedure illustrated in Fig. 2.4 to estimate the actual effect of tillage direction on water erosion, and particularly on gully initiation.

2.1.1.3 Unpaved Roads

Unpaved roads are a typical feature of agricultural catchments. Although roads also exist in rangelands and forest areas, they are very frequently used in agricultural catchments, as they serve the farmers for recurrent access to the fields during the growing season. Unpaved roads are compacted surfaces, and being less permeable to raindrops than soil-mantled areas, they act as a network of sealed lines. As a result, a dense road network may increase overland flow and consequent soil transport in agricultural areas.

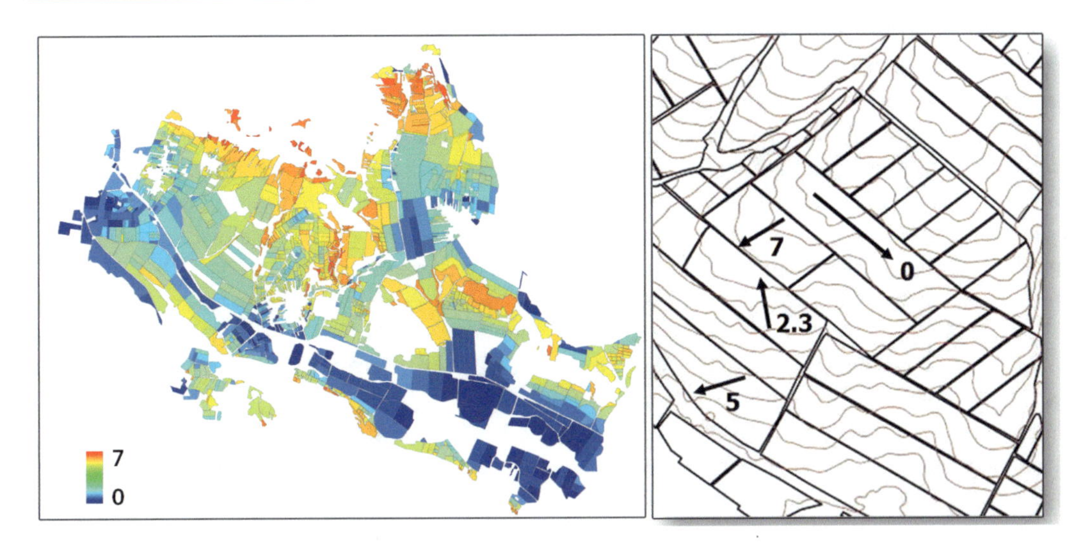

Fig. 2.4 Tillage direction quantification. On the right panel, we see the tillage direction in the fields (black arrows), and the elevation isolines illustrate the hillslope topographic orientation (aspect). The numbers are the score of tillage direction t in each field, computed by means of Eq. 2.1. The left panel is a map of tillage direction scores, as computed for the entire Harod catchment and aggregated using the operator mean to the field level. Lower scores are observed in the center of the catchment, in the flat areas, and higher values in the hilly areas in the northern and southern areas. Still high variability is observed in all fields depending on the tillage direction

Montgomery (1994) has found that such a road network in the upslope contributing area affects the topographic threshold for channel initiation (see in the next pages Eqs. 2.8–2.10). He also found that a smaller drainage area is required to form channeled flow in areas receiving road runoff. Based on the work of Dietrich et al. (1993), Montgomery (1994), suggested Eq. 2.4 to express the link between the discharge intercepted by roads q and *gully/rill* erosion. Montgomery (1994) did so by quantifying the Hortonian overland flow mechanism that occurs when rainfall intensity exceeds infiltration and depression storage capacity. Indeed, Horton runoff is not usually observed in humid areas, but being relatively sealed, unpaved roads provide conditions for Hortonian overland flow in both humid and dry areas.

$$a_{cr} = \frac{2 \cdot (\tau_{cr})^3}{q \cdot k \cdot v \cdot \rho_w^3 \cdot g^2 \cdot (\sin \theta)^2} + \left(\frac{T}{q}\right) \cdot \sin \theta, \quad (2.4)$$

where a_{cr} is the critical contributing area per unit contour length [m] required for incision; τ_{cr} is the critical sheer stress [F·A^{-1}; Pascal]; q is the rainfall rate in excess of the infiltration capacity of the surface generating the runoff [mm·s^{-1}]; k is a multiplier representing surface roughness [dimensionless]; v is the water viscosity parameter [Pa·s]; ρ_w is a parameter denoting water density [g·m^{-3}]; g is the earth gravity coefficient [m·s^{-2}]; θ is the hillslope gradient [^{0}]; and T is the transmissivity of saturated soil [m^3·s^{-1}].

The contribution of unpaved roads to runoff yield in agricultural catchments was also validated empirically. Ziegler and Giambelluca (1997) report that "hydraulic conductivity of unpaved roads is approximately one order of magnitude lower than any other land-surface type" in a mountainous agricultural catchment. They also found that roads generate runoff at an early stage of rainfall events and that despite the fact that they cover small areas, their contribution to the total runoff in the catchment is large during small rainfall events.

Roads, therefore, may act as major contributors to runoff yield in the catchment from their sealed surface. However, roads can also act as barriers to overland flow—due to their increased elevation and surrounding mounds which can divert or even halt overland flow. These two influential functioning features of roads are expressed in the geoinformatics methods used in this book, using GIS buffer zone operators (Fig. 2.5).

In Fig. 2.5, the flow direction is from east to west because the right side is higher than the left side. As a result, the overland flow that arrives from the eastern field is stopped by the road and the overland flow that arrives from the road drains into the western field. This is expressed in the multi-ring buffer with darker buffers simulating more water ponding in the road as a barrier case (due to water accumulation) and more soil loss in the case of road as a contributor (due to loss of water energy).

2.1.1.4 Cropping System

Cropping system is a term that refers to spatial and temporal changes of the crop type and the farmers' management of the fields due to these changes. The crop type—and in the long-term, the sequences of various crops that are grown in the fields by rotation—may substantially affect the resistance of the fields to erosion in two key aspects:

1. The different cultivation treatments between the crops and the different crop types initiate several chemical, biological, and physical processes, and therefore increase variation in susceptibility to erosion by changing soil properties (Page and Willard 1947).
2. Plants function as a shield for the soil's upper layer by absorbing part of the raindrops' kinetic energy, and the crop type can affect this shielding through various leaf and canopy characteristics. Closed canopies can prevent the development of a seal layer, thereby preventing the generation and flow of runoff water, while the opposite is true for open canopies and small leaves.

The second assertion is well evidenced in the literature: canopy cover [%] has been found to have an exponential effect on soil loss and soil detachment by splash (Gyssels et al. 2005). The differences in this respect between the various phases of growth—from small seedlings to full plant cover or between wide-leaf corn and wheat—may be very large.

2.1.2 Environmental Factors

2.1.2.1 Rainfall Characteristics

The entire process of soil detachment and consequent sediments transport is initiated by the raindrops impact which is affected by the kinetic energy [kJ] of the raindrops as they hit the soil surface. Variation in rainfall characteristics such as rainfall depth and its onset may also affect erosion process, since runoff generation and soil detachability also depend on soil infiltrability (Zhu et al. 1997) and antecedent moisture content (Ziadat and Taimeh 2013). However, many studies have found that rainfall intensity is the most influential factor on water erosion, especially in the case of heavy storms in semiarid to sub-humid areas (Knox 2001, Boer and Puigdefábregas 2005). In quantitative terms, raindrop intensity and its resulting impact on detachment have been found to depend on the following two measures: raindrop diameter D and raindrop fall velocity V, and are expressed as the sum of power law functions of D and V for each raindrop, based on empirical tests of soil detachment by raindrop impact (Eq. 2.5 from Salles et al. (2000)):

$$Er_{\alpha,\beta} = C_{\alpha,\beta} \cdot \sum_{i=1}^{n} D_i^{\alpha} \cdot V_i^{\beta}, \quad (2.5)$$

where $Er_{\alpha,\beta}$ is the rainfall erosivity; α and β are coefficients whose values depend on the rain parameter considered; and $C_{\alpha,\beta}$ is a coefficient, such that $Er_{\alpha,\beta}$ expresses a rainfall parameter. i is the running number of raindrops until n, D_i and V_i are the diameter and the fall velocity of the raindrop i, respectively. Aside from theoretical computations, rainfall intensity can be monitored

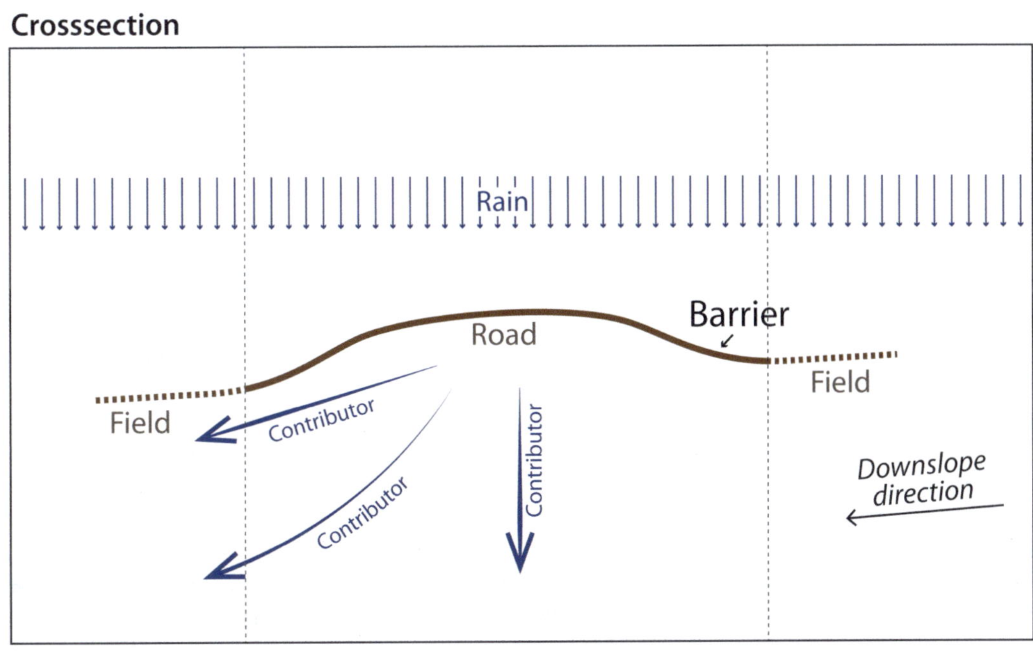

Fig. 2.5 The effect of an unpaved road on overland flow and soil erosion. The upper panel is a cross section that shows a road slightly higher than the surrounding fields. Under these conditions, the overland flow from the upper field is blocked by the road, which acts as a barrier and may cause water ponding in the edge of the field. Conversely, water accumulating on the road itself breaks downslope into the lower field, thereby acting as a contributor. In the lower panel, we see how the role of the road as a barrier and as a contributor is quantified using a buffer ring: On the right, the buffer shows the increased risk for water ponding; the left buffer shows the decreased risk for gully incision

to a high degree of certainty using widely accessible remote sensing means. Meteorological radar, for example, is used to estimate rainfall intensity using empirical parameters that link between the system reflectivity and rainfall intensity in a given grid-cell (Morin et al. 2003). Meteorological radar is often used in geoinformatics platforms along with other extrinsic factors to estimate water erosion risks (Svoray et al. 2015, Svoray and Ben-Said 2010).

2.1.2.2 Topography

Topography, which determines the geometry of water overland flow, is quantified in the existing literature using several indices (see Chap. 4 for more details). Among these measures, hillslope gradient S and the contributing area A (see Fig. 2.6) are the most used topographic factors in the study of water erosion risks—especially with regard to gulling and rill initiation and development in agricultural catchments (Torri and Poesen 2014). The importance of hillslope gradient and contributing area to predicting water erosion is well reflected in the large number of studies on the Topographic Threshold (TT)—a measure that will be further discussed in the next chapters.

However, note that the effect of topography is not always straightforward. In the case of *inter-rill* (or *sheet*) erosion, the effect of topography may be mediated by seal formation for example. Assouline and Ben-Hur (2006) have found that a more permeable layer was formed as hillslopes become steeper. Thus, for given rainfall characteristics, less runoff was produced for steeper hillslopes. This is counter-intuitive if the reader does not consider the seal formation, but more understandable once the spatial variation in permeability is taken into account.

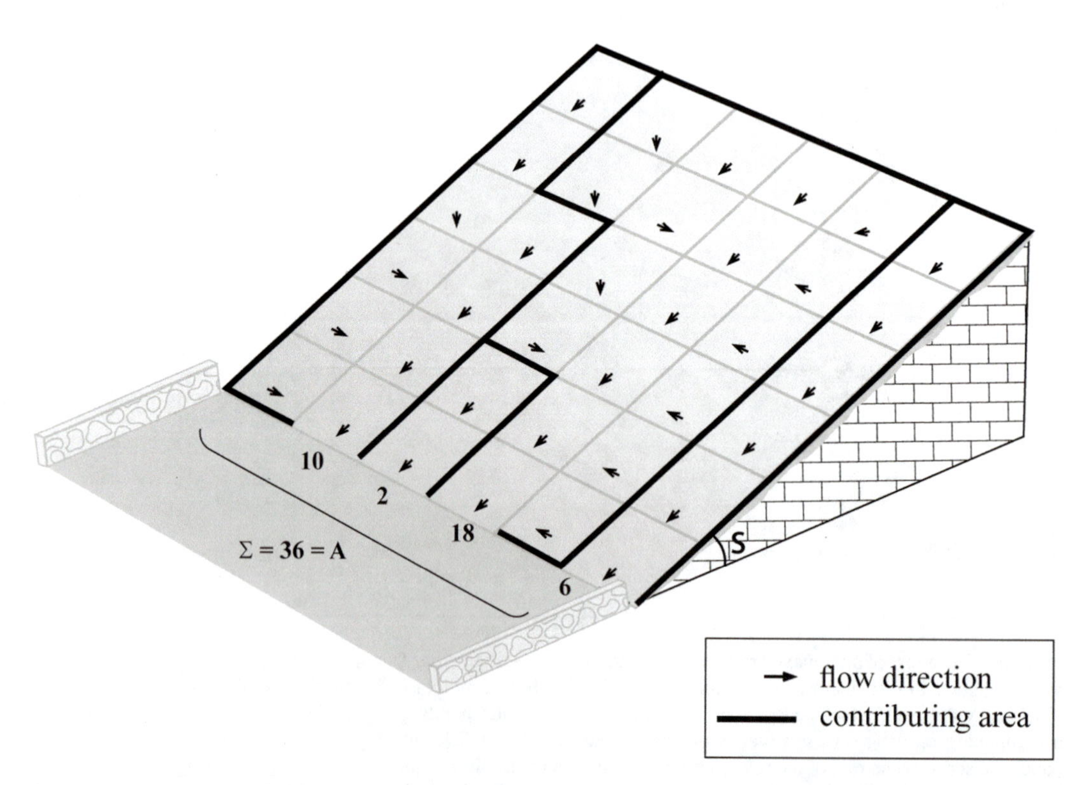

Fig. 2.6 The effect of hillslope gradient S and contributing area A on runoff development, flow and accumulation downslope. Each single grid-cell is assigned with a number of accumulating grid-cells depending on the flow direction determined by the steepest hillslope direction. In this simulation A_1 = 10 grid-cells, A_2 = 2, A_3 = 18, etc.. and the contribution of this entire hillslope to the channel is A = 36

2.1.2.3 Vegetation Cover

The rather dense vegetation layer in agricultural fields has an almost binomial effect on water erosion in these fields (yes/no runoff flow). The individual seedlings that grow gradually after germination in the field crops are major agents in restraining runoff development and soil loss, because the fields are densely sown, and in many of the crops the plants are high and maintain an entangled root system that, after germination and several weeks of growth, substantially halts runoff and subsequent erosion processes. This notion has been supported in many past empirical studies. Lessening of soil loss with growing vegetation cover and root mass has been found in previous studies to be exponential, for a review, see the work of Gyssels et al. (2005). Moreover, it has been found that for splash and *interrill erosion*, aboveground vegetation cover is the most influential vegetative parameter, whereas in the case of small ephemeral gullies and rills, aboveground and belowground biomass are of equal importance.

2.1.2.4 Parent Material

The parent material is the lithological layer that provides the soil's underlying infrastructure (see Fig. 1.1). This layer may affect hydrological and erosion processes mainly in shallow soils subjected to intensive erosion. Namely, those fields characterized by a thin soil profile, and when vegetation is absent, or negligible (Cerda 1999). In such agricultural fields, the impact of parent material on water erosion processes is expressed by the surface response to raindrops kinetic energy and impact, and by controlling infiltration, and increasing contributing areas due to sealed surfaces that may be very large in some bedrocks (Cerdà 2002). The observations of Cerdà which were acquired in semi-natural sites may also hold true for eroded agricultural fields that do not maintain the thick soil profile. The parent material is also influential in dictating some of the soil characteristics, which may affect the development of cracks in clayish soils with increased permeability and hydraulic connectivity —via preferential lines, for example, and the subsequent water flow.

2.1.2.5 Bioturbation

Bioturbation—the change in soil characteristics incurred by animals and plants—can be implicated in water erosion in agricultural catchments, for example, as a disturbance that may be used by the water to initiate a gully head. Changes of soil characteristics by animals can include digging and trampling, chemical changes due to fecal and urine activities, and even consumption of sediments. Such disturbances can be perpetrated, for example, by Günther's vole (*Microtus guentheri*), or any other rodent burrowing in the field. Mapping historical vole pit holes—or even termites (Breuning-Madsen et al. 2017) or isopods and porcupines activities (Yair 1995)—may indeed help to simulate factors affecting water erosion processes.

The erosion factors described so far are practically dictating the conditions that exert the main effect on the initiation, expansion, and consequences of water erosion processes in agricultural catchments. These factors may have a multi-factorial impact while their importance to erosion processes may differ temporally and spatially, and some factors may be of higher importance than others to erosion processes and consequences. Simulating these environmental and human factors with various geoinformatics tools is therefore a necessary step for understanding the processes and forms of water erosion in agricultural catchments.

2.2 Processes and Forms of Water Erosion in Agricultural Catchments

The literature, e.g.,Weil and Brady (2017), identifies five core types of hillslope water erosion processes: splash, sheet, rill, gully, and *piping* erosion. To these hillslope processes, we should add *valley* (or *stream*) erosion and *bank* erosion, but these two are less relevant to the hillslope agriculture studied in this book. This section reviews the basics of the first five water erosion processes and forms.

2.2.1 Splash Erosion

Splash erosion is a process that includes the breakdown and detachment of topsoil particles, and the transport of some of these, usually downslope, in some sort of a ballistic trajectory. Such a process may increase sediment yield in lower parts of a hillslope, but due to the fact that in splash erosion most of the material that moves include fine soil particles, splash erosion may also cause surface sealing downslope.

Splash erosion is usually caused by the impact of raindrops striking wet soils, which are more affected by repulsive hydration forces, and occur as the detached particles are transported further downslope through surface runoff (Brady and Weil 2016). Small craters are formed in the soil bed during the splash process, because the raindrops strike the soil particles with high energy.

The different shape and size of the formed craters were traditionally attributed to rainfall characteristics. The relationships between the raindrop energy and the splash crater characteristics were determined by Engel (1961), as described in Eq. 2.6:

$$D = K \cdot R \cdot \left(\rho \cdot V^2\right)^{\frac{1}{3}}, \qquad (2.6)$$

where D is the crater depth [cm], K is a constant that needs to be determined empirically for each site, R is the raindrop radius [cm], ρ is its density [$g \cdot cm^{-3}$], and V is the raindrop velocity [$m \cdot s^{-1}$]. However, subsequent research revealed that rainfall characteristics are not the only factor in determining the crater dimensions and the splash effects. Current studies show that under various field conditions, soil internal forces—such as electrostatic, hydration, and electromagnetic van der Waals—may also exert contribution to splash erosion, in addition to raindrop impact force (Hu et al. 2018). According to Huang et al. (2016), operating on the particle scale of soils, electrostatic and hydration forces are repulsive, and cause disaggregation of the soil particles, while van der Waals forces are attractive, reducing aggregate breakdown (Fig. 2.7).

The soil is not always exposed, and a dense vegetation layer can provide the soil surface with a shield that dramatically decreases the total amount of soil loss through splash erosion processes. Nonetheless, Eldridge and Greene (1994) have shown that splash erosion is an influential water erosion process in many agricultural fields. They have found that, under high rainfall intensity and exposed soil conditions, splash erosion

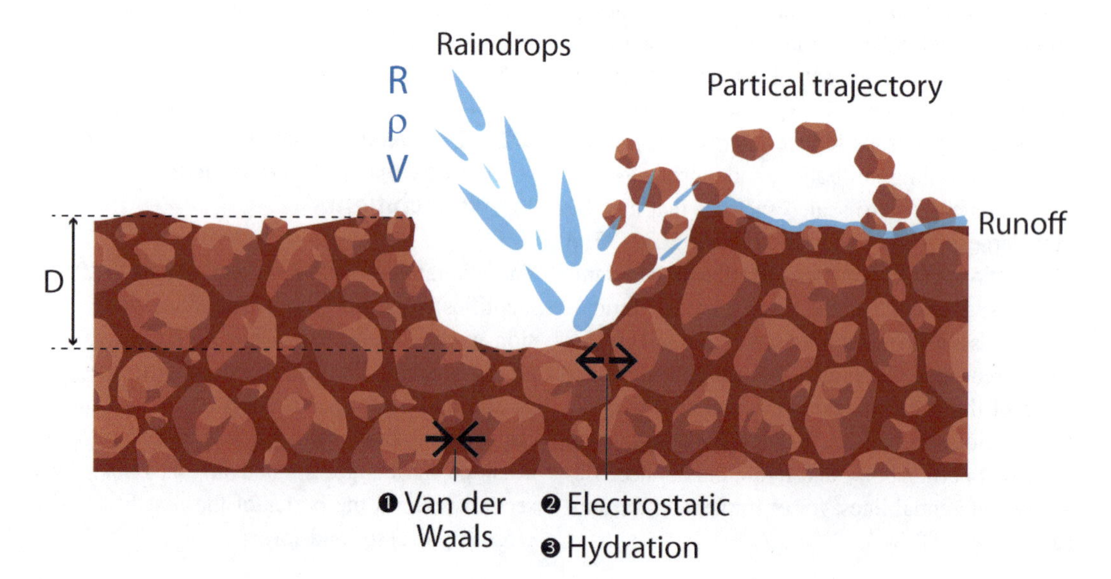

Fig. 2.7 Splash erosion in a theoretical flat agricultural field: R, ρ, and V stand for the raindrop radius, density, and velocity, respectively. The soil operating forces are the electrostatic and hydration (repulsive) and van der Waals (connective)

leads to repeated detachment, substantial transfer and loss of fine material, and high-value nutrients from agricultural fields.

2.2.2 Sheet Erosion

Sheet erosion is the process of soil removal and transportation downslope by water in an unchanneled formation, mostly due to wide and shallow overland flow. To enhance the distinction between channeled and unchanneled flow, sheet erosion is also occasionally referred to as interrill erosion. Several equations have been developed to quantify sheet erosion and they can be explored in the detailed review study of Wei et al. (2009). Generally, those quantifications stem from Eq. 2.7 (Flanagan and Nearing 2007), where D_i [kg·m^{-2}·s^{-1}] is the sheet erosion rate, K_i [kg·s·m^{-4}] is the sheet erodibility, I [mm·min^{-1}] is rainfall intensity, and q [m·s^{-1}] is the sheet runoff rate.

$$D_i = K_i \cdot I \cdot q \qquad (2.7)$$

Sheet erosion can cover large areas and is often considered as a uniform flow downslope—hence the "sheet" metaphor. However, surface roughness due to stones, mounds, and other objects can cause considerable differentiation in sheet flow structure.

Sheet erosion is a very common type of water erosion, while splash and sheet erosion occur simultaneously in many agricultural catchments although splash erosion usually dominates initially. In past empirical research, sheet erosion has been found to depend substantially on rainfall intensity [mm·min^{-1}], hillslope gradient [^{0}], and soil erodibility [kg·m^2·second] (Blanco-Canqui and Lal 2008a). One of the main threats in sheet erosion to farmers is that it can go unnoticed for a relatively long time, due to its characteristically shallow flow. However, the damage builds up, and may be noticed only after much of the soil profile is removed. Figure 2.8 illustrates the formation of sheet erosion, and the damage it causes at local and landscape scales of observation. Rock fractions from the C horizons are exposed across wide areas after the soil material between them is removed. The farmers clear the rock fractions before the sowing season—only to find other rock fractions being exposed in the following years, due to continual sheet erosion. The state of the soil illustrated in Fig. 2.8 is the outcome of sheet erosion over several decades of conventional cultivation in the Kibbutz Dalia wheat fields, just a few kilometers west of the Harod catchment. More often, sheet erosion may be found in fields with poorly consolidated soil material due to recent plowing and without the coverage of a protective vegetation layer.

2.2.3 Rill Erosion

The irregularities in soil roughness can direct the overland flow into small channels called rills. Small rills are usually eliminated by the farmer during the cultivation process—only to reform again in the following season, causing the loss of more soil from the field. Rill erosion has the character of a concentrated flow, which means much faster rates than the case of shallow flow that characterizes sheet erosion. The rill is initiated when the hydraulic shear of water flow (left side of Eq. 2.8) overcomes the soil resistance (right side of Eq. 2.8)—which is largely a function of soil erodibility (Shields 1936):

$$\tau_{cr} = \rho \cdot g \cdot R \cdot S > \theta \cdot g \cdot (\rho_s - \rho) \cdot D, \qquad (2.8)$$

where τ_{cr} is the critical shear stress below which there is no incision, ρ is the water density [g·cm^{-3}], g is gravitational acceleration [cm·s^{-2}], R is rill hydraulic radius [cm], S is the hillslope gradient [^{0}], θ is an empirical coefficient, ρ_s is the soil density [g·cm^{-3}], and D is the soil particle size [cm]. Begin and Schumm (1979) used empirical data to formulate Eq. 2.9 that links between the critical shear stress τ_{cr}, upslope contributing area, and hillslope gradient:

$$\tau_{cr} = (C \cdot \gamma) \cdot A^{rf} \cdot S, \qquad (2.9)$$

where C is a dimensionless coefficient to be determined empirically and γ is the weight per unit volume of the water [g·cm^{-3}], A is the

Fig. 2.8 Rock fragments from C-horizon are being exposed in the Kibbutz Dalia wheat fields after decades of continuous sheet erosion. Photographed by Benny Yaacobi, The Israeli Ministry of Agriculture

contributing area [m^2], *rf* is an empirical parameter, and *S* is the hillslope gradient [^{0}]. Since the 1970s, many—see the review by Torri and Poesen (2014)—have computed the combination of hillslope gradient and upslope contributing area required for erosion to occur, and have found the following relationship between the two factors to occur (Eq. 2.10):

$$S = a \cdot A^{-b}, \tag{2.10}$$

where *S* is the hillslope gradient [$m \cdot m^{-1}$], a is the empirical coefficient, *A* is the contributing area [ha], and b is the empirical coefficient (Desmet et al. 1999). Based on this theory, the catchment area and the hillslope gradient are proxies to the runoff volume, and channel head may initiate wherever a Topographic Threshold (TT) is exceeded. Previous studies have shown that the a and b coefficients may vary due to local clorpt conditions in the catchment and the method of data collection (Svoray and Markovitch 2009). After initiation occurs, the rill development to a certain width, length, and depth depends on the runoff water transport capacity.

2.2.4 Gully Erosion

Gullies are defined as "channels with a critical cross-sectional area larger than a square foot" (Hauge 1977), where a square foot is equal to 929 cm^2. Thus, rills and gullies differ only in size, and both are forms of small channels that flow along hillslopes. Gullies are usually classified into two main classes:

- *Permanent gullies*—that are too deep to be filled by the farmers during seasonal cultivation process (Foster 1986). Depth of permanent gullies was observed in several studies to be in the 0.5–30 m range (Poesen et al. 2003).
- *Ephemeral gullies*—which are simply smaller channels that can be erased by the farmers during the seasonal tillage process (Casali et al. 2006). A more detailed discussion of the differences between rills and ephemeral gullies is provided by Vandaele et al. (1996).

Gullies transport a very large amount of sediments in a wide range of agricultural environments (see Fig. 2.9), and therefore gullies act also as means for transferring water and soil from the upper parts of the catchment to the lower parts of it. Hence, permanent—and, to a lesser extent, ephemeral—gullies aggravate a wide range of off-site effects. As a means of transportation of large amounts of water, soil, chemical substances and biological information, gullies therefore increase landscape connectivity, and wreak considerable (if not most of the) damage (Poesen et al. 2003) to neighboring fields, roads, watercourses, and other properties.

2.2.5 Piping Erosion

Soil pipes are discrete preferential flow paths *underneath* the soil surface, formed parallel to the hillslope gradient (Wilson et al. 2018). Pipes are the product of soil pores, and are usually created and formed by various physical and biological processes. Piping is common, for example, in peatlands because erosion can occur under the peat surface (Li et al. 2018). The literature classifies soil pipes into two main groups: (1) vertically oriented preferential pathways, whose length is limited by the soil depth and (2) pipes that transport the soil material parallel to the hillslope gradient along preferential flow pathways.

Soil piping can be very intensive, and Eq. 2.11 represents a destructive feedback to the soil by the piping mechanism (Wilson et al. 2018).

$$q_s = \rho_d \cdot \frac{\partial R}{\partial t}, \tag{2.11}$$

where q_s is the sediment flux, which denotes the rate of soil material loss from the pipe walls, ρ_d is the bulk density, and R is the change of the pipe radius over time t. Once other factors, such as

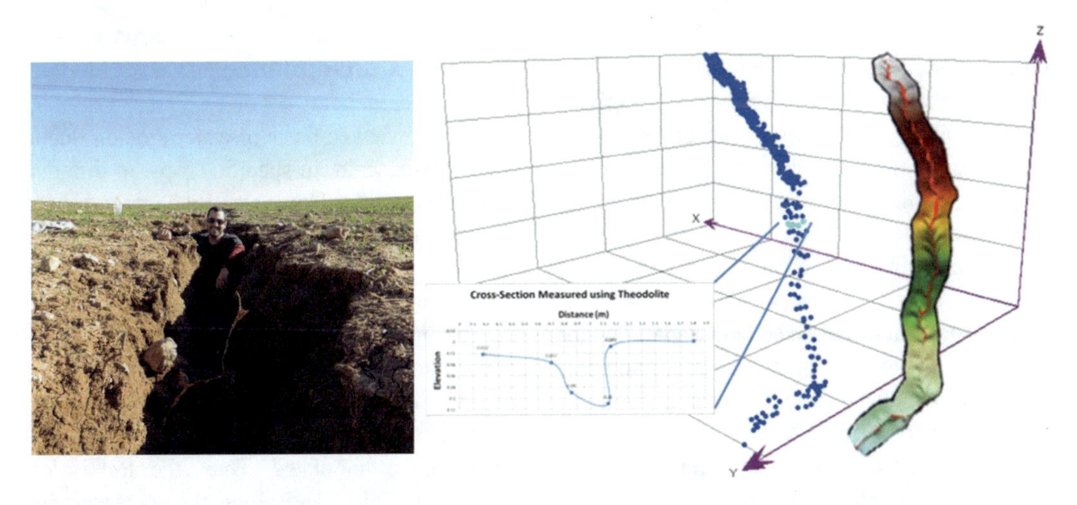

Fig. 2.9 An ephemeral gully developed in the Revadim catchment in the northern Negev region of Israel, between December 10 and 23, 2013. On the left—a photo of the investigated gully, with the MA student David Hoober in it illustrating its depth. The right panel shows the gully reconstruction using measurements from an electronic theodolite. Photo taken by Dror Zolta and the graph is unpublished data from Dudi Hoober's Master Thesis.

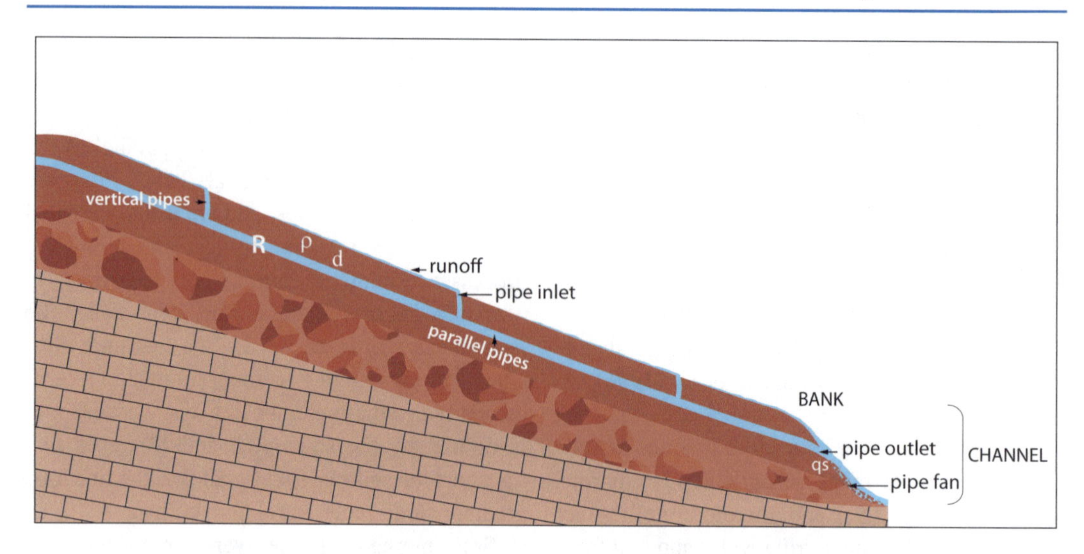

Fig. 2.10 Vertically oriented pipes and parallel pipes along a given hillslope. Note the pipeflow between the flute holes and the pipe outlet, with a small pipe fan of sediments at the river bank

hillslope gradient, are held constant (and therefore the partial derivative) the pipe radius grows with time, and the flow in the pipe can increase as a function of soil conditions and runoff interception. The increase in flow in the pipe results in increased water velocity, leading to increased yield in the pipe fan (Fig. 2.10).

As a result of the feedback presented in Eq. 2.11, hillslope parallel pipes substantially affect the soil loss processes for the hillslope as a whole, and should be treated differently from vertically oriented pipes (Sharma et al. 2010). Furthermore, the hydrological connectivity between soil cavities below the ground creates a network of preferential lines that transport the soil out of the field—sometimes without even the farmer's knowledge, as all these flow processes are hidden from view. Piping is a widespread phenomenon and has been observed in both natural and anthropogenic landscapes, under varying climatological, parent material, and soil settings (Bernatek-Jakiel and Poesen 2018).

To summarize Sect. 2.2., in water erosion processes, the soil surface in agricultural fields is transported by rainfall water, overland flow, and sometimes irrigation. Overland flow is the main driver of water erosion, but water, soil, and other substances can also be removed from the field through belowground flows. The water erosion process begins with a detachment of soil aggregates by raindrop with high kinetic energy in a splash process, followed by transport of detached particles in sheet or channeled flow, and deposition further downslope. This movement of water and soil has various consequences for agricultural fields which are detailed in the following section.

2.3 The Damage: On-Site and Off-Site Consequences

Chapter 1 reviewed costs of water erosion due to reductions in yield, in specific crops, in different parts of the world. However, the damage caused by water erosion to agricultural fields is far more than merely reductions in yield. The process of water erosion, from detachment to transportation and sedimentation, generates several consequences to the environment—both within the site, where the soil is removed, and to surrounding sites, where the soil and other substances are accumulated. Thus, the soil is lost from specific sites and along the transportation path; soil particles cover, pollute, and block various objects; and sedimentation occurs at unwanted locations in water bodies or inland areas. As a result, matters that are needed at the point of origin (including soil minerals, water, chemical and biological materials) are removed from it and appear elsewhere, where they are not needed.

Due to the movement and dynamics of various substances in water erosion processes, professionals usually distinguish between *on-site* and *off-site consequences*. This on-site/off-site distinction appears also in other disciplines such as economics when comparing workplaces.

With regard to water erosion, on-site damage usually denotes the physical reduction of the soil profile or damage to the topsoil layers—such as a reduction in availability of soil NPK (nitrogen, phosphorus, and potassium) nutrients. Such reductions in soil functionality decreases soil health, reducing plant productivity as well as yield quality and the soil's ability to provide other ecosystem services (Rinot et al. 2018). An important and unfortunate on-site characteristic is the loss of the soil itself—with the eroded material usually containing the finer fraction. In the face of such continuous erosion exertion over time, the remaining soil tends to lose its finer fraction and some of its water-holding capacity, which is crucial for plant productivity and other ecosystem services.

Off-site damage is usually caused by mass movement of soil material, pollutants, or other chemicals, from the site into adjacent sites/fields, thereby covering or blocking infrastructure—including, for example, road, bridges, and drainage ditches. The soil material transported outside the site can reduce the capacity of waterways and reservoirs; increase the risk of floods; and contaminate, poison, or bury other ecosystems—resulting in *eutrophication*, and infecting drinking or irrigation water sources with nitrogen and phosphorus. The soil can also build up on roads and cause accidents.

Figure 2.11 illustrates some examples of on-site and off-site consequences of water erosion in agricultural catchments.

There is another difference between the two types of consequences of water erosion that may affect soil conservation policy. Off-site consequences are usually visible, and in many cases noticed and treated. On-site consequences can at times be less visible, and therefore less monitored, and often neglected. As a result, on-site consequences—such as reduction in soil health (see Chap. 7)—build up over years and even decades, growing substantially in magnitude (Brady and Weil 2016). The on-site damage can emerge when it is too late, and the fields are at the point of no return, for cultivation purposes. Therefore, early warning and identification of the damage is crucial. In the following two sections, we detail and provide evidence from the literature of some of the more typical on-site and off-site consequences of water erosion. As will be shown in the next chapters, these are the kinds of consequences that the geoinformatics procedures provided in this book will tackle.

2.3.1 On-Site Consequences

The "site" is a tangible unit—a specific field, or a cluster of fields, or a particular topographic entity, such as a sub-catchment, a hillslope, or a hillslope catenary unit (see Chap. 4). On-site consequences are those related to the reduction in soil health at the site following substance losses. The loss of the soil minerals—namely, the reduction in soil depth—is the most commonly known on-site consequence, but it is not the only one. Table 2.1 reports a collection of studies on typical on-site consequences of soil erosion from around the world.

One example of on-site consequences was observed in the Swiss Plateau, where channeled and sheet erosion processes caused loss of up to 521 tons of soil from the uppermost layer of fields over a decade, at an annual average amount of 2.87 $ton{\cdot}ha^{-1}{\cdot}yr^{-1}$. The loss of soil material created depressions in the fields, where stagnating water accumulated, creating anaerobic conditions for the crop plants (Lemann et al. 2019). Soil loss can also occur in the wake of extremely destructive landslide processes (Vranken et al. 2013). Although landslides are rarer than channeled or sheet erosion, they can cause the loss of a huge amount of soil in a single catastrophic event. Landslides are also particularly dangerous, because in addition to the removal of the soil

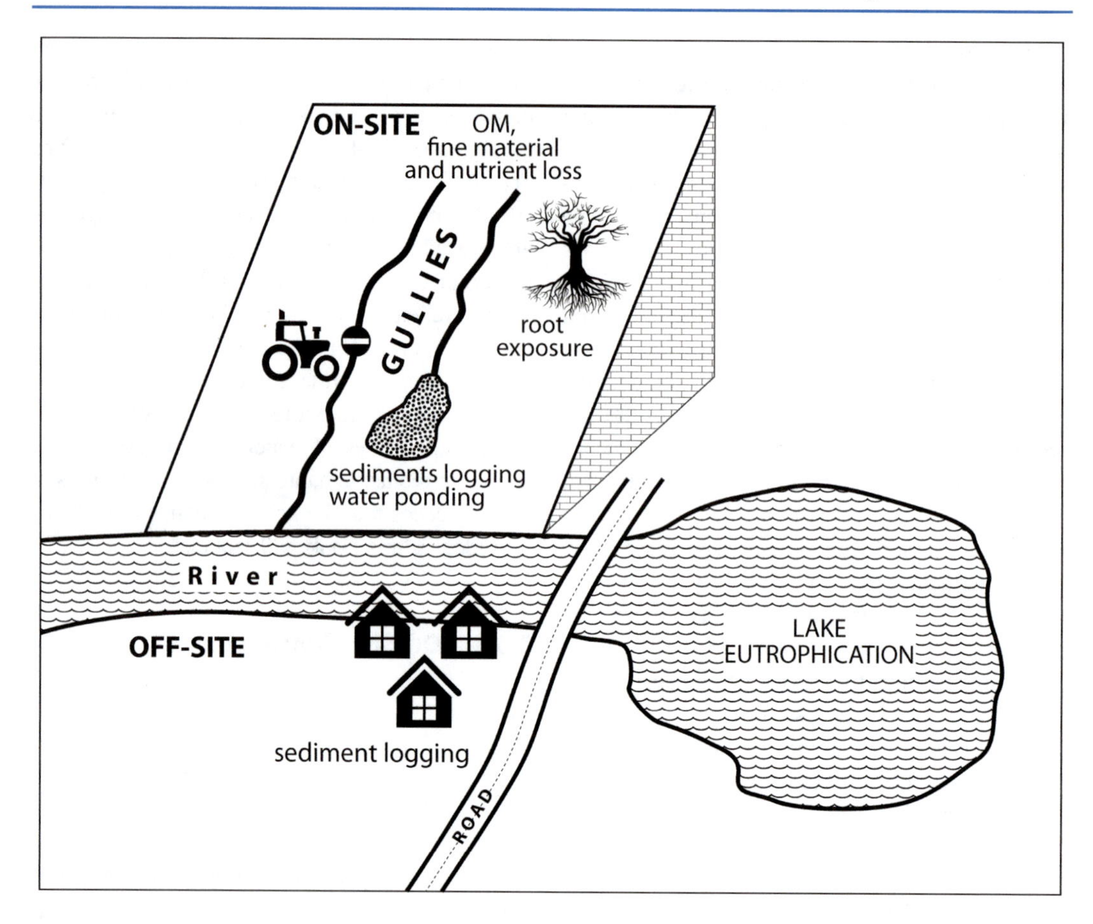

Fig. 2.11 A conceptual representation of on-site and off-site consequences of water erosion from a given site/field on a hillslope. The on-site consequences are mainly related to the reduction of soil health, while the off-site consequences pertain mainly to the accumulation of eroded mineral, chemical, and biological material in unwanted places. Note the on-site consequences can also, at times, include accumulation of material in unwanted places within the site itself as sometimes the runoff energy is insufficient to carry the material outside the site

Table 2.1 On-site consequences of water erosion in agricultural fields, in different environments

Source	Consequence	Climate	Mechanism	Method
Bell et al. (2011)	Soil compaction	Semi-humid	Mixed crop-livestock	Simulation model
Vranken et al. (2013)	Structural and functional damage	Humid	Landslide	Interviews
Fleshman and Rice (2014)	Seepage dam failures	Laboratory experiment	Piping erosion	Laboratory modeling program
Chen et al. (2015)	Dam failures	Laboratory experiment	Rainfall-induced	Flume simulator experiment
Ollobarren et al. (2016)	Reduction of soil health	Med	Water erosion	Chemical and physical analysis of soil sample
Lamandé et al. (2018)	Reduction of water storage and carbon sequestration	Humid	Agro-machinery	Field experiment
Lemann et al. (2019)	Stagnate waterSoil loss	Humid	Water erosion	Field experiment

itself from the site, the transported land mass involved can include—due to the magnitude of the event—agricultural structures, roads, electricity lines, and other networks at the site. Severe damage due to water erosion by landslides is dam failure—which can occur depending on factors such as dam geometry, the amount of inflow, hydraulic conductivity of the dam material, and riverbed conditions (Chen et al. 2015). In-site soil loss can also occur due to seepage that can lead to the development of subsurface flow—described earlier as piping—that may transport the soil out of the field in a belowground process that in many cases can hardly be observed by the farmers or remote sensing tools (Fleshman and Rice 2014).

Another on-site consequence is the effect of nutrients loss from the site, which substantially diminishes the quality of the remaining soil at the site, because nutrients are constituents used by the crop plants to produce biomass. Due to their importance, the loss of even a small amount of nutrients may have a large impact on the agricultural productivity in the site. For example, in the agricultural fields of Sicily, water erosion caused significant deterioration in soil quality. A soil quality index—based on chemical and physical soil variables, such as soil texture, bulk density, soil water content, pH, EC, carbon, nitrogen, and phosphorus content—diminished greatly, especially where a high degree of gully erosion was observed. The absence of these soil components was evident in the considerably reduced yield production (Ollobarren et al. 2016). The effect of water erosion on the various physical, chemical, and biological components of the soil is a key point to this book and it will be further discussed in Chaps. 7 and 8.

Another on-site consequence that has been frequently reported in the literature and has been found to affect agricultural cultivation is *subsoil compaction*, which can considerably increase the possibility of saturated runoff. Subsoil compaction is usually caused by agricultural machinery wheel load, due to the high pressure it imposes on the topsoil. The dense shallow seal layer reduces water storage and carbon sequestration (Lamandé et al. 2018), and as a result reduces yields. High shear forces are also caused by the tires of agricultural machinery, which can create small pits that may initiate gully or rill erosion. Raindrop impact can also create a dense layer and reduce hydraulic connectivity—especially when combined with other factors, such as livestock movement, which has been found to enhance the compaction effect. That is because animal treading may increase bulk density and reduce macro-porosity, and consequently the infiltration rate. Two experimental studies in southern Australia showed that soil compaction hampers root growth and reduces infiltration, which in turn may reduce grain yield by up to 10% (Bell et al. 2011).

The main danger in on-site consequences is the fact that the reclamation of soils that have lost their topsoil fertile layer is extremely difficult and expensive—far more than the consequence of common off-site damage, such as the clearance of infrastructure cover by soil or other substances.

2.3.2 Off-Site Consequences

Off-site consequences refer to the consequences of water erosion processes on the surroundings of a given site—such as neighboring agricultural fields, roads, or built areas. The most common off-site consequence is the transport of soil minerals, or other substances, from the site and their deposition on neighboring areas. This occurs as follows: Soil is lost from the fields and transported, usually downslope, crossing other fields, infrastructure, and built areas. Sediment and other matter are deposited on unwanted locations, causing expensive damage—or even the loss of human lives—when deposited on roads and creating a muddy surface. The actual damage to crops is caused by the accumulation of fine material on seeds and seedlings—effectively reducing the productivity of the eroded area. In the following season, the farmers simply level the soil, in anticipation of the next sediment flow to come and cover the seeds and seedlings again.

Another off-site consequence is eutrophication—which occurs when runoff water, rich with nutrients (especially nitrogen and phosphorous), arrives at a water body, causing accelerated increase in the primary production of lake algae and water plants, phytoplankton, and occasionally heavy metals, as well, from the field. Eutrophication decreases the oxygen needed for the lake fauna, reduces aesthetic enjoyment, and causes pollution to drinking water. It is most frequent in lakes that act as a base level to agricultural catchments due to over-fertilization, but can also occur in urban catchments due to organic matter from sewage systems. The costs due to the loss of function of the water body and maintenance activities (including filtering) needed to remediate the damage can be billions of dollars every year (Brady and Weil 2016).

Table 2.2 presents evidence from the literature on some of the most common off-site consequences in the world.

The following instances demonstrate the magnitude of the damage caused by off-site consequences. In the agricultural fields of northwest Switzerland, the amount of soil deposited off-site, next to the fields in question, was estimated to be over 50% of the total eroded soil—with 20% of this amount deposited in lakes and rivers, transportation networks, and ditches (Ledermann et al. 2010). Off-site soil deposits in agricultural catchments with water reservoirs can be even more costly. In the Missouri River Basin, the soil deposited off-site dramatically changed the level of sediments at the bottom of the reservoir, substantially reducing water storage capacity, and affecting the function and sustainability of the reservoir (Graf et al. 2010). Management programs to defend the dams and decrease the risk depend on damage predictions that may be erroneous and cannot be easily tested. A dam-break flood triggered in southern California water flows with velocity of 5 $m \cdot s^{-1}$ that caused structural failure to 41 wood-framed houses while almost half of them were completely washed out (Gallegos et al. 2009). Moreover, herbicides that are deposited off-site can adversely affect the aquatic ecosystem, since herbicides are poisonous chemical substances used to eliminate weeds. As such, even a small amount moving along random preferential lines may be more destructive to a stream or a lake than a large mass of soil deposit (Doppler et al. 2012). Reservoirs and lakes may also be contaminated by other agrochemicals, such as those

Table 2.2 Off-site consequences of water erosion in agricultural fields in different environments

Source	Consequence	Climate	Mechanism	Method
Ledermann et al. (2010)	Soil deposited on infrastructure	Humid	Muddy flood	Field survey
Graf et al. (2010)	Reduction of water storage capacity in reservoirs	Arid	Soil deposition in reservoir	USGS survey
Mansour et al. (2011)	Facilities damage	Worldwide	Slow moving landslides	Literature survey
Doppler et al. (2012)	Herbicide mobilization	Humid	Water transport	Field experiment
Gallegos et al. (2009)	Structural failure	Worldwide	High-velocity Urban Dam-break flood	Structural and hydraulic flood modeling
Gómez et al. (2014)	Contamination by agrochemicals of water bodies	Semi-arid	Water transport	Experimental plots
Cevasco et al. (2014)	Severe damage to infrastructure and facilities	Humid	Landslides	Field surveys, penetration, and permeability tests

from fertilizers, which may cause eutrophication and siltation of dams and rivers (Gómez et al. 2014). Landslides in agricultural fields can even cause off-site damage to neighboring settlements. A study by Mansour et al. (2011) found that damage to built areas and to infrastructure depends to large extent on the slide velocity or the accumulating displacement. Although movements as little as 100 mm may cause only moderate damage to urban communities, they can severely damage bridges. The off-site damage due to landslides can be even more extreme than on-site, as landslides can also kill. For example, thousands of shallow landslides—initiated by very high rainfall intensity—have caused flooding over wide areas. The water flow imposed severe damage to houses and to the infrastructure and 13 fatalities (Cevasco et al. 2014).

2.4 The Human Agent

Although on-site and off-site consequences of water erosion affect various populations at the national level, farmers are the ones who are most involved in water erosion processes and are the direct victims of their consequences. They pay the financial costs of reduction in yields due to soil degradation, and they are the ones who must perform the conservational actions needed to prevent and repair the on-site/off-site damage. As a result, farmers are torn between the financial desire to maximize yields and crop quality, and the need to behave sustainably to prevent long-term damage of water erosion. The authorities, who may be more aware of the long-term consequences of water erosion, make various efforts to recruit the farmers to engage in conservation, but for various reasons, as detailed below, this is not always possible. The farmers are therefore a key to any national program of conservational agriculture. This section attempts to explain why agricultural soils are not always treated well enough by the farmers, and the psychological barriers that may prevent farmers from cooperating with conservational efforts. Finally, this section reviews the actions that have been found to be effective in spurring farmers to adopt conservation practices.

2.4.1 Agricultural Soils as Social Traps

Non-conservational behavior in agricultural catchments is a special case of a general problem known as *the social trap* (Platt 1973). A social trap occurs when an individual or a group of individuals (in the case of soil conservation—the farmer, or group of farmers) act to obtain short-term gains (e.g., increase their yields), which in the long run may lead to a loss (reduction in soil health of the catchment, eutrophication in water bodies, etc.) for the group as a whole (the public). In other words, if farmers exploit the soil to the point of no return, the public can lose the ability to use those fields for agricultural uses, which may result in diminished food supply and food security. Social traps have already caused global disasters due to overuse of national resources—as in the case of overgrazing in the Sahel (Breman and Wit 1983) and the deterioration of rainforests in South America by excess logging of trees (Foley et al. 2007).

Social traps are a well-known problem in human behavior, and many studies, e.g., Balderacchi et al. (2016), have shown that individuals and groups frequently fall into a social trap behavior. In the specific case of combating soil erosion, as we shall see in the next section, there are many reasons—justifiable or otherwise—why a farmer would avoid sacrificing his own interests for the benefit of the public. In the case of soil conservation, the conflict between the benefit of the individual and the interests of the public at large is especially acute in countries where agricultural fields are government property that is leased to the farmer on an annual or decadal basis. In such instances, one notable explanation for social traps is the phenomenon known as "*the tragedy of the commons*" (Hardin 1998, Hardin 1968). According to Hardin, in a shared-resource system (such as publicly owned agricultural fields), individual users (e.g., farmers), acting independently according to their own self-interests, behave contrary to the common good of all users (the public), by depleting or spoiling that resource through their actions (non-conservational cultivation). Thus, if the soil is a common property and the farmer overuses it in a bid to maximize his profit, the main resource (the soil) is ultimately damaged (eroded, or even lost).

Hardin also suggested solutions: either privatize the land, or nationalize it and enforce conservation practices. In his view, the former is preferable, because enforcement of conservation practices by civil servants who are not assessed by their performance may be inefficient. But, as we have seen in the general case of social traps in soil conservation, even if the farmer owns the fields, he may face certain obstacles, for several reasons. Given the difficulties in recruiting farmers to engage in conservation practices due to deeply ingrained patterns of human behavior, and given the importance of this to conservation efforts as a whole, in the following two sections we shall try to address two questions: (1) what mental blocks, in theory, specifically may prevent farmers from taking part in conservation efforts and (2) what has been found in the literature to be successful in inducing farmers to act in a pro-environmental behavior.

2.4.2 The Psychological Barriers

The World Bank (2018) reports show that, in 2018, approximately 28% of the world's total employment was in agriculture (compared with approximately 44% in 1991). However, this average proportion varies greatly between countries. While, in 2018, again according to the World Bank data, the proportion of employees in agriculture was 1% in Israel, 1% in the US, 2% in Canada, 3% in Australia, 3% in the Euro area, and 6% in the Russian Federation; it stood at 72% in Somalia, 44% in India, 37% in Nigeria, 27% in China, and 21% in the Arab world. Farmers are therefore still a large group, and in agrarian countries they still constitute the majority. As with every human group, their motivation to action involves complex considerations, and may vary between individuals. The perception of the environment by this group and their will to perform sustainable agriculture may play a crucial role in soil conservation efforts, and the question is how to encourage such farmers to act in a pro-environmental manner, and to cooperate with governmental conservation initiatives?

Considerable research has been carried out in the field of environmental psychology on the psychological barriers that induce inaction in individuals. This has mainly focused on what may prompt individuals to mitigate (or, if this is not possible, to adapt to) climate change scenarios due to global warming. Among the most notable studies on the subject is that of Gifford (2011), who found that the willingness of individuals to engage in pro-environmental behavior —in our case toward soil conservation oriented behaviour—is an outcome of their perception of the environment (gathering information about it), appraisal of it (deciding whether it is good or bad), and attitude toward it (concern to actually protect it). In the face of the threat of water erosion, a farmer needs to weigh investing money in conservation efforts against short-term increased profits from yields. While conservation efforts may increase his profit in the long term (and possibly save the farm from irreparable soil degradation), it is too remote a prospect to observe on a daily basis.

However, even a good appraisal, a positive attitude, and concern about the consequences are not always translated into action (Gifford 2011). There are psychological obstacles that prevent concerned individuals (or farmers) from actually acting and investing to make a difference. When they are not concerned, the obstacles are even higher. The question that arises then is what are those psychological barriers that hamper or even prevent pro-environmental behavior, and how can those barriers be removed. Establishing a methodology to remove the psychological barriers facing farmers seeking to be pro-environmental is beyond the scope of this book, as it is a daunting task in its own right. But, for the benefit of the reader who may have to face the challenge of collaborating with farmers on national programs, we shall try to describe these barriers here as they relate to the special case of water erosion in agricultural fields. To do this, the next section relies entirely on the framework of seven psychological barriers to pro-environmental behavior against climate change that Gifford (2011) identified and dubbed the "*Dragons of Inaction*." Each of these "Dragons of Inaction" represents not a single barrier but a family of psychological barriers, as detailed below. The seven families of "Dragons of Inaction" are therefore categories of psychological barriers that may prevent farmers from acting in a pro-environmental fashion to conserve their soils.

2.4.2.1 Limited Cognition

Limited cognition refers to how individuals often fail to act rationally in decision-making due to inadequate thinking. With farmers, this may occur because the soil is a very complex system, and understanding the various threats it may face is difficult—even if you are a very experienced farmer who has made a living from the soil your entire life. Even scientists who spend their lives studying the soil system do not fully understand its various functions. Thus, if a given soil erosion problem—such as nitrogen leaking from the field, causing eutrophication in a remote lake—does not inflict direct and immediate damage to the farmer, she may be oblivious of it. Conversely, if the farmer is subjected to too much repetitive information and pressure by the government to conserve the soil, she may become indifferent to those attempts to encourage soil conservation activities. Other cognitive issues include: (1) *uncertainty* of the farmer about the actual threat posed by soil erosion—i.e., how real is the threat of soil degradation?; (2) typical *judgment discounting* of seemingly remote future risks; (3) *optimism bias*—i.e., the thought that if the soil has functioned well so far, it will continue to do in the future; and (4) *self-efficacy*, given that soil erosion is a huge infrastructure problem that is too large for the farmer to resolve, or do anything about.

2.4.2.2 Ideologies

This dragon includes worldviews such as capitalism, whereby the pursuit of a prosperous lifestyle spurs the farmer to engage in environmentally reckless behavior, because soil cultivation is only one component in the production line that can be replaced at will by other fields in exchange for money. Another philosophy/religious idea is the belief in a superhuman entity—such as Mother Earth—whom the farmer trusts will not abandon him and ensure that in the end, the soil erosion consequences are resolved. Another ideological barrier in the family of ideologies is the notion of *technosalvation*—whereby technology will solve every consequence of soil erosion and restore soil health through some newfangled invention. Last in the family of ideologies is the *justification* concept—which leads the farmer to believe that since the current status quo of soil health is OK, and the crops provide an adequate yield and livelihood, nothing needs to be changed.

2.4.2.3 Comparisons with Others

Among farmers, the adoption of conservation practices can be a function of the degree to which other farmers have done so. Thus, refusal to engage in good conservational practice may be justified on the grounds that neighboring farmers are not doing so, either. This dragon also includes norms that arise formally or informally within groups or settlements, and the false impression of perceived inequity—such as "I shall not use conservation tillage if they don't."

2.4.2.4 Sunk Costs

The Sunk Costs family of barriers concerns the financial investment involved in adopting conservational tillage. For example, if the farmer has recently purchased a conventional tillage plow, she would be loath to lose that investment by buying a new conservational plow. This may apply to purchased skills, as well: the expertise developed in using the old cultivation tools may act as a barrier to adopting new practices in operating more complex conservation tillage machines, and there is also the effect of a short-term profit on willingness to invest. The farmer is expected to be willing to carry the financial cost of soil aeration—which is necessary for biomass production and provides a direct profit—but less ready to pay the financial cost of soil conservation because it does not produce a profit in the short term.

2.4.2.5 Discredence

Discredence relates to when individuals think negatively about others. In the case of soil degradation, it occurs, for example, when farmers dismiss or discount the recommendations of field instructors. Farmers' negative views of experts can be due to distrust or simply a belief that the experts are wrong. Both negative views may lead to a refuse to follow the expert's advice. This state of affairs can arise from mistrust between farmers and experts over the described problem, which may prompt farmers to believe that the experts' prescribed program for soil remediation is inadequate, or statements such as "The experts don't see the field every day, so they don't know what is needed." Another barrier in this family is related to denial. Namely, the farmer may claim or even only believe that the erosion damage that his field experiences is not caused by the cultivation method that she applies. In some instances, this may even lead to resistance or rebellion against governmental conservation initiatives.

2.4.2.6 Perceived Risk

Perceived risk refers to what the farmer perceives might happen to her if she changes the cultivation method to a more conservational one. This includes: (1) functional risks about the effectiveness of conservation actions in cultivating the fields; (2) physical risks—such as dangers to workers, facilities, and yields; (3) financial risks regarding the question if the conservation efforts will pay back?; and finally (4) social risk of damage to one's reputation among colleagues and neighbors, if one is seen as collaborating with the authorities, or the loss of time and customers due to unnecessary actions.

2.4.2.7 Limited Behavior

Some farmers do indeed practice conservation tillage, but to a rather limited degree, which is insufficient and may not substantially contribute to soil conservation in the catchment, or even in their own fields. In these cases, while they may be attending conservation workshops or reading professional magazines, they are not fundamentally changing their cultivation methods. Limited behavior may occur due to (1) *Tokenism*, namely, the tendency of individuals to adopt low-cost actions that may have little or no impact on their actual goals, instead of more effective actions that have higher cost. Tokenism is theoretically based on the *low-cost hypothesis* (Diekmann and Preisendörfer 2003) that envisages that the intensity of effects of environmental concern on environmental behavior diminishes with increasing behavioral costs. Limited behavior can also occur due to (2) the *rebound effect* that states that after a positive step is taken (e.g., the acquisition of organic fertilization material), the gains made are diminished or erased through its overuse (e.g., spreading it all over the catchment). The theoretical basis for the rebound effect is the Jevons Paradox (Jevons 1865) and the Khazzoom–Brookes postulate (Brookes 1990, Khazzoom 1980).

Gifford's Seven Dragons of Inaction provide a holistic theoretical framework that has not yet been explored and validated empirically for soil conservation purposes. The Dragons of Inaction are cited here to help field instructors—who may use this book to introduce spatial soil conservation practices—to identify the specific psychological barriers facing farmers in their area, and to help farmers overcome these barriers and contribute more to conservation efforts.

The question now is what has actually been done in the field of soil conservation around the world, to motivate farmers to adopt conservational behavior? Are there any farmers who are willing to collaborate with national programs and to protect their own fields—and if so, why? The next section will attempt to answer these questions.

2.4.3 Evidence for Farmer's Conservation Actions

Since the cooperation of farmers is very much required to successfully implement conservational programs, large efforts have been made by various authorities around the world to encourage farmers to act pro-environmentally. As we shall see, many of these attempts involved financial inducements, but other incentives were also effective in converting farmers to the cause—including knowledge acquisition, enhanced socio-economic status, environmental awareness, and an ethical attitude toward the environment (Reimer and Prokopy 2014).

For example, De Graaff et al. (2008) found that the cost-effectiveness of soil and water conservation for the farming household proved to be a strong inducement for farmers to perceive conservation as worthwhile. They showed that if farmers saw a prospect of financial benefits—for example, through increased crop productivity, reduced manpower requirement, or higher profits—they were motivated to adopt, maintain, and pass on soil and water conservation practices. Due to the results in this research, the authors even recommended that the implementation of conservation practices will be conducted not in isolation, but always in conjunction with financial measures and other inducements that improve the prospect of increased future income for the farmer.

While financial aid may be a prerequisite for pro-environmental behavior, it may not be enough in and of itself. Intrinsic motivations have also been found useful to increase farmer's readiness to support conservation processes, from various reasons. Greiner and Gregg (2011) found that farmers were motivated by considerations of personal and family well-being, and to make decisions about soil conservation based on care-based ethics rather than simply in response to financial opportunities, imperatives, constraints, or social goals. Thus, ethics and altruism have been found to be important adoption factors. For example, Greiner and Gregg found that cattle grazing farmers across Australia's tropical savannas are driven in their actions by a very strong ethic that give rise to altruistic behavior, which in turn has resulted in visibly more efforts at soil conservation. Their findings have also shown large variability between farmers and between regions in the motivation to participate in conservation efforts and therefore they recommended to tailor conservation programs to harness the diverse aspirations and motivations of farmers in each case.

In a large study of ten European countries and 1000 farming households, Wilson and Hart (2000) found that environmental concern was as important as financial incentives in influencing farmers' participation in conservational schemes. Their results showed broadly similar participation patterns across countries, albeit with some differences between northern and southern Europe, and between arable farmers and grassland farmers. Accordingly, the authors stress that the financial imperative for participation must not disregard or eclipse the often equally important environmental concern, and the importance of demographic, socio-economic, and cultural background of farmers on their motivation to adopt conservational agrotechnology. They also recommended that these factors will be studied further in future research.

This indeed was carried out in a study of the factors affecting the awareness and adoption of conservation practices to reduce degradation and improve productivity among farmers in sub-Saharan Africa. The results of the study, which involved hundreds of participants, were very clear: factors such as the head of household's age, education, openness to professional help, and membership in farmer groups were critical in the farmer's increased awareness and pro-adoption decisions. This suggests that government policies should make effort to improve farmer's education, and increase land ownership, credit access, and social capital by group formations (Mango et al. 2018).

But simply adoption of conservational agrotechnology by the farmers is not the final goal. Even if a high degree of collaboration is achieved, the effectiveness of the conservation actions is actually the key factor. A study in Africa—a continent that suffers greatly from soil erosion consequences—involving a formal household survey, informal and focused group discussions, and field observations among a rural Ethiopian population, has revealed disturbing results: most of the farmers that took part in the conservation initiative did so unwillingly. The primary reason was the perceived ineffectiveness of the structures under construction. Other factors—such as low awareness, costs, and land tenure insecurity—were found to be less influential (Bewket and Sterk 2002).

Table 2.3 presents a further collection of very recent studies on the motivation of farmers in various forms of participation in conservation policy programs.

Table 2.3 and the detailed studies cited above reveal the following incentives that have been found to be effective in actually encouraging farmers to cooperate with conservational initiatives: financial inducements, group social support, involvement of more members of the household, and the farmers' knowledge and education. Age was another factor (young and less experienced farmers are more likely to adopt conservation measures), as were the size of the farm and the farming income (small-scale farmers and those with more farming income are significantly more likely to adopt). Notably, farmers are not only focused on profitability, but also concerned with environmental well-being. Land tenure also has a significant impact (farmers who own their land plots are consistently more likely to take up conservation than farmers tending to leased or borrowed)—as do ethical and altruistic motives.

In summary, the above studies show that some governmental institutions are able to overcome psychological barriers that farmers experience and successfully recruit them to the conservation cause. However, the chapter about the human agent is not intended to be a manual on how to persuade farmers to engage in soil conservation. Rather, the description of the theoretical psychological barriers formulated in the "Dragons of Inaction" by Gifford (2011) and the

Table 2.3 A summary of site locations and factors governing pro-environmental behavior by farmers in agricultural catchments around the world in the past decade

Source	Location	Contributing factors in take-up of conservation practices
Greiner and Gregg (2011)	Australia	Altruistic motives and ethics
McGuire et al. (2013)	Iowa, US	Activation of the farmers' conservationist identities in a group setting; social support
Roesch-McNally et al. (2018)	USA	Greater economic incentives and/or more diverse crop and livestock markets would be needed to achieve more widespread adoption of the practice
Sunny et al. (2018)	Bangladesh	Young farmers with less farming experience are more likely to adopt facilities. Small-scale farming, higher education, greater farming income, and more knowledge also significantly increased adoption. Most adopter farmers were concerned not only with profitability, but also with environmental well-being
Hamza et al. (2018)	Burkina Faso	Involvement of more members of the household as well the household head (male)
Mishra et al. (2018)	Kentucky, USA	Inadequate knowledge about sustainable farming and unfamiliarity with technology were significantly and negatively linked to adoption of sustainable practices
Harper et al. (2018)	Serbia	Type of partnership
Lawin and Tamini (2019)	Benin	Land tenure: conservation practices were consistently higher on owned plots than leased ones

empirical evidence from the literature on pro-environmental behavior of farmers is meant to demonstrate to the reader the challenges and potential in the field.

2.5 Summary

Water erosion processes—such as splash and sheet erosion, and channeled erosion (including rills, gullies, and piping)—occur in agricultural fields and create a great deal of damage. In agricultural environments, these processes may be enhanced by man-made factors such as cultivation methods, tillage direction, paved and unpaved roads, and plant rotation due to cropping systems. These human activities may also interact with environmental factors—such as rainfall intensity, parent material, topography, and bioturbation—that make the prediction of erosion risk and consequence even more challenging.

The consequences of water erosion in agricultural catchments are classified as either on-site consequences (relating to the loss of substance from the field to neighboring areas) and off-site consequences (the deposition of soil material at unwanted locations). The damage from water erosion depends markedly on the farmer's decisions regarding pro-environmental behavior. Therefore, understanding farmers' motivations and the barriers to their participation in national soil conservation programs is key to successful implementation of conservation initiatives and effective outreach. Several psychological barriers prevent farmers from adopting conservational tillage: limited cognition about the soil degradation problem; ideological worldviews that tend to preclude pro-environmental attitudes and behavior; comparisons with other individuals; sunk costs and behavioral momentum; discredence toward experts and authorities; perceived risks of change; and positive but inadequate behavior change. Some of the barriers can be removed through financial support, ethical encouragement, environmental concern, and education—making it possible to enlist farmers around the world to take up conservation.

The large and complex spatial and temporal variations in environmental and human-induced conditions in agricultural catchments give rise to multifaceted distribution patterns of processes, forms, and their consequences—across plot, hillslope, and catchment spatial scales, and at hourly, daily, and seasonal scales. Geoinformatics may provide a modern tool to unravel these complexities in agricultural catchments.

Review Questions

1. Make a list of the human-induced and environmental factors that affect water erosion in the agricultural catchment in your region. Rate the magnitude of effect of each of these factors, and order them from 1 to *n*. Base your rating on your own knowledge and on the literature.
2. Compute the effect of tillage direction in a northerly hillslope aspect direction (0°) in your field, assuming that its hillslope gradient changing gradually by 1° between 0° and 10°. In the next independent step, change the hillslope aspect between east 90° and west 270° using 10° interval. What would be the gradual effect?
3. A soil grid-cell has been found to be subjected to raindrops measuring 0.05 cm in radius; density of 1 $g{\cdot}cm^{-3}$; and velocity of 0.01 $m{\cdot}s^{-1}$. What is the crater depth created in cm, if the constant for this soil's characteristics in this area is 31.4?
4. Under which conditions do you expect your field to suffer from piping, and under which conditions is it expected to be more prone to gulling?
5. What is the difference between on-site and off-site consequences of soil erosion? Give four examples of each.
6. Explain the importance of hillslope gradient and contributing area in assessing the potential for gully incision.
7. What is meant by the statement, "Unpaved roads may act as contributors and barriers to runoff and erosion"? Where do they act as barriers, and where as contributors?

8. What are the factors affecting the critical shear stress in Eq. 2.8? Explain the left and right sides of the equation. Explain why do you think Shields (1936) chose them to predict incision? Also explain why, in your opinion, Begin and Schumm (1979) used hillslope gradient and contributing area as proxy for those variables.
9. Name the seven families of psychological barriers affecting pro-environmental behavior for soil conservation, and provide a brief explanation of each such family. With regard to each of these barriers, compare a collectivist group of farmers with an individualist group, and rate the level of influence each of the barriers would have on each group.
10. What are the factors documented in the literature as initiating collaborative actions in conservational programs? Which of these may be suitable to encourage farmers in your area to engage in pro-conservational behavior? Which of the aforementioned barriers (question 9) did they overcome?

Strategies to address the Review Questions

1. To answer this question, you need to select the human-induced and environmental factors that influence water erosion in your catchment of interest. This can be done as follows:
   - Figure 2.1 lists examples of human and environmental factors that affect water erosion in semiarid agricultural catchments. The human factors include cultivation method, tillage direction, unpaved roads, and cropping system, while environmental factors include rainfall characteristics, topography, vegetation cover, parent material, and bioturbation. Other factors can be suggested based on the conditions in the studied catchment.
   - Rating from most to least important factor can be done either using the methods described in Chap. 6 of this book or simply based on user experience, observations, and measurements in the studied catchment.
2. The answer to this question includes two parts: (1) compute t using Eq. 2.1 with S changed from 0° to 10° with 1° interval, and in step (2) compute t with β based on aspect changing between 90° and 270°. In the second step, S remains constant on any other value you choose. Higher hillslope values are expected to show higher effects. Note that $\cos^2 \beta$ varies between 0 and 1. The results show the relatively lower effect of hillslope gradient as a base to the function in this index and the gauss-like effect of changing distance between tillage direction and hillslope aspect.
3. The crater depth in the given grid-cell can be computed using Eq. 2.6:

$$D = 31.4 \cdot 0.05 \cdot \left(1 \cdot 0.01^2\right)^{\frac{1}{3}}, D = 0.073cm.$$

4. The difference between piping and gulling is described in Sect. 2.2.4–2.2.5. In a nutshell, gulling is aboveground channeled flow whereas piping is belowground channeled flow. Fields can suffer from piping due to high density of cracks and large pores, bioturbation, root activities, and drying and wetting processes. Gulling can occur due to irregularities in surface roughness and routing of sheet flow into specific locations. In locations where sheer stress is higher than the soil resistance, gully initiation occurs.
5. The topic of on-site and off-site consequences of water erosion is discussed in Sect. 2.3. Figure 2.11 illustrates the general differences between the two. In short, on-site consequences are mainly related to the reduction of soil quality, while the off-site consequences pertain mainly to the accumulation of eroded mineral, chemical, and biological material in unwanted places.
   - Examples for on-site consequences can be found in Table 2.1 and Sect. 2.3.1: organic matter loss, fine material and nutrients loss, roots exposure, soil loss, and damage to the topsoil layers.

- Examples for off-site consequences can be found in Table 2.2 and Sect. 2.3.2: covering or blocking infrastructure, reduce the capacity of waterways and reservoirs, increase the risk of floods, contaminate, poison, or bury other ecosystems.

6. Gulling occurs when water pressure is higher than soil resistance (Sect. 2.2.4). Equation 2.9 shows how hillslope gradient and contributing area can be used as a proxy to the water pressure on the soil and therefore could act as a proxy for assessing the potential for gully incision according to the sheer stress threshold by shields. Namely, S and A determine the critical shear stress below which there is no gully incision. The use of these two factors to predict gully incision is used by many studies since then and they are considered important in any erosion model.
7. The effect of unpaved roads on water erosion is described in Sect. 2.1.1.3 and illustrated in Fig. 2.5. Unpaved roads may act as contributors to runoff and erosion downslope by being less permeable to raindrops than soil-mantled areas, so they act as a network of sealed lines. As a result, a dense road network may increase overland flow and consequent soil transport in agricultural areas. Unpaved roads in lower parts of the hillslope may act as barriers to runoff and soil transport due to their increased elevation and surrounding mounds which can divert or even halt overland flow.
8. Equation 2.8 expresses the factors affecting critical shear stress in both sides of the equation:

   - Left-hand side represents the water stress based on: water density, gravitational acceleration, the gully hydraulic radius, and the hillslope gradient.
   - Right-hand side represents the soil resistance based on: soil density, particle size, gravitational acceleration, and an empirical coefficient.
   - Shields chose those factors to predict incision as they represent the critical shear stress beyond which incision occurs.
   - Begin and Schumm used hillslope gradient and contributing area as proxies for those variables to reduce the number of variables that need to be measured. Furthermore, the proxy variables can be measured more easily.

9. The seven psychological barriers for pro-environmental behavior by Gifford are discussed in Sect. 2.4.2.

   - Limited cognition—how individuals often fail to act rationally in decision-making due to limited thinking.
   - Ideologies—the pursuit of a prosperous lifestyle spurs the farmer to engage in environmentally reckless behavior.
   - Comparisons with others—refusal to engage in conservational practice because others do not do it.
   - Sunk costs—the financial investment involved in adopting conservational tillage is usually unprofitable in the short run.
   - Discredence—when individuals think negatively about recommendations of field instructors.
   - Perceived risk—what the farmer perceives might happen to him if he changes his cultivation methods to a more conservational one.
   - Limited behavior—farmers think about changing their cultivation methods but not actually do it.
   - The analysis of the effect of those barriers on farmers from collectivist and individualist societies requires some thought. For example, individualists may be affected by perceived risk, because they are expected to care about their own profit and cannot rely on help from other farmers. Farmers who work as individualists will be affected by limited behavior and ideologies, because they

may tend to adopt low-cost action. Farmers from collectivist societies will usually work as a team and will be mostly affected by comparison with others, discredence (because they think that as a team they know better) and sunk costs.

10. Section 2.4.3 describes in detail the conditions that encourage conservation actions by farmers around the world. Table 2.3 provides a few specific examples from different countries. These include (but not limited to, see more in Sect. 2.4.3): altruistic motives and ethics, group setting and social support, economic incentives, farmers age and education, strive for environmental well-being, involvement of more members of the household, type of partnership, and land tenure (conservation practices were consistently higher on owned plots than leased ones).

References

Abid M, Lal R (2009) Tillage and drainage impact on soil quality: II. Tensile strength of aggregates, moisture retention and water infiltration. Soil Tillage Res 103 (2):364–372. https://doi.org/10.1016/j.still.2008.11.004

Arvidsson J, Keller T (2011) Comparing penetrometer and shear vane measurements with measured and predicted mouldboard plough draught in a range of Swedish soils. Soil Tillage Res 111(2):219–223

Arvidsson J, Westlin H, Keller T et al (2011) Rubber track systems for conventional tractors—effects on soil compaction and traction. Soil Tillage Res 117:103–109

Assouline S, Ben-Hur M (2006) Effects of rainfall intensity and slope gradient on the dynamics of interrill erosion during soil surface sealing. Catena 66(3):211–220. https://doi.org/10.1016/j.catena.2006.02.005

Balderacchi M, Perego A, Lazzari G et al (2016) Avoiding social traps in the ecosystem stewardship: the Italian Fontanile lowland spring. Sci Total Environ 539:526–535. https://doi.org/10.1016/j.scitotenv.2015.09.029

Begin ZB, Schumm SA (1979) Instability of alluvial valley floors: a method for its assessment. Trans Am Soc Agric Eng 22:347–350

Bell LW, Kirkegaard JA, Swan A et al (2011) Impacts of soil damage by grazing livestock on crop productivity. Soil Tillage Res 113(1):19–29. https://doi.org/10.1016/bbj.still.2011.02.003

Bernatek-Jakiel A, Poesen J (2018) Subsurface erosion by soil piping: significance and research needs. Earth-Sci Rev 185:1107–1128

Bertolino A, Fernandes N, Miranda JPL et al (2010) Effects of plough pan development on surface hydrology and on soil physical properties in southeastern Brazilian plateau. J Hydrol 393(1):94–104

Bewket W, Sterk G (2002) Farmers' participation in soil and water conservation activities in the Chemoga Watershed, Blue Nile basin. Ethiopia Land Degrad Dev 13(3):189–200. https://doi.org/10.1002/ldr.492

Blanco-Canqui H, Lal R (2008a) Corn stover removal impacts on micro-scale soil physical properties. Geoderma 145(3):335–346

Blanco-Canqui H, Lal R (2008b) No-tillage and soil-profile carbon sequestration: an on-farm assessment. Soil Sci Soc Am J 72(3):693–701. https://doi.org/10.2136/sssaj2007.0233

Blanco-Canqui H, Lal R (2008c) Principles of soil conservation and management. Springer, London

Boer MM, Puigdefábregas J (2005) Assessment of dryland condition using spatial anomalies of vegetation index values. Int J Remote Sens 26(18):4045–4065. https://doi.org/10.1080/01431160512331338014

Borrelli P, Robinson DA, Fleischer LR et al (2017) An assessment of the global impact of 21st century land use change on soil erosion. Nat Commun 8(1):1–13. https://doi.org/10.1038/s41467-017-02142-7

Brady NC, Weil RR (2016) Theœ Nature and Properties of Soils, global edition of the 15th revised, edition. Pearson Education Limited, London

Breman H, Wit CT (1983) Rangeland productivity and exploitation in the Sahel. Science 221(4618)

Breuning-Madsen H, Kristensen JÅ, Awadzi TW et al (2017) Early cultivation and bioturbation cause high long-term soil erosion rates in tropical forests: OSL based evidence from Ghana. Catena 151:130–136. https://doi.org/10.1016/j.catena.2016.12.002

Brookes L (1990) The greenhouse effect: The fallacies in the energy efficiency solution. Energy Policy 18(2):199–201. https://doi.org/10.1016/0301-4215(90)90145-T

Casali J, Loizu J, Campo MA et al (2006) Accuracy of methods for field assessment of rill and ephemeral gully erosion. Catena 67(2):128–138. https://doi.org/10.1016/j.catena.2006.03.005

Cerda A (1999) Parent material and vegetation affect soil erosion in eastern Spain. Soil Sci Soc Am J 63 (2):362–368. https://doi.org/10.2136/sssaj1999.03615995006300020014x

Cerdà A (2002) The effect of season and parent material on water erosion on highly eroded soils in eastern Spain. J Arid Environ 52(3):319–337. https://doi.org/10.1006/jare.2002.1009

Cevasco A, Pepe G, Brandolini P (2014) The influences of geological and land use settings on shallow landslides triggered by an intense rainfall event in a coastal terraced environment. Bull Eng Geol Environ 73(3):859–875. https://doi.org/10.1007/s10064-013-0544-x

Chen SC, Chen CY, Lin TW (2015) Modeling of natural dam failure modes and downstream riverbed morphological changes with different dam materials in a flume test. Eng Geol 188:148–158. https://doi.org/10.1016/j.enggeo.2015.01.016

Cohen S, Willgoose G, Svoray T et al (2015) The effects of sediment transport, weathering, and aeolian mechanisms on soil evolution. J Geophys Res Earth Surf 120(2):260–274. https://doi.org/10.1002/2014JF003186

Dabney SM, Dalmo ANV, Yoder DC (2011) Effects of topographic feedback on erosion and deposition prediction. the American Society of Agricultural and Biological Engineers, St. Joseph. Int Symp Erosion Land Evol (ISELE) 18–21:18–21. https://doi.org/10.13031/2013.39282)

Daraghmeh OA, Jensen JR, Petersen CT (2008) Near-saturated hydraulic properties in the surface layer of a sandy loam soil under conventional and reduced tillage. Soil Sci Soc Am J 72(6):1728–1737. https://doi.org/10.2136/sssaj2007.0292

De Baets S, Poesen J, Meersmans J et al (2011) Cover crops and their erosion-reducing effects during concentrated flow erosion. Catena 85(3):237–244. https://doi.org/10.1016/j.catena.2011.01.009

De Graaff JD, Amsalu A, Bodnár F et al (2008) Factors influencing adoption and continued use of long-term soil and water conservation measures in five developing countries. Appl Geogr 28(4):271–280. https://doi.org/10.1016/j.apgeog.2008.05.001

Desmet B, De Vos A, Van de Velde H et al (1999) Relationship between semen morphology assessment according to strict criteria and the morphology of individual spermatozoa used for intracytoplasmic sperm injection. Hum Reprod 14:250. https://doi.org/10.1093/humrep/14.Suppl_3.250-a

Diamond JM (2005) Collapse : how societies choose to fail or survive. Centro de Investigaciones Sociológicas, London

Diekmann A, Preisendörfer P (2003) Green and greenback. Ration Soc 15(4):441–472. https://doi.org/10.1177/1043463103154002

Dietrich WE, Wilson CJ, Montgomery DR et al (1993) Analysis of erosion thresholds, channel networks, and landscape morphology using a digital terrain model. J Geol 101(2):259–278. https://doi.org/10.1086/648220

Doppler T, Camenzuli L, Hirzel G et al (2012) Spatial variability of herbicide mobilisation and transport at catchment scale: Insights from a field experiment. Hydrol Earth Syst Sci 16(7):1947–1967. https://doi.org/10.5194/hess-16-1947-2012

Eldridge DJ, Greene RSB (1994) Assessment of sediment yield by splash erosion on a semiarid soil with varying cryptogam cover. J Arid Environ 26(3):221–232. https://doi.org/10.1006/jare.1994.1025

Engel OG (1961) Collisions of liquid drops with liquids. US Department of Commerce, Office of Technical Services, Georgiana

Fielke JM (1999) Finite element modelling of the interaction of the cutting edge of tillage implements with soil. J Agric Eng Res 74(1):91–101

Flanagan DC, Nearing MA (2007) Water erosion prediction project (WEPP): Development history, model capabilities, and future enhancements. Amer Soc Agric Biol Eng 50(5):1603–1612

Fleshman MS, Rice JD (2014) Laboratory modeling of the mechanisms of piping erosion initiation. J Geotechn Geoenviron Eng 140(6):1–12. https://doi.org/10.1061/(ASCE)GT.1943-5606.0001106

Foley J, Gregory A, Marcos C et al (2007) Amazonia revealed: forest degradation and loss of ecosystem goods and services in the Amazon Basin. Front Ecol Environ 5(1):25–32. https://doi.org/10.1890/1540-9295(2007)5[25:ARFDAL]2.0.CO;2

Foster GR (1986) Understanding ephemeral gully erosion. Soil Conservation 2:90–125

Friedrich T (2000) Manual on integrated soil management and conservation practices 8:37–44

Fu B, Wang S, Liu Y, Liu J, Liang W, Miao C (2017) Hydrogeomorphic ecosystem responses to natural and anthropogenic changes in the Loess Plateau of China. Annu Rev Earth Planet Sci 45:223–243

Gallegos HA, Schubert JE, Sanders BF (2009) Two-dimensional, high-resolution modeling of urban dam-break flooding: a case study of Baldwin Hills. California. Adv Water Resour 32(8):1323–1335

Gifford R (2011) The dragons of inaction psychological barriers that limit climate change mitigation and adaptation. Am Psychol 66(4):290–302. https://doi.org/10.1037/a0023566

Gómez C, Pérez-Blanco D, Ramon B (2014) Tradeoffs in river restoration: Flushing flows vs. hydropower generation in the Lower Ebro River. Spain. J Hydrol 518:130–139. https://doi.org/10.1016/j.jhydrol.2013.08.029

Graf WL, Wohl E, Sinha T et al (2010) Sedimentation and sustainability of western American reservoirs. Water Resour Res 46(12). https://doi.org/10.1029/2009WR008836

Greiner R, Gregg D (2011) Farmers' intrinsic motivations, barriers to the adoption of conservation practices and effectiveness of policy instruments: empirical evidence from northern Australia. Land Use Policy 28 (1):257–265. https://doi.org/10.1016/j.landusepol.2010.06.006

Gyssels G, Poesen J, Bochet E et al (2005) Impact of plant roots on the resistance of soils to erosion by water: A review. Prog Phys Geogr 29(2):189–217. https://doi.org/10.1191/0309133305pp443ra

Hamza H, Auwal Y, Sharpson M (2018) Standalone PV system design and sizing for ahousehold in Gombe, Nigeria. Int J Interdis Res Innov 6(1):96–101

Hardin G (1968) The tragedy of the commons. Science 162:1243–1248

Hardin G (1998) Extensions of "The tragedy of the commons." Science 280(5364):682–683. https://doi.org/10.1126/science.280.5364.682
Harper JK, Roth GW, Garalejić B et al (2018) Programs to promote adoption of conservation tillage: a Serbian case study. Land Use Policy 78:295–302. https://doi.org/10.1016/j.landusepol.2018.06.028
Hauge C (1977) Soil erosion definitions. Calif Geol 30:202–203
Hernández-Hernández RM, López-Hernández D (2002) Microbial biomass, mineral nitrogen and carbon content in savanna soil aggregates under conventional and no-tillage. Soil Biol Biochem 34(11):1563–1570
Hillel D (2006) Introduction to soil physics. Academic Press, London
Hu F, Liu J, Xu C et al (2018) Soil internal forces contribute more than raindrop impact force to rainfall splash erosion. Geoderma 330(15):91–98
Huang M, Zettl JD, Lee Barbour S et al (2016) Characterizing the spatial variability of the hydraulic conductivity of reclamation soils using air permeability. Geoderma 262:285–293. https://doi.org/10.1016/j.geoderma.2015.08.014
Jevons W (1865) The coal question: an inquiry concerning the progress of the nation, and the probable exhaustion of our coal-mines. Macmillan and Co, London, Cambridge
Katra I (2020) Soil erosion by wind and dust emission in semi-arid soils due to agricultural activities. Agronomy (Basel) 10(1):89. https://doi.org/10.3390/agronomy10010089
Kettler TA, Lyon DJ, Doran JW et al (2000) Soil quality assessment after weed-control tillage in a no-till wheat–fallow cropping system. Soil Sci Soc Am J 64(1):339–346. https://doi.org/10.2136/sssaj2000.641339x
Khazzoom JD (1980) Economic implications of mandated efficiency in standards for household appliances. Energy J 1(4):21–40
Knox JC (2001) Agricultural influence on landscape sensitivity in the upper Mississippi river valley. Catena 42(2):193–224
Lal R (2001) Soil degradation by erosion. Land Degrad Dev 12(6):519–539. https://doi.org/10.1002/ldr.472
Lal R, Shukla MK (2004) Principles of soil physics. Marcel Dekker, New York
Lamandé M, Greve MH, Schjønning P (2018) Risk assessment of soil compaction in Europe—rubber tracks or wheels on machinery. Catena 167:353–362. https://doi.org/10.1016/j.catena.201805.015
Lawin KG, Tamini LD (2019) Land tenure differences and adoption of agri-environmental practices: evidence from Benin. The Journal of Development Studies 55(2):177–190. https://doi.org/10.1080/00220388.2018.1443210
Ledermann T, Herweg K, Liniger HP et al (2010) Applying erosion damage mapping to assess and quantify off-site effects of soil erosion in Switzerland. Land Degrad Dev 21(4):353–366. https://doi.org/10.1002/ldr.1008
Lemann T, Sprafke T, Bachmann F et al (2019) The effect of the Dyker on infiltration, soil erosion, and waterlogging on conventionally farmed potato fields in the Swiss plateau. Catena 174:130–141. https://doi.org/10.1016/j.catena.2018.10.038
Leys A, Govers G, Gillijns K et al (2010) Scale effects on runoff and erosion losses from arable land under conservation and conventional tillage: The role of residue cover. J Hydrol 390(3):143–154
Li C, Grayson R, Holden J, Li P (2018) Erosion in peatlands: Recent research progress and future directions. Earth Sci Rev 185:870–886
Li J, Li W, Yang X et al (2008) The impact of land use change on quality evolution of soil genetic layers on the coastal plain of south Hangzhou Bay. J Geogr Sci 18(4):469–482. https://doi.org/10.1007/s11442-008-0469-7
Ludwig B, Boiffin J, Chadœuf J, et al (1995) Hydrological structure and erosion damage caused by concentrated flow in cultivated catchments. Catena 25(1):227–252. https://doi.org/10.1016/0341-8162(95)00012-H
Mango N, Mapemba L, Tchale H et al (2018) Maize value chain analysis: a case of smallholder maize production and marketing in selected areas of Malawi and Mozambique. Cogent Bus Manag 5(1):1–15. https://doi.org/10.1080/23311975.2018.1503220
Mansour M, Morgenstern N, Martin C (2011) Expected damage from displacement of slow-moving slides. Landslides 8(1):117–131. https://doi.org/10.1007/s10346-010-0227-7
Martinez E, Fuentes JP, Silva P et al (2008) Soil physical properties and wheat root growth as affected by no-tillage and conventional tillage systems in a Mediterranean environment of Chile. Soil Tillage Res 99(2):232–244. https://doi.org/10.1016/j.still.2008.02.001
McGuire J, Morton L, Cast A (2013) Reconstructing the good farmer identity: Shifts in farmer identities and farm management practices to improve water quality. Agric Hum Values 30(1):57–69. https://doi.org/10.1007/s10460-012-9381-y
Mishra B, Gyawali B, Paudel K et al (2018) Adoption of sustainable agriculture practices among farmers in Kentucky, USA. Environ Manage 62(6):1060–1072. https://doi.org/10.1007/s00267-018-1109-3
Montgomery DR (1994) Road surface drainage, channel initiation, and slope instability. Water Resour Res 30(6):1925–1932. https://doi.org/10.1029/94WR00538
Morin E, Krajewski WF, Goodrich DC et al (2003) Estimating rainfall intensities from weather radar data. J Hydrometeorol 4(5):782–797.
Nearing MA, Polyakov VO, Nichols MH et al (2017) Slope–velocity equilibrium and evolution of surface roughness on a stony hillslope. Hydrol Earth Syst Sci 21(6):3221–3229. https://doi.org/10.5194/hess-21-3221-2017
Nosrati K, Haddadchi A, Collins A et al (2018) Tracing sediment sources in a mountainous forest catchment under road construction in northern Iran: comparison of Bayesian and frequentist approaches. Environ Sci Pollut

Res 25(31):30979–30997. https://doi.org/10.1007/s11356-018-3097-5

Ollobarren P, Capra A, Gelsomino A et al (2016) Effects of ephemeral gully erosion on soil degradation in a cultivated area in Sicily (Italy). Catena 145:334–345. https://doi.org/10.1016/j.catena2016.06.031

Page JB, Willard CJ (1947) Cropping systems and soil properties. Soil Sci Soc Am J 11:81–88

Pimentel D, Allen J, Beers A et al (1987) World agriculture and soil erosion. Bioscience 37(4):277–283

Platt J (1973) Social traps. Am Psychol 28(8):641–651. https://doi.org/10.1037/h0035723

Poesen J, Nachtergaele J, Verstraeten G et al (2003) Gully erosion and environmental change: importance and research needs. Catena 50(2–4):91–133. https://doi.org/10.1016/S0341-8162(02)00143-1

Reimer A, Prokopy L (2014) Farmer participation in U.S. farm bill conservation programs. Environ Manag 53 (2):318–332. https://doi.org/10.1007/s00267-013-0184-8

Rinot O, Osterholz WR, Castellano MJ et al (2018) Excitation-emission-matrix fluorescence spectroscopy of soil water extracts to predict Nitrogen mineralization rates. Soil Sci Soc Am J 82(1):126–135. https://doi.org/10.2136/sssaj2017.06.0188

Roesch-McNally GE, Basche AD, Arbuckle JG et al (2018) The trouble with cover crops: farmers experiences with overcoming barriers to adoption. Renewable Agric Food Syst 33(4):322–333

Salles C, Poesen J, Govers G (2000) Statistical and physical analysis of soil detachment by raindrop impact: Rain erosivity indices and threshold energy. Water Resour Res 36(9):2721–2729. https://doi.org/10.1029/2000WR900024

Schaeffer B, Stauber M, Mueller R et al (2007) Changes in the macro-pore structure of restored soil caused by compaction beneath heavy agricultural machinery: a morphometric study. Eur J Soil Sci 58(5):1062–1073. https://doi.org/10.1111/j.1365-2389.2007.00886.x

Schwen A, Böttcher J, von der Heide C et al (2011) A modified method for the in situ measurement of soil gas diffusivity. Soil Sci Soc Am J 75(3):813–821

Sharma R, Konietzky H, Kosugi K (2010) Numerical analysis of soil pipe effects on hillslope water dynamics. Acta Geotech 5(1):33–42. https://doi.org/10.1007/s11440-009-0104-5

Shields A (1936) Application of similarity mechanics and turbulence research for bed-load transport. California Institute of Technology, Pasadena, CA

Shmulevich I, Asaf Z, Rubinstein D (2007) Interaction between soil and a wide cutting blade using the discrete element method. Soil Tillage Res 97(1):37–50

Sunny F, Huang Z, Karimanzira T (2018) Investigating key factors influencing farming decisions based on soil testing and fertilizer recommendation facilities (STFRF)—a case study on rural Bangladesh. Sustainability 10(11):1–24. https://doi.org/10.3390/su101143
31

Svoray T, Ben-Said S (2010) Soil loss, water ponding and sediment deposition variations as a consequence of rainfall intensity and land use: a multi-criteria analysis. Earth Surf Proc Land 35(2):202–216. https://doi.org/10.1002/esp.1901

Svoray T, Levi R, Zaidenberg R et al (2015) The effect of cultivation method on erosion in agricultural catchments: Integrating AHP in GIS environments. Earth Surf Proc Land 40(6):711–725. https://doi.org/10.1002/esp.3661

Svoray T, Markovitch H (2009) Catchment scale analysis of the effect of topography, tillage direction and unpaved roads on ephemeral gully incision. Earth Surf Proc Land 34(14):1970–1984. https://doi.org/10.1002/esp.1873

Takken I, Govers G, Steegen A et al (2001a) The prediction of runoff flow directions on tilled fields. J Hydrol 248(1):1–13. https://doi.org/10.1016/S0022-1694(01)00360-2

Takken I, Jetten V, Govers G et al (2001b) The effect of tillage-induced roughness on runoff and erosion patterns. Geomorphology 37(1):1–14. https://doi.org/10.1016/S0169-555X(00)000593

The World Bank (2018) Employment in agriculture (% of total employment). https://data.worldbank.org/indicator/SL.AGR.EMPL.ZS?end=2017&start=1991 2018

Torri D, Poesen J (2014) A review of topographic threshold conditions for gully head development in different environments. Earth-Sci Rev 130:73–85. https://doi.org/10.1016/j.earscirev.2013.12.006

USDA (2011) National agronomy manual. United States Department of Agriculture (USDA).:500_1–509_1

Vandaele K, Poesen J, Govers G et al (1996) Geomorphic threshold conditions for ephemeral gully incision. Geomorphology 16(2):161–173. https://doi.org/10.1016/0169-555X(95)00141-Q

Vranken L, Van Turnhout P, Van Den Eeckhaut M et al (2013) Economic valuation of landslide damage in hilly regions: A case study from Flanders, Belgium. Sci Total Environ 447:323–336. https://doi.org/10.1016/j.scitotenv.2013.01.025

Wei H, Nearing MA, Stone J (2009) A new splash and sheet erosion equation for rangelands. Soil Sci Soc Am J 73(4):1087–1441. https://doi.org/10.2136/sssaj2008.0061

Weil R, Brady N (2017) The nature and properties of soils. 15th edn. Pearson, Essex, UK

Wilson GA, Hart K (2000) Financial imperative or conservation concern? EU farmers' motivations for participation in voluntary agri-environmental schemes. Environ Plan A 32(12):2161–2185. https://doi.org/10.1068/a3311

Wilson V, Wells R, Kuhnle R et al (2018) Sediment detachment and transport processes associated with internal erosion of soil pipes. Earth Surf Proc Land 43 (1):45–63. https://doi.org/10.1002/esp.4147

Yair A (1995) Short and long term effects of bioturbation on soil erosion, water resources and soil development in an arid environment. Geomorphology 13(1):87–99. https://doi.org/10.1016/0169-555X(95)00025-Z

Zhu TX, Cai QG, Zeng BQ (1997) Runoff generation on a semi-arid agricultural catchment: Field and experimental studies. J Hydrol 196(1):99–118. https://doi.org/10.1016/S0022-1694(96)03310-0

Ziadat FM, Taimeh AY (2013) Effect of rainfall intensity, slope, land use and antecedent soil moisture on soil erosion in an arid environment. Land Degrad Dev 24 (6):582–590. https://doi.org/10.1002/ldr.2239

Ziegler AD, Giambelluca TW (1997) Importance of rural roads as source areas for runoff in mountainous areas of northern Thailand. J Hydrol 196(1–4):204–229. https://doi.org/10.1016/S0022-1694(96)03288-X

Modeling the Erosion Process

3

Abstract

This chapter reviews physically based and empirical soil loss and deposition models, and their applications to quantifying water erosion processes. This review includes the equations and flowcharts of the CAESAR-Lisflood soil evolution model, the physically based WEPP model, and the Morgan–Morgan–Finney empirically based model. The principles of these models illustrate that topography (in particular, contributing area and hillslope gradient) is crucial to determine overland flow and water erosion processes, across all models. Rainfall depth (especially rainfall intensity) is found to be key to simulate soil detachment. Soil erosivity is expressed by texture and hydraulic conductivity. The models also highlight the importance of the effect of land use (especially cultivation method) on soil erosivity and consequent erosion, and the effect of crop cover, canopy interception, and even root layer, on the water erosion process. Finally, this chapter underlines the importance of spatial and temporal components in simulating the dynamics of the processes involved, for accurate model predictions.

Keywords

CAESAR-Lisflood · Empirical modeling · MMF · Physically based modeling · Water erosion factors · WEPP

The use of geoinformatics tools in general—and those presented in this book in particular—relies on the understanding of the developer the phenomena being studied. Information is needed about the important input variables, the resolution and type of spatial representation, the typical time intervals involved, the controls of spatial variation, and more. Examples of geoinformatics applications that require such a prior information include: *expert-based systems* to identify areas prone to erosion risks using a computer language; the application of pattern-recognition algorithms to identify flow path; the execution of classifiers to delineate potential rills and gullies dynamics; and the statistical and geostatistical analyses needed to map soil health.

Knowledge about water erosion processes comes from various disciplines in the earth sciences and can be quantitatively extracted from models that, in the past three decades, have played a major role in erosion studies (Borrelli et al. 2017; Batista et al. 2019). That is, by virtue of their following three capabilities: (1) erosion models provide a robust and validated tool for computing sediment and water yields; (2) they allow simulation of the temporal dynamics and spatial changes in these erosion outcomes; and (3) they enable simulations of future scenarios—such as climate and land use forthcoming changes—and their effect on water erosion processes, using heuristic mechanisms.

T. Svoray, *A Geoinformatics Approach to Water Erosion*,
https://doi.org/10.1007/978-3-030-91536-0_3

In view of these capabilities (and others, see Vereecken et al. 2016), spatially and temporally explicit soil erosion models can be used by developers of geoinformatics tools to acquire knowledge on soil erosion processes, and mainly soil loss and deposition, before developing and parameterizing those tools.

Some of the governing equations of water erosion processes have already been described in Chap. 2. However, those equations are limited to the quantification of specific sub-processes within the universal detachment–transportation–deposition soil erosion process. They are, therefore, insufficient for understanding soil erosion at the hillslope and catchment scales, which are most appropriate for soil conservation planning and practice. That is, because erosion risk assessment requires the synergy of additional equations to quantify erosion processes under various climatic, environmental, and man-made conditions, in high resolution. Such computations and analyses can be conducted through more comprehensive erosion models, as described in this chapter.

Spatially and temporally explicit soil erosion models are usually divided in the literature into the following three groups (De Vente and Poesen 2005):

Physically based modeling integrates physical principles of soil erosion through numerical simulation and mathematical equations. The use of generic physical principals increases model generalization. Therefore, culminating with successful estimation in various environments is relatively higher than the case of empirical models that are usually based on parameters that epitomize local conditions (Guisan and Zimmermann 2000).

Conceptual modeling is a representation of key concepts in the erosion process and the hypothesized relationship between them. As such, conceptual models can even be more general than physically based models; however, they do not provide a quantitative prediction in the output. Conceptual models are therefore used mainly to assist professionals with better understanding of erosion processes and mechanisms but not for explicit soil loss and deposition estimates.

Empirical modeling uses linear or non-linear functions to link directly, by data fitting, between independent variables such as rainfall intensity or hillslope gradient and dependent variables such as soil loss or runoff yield. Important components of empirical models are therefore site-specific coefficients to be determined using training data from field measurements or laboratory experiments.

Although the three groups of models appear to be very distinct, the differences between them are actually blurred. Physically based models may involve empirical parameterization, empirical or semi-empirical models may include physically based terms, while both empirically based and physically based models retain some sort of conceptualization.

Physically based and empirical models may have each disadvantages for quantitative predictions of soil loss and deposition, and yet many of them have been validated and widely used over the past four decades (Batista et al. 2019). This chapter reviews three typical water erosion models in detail: (1) the combined soil evolution and physically based model *CAESAR-Lisflood*; (2) the physically based *Water Erosion Prediction Project* (WEPP); and (3) the empirical *Morgan–Morgan–Finney* (MMF) model. Specifically, this chapter shall describe their input variables and modes of operation, and discuss the implications of their output data for soil erosion studies.

These specific three models were chosen to be described in this book for the following five reasons:

1. They are widely used and have been validated in several environments and continuously modified since they were originally proposed.
2. They represent different mathematical and computational approaches to modeling water erosion thereby allowing to learn from diverse scientific schools.

3. They are relatively simple, and as such have been applied in several environments around the world.
4. They were developed to simulate space–time dynamics at the hillslope and small-catchment scales, so they tackle challenges related to soil conservation planning and management.
5. They are thoroughly documented in the literature, so that they may be easily replicated by the readers of this book, for example, by means of heuristic research strategies.

As a basis for these three models, the next section will review a fourth model—the basic hillslope model—that represents the basic principles of sediment flux in a synthetic hillslope system. While such a model may not be applicable realistically in an agricultural catchment for harnessing water erosion, it can serve as an introduction for quantitative description of the key physical factors and the sediment balance in a hillslope erosion process over a long-range time scale.

3.1 Basics of a Hillslope—The Regolith Profile

Regolith is an unconsolidated solid material that builds up on the parent material bedrock, and comprises mostly dust, rock fractions, and soil (Ollier and Pain 1996). In theory, the regolith thickness in a bare (vegetation free) and homogenous hillslope system depends on the rates of: (1) regolith production through weathering of the parent material and (2) regolith transportation downslope depending on the topographic location and the hillslope gradient (Pelletier 2008). Accordingly, the thickness of regolith is expected to gradually increase from summit to footslope, due to downslope transportation of regolith material and, in particular, the fine material within it, namely, the soil (Fig. 3.1).

According to Pelletier (2008), the process of regolith production and transportation downslope can be described as follows. Theoretically, at the summit, there is no water or sediment movement, and the flux of these two components is assumed to be nil, because the summit is the dividing line between two drainage areas in opposite directions. Further downslope, water and soil particles movement is largely dependent on the topography—in particular, the hillslope gradient—so that the material is transported downslope due to the effect of gravitational forces.

In other words, in such a system, the regolith thickness at any given point is a function of the rate of regolith production due to bedrock weathering, and the point location on the hillslope, due to the gravitational effect. We further elaborate on the complex and crucial effect of relief on soil and water transportation on hillslopes in Chap. 4, where spatial data models are developed for the delineation of hillslope unit boundaries based on pedogenic characteristics of the hillslope catena (footslope, shoulder, summit, etc.).

For now, however, we assume that weathering forces along the hillslope and the regolith transportation downslope are uniform. In accordance with this assumption, and with the accumulated effect of time, the regolith profile is expected—and, in many erosion studies, is also observed—to be thin in the upper parts, and thick at the footslope. Furthermore, because the topography is first and foremost assumed to affect the direction of overland flow, many soil erosion studies also assume that the parts of the hillslope that accumulate a greater contributing area, which are usually the lower regions, near the channel network, are wetter than those along the upper part of the hillslope, whose contributing area is smaller—see, for example, Tarboton (1997).

These geometrical principles are not always a realistic reflection of overland flow continuity, especially in semiarid natural environments, given that vegetation patches may act as barriers to overland flow. Such barriers may result in low connectivity in overland flow and create different self-organized spatial patterns, such as mosaic-like structures of exposed areas (hydrological source) and vegetated patches (hydrological sinks) of runoff water and sediments with variable interconnection (Saco et al. 2007).

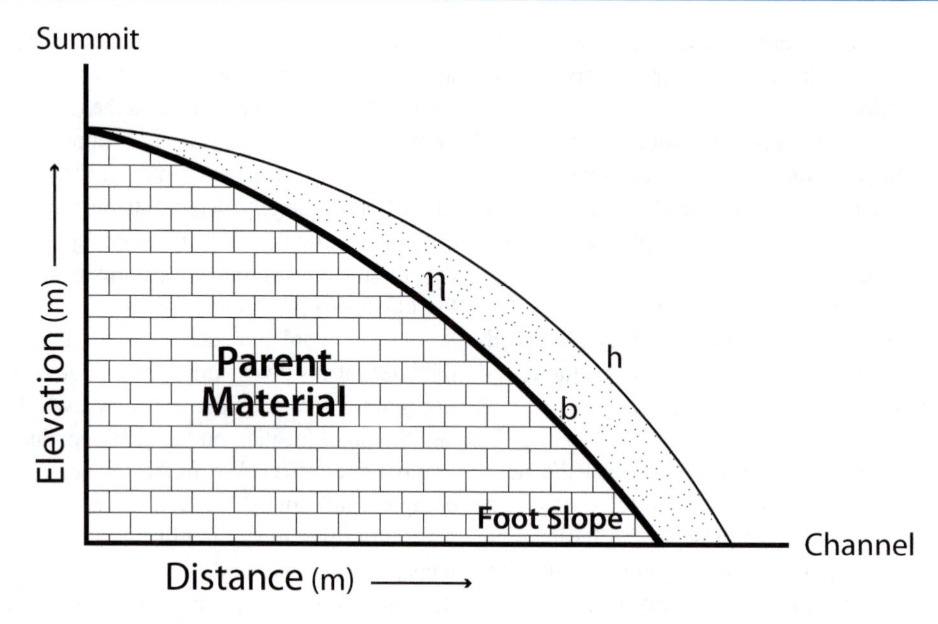

Fig. 3.1 A description of a synthetic hillslope with the bedrock weathering front *b*, the topographic surface *h*, and the regolith layer depth η. Water, as an agent, transforms the regolith from the summit area—or the interfluve—toward the footslope (Adapted from Pelletier (2008), J. D. Pelletier 2008 © Reproduced with permission of The Licensor through PLSclear)

Examples of such alterations in hydrological connectivity are *banded landscapes* which are comprised of alternating patches of bare soil and vegetation along the desert hillslopes (Moreno-de las Heras et al. 2012). The effect of hydrological sinks and banded landscapes should be considered—when relevant—in more complex water erosion simulations. For now, however, let us focus on the bare soil synthetic example while setting aside the effect of patchiness.

The long-term feedback between weathering front, topography, and regolith thickness in a continuous form is set out in Eqs. 3.1–3.4, from Pelletier (2008), based on Furbish and Fagherazzi (2001). According to these equations, topography controls the transportation and deposition processes, which control the regolith depths and dictate the rate of change in the weathering front. Equation 3.1 describes the change in the regolith depth over time, based on bedrock characteristics, sediment density, and topographic location along the hillslope (see also Fig. 3.1):

$$\frac{\partial h}{\partial t} = -\frac{\rho_b}{\rho_s} \cdot \frac{\partial b}{\partial t} - k \cdot \frac{\partial^2 h}{\partial x^2}. \quad (3.1)$$

The thickness change in the regolith layer *h* including soil and rock fractures with time *t* is the constant $-\frac{\rho_b}{\rho_s}$, where ρ_b is the parent material density and ρ_s is the sediment density—multiplied by the change of the parent material baseline *b* over time $\frac{\partial b}{\partial t}$, minus the constant k of the hillslope diffusivity, multiplied by the rate of change in topographic location $\frac{\partial^2 h}{\partial x^2}$, with *h* being the elevation representing the topographic surface, and *x* being the distance along the hillslope profile. The x^2 notation denotes a second-order derivative of the elevation, which quantifies the rate of change in hillslope gradient—or, in other words, the *surface curvature* (Pike et al. 2009).

Surface curvature may be defined as the degree to which the topographic surface deviates from a plane, by being either concave or convex: A negative value of $\frac{\partial^2 h}{\partial x^2}$ means that the hillslope is convex, and therefore flow is expected to slow down; a positive curvature value means that the given hillslope is concave, whereby the flow is accelerated. A hillslope of a zero-curvature value is assumed to be straight.

Equation 3.2 describes the rate of change in the weathering front $\frac{\partial b}{\partial t}$ over time which is actually the rate of regolith production from the bare parent material P_0 multiplied by the exponent of the ratio between the regolith depth h at a given moment, and the regolith depth at $t = 0$.

$$\frac{\partial b}{\partial t} = -P_0 \cdot e^{\frac{-h}{h_0}}. \qquad (3.2)$$

The regolith profile thickness is therefore described as the dynamic difference between the topographic conditions, and the weathering of the parent material. Thus, the rate of change of regolith depth over time is the difference between the rate of change in parent material over time, multiplied by the ratio of the parent material to sediment density, and the second derivative, namely, the curvature of the topographic profile, multiplied by the hillslope diffusivity.

According to Nishiizumi et al. (1997), reduction in parent material is most pronounced on bare hillslopes and decreases exponentially with increasing regolith depth. Consequently, Eqs. 3.1–3.2 can be developed into Eq. 3.3, where the exponent in Eq. 3.2 is reversed to *ln* and merged in Eq. 3.1 to extract the regolith profile thickness.

$$h = h_0 \cdot ln\left(\frac{\rho_b}{\rho_s} \cdot \frac{P_0}{k} \cdot \frac{1}{-\frac{\partial^2 h}{\partial x^2}}\right). \qquad (3.3)$$

Similarly, Eq. 3.4 represents the rate of change over the distance where $h = h_0$.

$$\left.\frac{\partial^2 h}{\partial x^2}\right| \text{bare} = -\frac{\rho_b}{\rho_s} \cdot \frac{P_0}{k}. \qquad (3.4)$$

Equations 3.1 through 3.4 illustrate the role of surface geometry or topography and parent material in the very basic sediment transportation process on a hillslope system. In this regolith profile, the fine soil material accumulates over time. As more soil builds up on the regolith layer, and plants and agricultural activities begin to affect the erosion process, it becomes more complex, due to a multitude of factors that either negate or enhance the basic impacts of topography and parent material. Furthermore, to simulate more realistic erosion processes under the conditions of active agricultural environments; the effect of factors such as rainfall intensity, duration and depth, plant transpiration, evaporation, and overland and below surface water flow—vertical and horizontal—should be considered. This may require a more careful treatment for erosion simulation than Eqs. 3.1–3.4. However, because the regolith profile equations are dynamic, a range of regolith depth changes can be simulated—such as heuristic studies of the effect of topographic gradient and parent material density in predicting soil infrastructure development under various conditions of parent material and topography.

3.2 Landscape Evolution Models

3.2.1 Background

Landscape Evolution Models (LEM) simulate space–time dynamics in surface elevation by quantifying hillslope and catchment scale erosion and deposition processes (Hancock et al. 2017). By doing so, LEMs consider more complex and realistic environmental and climatic effects on soil erosion than the basic hillslope model in Eqs. 3.1–3.4, while also relying on physical principles. The LEMs are applied to *DEMs* (Digital Elevation Models) as the input spatial data, over time iterations ranging from tens to thousands of years (Hancock et al. 2015). By doing so, they allow to predict landscape evolution in hillslopes and catchments at high spatial resolution which may vary from meters to kilometers (Coulthard and Van De Wiel 2017). The model strategy is executed by routing simulated overland flow across the DEM grid-cells, and changing the elevation in each grid-cell to represent fluvial and hillslope erosion and deposition processes.

A general LEM equation that can be used to estimate the rate of change of surface elevation over time $\frac{\partial h}{\partial t}$ is summarized in Eq. 3.5 (Braun 2019, 2018):

$$\frac{\partial h}{\partial t} = U - K_f \cdot A^m \cdot S^n + K_D \cdot \nabla^2 \cdot h, \quad (3.5)$$

where h is the topographic elevation and t is the time, U is the tectonic uplift, and the effect of fluvial morphology is expressed by the amount of water flow in the river network relying on the upslope contributing area A and the hillslope gradient S with the parameters m and n as exponents that should be determined empirically, based on local conditions. In addition, the parameters K_f and K_D are more complex, and according to Braun (2018) they depend on the effect of rainfall and overland flow characteristics, parent material physiognomies, canopy cover, and other factors. Finally, ∇^2 is the diffusion parameter. The central term on the right side of Eq. 3.5—$K_f A \cdot^m \cdot S^n$—quantifies the long-range processes that link together various parts of the landscape through erosion and deposition processes, while the rightmost term—$K_D \cdot \nabla^2 \cdot h$—represents the short-range processes that smoothen the landscape. Equation 3.5 is indeed one dimensional, however, as was mentioned before, the use of a DEM allows for application of the computations over an entire catchment and offers analysis at a landscape scale (Cohen et al. 2010).

DEMs therefore play a major role in the upscaling process from point to catchment or hillslope. They have been used in the literature as input data to LEMs in two formats: (1) matrices with square grid-cells of equal spatial resolution, which allow the user to exploit the geometric characteristics of raster DEMs (Cohen et al. 2015) or the less widespread version and (2) an object-oriented framework based on adaptive irregular network that represents more active parts of the catchment at higher resolution, and less active parts at lower resolution (Tucker et al. 2001b).

Figure 3.2 illustrates the spatio-temporal dynamic concept of the more common, matrix DEM-based LEM, and the use of temporal layers of raster data models to simulate alterations in surface elevation during several time steps t. Such a computational platform provides a useful tool in predicting landscape evolution between loss and deposition of surface material.

The early LEMs were developed in the mid-1970s, and became more sophisticated over the 1980 and 1990s with the increase in computation power (Coulthard 2001). LEMs simulate key surface processes—such as sediment transport and deposition, landslides, regolith production, rockfall, and debris flows. One notable characteristic of LEMs is the computational platform described in Fig. 3.2, which incorporates spatio-temporal analysis that is so crucial in the study of water erosion processes, especially in the hillslope and catchment scales. Within this spatio-temporal framework, among the most relevant LEM developments to soil erosion studies in agricultural environments are the short-range Soil Evolution Models (SEMs)—a specific branch of LEMs that will be further discussed later in this chapter.

Coulthard (2001) surveyed four of the most used LEMs that featured similar principles but different methodologies: SIBERIA (Willgoose et al. 1991), GOLEM (Tucker and Slingerland 1994), CASCADE (Braun and Sambridge 1997), and CHILD (Tucker et al. 2001a). All of these LEMs, he found, are successful in surface elevation predictions at different spatio-temporal scales of observation. However, CASCADE and GOLEM are more effective for predicting longer periods and lower spatial resolution, and SIBERIA and CHILD are more suited for shorter periods and higher spatial resolution. CHILD and GOLEM offer modeling of fluvial processes in greater detail, while SIBERIA offers more successful modeling and predictions of hillslope processes. Among all four, CASCADE is the only one to accommodate tectonic movements. All of the above means that models need to be fitted to the specific environmental conditions in each site. For example, GOLEM and SIBERIA are more flexible models that can represent various hillslope processes including weathering effects. Therefore, they can specifically fit to quantify the unique runoff mechanisms of dry environments. However, if the model needs to be

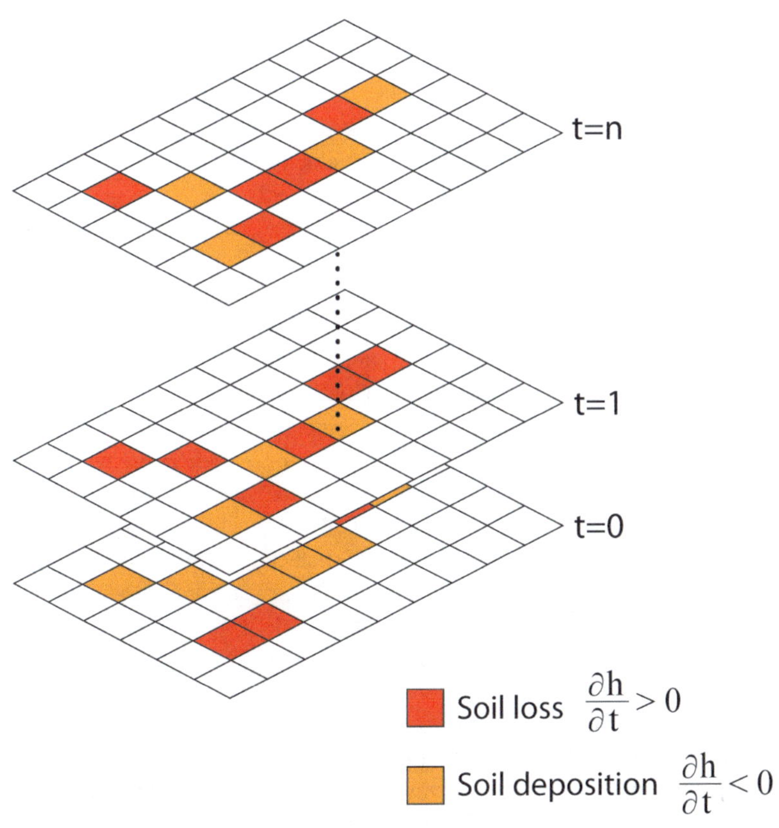

Fig. 3.2 The basic space–time structure of matrix DEM-based LEM. Grid-cells are assigned with values based on physically based simulations of erosion and deposition processes, such as those described in Eq. 3.5. Different t values denote time intervals ranging from minutes in short-term soil evolution models to hundreds of years in LEMs. The outcomes are therefore patterns of soil erosion (negative values) and deposition (positive values)

applied to convoluted channel networks, CASCADE and CHILD, which are based on irregular mesh data model, can be more suitable.

LEMs are indeed very useful for long-term landscape analysis, such as topographic change and tectonic uplift, but to study high-resolution pedogenic processes we need the aforementioned SEMs. SEMs simulate soil formation and loss based on quantification of short-term soil and hydrological processes using raster-based DEM data model (Cohen et al. 2010, 2015). While in LEMs surface elevation is predicted, in SEMs, it is the soil profile thickness that is predicted. However, the roots of the model and the modeling principles of SEMs stem from their antecedents, the LEMs. The LEM CHILD model, for example, was modified to simulate gully development using a three-dimensional soil-LEM framework (Flores-Cervantes et al. 2006). Another SEM, the mARM3D was developed to couple soil profile evolution and detailed hydrological processes to predict soil thickness (Cohen et al. 2010). The Integrated Landscape Evolution and Soil Development model (MILSED) was developed primarily to simulate pedogenetic processes (Vanwalleghem et al. 2013). The tRIBS-erosion model is a spatially distributed TIN-based real-time integrated basin simulator (Ivanov et al. 2004), coupled with hillslope and channel erosion processes. tRIBS represents large number of important and advanced soil erosion components but it requires large computational resources (Francipane et al. 2012).

3.2.2 The CAESAR-Lisflood

Among the SEMs, the commonly used CAESAR-Lisflood (CL) is aimed to predict soil thickness in temporal scales between hours and thousands of years. CL quantifies water erosion processes in relatively small catchments by combining the geomorphic CAESAR model

(Cellular Automaton Evolutionary Slope And River) with the Lisflood-FP 2d hydrodynamic flow model (Bates et al. 2010).

CAESAR was developed in the late 1990s (Coulthard et al. 1998, 2002) as a simulator of fluvial and hillslope complex—and sometimes non-linear—processes. The model is also aimed to reveal positive and negative feedbacks between forms and processes. A calibrated CAESAR model was able to predict suspended sediment and bedload yields, corresponding with field measured data in both volume and timing (Coulthard et al. 2012). The use of a *cellular automata by CAESAR ensures high fidelity of representation of neighborhood effects, and an efficient spatio-temporal* data model platform. In a nutshell, cellular automata procedures use a series of mathematical or logical rules—applied to each raster grid-cell and its immediate D8 neighborhood, known also as the Moore neighborhood—to govern the behavior of the entire hillslope or catchment soil evolution (Bastien et al. 2002). The rules are representations of spatial processes and their synergy allows quantifying non-linear response of processes and feedbacks in water erosion (Van De Wiel et al. 2007).

The other half of CL, LISFLOOD-FP, is a physically based flood model that also uses a raster-based DEM as an input data model to represent explicitly the surface and subsurface water flow discharge variation in the catchment. LISFLOOD-FP is specifically designed for channel hydraulic routing problems while minimizing the floodplain hydraulics representation, to achieve acceptable predictions in relation to measured data (Bates et al. 2010). The model computations are based on the one-dimensional shallow-water equations that were fitted for use in two-dimensional representation where flows in the longitudinal and latitudinal directions are merged. The outcome model can be solved explicitly at reasonable computational cost, and, despite its complexity, it was verified successfully against several independent sources.

CL, as a synergized model between these two components, was validated and was noted for its efficiency. Specifically, to our aim in this book, CL has been used to simulate the development and dynamics of ephemeral gullies in agricultural catchments, using spatially and temporally explicit topographic, soil and rainfall intensity data (Hoober et al. 2017).

3.2.3 The Model Operation

Generally, CL is applied by simulating water routing across a DEM, to quantify the processes that result in a change in soil depth in each of the DEM grid-cells. These processes are driven by the geomorphological processes simulated in CAESAR, with temporal iterations. The outcomes of these model iterations are the elevation changes in the grid-cells at each time step t, in accordance with the rainfall data and environmental conditions. Thus, CL operates on a cell basis using four data layers:

1. An original DEM layer with elevation data of the catchment before the erosion/deposition process began.
2. Hourly rainfall rate from a meteorological station or a meteorological radar.
3. Data on soil particle size fraction.
4. A parent material layer.

The processing of these four data layers in CL is applied in four steps, as set out in Fig. 3.3.

In the first step, the model simulates the hydrologic fluxes (infiltration-excess, overland and subsurface flows, infiltration, exfiltration, evapotranspiration, and water flow in the channel), and explicit ground and surface–water interactions. This is done using the commonly used TOPMODEL equations (Beven and Kirkby 1979). Runoff simulations consider soil moisture storage and rainfall rate based on saturated flow assumptions.

Secondly, runoff flow is simulated as a discharge, using the LISFLOOD-FP hydrodynamic flow computations (Bates et al. 2010), which resolve a reduced form of the shallow-water equations (Saint–Venant 1871) using a numerical scheme. As an approximation of the diffusion wave, LISFLOOD-FP computes water flux in X- and Y-directions, using Manning's equation that allows to estimate empirically—based on

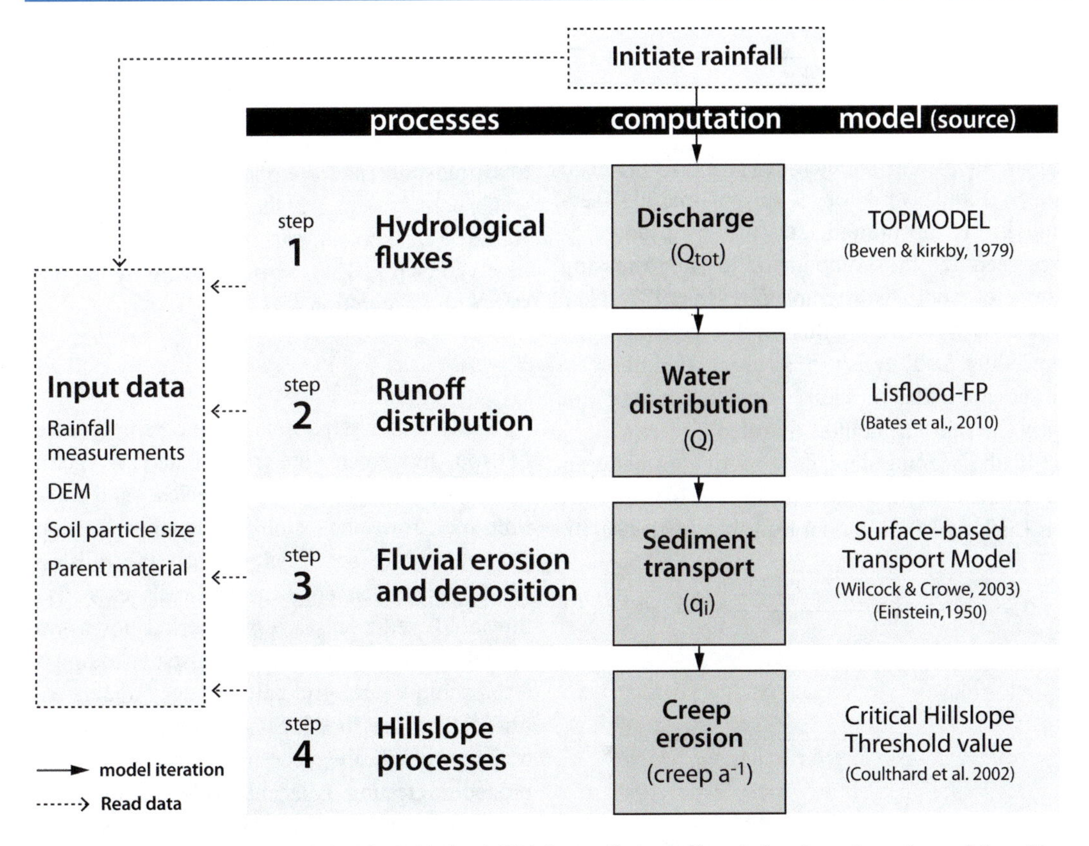

Fig. 3.3 The main four steps of CAESAR-Lisflood (CL) for predicting soil evolution through erosion and deposition processes. See the text below for details

channel geometry and friction coefficient—the mean velocity of water flow in the channel. In other words, the computations are based on the surface roughness, hydraulic radius, and channel inclination. This allows using simple equations to compute fluxes explicitly, so the computational costs are feasible.

Thirdly, Manning's equation is used to estimate water thickness, and the flow velocity is calculated by dividing water flux between grid-cells. Shear stress threshold is computed based on the flow depth and velocity. Then shear stress is used to compute fluvial erosion based on the sediment transport equation (Wilcock and Crowe 2003). Sediment transport is simulated by characterizing the gully bed, the stratigraphic unit, and the parent material layer (Van De Wiel et al. 2007).

Fourth, hillslope processes are represented as a mass movement that occurs when a hillslope threshold is exceeded, and soil creep as a function of a threshold of the hillslope gradient.

Equations 3.6–3.12 (Coulthard et al. 2013, 2002) are used to apply the operational steps of CL, as illustrated in Fig. 3.3.

As a first step, in Eqs. 3.6–3.7, a version of the physically realistic but parametrically simple, rainfall-runoff TOPMODEL (Beven and Kirkby 1979) is executed to simulate surface, and subsurface, discharge Q_{tot} through the expression of variation in soil moisture storage j_t:

$$Q_{tot} = \frac{m}{T} \cdot \log\left(\frac{(r - j_t) + j_t \cdot e^{\left(\frac{r \cdot T}{m}\right)}}{r}\right), \quad (3.6)$$

$$j_t = \frac{r}{\left(\frac{r-j_{t-1}}{j_{t-1}} \cdot exp\left(\left(\frac{(0-r) \cdot T}{m}\right) + 1\right)\right)}, \quad (3.7)$$

where m is a parameter that controls the soil water storage variation, T is the timestep [seconds], r is the rainfall rate [m · h^{-1}], and j_t is extracted by the computation of a decreasing curve of flood hydrograph (Beven 1977). The model is dynamic in time and j_t considers the preceding soil moisture storage conditions as observed in the previous iteration j_{t-1} and not only the rainfall input at timestep T.

In the second step, flow routing Q between grid-cells is computed using Eqs. 3.8–3.9, based on Lisflood-FP 2d model by Bates et al. (2010).

$$Q = \frac{q - g \cdot h_{flow} \cdot \Delta t \cdot \frac{\Delta(h+z)}{\Delta x}}{\left(1 + g \cdot h_{flow} \cdot \Delta t \cdot n^2 \cdot |q| \cdot h_{flow}^{-10/3}\right)} \Delta x, \quad (3.8)$$

$$\frac{\Delta h^{i,j}}{\Delta t} = \frac{Q_x^{i-1,j} - Q_x^{i,j} + Q_y^{i,j-1} - Q_y^{i,j}}{\Delta x^2}, \quad (3.9)$$

where q is the flux between grid-cells from the previous iteration [m^2 · s^{-1}]; g is acceleration due to gravitation forces [m · s^{-2}]; h is the soil depth [m]; and h_{flow} is the maximum water thickness between two grid-cells (a threshold value with a default = 0.00001 m) and is being used to prevent the flow model from trying to move water when there are small gradients between grid-cells and t is time [seconds]. z is the grid-cell elevation [m], x and y are the grid-cell width and length [m], n is the Gauckler–Manning roughness coefficient [s · m$^{-0.3}$]. After the discharge across the grid-cell boundaries is computed—using i and j grid-cell coordinates—the grid-cell water depth h is updated.

In the third step, sediment transport rate q_i is computed in Eq. 3.10 based on Wilcock and Crowe (2003).

$$q_i = \frac{F_i \cdot U_*^3 \cdot W_i^*}{(s-1) \cdot g}, \quad (3.10)$$

where F_i is the fractional volume of the ith sediment in the active layer; U_* is a function of shear velocity [cm · s^{-1}]; W_i^* is a multifaceted expression that relates fractional transport rate to total transport rate; s is the ratio of sediment to water density; and g is the power of gravitation. Rates of transport q_i can be converted into volume values [cm^3], V_i, by multiplying by the time step t of the iteration (Eq. 3.11).

$$V_i = q_i \cdot dt. \quad (3.11)$$

In the fourth step, downslope creep processes of soil movement are represented as gentle downward transport of soil particles—and sometime rock fractions—along a moderate hillslope gradient. More specifically, when the hillslope gradient between adjacent grid-cells exceeds a threshold, sediments are transported downslope until they reach a relatively flat (also depending on a threshold value) grid-cell. Because a slide at a grid-cell at the footslope may initiate sediment transport uphill, the creep model uses an iterative procedure testing adjacent grid-cells until soil transport is stopped. Therefore, soil creep as a function of the surrounding area a^{-1} is computed between each grid-cell at every temporal iteration t according to Eq. 3.12 where S is the hillslope gradient [°] and D_x is the horizontal spacing or size of grid-cell [m] (Coulthard et al. 2002):

$$Creep \cdot \left(a^{-1}\right) = \frac{S \cdot 0.01}{D_x}. \quad (3.12)$$

To fit the CL simulation to a short-term agricultural catchment analysis, one must calibrate the input parameters. The values assigned by Hoober et al. (2017) for a small-catchment scale, single season simulation of gully development in an agricultural catchment in central Israel are as follows (within brackets): h—soil thickness (0.1 m), x, y—DEM grid-cell size (2 · 2 m^2), lateral erosion rate (0.0005), n—Manning's roughness coefficient (0.0396), maximum erode limit (0.01 m), minimum discharge (0.02 Q • m^{-3}), T—rainfall time step [hour], and water depth threshold [m]. Computation of these steps using spatial data models yields an outcome of model output that can predict grid-cells where soil is eroded (lost) or deposited at each time step.

3.2.4 The Model Output

The model output of CLs is a collection of layers representing the change in soil thickness, in each grid-cell, as a function of soil loss and deposition processes. As in the basic dynamic regimes of SEMs—specifically, the CL—the model output can be applied to analyze the development of soil loss and deposition damage, as well as the study of the most influential parameters. While CL models have indeed been applied at the decadal and even the millennial scales (see, for example, Hancock et al. (2011), Coulthard et al. (2013)), recent studies have adapted them to the short-range modeling strategy. For example, in a recent seasonal analysis of real-world conditions of an agricultural catchment in Israel's central region, CL output predictions were found to be accurate and effective in studying the development of ephemeral gullies in wheat fields (Hoober et al. 2017). The authors have calibrated the CL for hourly rainfall characteristics and to a catchment area of 0.37 km^2. The development of ephemeral gullies was simulated for two consecutive rainfall seasons in a grid-cell resolution of $2 \cdot 2$ m^2 using electronic theodolite measurements (see model output in Fig. 3.4).

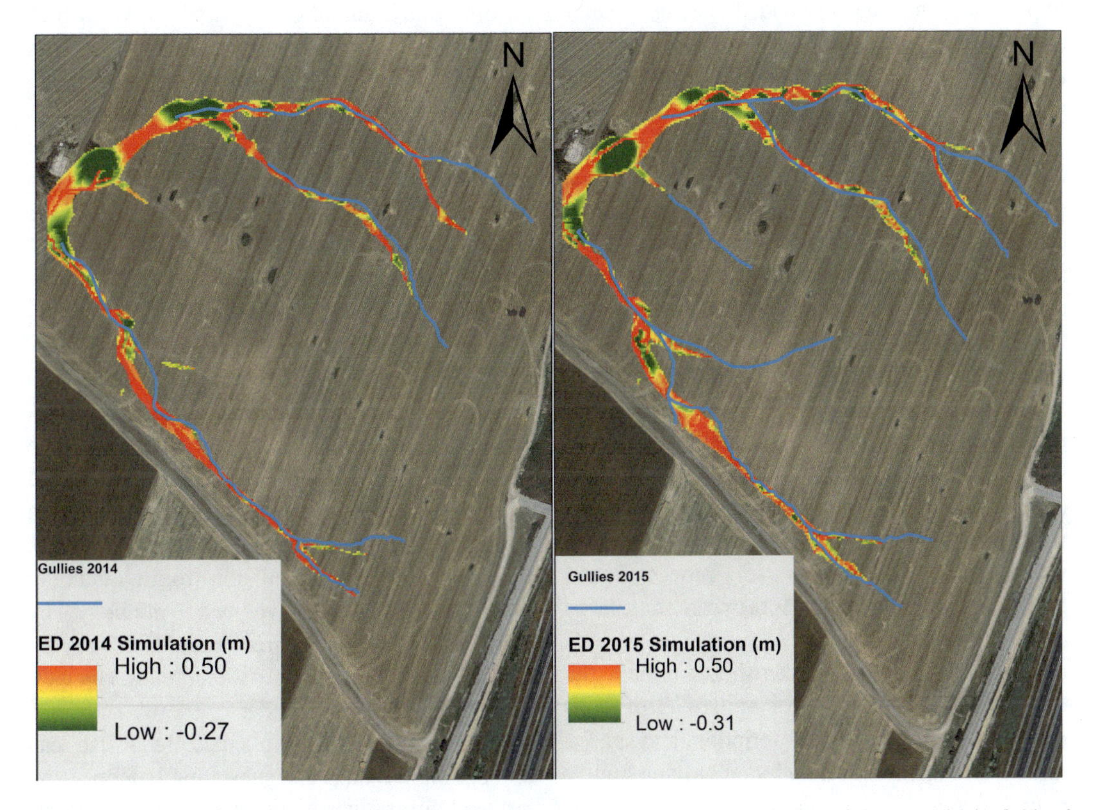

Fig. 3.4 Elevation Difference (ED) at the end of a CL simulation for the 2014 and 2015 rainfall seasons in the fields of Kibbutz Revadim, Israel. The blue lines represent ephemeral gullies that were developed each year and were delineated using DGPS. Positive values (red grid-cells) represent erosion; negative values (green grid-cells) are deposition (*Source* Hoober D, Svoray T, Cohen S, Using a landform evolution model to study ephemeral gullying in agricultural fields: The effects of rainfall patterns on ephemeral gully dynamics. Earth Surface Processes and Landforms 42(8):1213–1226. With permission from Wiley, copyright © 2017 John Wiley & Sons, Ltd). A full simulation of the model runs can be seen at: https://www.youtube.com/watch?v=DZ2x53p0fyk

Two observations arise from the model output illustrated in Fig. 3.4. First, there is a clear spatial correspondence between the blue lines (the ephemeral gullies as measured and delineated in the field using high-resolution differential GPS) and the predictions of the dynamic erosion-deposition processes by the CL. In other words, the CL provides accurate geometrical identification of the location and length of the ephemeral gullies. CL runs can also show if the gullies repeat in different seasons, under different rainfall intensity regimes using change-detection analysis (see Sect. 5.4.4.2). Second, the model output clearly demonstrates how erosion (red) and deposition (green) processes can vary within the ephemeral gully boundaries, depending on the channel's water energy and carrying capacity. Importantly, the data is provided at a high spatio-temporal resolution that is difficult to measure in the field with traditional tools, for logistical reasons.

In conclusion, CL is a SEM that can provide us with an accurate tool for predicting areas at risk and the rate of channeled erosion, at a temporal resolution that cannot practically be achieved via field measurements. CL model outputs have been compared with those of SIBERIA—a LEM that has been modified to be used as a SEM (Hancock et al. 2010). CL is more suitable for temporally explicit simulations of individual events, whereas SIBERIA is more generalized in time and uses mean temporal erosion rates. Both model predictions compared well with field data—however, a small difference was observed between the soil processes and modeled catchments after 1000 years, and large differences were observed at simulation periods of $\sim$ 10,000 yr. CL has been shown to have the upper hand with regard to simulations of sediment transport, grain size description, and space–time hydrological modeling by employing hourly rainfall depth.

3.3 Water Erosion Prediction Project (WEPP)

3.3.1 Background

WEPP is a process-based water erosion model that was developed to estimate soil loss and deposition by researchers of the Agricultural Research Service of the U.S. Department of Agriculture (USDA). WEPP was specifically created to help soil and water conservation efforts in rangelands, forest areas, and croplands (Flanagan and Nearing 2007). The hydrological and soil processes modeled by WEPP include infiltration and overland flow, raindrop and flow detachment, and soil transport and deposition. In addition, WEPP simulates the effects of residue decomposition and of crop plants' growth, which are of key importance in agricultural areas with regard to, for example, crop-rotation activities. In spatial terms, the original WEPP computations provided estimates along a hillslope profile. At a later stage, however, WEPP was further developed into the GeoWEPP platform, to include spatially explicit soil loss and deposition estimates, off-site and on-site, using GIS layers.

WEPP is a widely held soil erosion model, used by both researchers and practitioners. Since its inception in the mid-1980s, it has been validated and parameterized in many research studies in the US and other countries. One up-to-date application of the WEPP is the study of the effect of climate change scenarios on erosion yields in a sub-tropical environment (Anache et al. 2018). Overland flow and soil erosion from plots under different land uses have been predicted by WEPP, using General Circulation Models as the weather engine to simulate the climatic changes. The WEPP results under various scenarios showed how carefully planned land use can reduce soil loss rates and substantially reduce soil vulnerability to worsening in rainfall conditions.

So, the WEPP can be applied for practical uses of soil conservation not only by estimating areas prone to on- and off-site consequences but also by envisaging the effect of farmer's manipulation on erosion risks.

Yet, we ought to be aware that similarly to the case of the SEMs, or any other spatial model application to real-world conditions, the availability and quality of the input data are crucial to the model performance. For example, while using low-resolution DEMs, such as the case of contour-based DEMs, we can expect relatively erroneous topographic representation which may decrease the certainty of the model predictions. Alternatively, modern high-resolution LiDAR data may be more reliable. This is naturally true also for the availability of soil properties and climate data. Also, similarly to the case of the SEMs, WEPP estimates soil loss and deposition and does not suggest any estimations of the consequences of water erosion to the chemical or the biological components of the soil.

3.3.2 Model Operation

The original WEPP model operation for a hillslope cross section is available through the software tool from Purdue University (at: https://milford.nserl.purdue.edu/wepp/weppV1.html). The spatial GeoWEPP GIS version on a catchment scale is available from the State University of New York (SUNY), Buffalo, New York, at: http://geowepp.geog.buffalo.edu/.

Similar to other soil erosion models, WEPP uses input data of climatic, topographic, and soil- and human-induced conditions as the infrastructure in the simulation of the governing hydrologic and soil processes (Fig. 3.5). Based on this input data, the strategic approach of WEPP, scientifically and operationally, is to distinguish between (and, subsequently, to combine) the channeled overland flow from the rill area and the shallow depth flow of the sheet/interrill area.

In the case of channeled flow, shear stress is quantified by means of theory of rill hydraulics. That is, critical shear stress is set as a threshold and if the threshold value is crossed, a simplified transport equation is applied to calculate the transport capacity.

Thus, the model developers assume that soil detachment is initiated once the value of hydraulic shear stress > value of critical shear stress, and soil is transported downslope when sediment load is smaller than the rill's transport capacity. Alternatively, soil deposition occurs when sediment load > transport capacity.

In the model operation, the sediments of the sheet or interrill area are delivered into the rill network. They are then either deposited into the rill bottom or transported by the channeled flow downslope to the river network. Sediment delivery from the interrill area is not only a function of the square of rainfall intensity, but also of the hindrance effects of vegetation canopy cover, crop residue cover, and interrill soil erodibility. WEPP modeling can be therefore used to map soil loss, including consideration of spatial variability in hillslope gradient and contributing area, surface roughness, soil properties, and land use/land cover conditions in the catchment/hillslope mapped. The entire erosion framework is quantified by the WEPP equations, which are described in detail in many papers, e.g., Huang et al. (1996) and Yu (2003). Rather than repeat the entire system of equations here, we shall only indicate the main ones, based on the description by Nearing (1989).

The original WEPP equations operate along a well-defined hillslope profile and require detailed input data from the user on (Pieri et al. 2007): rainfall depth and temperature characteristics

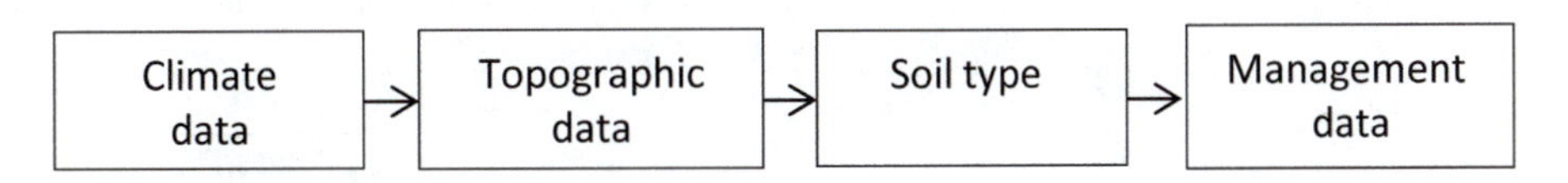

Fig. 3.5 Flowchart of model operation of WEPP. Climate data is imported from a weather station, then the user enters the hillslope characteristics and defines the soil type and management tools

(climate data), a DEM (topographic data), soil properties including texture, organic matter percentage, cation exchange capacity, and soil albedo (soil type), as well as various crop information inputs such as canopy cover, biomass energy ratio, root-to-shoot ratio, senescence period, in-row plant spacing, harvest index and tillage intensity, ridge interval, and mean tillage depth (management data).

Modeling such detailed effects on the erosion processes may be very accurate but may also bear a certain amount of noise and even errors in the predictive ability, due to data limitations. As further discussed in the following section, the empirical-parsimonious models bypass specific effects on the modeling effort by using data-fitting approaches.

The first step in WEPP computation begins with quantifying the change in sediment load along the hillslope profile. The model computes the soil loss unit area of the hillslope based on the erosion rate in both the rill and interrill parts of the hillslope, as described in the sediment continuity equation (Eq. 3.13):

$$\frac{dG}{dx} = D_f + D_i, \tag{3.13}$$

where G [kg · s^{-1} · m^{-1}] is the sediment load; x [m] is the distance downslope along the hillslope profile; D_f [kg · s^{-1} · m^{-2}] is the rill erosion rate; and D_i [kg · s^{-1} · m^{-2}] is the interrill erosion rate. D_f, is positive for detachment and negative for deposition, while D_i is considered independent of x. Both erosion rates are computed on a per rill area basis, and consequently G is solved on a per unit rill width basis. In the end of the process, soil loss is computed per hillslope unit area.

The rill erosion rate D_f is controlled by the flow detachment capacity D_c [kg · s^{-1} · m^{-2}], and the sediment transport capacity within the rill T_c [kg · s^{-1} · m^{-1}], which is controlled by the rill geometrical dimensions (Eq. 3.14):

$$D_f = D_c \cdot \left[1 - \frac{G}{T_c}\right]. \tag{3.14}$$

The detachment capacity D_c is computed using Eq. 3.15, whereby K_r[s · m−1] is the soil erodibility parameter. The equation is based on a threshold value where the hydraulic shear stress τ_f[Pa] is higher than the critical shear stress τ_c[Pa]:

$$D_c = K_r \cdot \left(\tau_f - \tau_c\right). \tag{3.15}$$

If the shear of the soil is higher than the critical shear, the rill detachment is considered zero. Soil deposition is calculated when sediment load, G, exceeds the sediment transport capacity, T_c. Accordingly, for deposition, Eq. 3.16 is used to compute the rill erosion rate D_f:

$$D_f = \left[\frac{V_f}{q}\right] \cdot [T_c - G], \tag{3.16}$$

where by V_f [m · s^{-1}] is the effective fall velocity for the sediments and q [m^2 · s^{-1}] is the flow discharge.

The following three variables are computed in the WEPP hydrological component which generates breakpoint rainfall information and runoff hydrographs: (1) the effective overland flow duration t_r [seconds]; (2) the effective rainfall intensity I_e [m · s^{-1}]; and (3) the peak runoff rate P_r [m · s^{-1}]. To transport the dynamic information into steady-state terms for the erosion equations, P_r is assigned the value equal to that of the peak runoff on the hydrograph and t_r (Eq. 3.17) is assigned as the time required to produce a total runoff volume equal to that given by the hydrograph with a constant runoff rate of P_r:

$$t_r = \frac{V_t}{P_r}, \tag{3.17}$$

where V_t [m] denotes runoff depth per rainfall occurrence and I_e is computed independently using Eq. 3.18:

$$I_e = \sqrt{\left[\frac{\left(\int I^2 \cdot dt\right)}{t_e}\right]}, \tag{3.18}$$

where I is the rainfall intensity [m · s^{-1}], t is the time, and t_e denotes time duration that rainfall rate > infiltration rate and creates a water surplus.

As a next step, the WEPP operating functions compute the geometrical characteristic of the rill cross section. The rill width w [m] is calculated based on the relationship in Eq. 3.19 with c as an empirical parameter, Q_e [m^3 · s^{-1}] as the flow discharge at the bottom of the profile, and d as an empirical parameter.

$$w = \mathrm{c} \cdot Q_e^{\mathrm{d}}. \tag{3.19}$$

Q_e is determined by Eq. 3.20 with again, P_r [m · s^{-1}] as the peak runoff rate, L [m] as the hillslope length, and R_s [m], the rill spacing, is the distance between the flow channels:

$$Q_e = P_r \cdot L \cdot R_s. \tag{3.20}$$

Equation 3.21 is used to compute τ_{fe}[Pa] which is the shear stress that is computed with γ [kg · m^{-2} · s^{-2}] as the water weight, S is the mean hillslope gradient [dimensionless], R is the hydraulic radius [m], f_s is the soil friction factor, and f_t is the total rill friction factor.

$$\tau_{fe} = \gamma \cdot S \cdot R \cdot \frac{f_s}{f_t}. \tag{3.21}$$

Next, T_c the transport capacity [kg · m^{-1} · s^{-1}] is computed in Eq. 3.22 based on a transportation coefficient:

$$T_c = k_t \cdot \tau_f^{1.5}, \tag{3.22}$$

where K_t is the transport parameter and $\tau_{\mathrm{f}}^{1.5}$ is the hydraulic shear stress [Pa]. WEPP calculations are applied using non-dimensional equations. As a next step, the model redimensionalizes the final solution. Conditions at the footslope of a uniform hillslope along a given profile are used to normalize the erosion equations. To do so, the hillslope is divided into homogenous units, and each unit is modeled as an independent entity, with added inflow of water and sediment from the upslope homogenously modeled unit.

WEPP is therefore a predictive model of water balance including surface and subsurface overland runoff computation and evapotranspiration, soil detachment and deposition, and sediment delivery, which are all computed along those strips of a hillslope profile, while the output data of a WEPP model run is projected on a hillslope profile, as shown in Fig. 3.6. The technique of prediction through hillslope strips may be a shortcoming of WEPP, since practical conservation agriculture is increasingly based on spatial assessments that are linked to geographical information systems at the catchment scale.

An important advantage of WEPP was to enable catchment scale analysis—and to that end, the aforementioned GeoWEPP was developed. The GeoWEPP is a geo-spatial interface for WEPP, that its predicted output includes a spatially explicit GIS layer, rather than the WEPP profile in Fig. 3.6. GeoWEPP uses DEMs and other GIS layers for spatially explicit soil and water conservation. The data model is the subcatchment with a single soil and land use characteristic per unit (Renschler 2003). It is applied using the following modules to extrapolate the environmental data into the spatial domain:

1. Generation of interpolated climate data.
2. DEM acquisition, channel network extraction, overall watershed delineation, and subcatchment.
3. Delineation and flow path computation—while the terrain analysis is conducted by means of topographic parameterization (TOPAZ) software by (Garbrecht and Martz 1999).
4. Execution of WEPP computations.
5. Processing the WEPP runoff, soil loss, and sediment yield outputs for spatial display on a cell basis.

The transformation of WEPP simulation of a profile into the spatially explicit GeoWEPP maps involves the scale transformation issue. It is therefore important to express scale issues in both the processes simulated and the data measured in the field. Some of the model parameterization and the processes simulated may stem

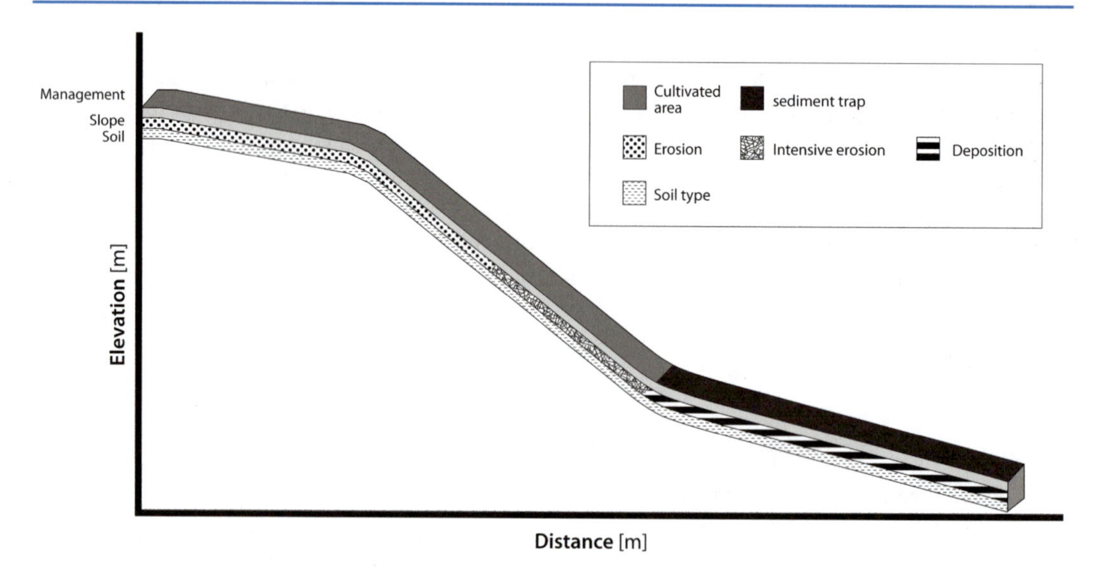

Fig. 3.6 A typical outcome profile of WEPP calculations. The lower layer carries the soil data, the intermediate layer represents the processes (erosion/deposition) and the upper layer is the land use/cover. Applied using synthetic data

from the original model and therefore incautious use of them in different scale or under different environmental conditions may cause inaccuracies in predictions.

3.3.3 The Model Output

The GeoWEPP links the original WEPP equations with GIS formalisms, to directly model catchment scale erosion and deposition, on-site and off-site. The GIS platform allows the WEPP to consider extrinsic and intrinsic factors (see Chap. 1) and their variations in space and time in greater detail. Thus, the GeoWEPP may enable prediction of soil erosion consequences at higher resolution, for further treatment through conservation actions.

For example, a GeoWEPP platform has been used to assess the effect of various soil cultivation and crop management strategies at the field-scale resolution, on runoff development and yield, and consequent soil loss, in a 1990-hectare catchment in southeastern Brazil, with sugarcane, forest, and pasture land uses (De Jong et al. 2005). The predicted results were plotted in a GIS software application, for decision-making regarding conservational land cultivation.

An application of the GeoWEPP using synthetic climate data of the entire Harod catchment in Israel (Fig. 3.7) and a more focused one to the Yehezkel catchment (Fig. 3.8) illustrate the spatially explicit nature of the output layer. The GeoWEPP output layers of both on-site and off-site consequences are divided into categories of soil loss and deposition. The formers are defined by the software manual (http://geowepp.geog.buffalo.edu/wp.content/ploads/2013/08/GeoWEPP_for_ArcGIS_9_Manual.pdf) as follows: the on-site assessment layer quantifies the amount of erosion or deposition occurring in each grid-cell of the raster output (because in a raster data model, the basic unit is the grid-cell). The off-site layer quantifies the amount of sediment that leaves the catchment hillslopes (represented by grid-cells), and arrives at the corresponding catchment outlet. The model determines which of the sediments will leave the site, and which will remain, to calculate the transport capacity.

The upper panel of the map in Fig. 3.7 shows the on-site damage in the Harod catchment with higher levels of soil loss in reddish hues, lower levels in greenish hues. We can clearly see the effect of the surrounding higher elevation areas on the catchment—Mt. Gilboa to the south and Hamoreh Hill to the north—resulting in high soil

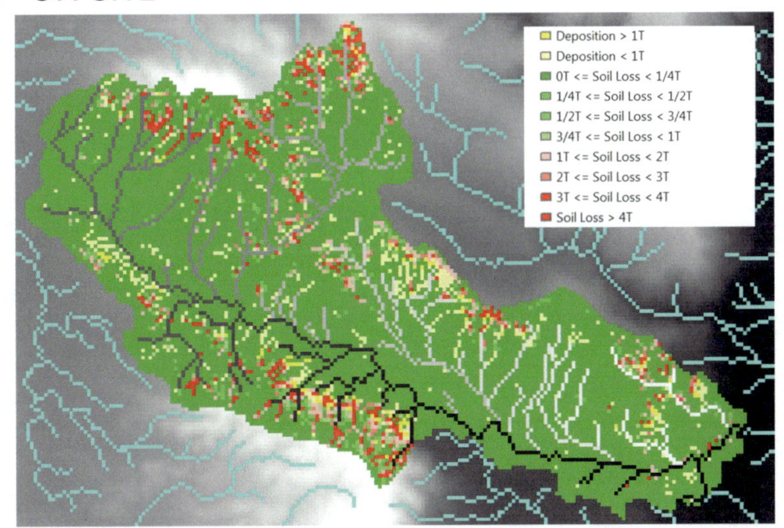

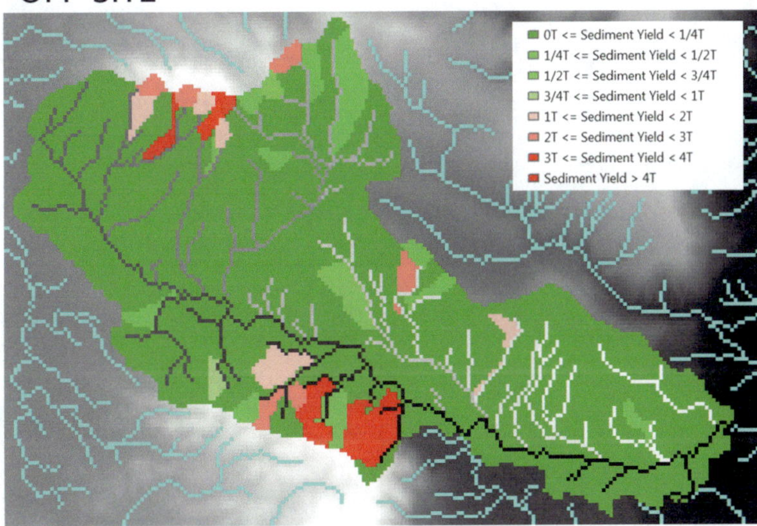

Fig. 3.7 A GeoWEPP output of the Harod catchment, based on synthetic data, shows the on-site soil loss (upper panel) and off-site sediment yield (T = tons) (lower panel). The blue elongated lines are the channel network, while the gray lines within the catchment show the flow accumulation

loss, mainly due to large contributing area and steep hillslope gradients. These areas of soil loss are also evident in the off-site layer, with the corresponding catchments showing high levels of sediment yield. The on-site layer also shows that some of the soil material is deposited in neighboring grid-cells to the grid-cells of high degree of soil loss—as reflected in the relatively large number of yellow grid-cells next to red ones. This may suggest that transport capacity is relatively low in several subcatchments under the simulated rainfall intensity, resulting in depositing of soil and on-site damage.

However, several grid-cells with deposited sediments can be observed near the center of the catchment, in the vicinity of the main channel—meaning that this soil mass has reached the river, and is being transported by the channel system out of the catchment. In addition, large parts of the catchment in the center part are in green—indicating that they are stagnant, because they are relatively flat, or of very minor low hillslope gradient.

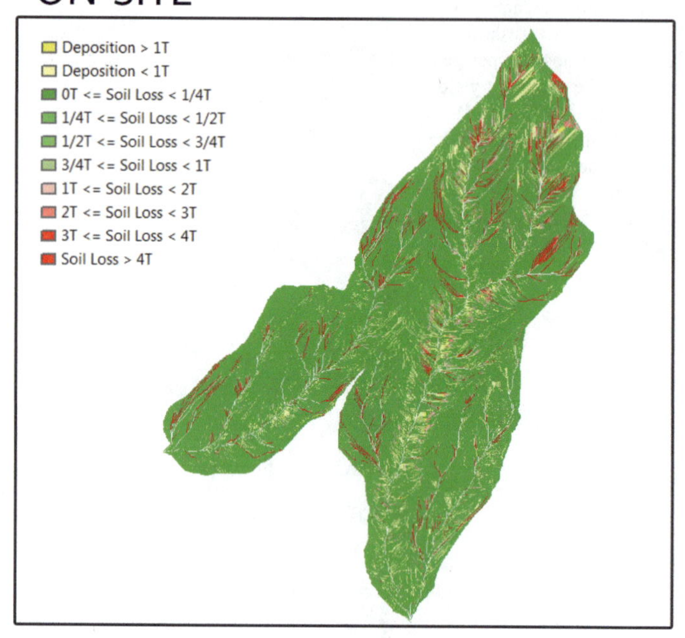

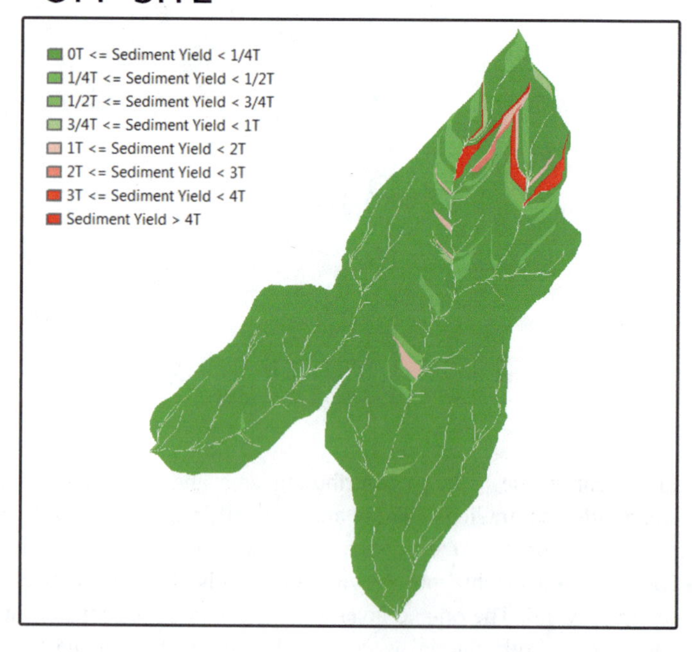

Fig. 3.8 The GeoWEPP output on-site and off-site prediction layers, as applied to the Yehezkel catchment (T = tons)

The case of the Yehezkel catchment in Fig. 3.8 shows similar predictions, at a higher spatial resolution. The northern part of the catchment—Hamore Hill—is characterized by areas of steep hillslope gradient. In this area, we see reddish hues of the clusters of grid-cells of areas that suffer high soil loss on-site (upper panel). This is also manifested in the GeoWEPP predictions to these grid-cells with high sediment yield off-site reaching the outlet (lower panel). In the lower sections, some grid-cells are also predicted to feature high levels of on-site soil loss, but they are not expressed as large damage in the off-site at the outlet (only a single pink sub-catchment). More generally, Fig. 3.8 shows that the Yehezkel is a catchment that requires considerable attention by soil conservationalists, because large parts of it suffer soil loss—particularly in the form of on-site damage. The figure clearly shows how some of the soil lost is deposited in the fields, creating sediment deposition damage in unwanted areas. This can be clearly seen in the upper panel, as a large number of yellow grid-cells having evolved along the central large gully.

Figures 3.7 and 3.8 are two simple representations of the use of GeoWEPP. Other studies have taken the physically based representations of erosion simulated by WEPP further. Wu et al. (2018) showed that the WEPP approach of using straight, parallel rills is inaccurate for simulations of spatio-temporal rill network dynamics, and suggested a modification using key rill network characteristics to improve predictive ability of the model and more realistic representation of the rill network. Experimental plots, subjected to three successive rainfall events, showed that the original WEPP approach of simulating rill erosion led to errors up to 30% larger than the new approach proposed by the authors.

However, the relative simplicity of the WEPP equations has not limited its application to various environments around the world: WEPP and its successors have been applied to many agricultural catchments with high degree of certainty. While it can be limited in its ability to express subtleties—such as tillage direction, or distance from the road, as a factor in contributing or halting runoff flow—WEPP can be, in fact, very powerful in physical reconstruction of rill and interrill sediment yield.

To sum up the results in Figs. 3.7 and 3.8, the GeoWEPP output might offer two maps of consequences for catchment scale analysis of erosion damages—of areas that are most at risk in the agricultural fields, and those that are not in need of attention. The use of such a GeoWEPP infrastructure in a *decision-support* system can entail one of three possible activities: identification of areas under threat, to assign immediate soil conservation activities; delineation of areas with no erosion activities even under severe rainfall intensity scenarios, to avoid allocating soil conservation resources to those areas; and, finally, demarcation of areas under mild erosion threat, to allocate further research to those areas. But, at any rate, the outcome of the WEPP and the GeoWEPP is all about soil loss and deposition while they provide no information about changes in other soil properties due to the erosion process.

3.4 Morgan–Morgan–Finney (MMF)

3.4.1 Background

CL and WEPP are complex physically based soil erosion models that require detailed input data, apply rather complex physically based computations, and can operate at the scale of individual rainfall events and high spatial resolution (meter and even sub-meter grid-cell resolution). A successful implementation of such models therefore requires an accurate representation of land use by GIS layers. These include: agricultural cultivation methods, measurements of soil properties, characteristics of vegetation strata, and topographic information. Thus, models such as the CL and the WEPP can provide vital information about the sub-processes of soil erosion but may also be prone to possible errors, due to data limitations. For example, rills and gullies delineation by WEPP can limit the accuracy of predicted sediment and water amounts due to naive representations of these surface entities, given the generalized 45° flowdirection angle by the matrix structure of raster data, which can result in unrealistic parallel channels (Svoray 2004).

Accordingly, based on the well-known *principle of Occam's Razor*, in some circumstances, we may need to use more simple empirical models. Occam's Razor principle—formulated in the fourteenth century by William of Ockham, and frequently used in contemporary heuristic and problem-solving studies—states that entities (or, in our case, model components) should not be multiplied unnecessarily. With regard to erosion models, this means that without the necessary means (data or even profound understanding of the governing processes), complex process-based models may be limited in the realism of their predictions of soil and water quantities, simpler models may be better at predicting erosion consequences. That explains why, in the soil erosion literature, we find many papers that develop and apply empirical and parsimonious models that allow analysis of observed data with a small number of variables, yet still have explanatory power. Sometimes the "Keep It Simple" approach provides us with more certain information (Svoray and Benenson 2009; McBratney et al. 2003).

In other words, empirical models can provide an alternative or supplementary source of information to complex process-based ones, because they are tailored to local conditions through data fitting; set out to represent fewer variables; and may make it possible to bypass detailed simulations of processes that we do not fully understand, or cannot fully quantify. To this end, empirical modeling relies on laboratory experiments and/or field measurements to extract model parameters through statistical analyses—such as simple regression models—between measured data, such as rainfall intensity and modeled physical parameters, such as raindrop energy.

The Morgan–Morgan–Finney (MMF) empirical model of water erosion (Morgan 1984, Morgan 2001) is a good case in point. Used over the past four decades to study water erosion—thanks to its ability to estimate annual soil loss, while retaining simplicity offering prediction at relatively high certainty—MMF can be used at the plot, hillslope, or small-catchment scales. Although spatially and temporally explicit applications of MMF are comparatively rare, and —as will be demonstrated below—may encounter difficulties, MMF has been applied successfully in various environments around the world. Furthermore, Tan et al. (2018) compared the performance of eight spatially explicit soil erosion models in hundreds of catchments in five different countries and the authors have found the MMF to be the best predictor of continental scale sediment yield at the scale of square kilometers. MMF provided the most accurate predictions—according to the authors—because of a more realistic representation of runoff-driven erosion and sediment transport capacity.

MMF has been validated against field measurements in many studies. To begin with, when it was first launched in the 1980s, the model was subjected to thorough validation procedure against field measurements from 67 sites, across 12 countries. It has also undergone sophisticated sensitivity analyses, in an effort to explore the effect of input parameters on soil loss assessments—in particular, the effect of shifting cultivation on soil loss—to provide recommendations on the adoption of soil conservation practices (Morgan et al. 1984).

From an operational point of view, and similarly to other erosion models, the MMF comprises two model components: water and sediment. Based on the general erosion theory (Chap. 2), it computes soil erosion quantities from the stage of soil detachment, through to the stage of transportation of soil downslope by runoff flow. The raindrop energy involved in detaching the soil particles by splash and the runoff volume are estimated in the water component. In the following section, we describe the steps involved in the model operation of MMF, based on the description by Morgan et al. (1984).

3.4.2 Model Operation

As a general framework, MMF models the movement of water and sediments from the source areas in the hillslopes to the channel network (Zema et al. 2020). The basic units for water and sediment transport are usually polygons of lumped homogenous hillslope characteristics. In other words, the studied area is divided into spatially uniform units based on soil properties, land use category, and topographic characterization.

In an MMF application, the entire erosion process in each site is based on computations of the sediment and water budgets for each of the aforementioned uniform unit while those materials are routed to adjacent downslope units based on the steepest hillslope gradient. This is usually applied such that each hillslope unit receives water and sediments from only a single upslope unit, and accordingly this unit transports the water and sediments to only a single downslope unit (Morgan and Duzant 2008).

The following operating functions for the water component in the MMF are relatively simple and were chosen by Morgan et al. (1984) from the literature for their "predictive ability, simplicity, and ease of determination of their input parameters." The water component is formulated using Eqs. 3.23–3.26 (Morgan et al. 1984):

$$E = R(11.9 + 8.7 \cdot log\ I), \tag{3.23}$$

$$Q = R \cdot e^{\left(\frac{-R_C}{R_O}\right)}, \tag{3.24}$$

$$\text{where } R_c = 1000 \cdot M \cdot \gamma \cdot Dr \cdot (E_t/E_o)^{0.5}, \tag{3.25}$$

$$\text{and } R_o = \frac{R}{R_n}, \tag{3.26}$$

where E is the kinetic energy of the rainfall [J · m^{-2}] and R is annual rainfall depth [mm]. I is a typical value for rainfall intensity of an erosive rainfall event [mm · h^{-1}] and Q is the depth of the overland flow [mm]. According to Morgan et al. (1984), R triggers soil particles detachment and consequent transport capacity. Q is highly dependent on rainfall, soil characteristics, and the evaporation rate that influences the soil water content and soil saturation. R_c is a critical daily rainfall value for runoff occurrence and R_o is the mean rainfall amount per day [mm]. M is the soil water content at field capacity [%]; γ is the bulk density of the topsoil layer [Mg · m^{-3}]; D_r is topsoil rooting depth [m]; and R_n is the number of rain days in the year [mm]. Thus, we can see the importance that the model attaches to the initial soil water conditions.

Equations 3.23–3.26 exhibit the empirical approach of MMF, by extracting energy values from measured intensity values using empirical coefficients. The slope and intercept coefficients in Eq. 3.23 may be extracted for other conditions (sites) using a regression model. These coefficients may vary with site conditions, thereby replacing more complex physically based representations of the rainfall energy role in the detachment process.

Accordingly, the sediment component—including the rate of splash detachment F [kg · m^{-2}] and soil transport capacity of the overland flow G [kg · m^{-2}]—is formulated using relatively simple equations and coefficient derivation, as in Eqs. 3.27–3.28, respectively (Morgan et al. 1984):

$$F = K \cdot (E \cdot e^{-a \cdot p})^b \cdot 10^{-3}, \tag{3.27}$$

where K is the soil detachability index [g · J^{-1}], that is, the detached soil mass per rainfall energy unit; a is an exponent to be determined empirically; p is the % rainfall contributing to interception and stemflow; and b is an exponent to be determined empirically.

$$G = C \cdot Q^d \cdot (\sin S) \cdot 10^{-3} \cdot I, \tag{3.28}$$

where C is the cover crop management factor; Q is the amount of the overland flow [mm]; the local parameter d is an empirical exponent, which is to be determined empirically against measurements; and S is the hillslope gradient [°] and I again is the rainfall intensity [mm · h^{-1}].

The sediment transport component assumes, based on theoretical grounds (Chap. 2), that the erosion process is initiated by the detachment of particles from the topsoil, due to raindrop impact. Transport of these particles subsequently occurs by runoff water flow. The rainfall energy required for splash detachment and runoff volume is, as was previously shown, estimated in the water component—namely, the water and sediment components are combined, as in real-world processes. However, despite this seemingly complex feedback, all five model equations are simple, and their solution for real-world conditions relies on data fitting. This requirement is basically the shortcoming of the MMF, where the advantage of simplicity comes at the price of having to calibrate the model for every site (or nearly every site—because in some circumstances coefficients may be generalized for sites with similar environmental conditions).

Figure 3.9 is the original flowchart of the MMF model components, adapted from Morgan et al. (1984). The flowchart demonstrates the entire model conceptual framework that is the basis for Eqs. 3.23–3.28. The left side of the flowchart illustrates the factors that lead to soil detachment, based on rainfall energy computations from rainfall intensity and volume and the effect of soil characteristics on detachability and vegetation layer on rainfall interception. The right side of the flowchart represents the factors that affect soil transport capacity—including soil moisture storage, overland flow characteristics, crop type, and conservation management.

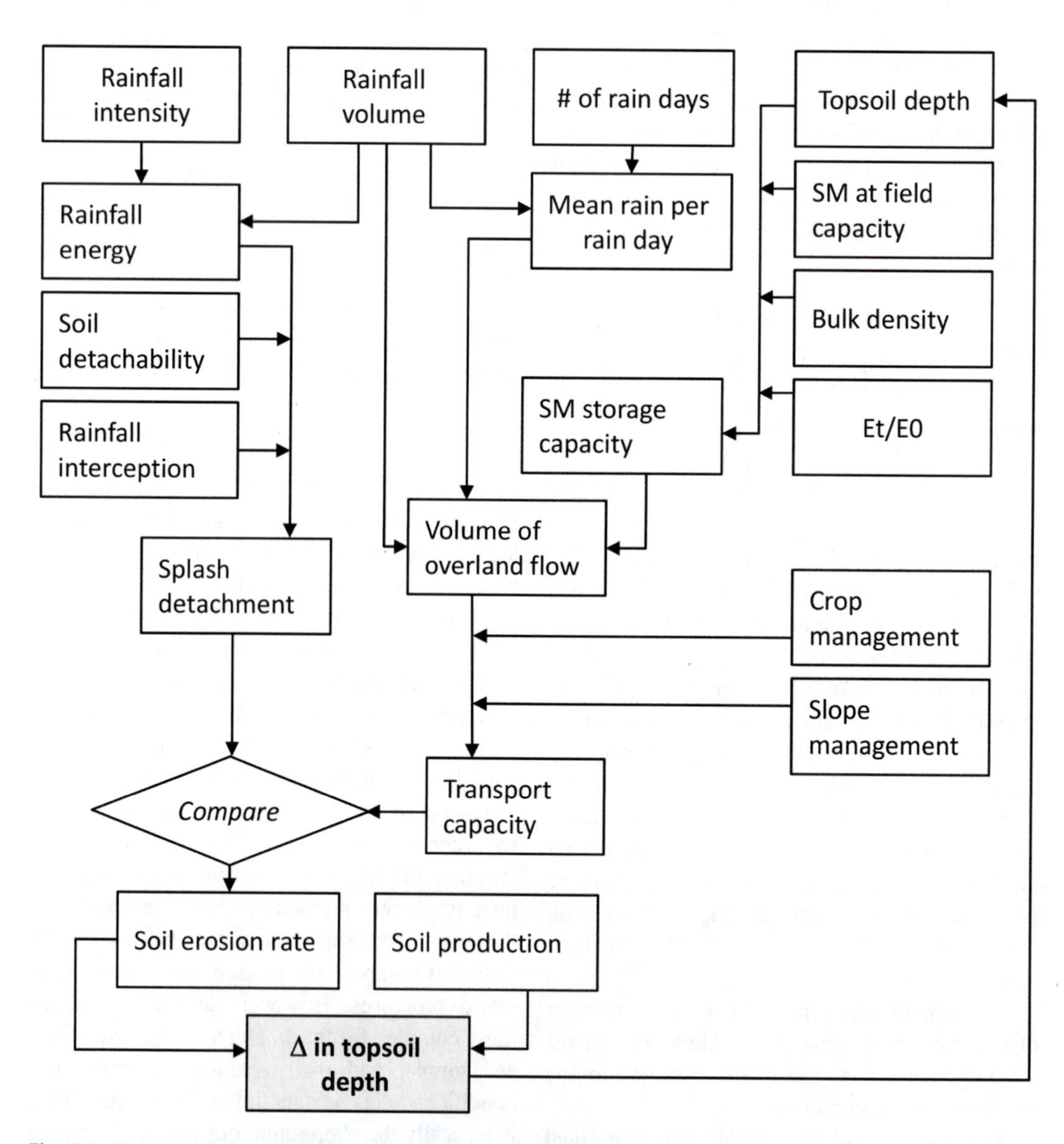

Fig. 3.9 The flowchart of the MMF model operation, from input data sources to topsoil change. From a bird's-eye view, the model compares the sediment transport rate with soil detachment to predict soil erosion rate, which is considered along with soil production to predict variation in soil depth. Reprinted from: Morgan, Morgan, Finney, predictive model for the assessment of soil erosion risk. Journal of Agricultural Engineering Research 30:245–253 (1984), with permission from Elsevier

The results of the computations of the left and right side of the flowchart model components are compared in the following stage of the model application, to determine the soil erosion rate (i.e., the soil loss). The difference between soil loss and soil production is then translated into the change in soil depth—which provides the topsoil depth variable for the next model iteration. However—as we noted from the field data provided in Chap. 1—soil production is usually dozens or hundreds of percent slower than the erosion rate.

3.4.3 Further Development and Model Application

For many decades, soil erosion studies have been based on data measured in situ using various gauging tools (Fernández-Raga et al. 2017; Hudson 1993). The sampling effort along the years included permanent instrumentation of experimental catchments for long-term measurements. Examples for such catchments are the Reynolds Creek Station in Idaho (Slaughter et al. 2001), and the Walnut Gulch Experimental Watershed in Arizona (Keefer et al. 2008), where long-term hydro-meteorological, rainfall and runoff and soil erosion/sedimentation data is collected over decades.

The climatic, hydrological, and soil data that has been measured in the experimental sites around the world was published widely and have been made available to the scientific community for the extraction of the MMF (and other models) coefficients. This has allowed the MMF to be applied to real-world conditions at various locations—which has indeed occurred, leading in more recent studies to further development of the original six equations (Eqs. 3.23–3.28). Such developments were used to apply the MMF to model soil erosion under even more complex conditions, while maintaining its empirical character. For example, a follow-up work to the original MMF, by Morgan (2001), provides the MMF with enhanced capabilities to simulate soil particle detachment by raindrop impact, in a manner that accounts for plant canopy stature and leaf drainage effect on detachment. Accordingly, the effect of raindrop impact can be quantified in different crop types. Morgan (2001) has also added a model component that expresses soil detachment by the overland flow itself and not only by direct or intercepted raindrop. Such modifications provide more detail on the detachment phase and advance the MMF simulation of the sediment source, yield, and delivery to streams. These all further advance the ability to apply MMF in a more operational manner to support soil conservation management.

In a later work, Morgan and Duzant (2008) have advanced MMF even further by adding more detailed crop cover parameters to the model simulations. They also changed how the model deals with sediment deposition and incorporated particle size selectivity in the entire process. The additional model parameters include very detailed data on the crop canopy cover, stem characteristics and the plant stature, wetting front, and a new modeling component of deposition, by accounting for particle settling velocity, flow velocity, flow depth, and hillslope length. This data was then used to simulate the effect of agricultural crop plants on both runoff and sediments, including fine processes such as the effect of interception, compared with throughfall. Thus, with time, the simple empirical models have also become more complex. The modeling of crops effect on erosion processes was applied by Morgan and Duzant using the following six steps:

Step 1: The crop cover contribution was expressed through the evapotranspiration ratio ($\frac{E_t}{E_0}$), using an empirical relationship that was established by Kirkby (1976).

Step 2: The effect of canopy interception was expressed in splash detachment modeling by altering the original rainfall energy component.

Step 3: The model expresses the different rooting depth, specifically to distinguish between crops and orchards.

Step 4: The model quantifies the increase in topsoil root layer due to different crop-rotation strategies and the transformation of living plants into humus.

Step 5: The model computes the ratio of soil loss under a given crop cover management factor—that combines C and P factors of the USLE (see Chap. 1, Eq. 1.7)—to soil loss from bare soil.

Step 6: The runoff transport capacity is determined with experimental support, and is a function of the runoff volume, the hillslope gradient, and the vegetation canopy cover.

Thus, the number of processes has increased overall, but the modeling approach remained empirically based.

A recent development of the MMF is the attempt to compute temporal dynamics of soil loss at the scale of daily iterations. This temporal approach has been developed to extend the original MMF strategy of simulation of annual budgets (Choi et al. 2017). In it, the authors present a daily-based MMF to estimate overland flow and soil spatial arrangement, under varying field conditions. The main simulated processes are similar to those in the original MMF but, by using revised equations, temporal dynamics are modeled based on daily time steps. Sensitivity analyses have shown that the rationale for using these input parameters is consistent with previous versions of MMF, and evaluations of predictive ability in two fields in South Korea have shown acceptable results for water and soil loss estimation (Fig. 3.10).

The graphs in Fig. 3.10 show the output of the MMF model, compared with data measured in two agricultural fields in South Korea. The results of Choi et al. (2017) show good model performance, but note that the analysis was applied on a field basis (aggregated data), and not in a spatially explicit representation as in the case of the GeoWEPP and in the various versions of SEMs. The difference observed in the predictive abilities between the two fields was explained by the authors by the effect of local conditions.

The aggregated data representation may be a limitation of the model. Although rainfall intensity, topography, plant cover, and land management can vary greatly in space and time, and these variations may greatly affect erosion in different parts of the field, only relatively few studies have embedded MMF in a GIS software to study water erosion in the catchment scale. The reason for this scarcity may be the need to extract coefficients that may need to be non-stationary, namely, the coefficients may need to vary in space due to variation in the relationships between the measured and modeled variables. This variation may be caused by a non-stationary relationship between different variables such as the rainfall-runoff relationships that may depend largely on surface vegetation cover or the effect of non-stationary land use change that requires the use of statistical tools such as the Geographically Weighted Regression (GWR—Fotheringham et al. (2002)) to quantify spatial non-stationarity and local coefficients in simulating land use and land cover changes (Wang et al. 2019). Due to the important role that empirical coefficients play in the MMF, its application may suffer limitations in prediction although this question was not studied in full yet.

Despite the limitation of extracting empirical coefficients in a domain that may be non-stationary, several recent studies have applied the MMF in a GIS environment to predict soil erosion over wide regions. In a study by Tesfahunegn et al. (2014), the loosely coupled MMF-GIS model was applied to the Mai-Negus agricultural catchment in northern Ethiopia. The authors used input raster (and rasterized) data for the MMF model application—including rainfall [mm], land use, DEM for hillslope layer extraction, soil texture, soil moisture content at field capacity [$\%w \cdot w^{-1}$] (% mass of water/mass of soil at $\Psi_{FC} = -33$ kPa, Assouline, and Or 2014), soil detachability index [$g \cdot J^{-1}$], bulk density of soil [$M \cdot g \cdot m^{-3}$], cohesion of soil surface [kPa], soil moisture storage capacity [m], effective hydrological top soil depth [m], and ratio of actual to potential evapotranspiration. The output maps of the Tesfahunegn et al. (2014) MMF-GIS predictions include rate of soil detachment by raindrop impact, and by runoff impact (two different layers), total detachment as a third layer, and soil transport capacity of overland flow. Another example is the application of spatial MMF to the area of the Narmada River

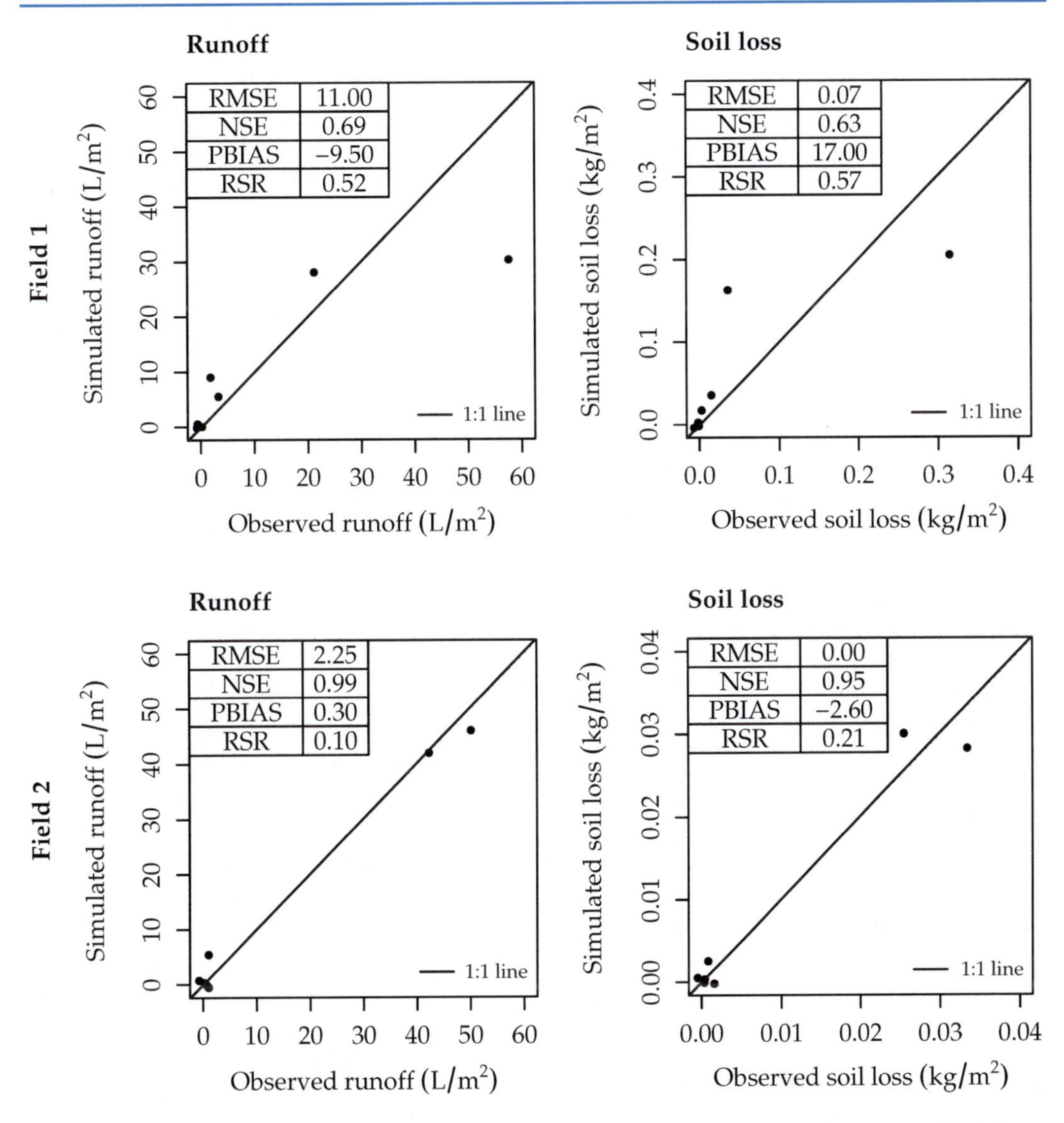

Fig. 3.10 Comparison between predicted and observed runoff and soil Loss (SL) for two potato fields in South Korea. The model predictions are from the daily version of MMF. Model predictive ability was assessed using the common RMSE, the Nash–Sutcliffe efficiency coefficient (NSE), percent bias (PBIAS), and the RMSE-observation standard deviation ratio (RSR), reproduced from Choi et al. (2017). The simulations in Field 2 (lower panel) are more in line with observed data than the simulations in Field 1

catchment in Madhya Pradesh, India, using soil maps, satellite images, and ASTER DEM, as well as rainfall data (Mondal et al. 2018). The authors compared the spatial MMF with the RUSLE (see Chap. 1), and found that the RUSLE was more suitable, with only a ~5% difference in the values of soil erosion from the observed data, whereas MMF yielded a very large difference of ~40%.

In summary, MMF is an empirical erosion model that uses simplifications in the operating functions, while relying heavily on empirical coefficients that may suffer from spatial non-stationarity effect. The various versions of MMF have been found to reliably predict the effect of detachment and transport capacity on water erosion at the annual scale—but also, as we have seen recently, at the hourly time scale. It must be borne in mind, however, that the MMF and the other empirical models do not aim to quantify the precise physical processes that cause soil erosion, but to provide a proximate prediction of soil loss. Once the appropriate coefficients, preferably non-stationary coefficients with GWR, for example, are obtained, and the model predictions are validated against independent measured data, the user can expect reliable changes in topsoil depth. Note, however, that more recent versions of the MMF aspire to simulate the effect of an increasing number of environmental variables (such as crop type) and hydrological processes (such as detachment by runoff) on soil loss. Such a trend of increasing model complexity may turn the MMF into a more multifaceted model, thereby excluding it from the group of simple and parsimonious models.

3.5 Summary

Scientists use physically based and empirical models as computational tools to predict soil loss and deposition, splash detachment, and overland flow rates in agricultural catchments. The erosion models—in particular, the physically based ones—provide a heuristic mechanism for studying the underlining processes that create observed water erosion phenomena such as pipes, rills, gullies, and the traces of sheet erosion, under various climatic, environmental, and human-induced conditions. The heuristic study is usually applied by independently controlling the climatic, environmental, and human-induced factors within the process.

Erosion models may complement field studies by allowing the exploration of the effect of extreme events—that may have not even yet occurred and thus were not measured—on soil erosion and deposition processes. Such extreme events may include, for example, (1) an outstandingly large increase in rainfall intensity and its consequences on splash detachment; (2) the effect of massive exposure of vegetation strata—due to abandon of agricultural areas—on overland flow; and (3) a major increase in unpaved roads coverage in the catchment that may cause an increase runoff development and yield.

The erosion models are usually applied to mimic the governing processes of soil erosion using two main components: hydrological and sediment. The hydrological component includes overland flow generation and development in space with a focus on the transport capacity and the sediment component includes particle detachment and bed load or suspended load transport and deposition. Yet, another process that caught less attention in the literature but was also modeled in the aforementioned models is the diffusive creep that involves a coupling between changes in topography, soil transport rates, formation, and depth (Furbish and Fagherazzi 2001).

Spatially and temporally explicit input data is crucial to providing high-certainty predictions to these erosion models, because the earth surface is not homogenous, and local climatic, human-induced, and environmental conditions may affect the erosion process from detachment to deposition. The driving forces of the earth surface heterogeneity are the intrinsic and extrinsic factors (see Chap. 1) that operate at various scales and need to be considered to enable educated mitigation and management of water erosion by farmers. While the soil erosion models can indeed assist with unraveling the surface heterogeneities, they differ by the simulation methods they use of the erosion process, as well as in the data input and model calibration.

Fundamentally, the mechanism of regolith production and transport downslope in a synthetic hillslope is quantified by the rate of weathering and the topographic location (Pelletier 2008). However, a more realistic simulation of water erosion processes in complex agricultural catchments should consider more factors and processes. Three of the more detailed erosion models were described in this chapter.

The CAESAR-Lisflood uses an initial layer of DEM and simulates changes in the output grid-cells elevation using cellular automata technique and physically based equations of erosion and deposition processes. The WEPP is a process-based model that uses physical equations to quantify the change in sediment load along a hillslope profile, by computing the soil loss unit area based on the erosion rate in both the rill and interrill parts of the hillslope, using the sediment continuity equation. Finally, the MMF is an empirical model that uses coefficients and data-fitting techniques to solve the effect of climatic and soil factors on water erosion and deposition outcomes, thereby bypassing the need to solve complex physically based equations. For example, the MMF associates the kinetic energy of the rainfall drop with the rate of splash detachment and the runoff volume with transport capacity. Clearly, LEMs and particularly the CAESAR-Lisflood and WEPP require relatively low calibration effort compared with empirically based models such as the MMF that base their operating functions on linear or non-linear relationships extracted at a specific site or cluster of sites, using model coefficients.

There is also a difference between these models in terms of their spatial representation, which is crucial to geoinformatics modeling. The GeoWEPP and the spatially explicit applications of MMF are based on local operators, namely, computation of the erosion and runoff equations is applied on a single grid-cell basis. LEMs are more unique, because they are based on cellular automata techniques that allow computation that includes the effect of the neighboring grid-cells —at high temporal (hours to days) and spatial (sub-meter) resolution. These capabilities of cellular automata modeling may be especially useful when studying water erosion processes at the hillslope and catchment scales and the effect of neighboring grid-cells is of prime importance for soil loss and deposition predictions.

Li et al. (2017) compared 11 soil erosion models for the Loess Plateau in China based on their prediction accuracy, process representation, data requirements and calibration needs, and potential for use in scenario studies. Their results show that prediction accuracy does not increase with model complexity. Empirical models are useful for a quick assessment of soil erosion rates and sediment yields over an area, while process-based models are successful in detailed soil erosion and sediment yield assessments. Both WEPP and MMF produced satisfactory predictions for the Loess Plateau. WEPP is a powerful tool for simulating event/annual soil erosion rates at the hillslope scale and continuous sediment yield at the small-catchment scale. MMF is most promising for accurate annual sediment yield predictions for medium to large catchments. Further improvement of soil erosion models should concentrate on enhancing the quality of data for model implementation, improving process representation, comparing similar models, and implementing fundamental and spatial testing.

The principles described in the four models reviewed in this chapter illustrate that topography (especially contributing area and hillslope gradient) is a crucial determinant of overland flow and water erosion processes, in all models. Rainfall depth and mainly rainfall intensity are found important for simulating soil detachment. Soil erosivity is expressed by texture and hydraulic conductivity. The models also highlight the importance of the effect of land use (especially cultivation method) on soil erosivity and consequent erosion and the effect of crop cover, canopy interception, and even root layer on the water erosion process. All these enhance the importance, efficiency, and accuracy of spatial and temporal dynamic soil erosion models in predicting soil loss and deposition.

Review Questions

1. What are the important governing factors to predict soil loss and deposition based on the four models presented in Chap. 3? Make a list to each model separately and outline the overlapping variables.
2. What is the difference between the four models presented in this chapter with regard to:

    - Spatial data models used?
    - Temporal iterations?
    - Governing equations?
    - Predicted variables?

4. Which of the models you think best fit to the catchments in your area? Why?
5. In a field with hillslope length L of 100 m, peak runoff rate P_r of 1 m · s^{-1}, and a distance between flow channels of 10 m, what is the rill width, according to WEPP if c and d coefficients equal to 14 and -0.41, respectively?
6. A field is subjected to crop cover management factor C of 0.05; the amount of overland flow Q of 2 mm; the local coefficient d extracted as 20; I, the rainfall intensity is 10 mm · h^{-1}; and S is hillslope gradient of 2.2^0. What is the soil transport capacity of the overland flow G [kg · m^{-2}], according to MMF?
7. What is the principal climatic property that affects water erosion in agricultural catchments? How it is expressed mathematically in CAESAR-LISFLOOD, in WEPP, and in MMF?
8. Explain why ancient farmers saw terrace farming as beneficial in combating soil erosion.
9. What is meant by spatially and temporally explicit modeling? Why is it so important in water erosion modeling? What is the necessary data used to apply such modeling effort?
10. What is a *parsimonious model*, and how does it contribute to water erosion modeling, despite its simplicity?

Strategies to address the Review Questions

1. To answer question 1, you need to create a two-dimensional table that covers the following five models: simple hillslope, LEM, SEM, WEPP, and MMF. The governing factors can be listed from the equations: simple hillslope in Eqs. 3.1–3.4; LEM in Eq. 3.5; SEM (CAESAR-LISFLOOD) in Eqs. 3.6–3.12; WEPP in Eqs. 3.13–3.22; and MMF in Eqs. 3.23–3.26. Then you mark the overlapping factors in the different models: these are the factors that the developers agree on their importance to quantify soil loss and deposition.
2. You can compare the models by using a comparative table. The five models may be placed in the rows and the following four measures the columns:
   - Spatial data: compare between the spatial resolution used in the models between pixel based (individual data) and catchment based (lumped data).
   - Temporal iterations: compare the temporal resolution of the models between minutes, hours, years, etc.
   - Governing equations: classify the models between physically based, semi-empirical, and empirical.
   - Predicted variables: LEM predicts the rate of change of surface elevation over time. SEM predicts the change in soil thickness, as a function of soil loss and deposition processes. WEPP model's output is erosion and deposition level in physical units, on-site and off-site. Finally, the MMF model output is the Δ in soil thickness.
3. To answer this question, I suggest to execute three models: CAESAR-LISFLOOD, WEPP, and MMF to the catchment of interest. This will allow to explore the model's certainty for your site.
4. The rill width, w, can be computed using Eq. 3.19. To compute Q_e^d we can use Eq. 3.20 with: $P_r = 1\text{m} \cdot \text{s}^{-1}$, $L = 100\text{m}$, and $R_s = 10\text{m}$.

$$Q_e = 1 \cdot 100 \cdot 10 = 1000\text{m}^3 \cdot \text{s}^{-1}$$

$$w = 14 \cdot 1000^{-0.41} = 0.82\text{m}.$$

5. The soil transport capacity of the overland flow, G, can be computed using MMF. The data provided in question 5 can be plugged in Eq. 3.28 to do so:

$$G = 0.05 \cdot 2^{20} \cdot (\sin 2.2) \cdot 10^{-3} \cdot 10$$
$$= 19.6\text{kg} \cdot \text{m}^{-2}.$$

6. The principal climatic property that affects water erosion in agricultural catchments in all models is rainfall:

CAESAR-LISFLOOD: in Eqs. 3.6–3.7—rainfall rate r.
WEPP: in Eq. 3.18—rainfall intensity I.
MMF: in Eq. 3.23 and Eq. 3.27—rainfall kinetic energy E.

7. Terracing allows to "break" the runoff continuity and thus runoff water accumulation into smaller plots and as a result reduces the runoff transport capacity. Most likely, ancient farmers attempted to trap the water in the small plots (the terraces) and use them for crop irrigation rather than as sediment transport agents.
8. Spatially and temporally explicit models combine water and sediment simulations with layers of environmental variables to quantify soil loss/deposition for each landscape unit—be it a pixel, a hillslope catena unit, a sub-catchment, or other. The amount of soil loss/deposition is therefore explicitly quantified by the model, and the effect of changing conditions on water erosion can be calculated. Spatially and temporally explicit modeling is important for water erosion studies because the soil is a highly heterogenous medium in various scales and conditions for erosion change vastly in space and time. The necessary data for spatially and temporally explicit modeling is a detailed GIS of environmental and human-induced conditions.
9. Parsimonious models are empirical models that are based on data-fitting approach and are being used in water erosion studies to predict soil loss/deposition mainly for explanatory uses. The aim of the developer is to use minimum number of variables (an idea that stems from Occam's razor principal, Sect. 3.4.1).

References

Anache JAA, Flanagan DC, Srivastava A et al (2018) Land use and climate change impacts on runoff and soil erosion at the hillslope scale in the Brazilian Cerrado. Sci Total Environ 622–623:140–151. https://doi.org/10.1016/j.scitotenv.2017.11.257

Assouline S, Or D (2014) The concept of field capacity revisited: defining intrinsic static and dynamic criteria for soil internal drainage dynamics. Water Resour Res 50:4787–4802. https://doi.org/10.1002/2014WR015475

Bastien C, Dupuis A, Masselot A et al (2002) Cellular automata and lattice Boltzmann techniques: an approach to model and simulate complex systems. Adv Complex Syst 5(2n03):103–246. https://doi.org/10.1142/s0219525902000602

Bates PD, Horritt MS, Fewtrell TJ (2010) A simple inertial formulation of the shallow water equations for efficient two-dimensional flood inundation modelling. J Hydrol 387(1):33–45. https://doi.org/10.1016/j.jhydrol.2010.03.027

Batista P, Davies J, Silva M (2019) Earth science reviews. Earth Sci Rev 197. doi:rg/https://doi.org/10.1016/j.earscirev.2019.102898

Beven KJ (1977) Hillslope hydrographs by the finite element method. Earth Surface Process 2(1):13–28. https://doi.org/10.1002/esp.3290020103

Beven KJ, Kirkby MJ (1979) A physically based, variable contributing area model of basin hydrology/Un modèle à base physique de zone d'appel variable de l'hydrologie du bassin versant. Hydrol Sci Bull 24(1):43–69. https://doi.org/10.1080/02626667909491834

Borrelli P, Robinson DA, Fleischer LR et al (2017) An assessment of the global impact of 21st century land use change on soil erosion. Nat Commun 8(1):1–13. https://doi.org/10.1038/s41467-017-02142-7

Braun J (2018) A review of numerical modeling studies of passive margin escarpments leading to a new analytical expression for the rate of escarpment migration velocity. Gondwana Res 53:209–224. https://doi.org/10.1016/j.gr.2017.04.012

Braun J (2019) Modelling landscape evolution: from simulation to inspiration (MAL2): arthur holmes medal lecture. In: EGU2019 conference, 7–12 April 2019, Vienna

Braun J, Sambridge M (1997) Modelling landscape evolution on geological time scales: a new method based on irregular spatial discretization. Basin Res 9(1):27–52. https://doi.org/10.1046/j.1365-2117.1997.00030.x

Choi K, Arnhold S, Huwe B et al (2017) Daily based Morgan–Morgan–Finney (DMMF) model: a spatially distributed conceptual soil erosion model to simulate complex soil surface configurations. Water:1–21. https://doi.org/10.3390/w9040278

Cohen S, Willgoose G, Hancock G (2010) The mARM3D spatially distributed soil evolution model: three-dimensional model framework and analysis of hillslope and landform responses. J Geophys Res 115(F4). https://doi.org/10.1029/2009JF001536

Cohen S, Willgoose G, Svoray T et al (2015) The effects of sediment transport, weathering, and aeolian mechanisms on soil evolution. J Geophys Res Earth Surf 120(2):260–274. https://doi.org/10.1002/2014JF003186

Coulthard TJ (2001) Landscape evolution models: a software review. Hydrol Process 15(1):165–173. https://doi.org/10.1002/hyp.426.abs

Coulthard TJ, Hancock GR, Lowry JBC (2012) Modelling soil erosion with a downscaled landscape

evolution model. Earth Surf Proc Land 37(10):1046–1055. https://doi.org/10.1002/esp.3226
Coulthard TJ, Kirkby MJ, Macklin MG (1998) Non-linearity and spatial resolution in a cellular automaton model of a small upland basin. Hydrol Earth Syst Sci 2:257–264
Coulthard TJ, Macklin MG, Kirkby MJ (2002) A cellular model of Holocene upland river basin and alluvial fan evolution. Earth Surf Proc Land 27(3):269–288. https://doi.org/10.1002/esp.318
Coulthard TJ, Neal J, Bates P et al (2013) Integrating the LISFLOOD-FP 2D hydrodynamic model with the CAESAR model: Implications for modelling landscape evolution. Earth Surface Processes and Landforms 38:1897–1906. https://doi.org/10.5281/zenodo.321820
Coulthard TJ, Van De Wiel MJ (2017) Modelling long term basin scale sediment connectivity, driven by spatial land use changes. Geomorphology 277:265–281. https://doi.org/10.1016/j.geomorph.2016.05.027
De Jong V, Sparovek G, Flanagan DC et al (2005) Runoff mapping using WEPP erosion model and GIS tools. Comput Geosci 31(10):1270–1276. https://doi.org/10.1016/j.cageo.2005.03.017
De Vente J, Poesen J (2005) Predicting soil erosion and sediment yield at the basin scale: scale issues and semi-quantitative models. Earth Sci Rev 71(1–2):95–125
Fernández-Raga M, Palencia C, Keesstra S et al (2017) Splash erosion: a review with unanswered questions. Earth Sci Rev 171:463–477
Flanagan DC, Nearing MA (2007) Water erosion prediction project (WEPP): development history, model capabilities, and future enhancements. Am Soc Agricult Biol Eng 50(5):1603–1612
Flores-Cervantes J, Istanbulluoglu E, Bras RL (2006) Development of gullies on the landscape: a model of headcut retreat resulting from plunge pool erosion. J Geophys Res 111(F1). https://doi.org/10.1029/2004JF000226
Fotheringham AS, Brunsdon C, Charlton M (2002) Geographically weighted regression. Wiley, Sussex
Francipane A, Ivanov VY, Noto LV et al (2012) tRIBS-Erosion: a parsimonious physically-based model for studying catchment hydro-geomorphic response. Catena 92:216–231. https://doi.org/10.1016/j.catena.2011.10.005
Furbish DJ, Fagherazzi S (2001) Stability of creeping soil and implications for hillslope evolution. Water Resour Res 37(10):2607–2618. https://doi.org/10.1029/2001WR000239
Garbrecht J, Martz LW (1999) An overview of TOPAZ: an automated digital landscape analysis tool for topographic evaluation, drainage identification, watershed segmentation and subcatchment parameterization. Agricultural Research Service, El Reno, Oklahoma, USA, report no. GRL, 99–1.
Guisan A, Zimmermann NE (2000) Predictive habitat distribution models in ecology. Ecol Model 135(2–3):147–186. https://doi.org/10.1016/s0304-3800(00)00354-9
Hancock GR, Coulthard TJ, Martinez C et al (2011) An evaluation of landscape evolution models to simulate decadal and centennial scale soil erosion in grassland catchments. J Hydrol 398(3):171–183. https://doi.org/10.1016/j.jhydrol.2010.12.002
Hancock GR, Lowry JBC, Coulthard TJ et al (2010) A catchment scale evaluation of the SIBERIA and CAESAR landscape evolution models. Earth Surf Proc Land 35(8):863–875. https://doi.org/10.1002/esp.1863
Hancock GR, Lowry J, Coulthard TJ (2015) Catchment reconstruction — erosional stability at millennial time scales using landscape evolution models. Geomorphology (amsterdam, Netherlands) 231:15–27. https://doi.org/10.1016/j.geomorph.2014.10.034
Hancock GR, Verdon-Kidd D, Lowry JBC (2017) Soil erosion predictions from a landscape evolution model – an assessment of a post-mining landform using spatial climate change analogues. Sci Total Environ 601–602:109–121. https://doi.org/10.1016/j.scitotenv.2017.04.038
Hoober D, Svoray T, Cohen S (2017) Using a landform evolution model to study ephemeral gullying in agricultural fields: the effects of rainfall patterns on ephemeral gully dynamics. Earth Surf Proc Land 42 (8):1213–1226. https://doi.org/10.1002/esp.4090
Huang C, Laflen J, Bradford J et al (1996) Evaluation of the detachment-transport coupling concept in the WEPP rill erosion equation. Soil Sci Soc Am J 60 (3):734–739. https://doi.org/10.2136/sssaj1996.03615995006000030008x
Hudson N (1993) Field measurement of soil erosion and runoff. Food and Agriculture Organization of the United Nations, Italia.
Ivanov VY, Vivoni ER, Bras RL et al (2004) Catchment hydrologic response with a fully distributed triangulated irregular network model. Water Resourc Res 40 (11). https://doi.org/10.1029/2004WR003218
Keefer T, Moran M, Paige G (2008) Long-term meteorological and soil hydrology database, Walnut Gulch Experimental Watershed, Arizona, United States. Water Resourc Res 44(5). https://doi.org/10.1029/2006WR005702
Kirkby MJ (1976) Tests of the random network model, and its application to basin hydrology. Earth Surface Process 1(3):197–212. https://doi.org/10.1002/esp.3290010302
Li P, Mu X, Holden J, Wu Y, Irvine B, Wang F, Gao P, Zhao G, Sun W (2017) Comparison of soil erosion models used to study the Chinese Loess Plateau. Earth Sci Rev 170:17–30
McBratney AB, Mendonça Santos ML, Minasny B (2003) On digital soil mapping. Geoderma 117(1):3–52. https://doi.org/10.1016/S0016-7061(03)00223-4
Mondal A, Khare D, Kundu S (2018) A comparative study of soil erosion modelling by MMF, USLE and RUSLE. Geocarto Int 33(1):89–103. https://doi.org/10.1080/10106049.2016.1232313
Moreno-de las Heras M, Saco P, Willgoose G et al (2012) Variations in hydrological connectivity of Australian

semiarid landscapes indicate abrupt changes in rainfall-use efficiency of vegetation. J Geophys Res: Biogeosci 117(G3). https://doi.org/10.1029/2011JG001839

Morgan D, Morgan R, Finney H (1984) A predictive model for the assessment of soil erosion risk. J Agric Eng Res 30:245–253. https://doi.org/10.1016/S0021-8634(84)80025-6

Morgan R (2001) A simple approach to soil loss prediction: a revised Morgan–Morgan–Finney model. Catena 44(4):305–322. https://doi.org/10.1016/S0341-8162(00)00171-5

Morgan R, Duzant J (2008) Modified MMF (Morgan–Morgan–Finney) model for evaluating effects of crops and vegetation cover on soil erosion. Earth Surf Process Landforms 33(1):90–106. https://doi.org/10.1002/esp.1530

Nearing MA (1989) A process-based soil erosion model for USDA-water erosion prediction project technology. Trans ASAE 32(5):1587–1593. https://doi.org/10.13031/2013.31195

Nishiizumi K, Finkel RC, Dietrich WE et al (1997) The soil production function and landscape equilibrium. Nature 388(6640):358–361. https://doi.org/10.1038/41056

Ollier C, Pain C (1996) Regolith, soils and landforms. Wiley, Chichester

Pelletier JD (2008) Quantitative modeling of earth surface processes. Cambridge University Press M.U.A

Pieri L, Bittelli M, Wu JQ et al (2007) Using the water erosion prediction project (WEPP) model to simulate field-observed runoff and erosion in the Apennines mountain range. Italy. J Hydrol 336(1):84–97. https://doi.org/10.1016/j.jhydrol.2006.12.014

Pike R, Evans I, Hengl T (2009) Geomorphometry: a brief guide. In: Hengl T, Reuter H (eds) Geomorphometry. Elsevier, London

Renschler CS (2003) Designing geo-spatial interfaces to scale process models: the GeoWEPP approach. Hydrol Process 17(5):1005–1017. https://doi.org/10.1002/hyp.1177

Saco PM, Willgoose GR, Hancock GR (2007) Eco-geomorphology of banded vegetation patterns in arid and semi-arid regions. Hydrol Earth Syst Sci 11 (6):1717–1730. https://doi.org/10.5194/hess-11-1717-2007

Saint-Venant A (1871) Théorie du mouvement non permanent des eaux, avec application aux crues des rivières et a l'introduction de marées dans leurs lits. Comptes Rendus De L'académie Des Sciences 73 (147–154):237–240

Slaughter C, Marks D, Flerchinger G et al (2001) Thirty-five years of research data collection at the Reynolds Creek Experimental Watershed, Idaho. United States. Water Resourc Res 37(11):2819–2823. https://doi.org/10.1029/2001WR000413

Svoray T (2004) Integrating automatically processed SPOT HRV Pan imagery in a DEM-based procedure for channel network extraction. Int J Remote Sens 25 (17):3541–3547. https://doi.org/10.1080/01431160410001684992

Svoray T, Benenson I (2009) Scale and adequacy of environmental microsimulation. Ecol Complex 6:77–79. https://doi.org/10.1016/j.ecocom.2018.05.002

Tan Z, Leung L, Li H et al (2018) Modeling sediment yield in land surface and Earth system models: Model comparison, development, and evaluation. J Adv Model Earth Syst 10:2192–2213. https://doi.org/10.1029/2017MS001270

Tarboton DG (1997) A new method for the determination of flow directions and upslope areas in grid digital elevation models. Water Resour Res 33(2):309–319. https://doi.org/10.1029/96wr03137

Tesfahunegn GB, Tamene L, Vlek PLG (2014) Soil erosion prediction using Morgan-Morgan-Finney Model in a GIS environment in northern Ethiopia catchment. Appl Environ Soil Sci 2014:1–15. https://doi.org/10.1155/2014/468751

Tucker G, Lancaster S, Gasparini N et al (2001) The channel-hillslope integrated landscape development model (CHILD). In: Doe W (ed) Harmon R. Landscape erosion and evolution modeling Kluwer, New York, pp 349–388

Tucker G, Lancaster S, Gasparini N et al (2001) An object-oriented framework for distributed hydrologic and geomorphic modeling using triangulated irregular networks. Comput Geosci 27(8):959–973. https://doi.org/10.1016/S0098-3004(00)00134-5

Tucker G, Slingerland R (1994) Erosional dynamics, flexural isostasy, and long-lived escarpments: a numerical modeling study. J Geophys Res Solid Earth 99(B6):12229–12243

Van De Wiel MJ, Coulthard TJ, Macklin MG et al (2007) Embedding reach-scale fluvial dynamics within the CAESAR cellular automaton landscape evolution model. Geomorphology 90(3):283–301. https://doi.org/10.1016/j.geomorph.2006.10.024

Vanwalleghem T, Stockmann U, Minasny B et al (2013) A quantitative model for integrating landscape evolution and soil formation. J Geophys Res Earth Surf 118 (2):331–347. https://doi.org/10.1029/2011JF002296

Vereecken H, Schnepf A, Hopmans JW et al (2016) Modeling soil processes: review, key challenges, and new perspectives. Vadose Zone J 15(5):1. https://doi.org/10.2136/vzj2015.09.0131

Wang H, Stephenson S, Qu S et al (2019) Modeling spatially non-stationary land use/cover change in the lower connecticut river basin by combining geographically weighted logistic regression and the CA-Markov model. Int J Geogr Inf Sci 33(7):1313–1334. https://doi.org/10.1080/13658816.2019.1591416

Wilcock PR, Crowe JC (2003) Surface-based transport model for mixed-size sediment. J Hydraul Eng 129 (2):120–128. doi:2(120)
Willgoose G, Bras RL, Rodriguez-Iturbe I (1991) A coupled channel network growth and hillslope evolution model: 1. Theory. Water Resourc Res 27 (7):1671–1684. https://doi.org/10.1029/91WR00935
Wu S, Chen L, Wang N et al (2018) Modeling rainfall-runoff and soil erosion processes on hillslopes with complex rill network planform. Water Resour Res 54 (12):1–17. https://doi.org/10.1029/2018WR023837
Yu B (2003) A unified framework for water erosion and deposition equations. Soil Sci Soc Am J 67 (1):251–257. https://doi.org/10.2136/sssaj2003.2510
Zema DA, Nunes JP, Lucas-Borja ME (2020) Improvement of seasonal runoff and soil loss predictions by the MMF (Morgan-Morgan-Finney) model after wildfire and soil treatment in Mediterranean forest ecosystems. Catena 188:104415. https://doi.org/10.1016/j.catena.2019.104415

4 Spatial Variation in Soils

Abstract

Soil variation can be modeled either by dividing the space into discrete units, or by quantifying autocorrelation in space from known points, using geostatistical tools. Four discrete units are discussed in this chapter: the pedogeomorphological unit of the hillslope catena, the subcatchment unit, the parcel (an agricultural field) representing data aggregation, based on administrative units, and the raster grid-cell, or pixel unit. As an alternative to these four discrete units, a set of continuous data analysis tools can be applied to map soil properties from point measurements, using geostatistical equations. The arsenal of geostatistical tools described here includes: sampling procedures, variogram envelopes and Moran's I analyses for testing spatial autocorrelation, and the various kriging interpolation techniques (with or without ancillary data). Geostatistical models are advantageous for several reasons: they can improve sampling strategy, introduce and quantify the Tobler Law in a bid to understand patterns of variation in space, and offer an account of the roles of various climatic and environmental factors in water erosion processes. However, they may also suffer from a large degree of noise, and not all variables are spatially continuous. In this chapter, we focus on the geoinformatics data models needed for water erosion studies. We present the procedures for applying spatial data models in a discrete fashion, as well as the equations necessary for soil sampling, autocorrelation analysis, and interpolation. These methods make it possible to map the spatial variation in the catchment characteristics that are needed for the water erosion data analyses, with relatively high certainty.

Keywords

Autocorrelation · Catchment analysis · Geostatistics · Interpolation · Sampling · Terrain characterization

Recurrent uses of geoinformatics in water erosion studies for various applications—such as prediction of soil loss risks, mapping of soil quality after an erosion event, and spatial decision-making in soil conservation—require accurate and reliable quantification of spatial variation in soil properties. That is because soil properties markedly affect soil erosivity, and consequently play a major role in erosion processes, such as soil detachment and soil transportation. Spatial variation in soil properties is also an indicator of the effect of extrinsic and intrinsic factors on water erosion, and for this reason, quantification of the variability in soil properties has been successfully merged in heuristic studies of water erosion in various agricultural environments (Annabi et al. 2017; Uyan 2016; Fan et al. 2016; Svoray et al. 2015a).

T. Svoray, *A Geoinformatics Approach to Water Erosion*,
https://doi.org/10.1007/978-3-030-91536-0_4

Spatial variability in soil properties is usually very large in agricultural catchments. Although one might assume that soil properties in agricultural fields—that are habitually located in valleys—are spatially homogenous due to the reduced topographic effects; even this is not always true. Due to the effect of cultivation, fertilization, and other agricultural activities, soil properties can vary largely also in flat areas (see Sect. 1.1.2 on causes of spatial variation in soils due to the effect of extrinsic and intrinsic factors). Agricultural fields situated in sharp topography, or in areas with disparities in other clorpt components, can produce even higher variation of soil properties. Indeed, research shows that spatial variation in agricultural catchments occurs across all spatial scales—from pores to entire regions (Mulder et al. 2016).

Soil variation at the macroscopic scale includes characteristics such as catchment hydrology, parent material alterations, discrepancy in cultivation methods due to societal and cultural factors, and regional physiography. At the mesoscopic scale, soil variability is usually associated with the hillslope catena features that are determined by soil and hydrological processes—such as the continuity of overland flow, the pattern of vegetation growth, and surface sealing. The microscopic scale is associated with soil properties variations within the soil pedon (a representative three-dimensional soil sample of the characteristics of the soil horizons), and the interactions of intrinsic mineral-organic components, through physical, chemical, and biological processes.

The irregularities in soil properties should be considered when attempting to compute model coefficients, in a bid to simulate surface processes such as soil crusting, splash detachment, and the formation of preferential flow lines. Thus, one must combine knowledge about soil variation from soil physics, pedology, and hydrology, at the micro-, meso-, and macroscales of observation, with data assimilation from soil surveys that quantify soil erosion processes (Campbell 1979; Lin 2003).

Indeed, spatial variation in soil properties has been studied extensively by researchers, using: (1) terrain analysis and morphometric computational tools (Hengl and Reuter 2009); (2) a range of quantitative techniques from the scientific discipline of geostatistics (McBratney et al. 2000, 2003); and (3) quantification of intrinsic soil properties, using non-linear system complexity research tools (Phillips 2017; Malanson 1999).

In these three schools of studies, two main data models have been used to quantify spatial variation in soils. The first relies on a division of the studied site into discrete units, within which soil properties are assumed to maintain relatively low variability. In agricultural catchments, these units might be: (1) hillslope and subcatchment units; (2) parcels of agricultural fields; and (3) grid-cells or image pixels derived from remotely sensed data or other sources. The second data modeling approach is based on the premise that soils are collections of continuous variables that make it possible to estimate the soil properties in unobserved areas (Malone et al. 2018). These are the various methods of interpolation from point to area, using point measurement layers (Oliver and Webster 2014). Accordingly, the present chapter is divided into two sections: the first (4.1) surveys procedures of mapping discrete units as a basis for spatial data models, based on clorpt characteristics and mainly topography, human activity, and climate (rainfall), while the second Sect. (4.2) describes the wide range of geostatistical methods that exist in the study of the continuity of soil variation.

4.1 Discrete Spatial Units

Representation of spatial objects using discrete polygons is a well-established paradigm in the geographical sciences. Discrete cartographic units, or irregular polygons, are used in choropleth maps to aggregate various data from point measurements. Data such as family income that were measured in a sample of households are aggregated using zonal statistics into tangible units—such as city boundaries or statistical parcels—to create a categorical map composed from discrete areal units (Boots and Csillag 2006). Discrete units were also often used by pedologists for soil mapping in various parts of the world—first in pen-and-paper drawings, and subsequently with computerized tools (Hewitt 1993).

In the historical maps—with the discrete units usually being the soil formation—ancillary data such as airphotos, geological transects, soil profiles, vegetation type, and terrain maps were used to infer variation in soil properties—both within and between the soil formations. The mapping process was usually subjected to the interpretation of professional surveyors, who used their experience, knowledge, and understanding to integrate the clorpt factors to evaluate the variation in soil properties. Such a mapping procedure assumes that the variation between soil properties within the discrete unit is relatively minor and may be considered negligible, and that critical transitions in soil properties occur at the polygon boundaries.

According to the discrete units approach, if the entire study area is an agricultural catchment, it can be divided into sub-units of similar clorpt model characteristics, while studies are usually focused on a subset of components of the clorpt model. One very notable example of such a clorpt component is the highlighting of topography (or relief). The use of topography as a principal clorpt component leads to a division of the catchment into hillslope units, or in a coarser scale, into subcatchment units. But topography is not the only clorpt component to be used for catchment fragmentation: agricultural catchments are also divided into land use units—specifically, into units based on agricultural usage, delineated by the boundaries of the field parcels that represent the human effect on the soil system. A third classic division method is the discretization of a given site into raster-based grid-cells or image pixels, based on, for example, remotely sensed data of vegetation or rock coverage.

Section 4.1 reviews these three discrete unit approaches. First is the topographic approach, including the hillslope catena pedogeomorphic analysis and the subcatchment divisions. Second is the field-based approach, which is related to cultivation and conservation processes that are usually applied at the field scale, due to decisions made by the farmer. Third is the image pixel approach, which is demonstrated here using meteorological radar data.

Before we proceed into a detailed description of the various methods of division of the study site into discrete units, it is important to note that despite its widespread use, the discrete unit approach has been criticized in past studies, on the grounds that spatial soil variation is occasionally more continuous than categorical, and that the boundaries between the discrete units are, in most cases, indeterminate (Malone et al. 2018; Nettleton et al. 1991). Indeed, some have gone as far as to say that the use of boundaries to represent clear demarcations between discrete units is downright false. In an introductory chapter to a book on geographic objects with indeterminate boundaries, Burrough et al. (1992) showed that the boundaries of discrete soil units are zones of confusion—rather than lines—between overlapping soil properties, which may exhibit a gradual change in soil properties, rather than crisp boundaries between soil formations.

4.1.1 The Topographic Approach

4.1.1.1 The Hillslope Catena

Arable lands on earth are situated in large valleys with deep and fertile soils. However, to maximize the use of the land, most of the hillslopes in agricultural catchments are also cultivated. Consequently, many agricultural fields are located on mild or even sharp topography, and therefore a common approach is to map the spatial occurrence of soils in discrete units along such hillslopes.

On a simple hillslope, water and sediments are transported downslope due to gravitational forces (see Sect. 3.1). If topography is one of the major factors affecting the transportation process, one can, at least in theory, divide the cultivated hillslopes into discrete units based on three simple topographic variables: (1) the upslope contributing area A; (2) the hillslope gradient S; and (3) the hillslope curvature C. The contributing area is defined as all the grid-cells that inflow into any given grid-cell in the studied site. It is computed in Eq. 4.1 using the sum of the areas a_i of the grid-cells i to n that flow toward the given grid-cell:

$$A = \sum_{i=1}^{n} a_i. \tag{4.1}$$

The contributing area A can be computed using the predetermined flow direction in each DEM grid-cell, with the *eight flowdirection matrix (D8)* methodology (O'Callaghan and Mark 1984).

Based on D8, we assume that water flows from every given grid-cell i to one of eight neighbors in the raster data model, using the steepest hillslope gradient premise. In other words, the algorithm computes the hillslope gradient S from the grid-cell in the center of the D8 neighborhood to each of the surrounding eight grid-cells, and the water flows to the steepest hillslope gradient, based on Eq. 4.2:

$$S_{D8} = max_{i=1..8} \frac{z_8 - z_i}{x_i}, \tag{4.2}$$

where S_{D8} is the steepest hillslope gradient downslope, Z_8 is the topographic elevation of the center grid-cell, and Z_i is the topographic elevation at the D_8 surrounding grid-cells. X_i is the distance between the center grid-cell and each of the D8 surrounding grid-cells. Computation of X_i must consider that the distance between cardinal neighbors is the grid-cell resolution in meters multiplied by 1 but the same distance in meters is multiplied by $\sqrt{2}$ for the diagonals. Thus, for example, a 30 m resolution DEM, the cardinal X_is will be 30 m and the diagonals 42 m.

The D8 approach was further advanced in the late 1990s by David Tarboton to account for the inherent premise that water flows from one place to another at an angle of 45°, due to the data model structure of a grid-cell in the DEM being a raster matrix. In the algorithm D-infinity (Tarboton 1997), the contributing area is composed of meaningful fractions of the contributing grid-cells while the author developed a procedure for computing contributing areas and flow direction, by directing the flow downslope along the steepest hillslope gradient on the eight three-cornered planes located around the center of each grid-cell of the raster layer. The contributing area is then computed by assigning the flow between every couple of grid-cells, based on the distance from the direct route to the grid-cell downslope. Tarboton (1997) procedure is available to users as the TauDEM (Terrain Analysis Using Digital Elevation Models) software for spatially explicit computation and analysis of drainage structure and water flow from DEMs, at: https://hydrology.usu.edu/taudem/taudem5/index.html.

The hillslope curvature C, in Eqs. 4.3–4.4, is the second surface elevation z derivative to determine the rate of change of the hillslope gradient in a given direction (x as the longitude and y being the latitude), and to some extent controls the spatial distribution of soil loss rates (Hurst et al. 2012).

$$C_x = \frac{\partial^2 z}{\partial x^2}, \tag{4.3}$$

and similarly in the y-direction:

$$C_y = \frac{\partial^2 z}{\partial y^2}. \tag{4.4}$$

When we map the hillslope units based on these three variables (A, S, and C), we expect each of such a hillslope sub-unit—whose boundaries will be defined later—to have unique soil properties, due to the different erosion and deposition processes that they experience. Such an approach of giving a substantial weight to topography in estimating spatial occurrence of soil properties has been widely used in the soil sciences and geoinformatics literature. Zhu et al. (1997), for example, developed a soil-land inference model (SoLim) to map soil types, based on the following topographic characteristics: surface elevation, hillslope orientation (aspect), hillslope gradient, and hillslope curvature—yet despite the importance the authors have attributed to topography, they also mapped two other clorpt model components: parent material and vegetation cover.

The link between the location on the hillslope, erosion, and deposition processes, and spatial occurrences of soils, has a theoretical basis, which has given rise to several hypotheses and basic terms—such as the soil-landscape paradigm and the hillslope catena (Conacher and Dalrymple 1977). According to the soil-landscape paradigm (Hudson 1992), when other clorpt factors (parent material, climate, organisms, and time) can be treated as uniform at the hillslope scale, or if their variation in the hillslope can be ignored, spatial variation of soil properties may be dominated by terrain characteristics. This idea leads to the hillslope catena hypothesis that states that soils develop along the hillslope catena based on the effect of water flow over the landscape. According to this notion, soil characteristics can be estimated along the hillslope catena—and indeed, several studies have developed geomorphometric approaches for landscape scale analysis of soil heterogeneity (Moore et al. 1993). Furthermore, the literature provides empirical evidence in support of this theoretical notion. Soil texture, for example, as an important property in determining hydraulic conductivity, infiltration, and water erosion, has been found to depend strongly on surface elevation, relative hillslope position, and vertical distance to channel networks (Kokulan et al. 2018).

The hillslope catena—or to use its other name, the soil chain—refers to a series of soil units along the hillslope profile that differ from one another, but have evolved under similar climatic, parent material, biological, and human-induced conditions. Thus, the difference between soils and soil properties is caused (and therefore can be estimated) by their location along the hillslope profile. The reason for this is that, due to the heterogeneous structure of the hillslope profile, the hillslope units differ in detachment, transportation, and sedimentation processes, and consequently in soil properties, such as soil depth, texture, structure, nutrient distribution, and others. The location on the hillslope unit can also affect vegetation productivity, due to variations in water availability, soil depth, and nutrients supply (Milne 1936).

Based on observed variation in soil properties between the hillslope units and the hillslope catena hypothesis, several models—such as the nine-unit soil-landscape model (Conacher and Dalrymple 1977)—have been developed to estimate soil properties along a given hillslope profile. The Conacher and Dalrymple model simulates a three-dimensional geomorphometric structure that extends from the top of the hillslope (the summit) to the channel network. Each point on such a structure can be assigned with one of nine units, and each unit is expected to have distinct soil properties—depending on the effect of sediment and water flow along the hillslope topography. The Conacher and Dalrymple model is based on the following three assumptions: (1) hillslope landscapes have characteristic hillslope and sub-hillslope forms; (2) surface processes, such as overland flow and sediment transport are mostly, but not solely, controlled by topographic features; and (3) of all the clorpt components, hillslope units are the most efficient indicators of pedogenetic processes, and consequently of soil properties, as well.

Conacher and Dalrymple's hillslope model was subsequently advanced by Park et al. (2001) into a computer program for characterizing landform components using a DEM as an input layer. Park et al. developed a landscape delineation procedure based on the theoretical idea of relations between landscape and soils designed by Conacher and Dalrymple. Specifically, they reduced the Conacher and Dalrymple nine landscape units down to six units: "Interfluve; shoulder (seepage slope and convex creep slope); backslope (free face and transportational midslope); footslope (colluvial footslope); toeslope (alluvial toeslope); and channel (channel wall and channel bed)." They did so because they found that three of the original nine landscape units are either too specific while comparing the hillslope processes, for example, seepage

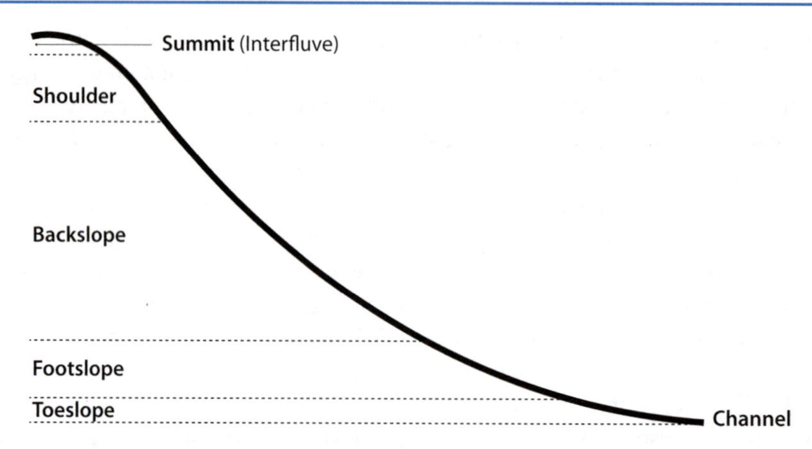

Fig. 4.1 A conceptual draw of a simple hillslope catena from the summit to the channel network. Note that in more complex hillslopes (as most realistic cases are), the units are not necessarily ordered from top to bottom: footslope, for example, might appear at the bottom of the shoulder, and above the backslope unit. The characterization of the units is based on the unit hillslope gradient and cumulative runoff from the upper unit. At larger scales, such as a mountain ridge catena, these units may provide the infrastructure for different soil types, due to microclimatic differences

hillslope versus convex creep hillslope, or due to the fact that the units (i. e., channel wall and channel bed) are too similar to be separable the most currently available DEMs. Figure 4.1 illustrates these six hillslope units on a theoretical hillslope.

More details on the characteristics of the hillslope units as depicted in Fig. 4.1 are provided in Table 4.1. Note the link between the water erosion processes and the landscape unit. These processes link between the hillslope location and soil properties. As we shall see further below in this chapter, this spatial relationship between topography, erosion processes, and soil properties along a topographic sequence may be further used to gain a better understanding of, and to estimate, water erosion processes in agricultural catchments, using readily available DEMs.

The algorithm proposed by Park et al. (2001) uses continuity equations to describe the distribution of the soil particles over hillslopes and can be implemented using many geoinformatics software tools. Moreover, it has been validated against field data, and its application on a catchment in Wisconsin showed similar results with soil survey based on field data. Park model has also been validated at the Lehavim LTER in Israel (Svoray and Karnieli 2011), and has made it possible to estimate the aboveground net primary production and biomass yields of herbaceous vegetation, with high degree of certainty, from the Satellite Pour l'Observation de la Terra (SPOT) XS images using NDVI.

Based on the associations between hillslope unit morphology and the dominant erosion processes described in Table 4.1, Park et al. (2001) suggest quantifying the hillslope unit locations using Eqs. 4.5–4.9 and the model operation is illustrated in Fig. 4.2.

The model begins with computing the change in soil thickness p over the time duration t, with the volume of soil material μ which is the outcome of weathering of bedrock with a given volume (as described in Sect. 3.1, based on the simple hillslope dynamics)—where W is the weathering rate, S is the mean sediment transport, and x is the distance of the given location from the hillslope summit (Carson and Kirkby 1972).

$$\frac{\partial p}{\partial t} = \mu \cdot W - \frac{\partial S}{\partial x}. \tag{4.5}$$

Accordingly, the mean rate of transport of soil particles S through diffusive hillslope processes, such as splash-wash, creep, and subsurface movement of coarse particles from weathering,

Table 4.1 Soil landscape units along the hillslope catena, with dominant water erosion processes and expected surface characteristics in each unit. Reprinted from: Conacher & Dalrymple, The nine-unit land surface model and pedogeomorphic research. Geoderma 18(1):127–144, (1977) and from Park, McSweeney, & Lowery, Identification of the spatial distribution of soils using a process-based terrain characterization. Geoderma 103(3):249–272, (2001), with permission from Elsevier

Landscape units Park et al. (2001)	Landscape units Conacher and Dalrymple (1977)	Typical surface morphology	Dominant processes
Summit (Interfluve)	Interfluve; Seepage slope	Flat, smooth surface (Modal Slope 0–1°) Wide spatial extent (>30 m in two directions)	Vertical leaching; Limited lateral surface and subsurface flow; Water retention
Shoulder	Seepage slope; Convex creep slope	Convexity narrow drainage divide irregular surface (Interfluve < modal slope < backslope)	High intensity of surface erosion and subsurface eluviation; Mechanical removal (soil creep)
Backslope	Fall face; Transformational midslope	Modal slope (1–45°); Straight middle Slope segment; Smooth surface	Transportation of large amount of materials downslope by flow, slump, slide, raindrop impact, surface wash and cultivation
Footslope	Colluvial footslope	Concavity; Water seepage	Colluvial deposition from upslope; Episodic slope failure; Accumulation of throughflow
Toeslope	Alluvial toeslope	Flat; Surface water; Saturation; Adjacent to stream channel	Alluvial deposition of soil material from upvalley; Water saturation by ground water
Channel	Channel wall; Channel bed	Channel	Fluvial processes by a channel flow

can be substituted, on a transport-limited hillslope, by transportation capacity T_h of soils at a given grid-cell:

$$S = T_h = \mathrm{k} \cdot f(x)^{\mathrm{m}} \cdot g(x)^{\mathrm{l}}, \quad (4.6)$$

where k is a local coefficient to be determined empirically, *f(x)* is the distance of x from the hillslope summit and m is a local coefficients to be determined empirically, *g(x)* is the elevation change with position on the hillslope profile, and l is a local coefficients to be determined empirically in each new environment studied.

In Eq. 4.7, the upslope contributing area A_s is computed for any grid-cell i in the catchment. This is usually done using the flow accumulation matrix based on the D8 algorithm (O'Callaghan and Mark 1984) or TauDEM (Tarboton 1997)—both of which are readily available in GIS software tools such as ARCGIS or a stand-alone software. Another option to compute A_s is SAGA GIS that provides flow accumulation algorithms, including D8, D-Infinity, and several others: http://www.saga-gis.org/saga_tool_doc/2.3.0/ta_hydrology_0.html.

$$A_s = \left(\frac{1}{b}\right) \sum_{i=1}^{n} \rho_{\mathrm{i}} \cdot A_i, \quad (4.7)$$

where b is the contour width based on the original spatial resolution of the DEM grid-cell; n is the number of grid-cells in the upslope contributing area cluster; ρ is the runoff coefficient weight depending on the surface cover such as roads, surface sealing, and vegetation strata; and A is the area of a grid-cell in the DEM.

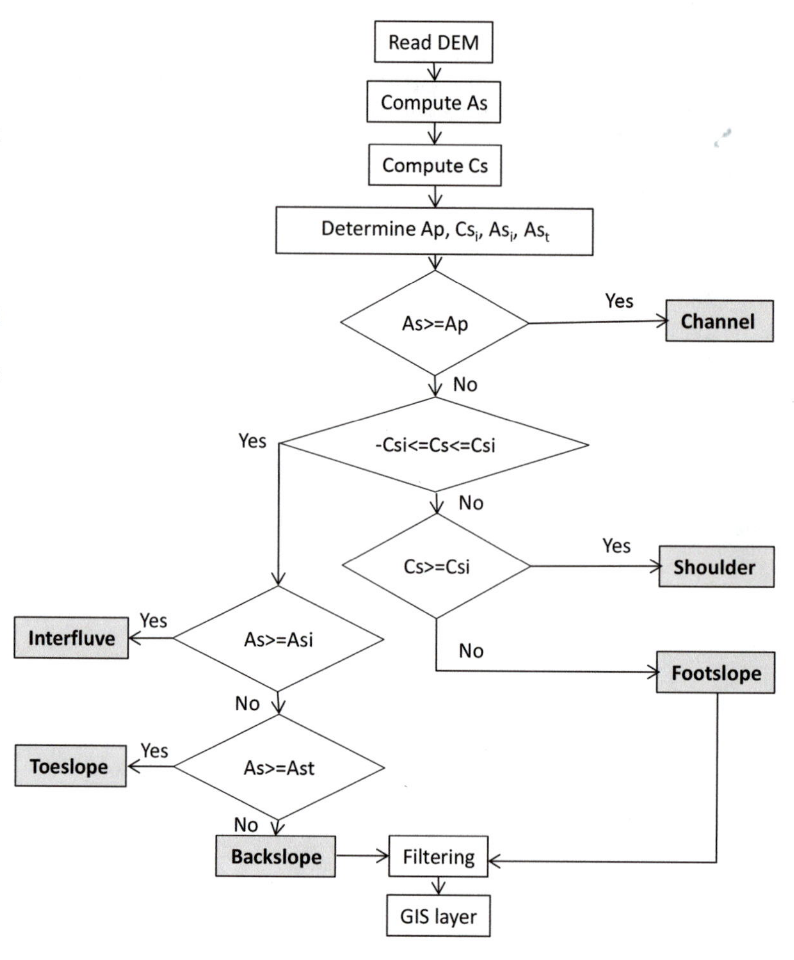

Fig. 4.2 A flowchart to assign each of a DEM grid-cells to one of six catenary units, based on the grid-cell curvature C_s and contributing area A_s. C_{si} is a buffer for C_s—to identify interfluve, backslope, and toeslope from the convex shoulder and concave footslope—while A_p is a critical value of A_s for channel initiation, and A_{si} and A_{st}, respectively, are values that differentiate the interfluve and toeslope from the backslope. Reprinted from: Park, McSweeney, and Lowery, Identification of the spatial distribution of soils using a process-based terrain characterization. Geoderma 103(3):249–272, (2001), with permission from Elsevier

The surface curvature C_s is then computed, where z_i is the elevation of the ith grid-cell and z_n is the elevation of the surrounding grid-cells of this ith grid-cell; n is the total number of surrounding grid-cells in the specific run; and d is the horizontal distance between the two model points. d_{in} is therefore the horizontal distance between the ith grid-cell and its neighbors:

$$C_s = \frac{\left(\sum_{i=1}^{n} \frac{(z_i - z_n)}{d_{in}}\right)}{n}. \tag{4.8}$$

The transportation capacity T_h, also denoted as the Terrain Characterization Index (TCI), is the surface curvature C_s multiplied by log A_s:

$$T_h = TCI = C_s \cdot log(A_s). \tag{4.9}$$

To delineate the hillslope units (based on the Park et al. (2001) algorithm in Fig. 4.2), the channel network should be extracted from the DEM. A detailed example of a channel network extraction procedure is presented in Chap. 5.

Apart from the solution of Eqs. 4.5–4.9, the Park et al. algorithm requires four criteria. The C_{si} is a range around C_s, used to differentiate three hillslope units: interfluve, backslope, and toeslope from the other hillslope units: the convex shoulder and concave footslope. A_{si} denotes the contributing area to the interfluve and A_{st} is the contributing area to the toeslope. These are threshold values for differentiating the interfluve and toeslope from the backslope. A_s is an empirical threshold of contributing area value for channel incision and Ap is a critical value of contributing area for channel incision. Whereas when $A_s > A_p$, the grid-cell is likely to become part of the channel network. Park et al. (2001) assigned C_{si} as 0.25·standard deviation of C_s; $(\log 10 \cdot A_s) \cdot A_{si} = 2.5$; $(\log 10 \cdot A_s) \cdot A_{si} = 3.5$; and $(\log 10 \cdot A_s) \cdot A_p = 4$. However, to fit these threshold values to your area, you may need several trials against a validated database.

The theoretical foundations of the Park et al. (2001) model also make it very general. Due to the fact that the spatial computations—both discrete and continuous—rely on well-established hillslope processes, the Park et al. (2001) model is applicable to estimate soil distribution in other hillslopes. Figure 4.3 is an output file of the application of the Park et al. (2001) model, to map hillslope catenary units using a 25 m resolution DEM of the study area of the Yehezkel catchment. Note that despite the high "salt-and-pepper" variation, the six hillslope units can be observed clearly: although the hillslope units are not uniform, as conceptually illustrated in Fig. 4.1, the gullies can be very clearly identified. The interfluve areas (summit) are equivalent to the gully lines in most cases, with all other units dissociating between them in the more developed hillslopes, while the toeslopes are rarely identified because they are relatively narrow units that cannot be identified in coarse resolution DEMs. In fact, the shoulders are clearly identified and show overestimation in the upper part of the catchment.

The quantitative definition of the hillslope catena is therefore based on primary and secondary topographic variables (Wilson and Gallant 2000)—which once again underline the importance of topography on hillslope erosion, as was already shown in the models described in Chap. 3. The primary variables are the hillslope gradient and curvature, and contributing area, as detailed above.

The secondary topographic variables can be even more informative, and they include the Topographic Wetness Index—TWI (Barling

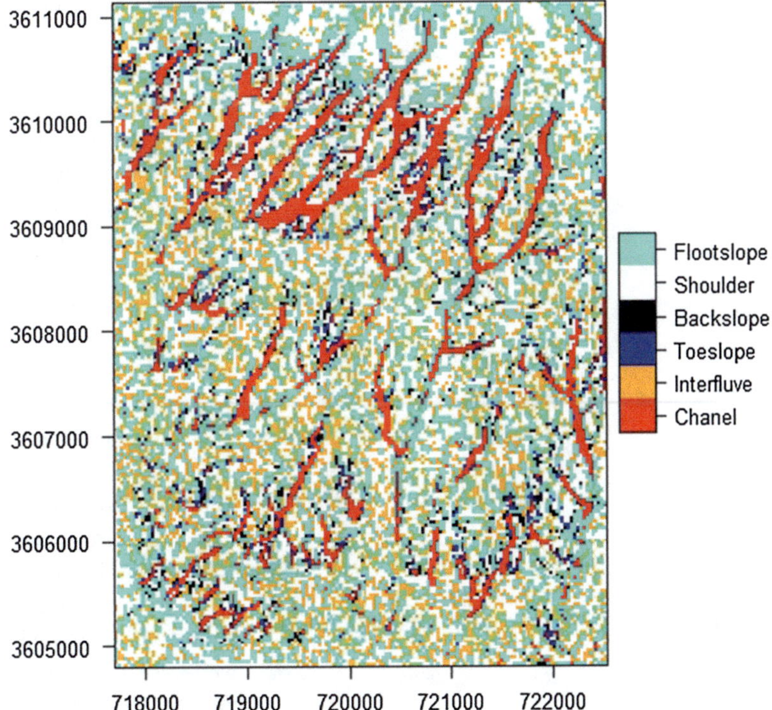

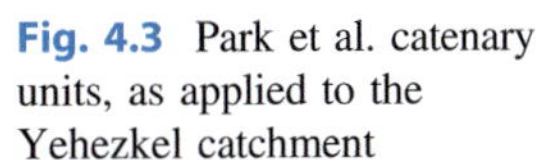

Fig. 4.3 Park et al. catenary units, as applied to the Yehezkel catchment

et al. 1994)—as computed in Eq. 4.10, to estimate how potentially wet a given area is.

$$TWI = ln \cdot \left(\frac{A_s}{\tan S}\right), \quad (4.10)$$

where A_s is the contributing area [m^2] and S is the hillslope gradient [°]. Thus, when the contributing area is large and the surface is flat, the water accumulates in the grid-cells—whereas in grid-cells where the hillslope gradient is relatively large, even if water does flow there from other grid-cells, it is transferred downslope. Figure 4.4 shows the variation in potential wetness as a result of topography, due to variations in hillslope gradient and topographic area.

TWI is not only a measure of ecosystem function and water available for plants. Based on DEM calculations, the *TWI* has been used to locate areas at risk of soil degradation, due to dynamic saturation process. That is because the *TWI* can be used to locate areas of potential saturation that under heavy rainfall events can fulfill the potential and create ponding and disastrous standing water (Kok and Kim 2019).

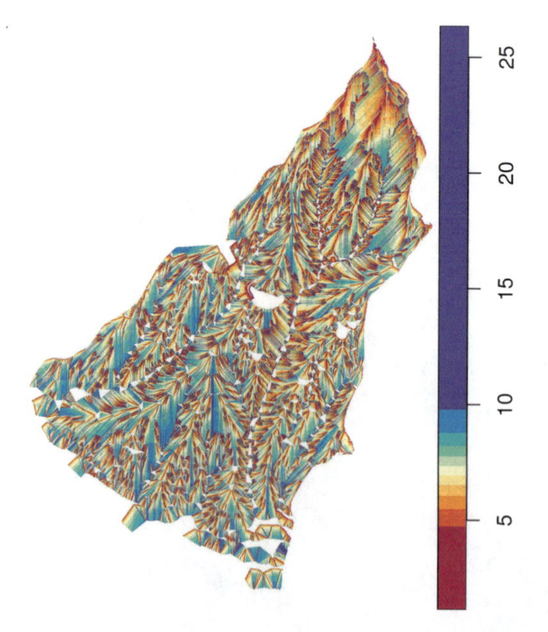

Fig. 4.4 Topographic Wetness Index (TWI) computed for the Yehezkel catchment. Bluish colors represent higher TWI values, and brown colors indicate lower values. Note that the color scale intervals applied is "quantile"

A second useful secondary topographic variable is the Sediment Transport Capacity Index Tc that was aforementioned. The use of the hillslope gradient and contributing area, normalized to certain specific coefficients, can also be used to calculate soil loss potential variation in space with the sediment transport capacity index proposed by Wilson and Gallant (2000) in Eq. 4.11:

$$T_{ci} = \left(\frac{\sum_{i=1}^{n} A_{si} \cdot \mu_i}{22.13}\right)^{0.6} \cdot \left(\frac{\sin S_i}{0.0896}\right)^{1.3}, \quad (4.11)$$

where A_s is the contributing area to the ith grid-cell and n is the total number of contributing grid-cells to the cell i. μ_i is a coefficient to be determined empirically. The value of μ_i is assigned 0 when no rainfall excess is generated in the grid-cell, 1 when all rainfall in the grid-cell is subject to direct surface runoff, and may be transferred to the next grid-cell where the flow is directed. Thus, μ_i controls the runoff/infiltration ratio, depending on the soil's sealing properties. S is the hillslope gradient. Figures 4.4 and 4.5 show an application of the two secondary variables, T_{ci} and *TWI*, respectively, to the Yehezkel study area.

We summarize the hillslope section by highlighting three basic topographic measures (contributing area, hillslope gradient, and hillslope curvature) that may be linked to hillslope erosion processes, and therefore act as basic hillslope units in the effort to map water erosion in agricultural catchments. These basic hillslope units have been linked to soil properties distribution and water processes and can be used to aggregate soil data. As a result, the hillslope sub-units can be used to unravel the real-world complexity, and to estimate soil properties in unvisited locations.

4.1.1.2 Catchment Variables

It is hard to generalize—using simple numerical equations—the effect of hydrologic and geologic characteristics of a catchment on its soil distribution and crop productivity (Horton 1932). One well-known ubiquitous characteristic of catchments, however, is their hierarchical topographic

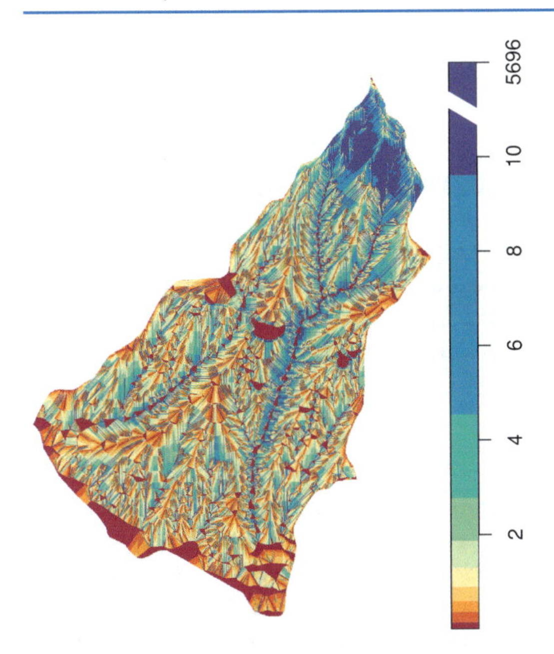

Fig. 4.5 T_{ci} values of the Yehezkel catchment, as computed based on Eq. 4.11. Note that the color scale intervals applied is "quantile"

structure that conveys water flow from the upper parts of the catchment to the lower ones, until the runoff water finally drain into a single outlet. The robust flow of runoff water downslope and the topographic structure of the catchments allows their division into appropriate physically based discrete units for soil conservation and drainage practices—the subcatchments (Downs et al. 1991). Due to the heterogeneity of clorpt components within and between catchments (Lin 2003), one common practice in soil conservation and drainage planning is to divide the catchment into the basic units of subcatchments, and to quantify their respective clorpt characteristics.

Among the most common methods of dissecting a catchment into subcatchments is the stream ordering approach. This allows a systematic division of the catchment based on hierarchical topographic structure of streams and subcatchments, and their labeling and ordering, to suit. The labeling is important, because the amount and energy of water and sediment flow is much lower upstream than downstream toward the outlet. The Stream Ordering Index (SOI) developed by Horton (1945), and subsequently modified by Strahler (1952), is an index of the level of branching in a channel network, and a key method of characterization of catchments. Horton–Strahler order is designed to account for catchment morphology and it lays the basis for computing important variables of catchment structure such as drainage density and bifurcation ratio. The topological ordering of the streams in a given catchment is based on their distance from the outlet, and their location within the channel network. Stream order is a key factor in the understanding of connectivity between different parts of the catchment—particularly with regard to water overland flow and erosion processes.

In practical terms, the Horton–Strahler algorithm works as follows (Tarboton et al. 1991): all branch-free streams—first-order streams—are assigned an order of 1. Along the channel network, if two streams with the same order (first-order streams and first-order streams for example) meet, the stream order from the point of meeting (node) onward increases (the next stream will become second-order stream in this example). When two streams of different orders meet, there is no increase in stream order, but the next stream downslope will be assigned with the order of the higher ordered two converging streams. Thus, the node that connects a second-order stream and a third-order stream yields a third-order stream. Figure 4.6 illustrates a stream ordering of sample data based on the Horton–Strahler Stream Ordering Index. Note that two first-order channels on the left side of the network intersect and create a second-order channel but when this second-order channel intersects with a third-order channel, the outcome channel remains a third-order channel (Fig. 4.6).

Each of the streams in the channel network in Fig. 4.6, and every other network, has a confined contributing area, usually denoted as a subcatchment. The subcatchments, which make up the substructure of the agricultural fields in every catchment, can be classified by the reader using the Horton (1932) framework based on the following factors: (1) morphology that depends on the topography of the constituent landforms of the subcatchment; (2) the physical properties of the soils within the subcatchment; (3) the

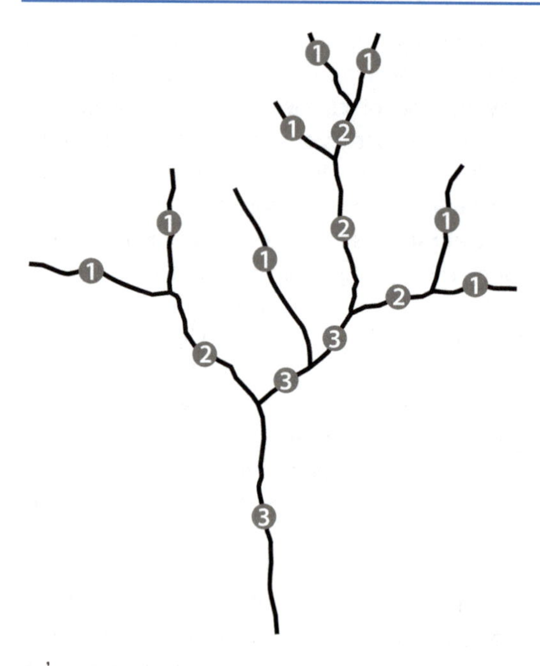

Fig. 4.6 Demonstrated over small part of the Harod catchment, top-down stream ordering based on Horton–Strahler numerical measure of branching complexity of a mathematical tree. Note that stream number increases as the distance from the outlet decreases. This means that first- and second-order streams flow is lower than 3 and 4, etc

geological characteristics of the subcatchment that affect its surface processes and ground water conditions; (4) vegetation growing within the subcatchment (cultivated or natural); and (5) climatic hydrologic factors (such as rainfall and evaporation). Clearly, the subdivision of a catchment into subcatchments with classified clorpt model components may have important ramifications for analyzing the intensity of water erosion processes and the consequent conservation actions.

As a means of catchment analysis, Horton suggested several measures for quantifying the differences between the subcatchments in a catchment, as shown in Eqs. 4.12–4.14. Among these, the form factor F represents the ratio between the subcatchment area A and the length of the streams contained in the catchment L squared.

$$F = \frac{A}{L^2}. \tag{4.12}$$

The form factor F can be associated with the stream's flood regime. Similarly, the compactness factor C of a subcatchment represents the similarity of the subcatchment perimeter P to a circle—with unity (C = 1) representing a circular subcatchment, where long catchments may represent higher transmission losses.

$$C = \frac{P}{2 \cdot \sqrt{\pi \cdot A}}. \tag{4.13}$$

Horton (1932) also found that the mean hillslope gradient and mean surface elevation could provide important information about the water flow in the catchment and the streams. Another important measure of the intensity of water erosion processes in a given catchment is the drainage density DD which quantifies the length of channeling versus the catchment area:

$$DD = \frac{\sum_{i=1}^{n} L}{A}, \tag{4.14}$$

where i is denoted as the given channel between two channel network intersections (i.e., of a given order) and n is the total number of channels within the catchment. ΣL is the length of the channels [m] within the subcatchment of a given order and A is the contributing area [m^2]. Drainage density is a basic measure at the landscape scale, and is recognized as a transition point between various scales of observation (Tarboton et al. 1992).

With the increased use of DEMs in geoinformatics, these descriptive factors of subcatchments have been automated, to determine the drainage network and subcatchment areas. Several algorithms have been executed, such as identification and treatment of depressions, flowdirection analysis in flat areas, the assemblage of flowdirections into flow vectors, delineation of subcatchmnet boundaries, computing flow accumulation in the subcatchment on a grid-cell basis, and the extraction of the entire channel network (Martz and Garbrecht 1992).

Catchment measures are therefore analytical tools with a physical basis, involving geometrical measures of the subcatchments and streams. The delineation of subcatchments may be used to aggregate topographic characteristics from DEM grid-cells or surface cover data (vegetation cover, for example) from remote sensing imagery into more hydrologically meaningful units. The subcatchments maintain strong associations with physical characteristics at the catchment scale and are equivalent to the hillslope units that are based on soil and hydrology processes along the hillslope catena.

One example of a subdivision of the Harod catchment into subcatchments using a polygonal data model is presented in Fig. 4.7. The classified data of the subcatchments in the illustration represents the mean values of the subcatchment characteristics that were extracted from the DEM data: surface elevation, flow accumulation, hillslope gradient, and topographic wetness index. This classification may be a useful means of analyzing the catchment characteristics and of planning priorities in soil conservation actions.

The results in Fig. 4.7 show a clear pattern of the topographic characteristics that may be explained by the hydrological processes. The upslope contributing area (mean flow accumulation layer) in the Harod subcatchments (Fig. 4.7a) increases in line with the Horton–Strahler number, and the main stream at the center of the Harod catchment is clearly identified by the black subcatchments with high flow accumulation. The higher order subcatchments are more grayish in color, and some are white. The hillslope gradient (Fig. 4.7b) is less affected by the topographic location in the catchment and varies more randomly within the catchment. The hillslope gradient layer is, however, a very important layer in soil erosion analyses, as hillslope gradient has been found to greatly affect water erosion processes. In such layers, subcatchments with a steep hillslope gradient can quickly be identified as prone to soil loss and become the focus for further exploration and perhaps priority for conservation treatments. The mean surface elevation map (Fig. 4.7c) simply illustrates the general trend of the Harod catchment from west to east, with elevation change from +400 m AMSL to −143 in the Jordan River. The Wetness Index (Fig. 4.7d)—representing subcatchments with water ponding risks—shows a similar trend to that of the hillslope gradient, but apparently the addition of the upslope contributing area creates spatial variation within relatively uniform hillslope gradient units, such as those in the southern and northwestern sections.

The Harod subcatchments in Fig. 4.7 were identified and delineated using the r.watershed tool of the GRASS GIS 7.6.1svn software plug-in embedded in the GIS software QGIS. This tool was originally designed to calculate several hydrological parameters—specifically, the RUSLE factors (see Chap. 1). The DEM used as an input layer was a $30 \cdot 30\ m^2$ spatial resolution raster, extracted from elevation points measured by a certified surveyor in the Harod catchment at sub-meter resolution, and supplied by the Israeli Ministry of Agriculture. The minimum size of the external subcatchment is determined in the r.watershed tool, using a threshold value provided by the user. The decision regarding the minimum size of the catchment naturally affects the number of subcatchments, and should be determined empirically.

Figure 4.8 illustrates how the number of subcatchments depends on the minimum size of the smallest subcatchment in our data analysis of the Harod catchment, as arising from our simulations. Note that when threshold values are too low, run time increases, and interpretation of the output subcatchments is more difficult.

The user can decide on the minimum size—and resulting number of catchments—based on the scale they want to analyze the erosion processes in the catchment. In the case of Fig. 4.7, we used a 1000 threshold value—resulting in 163 subcatchments to the Harod catchment (Fig. 4.8).

Sections 4.1.1.1–4.1.1.2 show possible data aggregation and landscape units in the hillslope and the catchment scale based on topographic analysis of the agricultural areas, due to

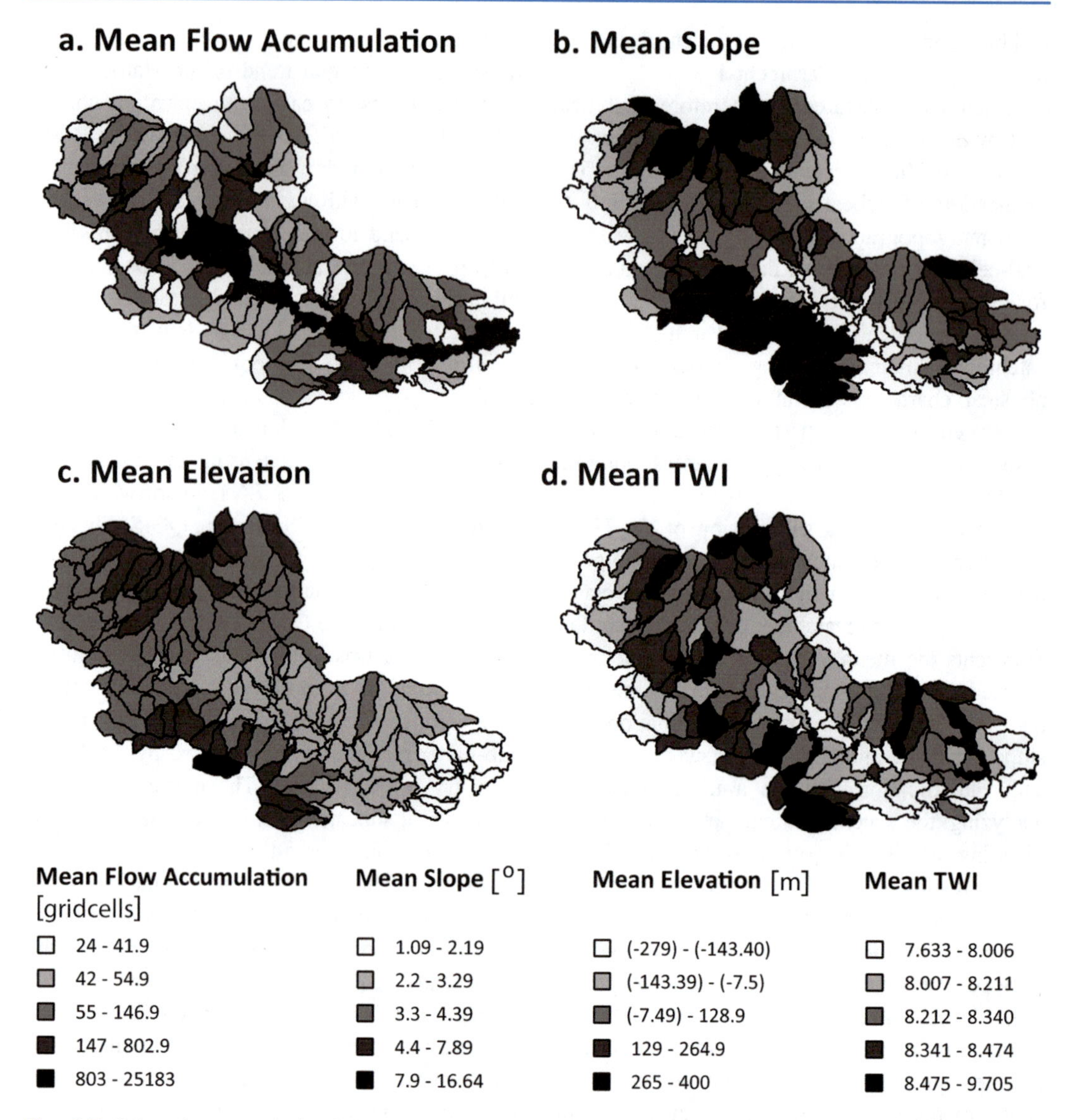

Fig. 4.7 Subcatchment analysis of the Harod catchment, represented by mean values of catchment variables, as computed from the DEM covering the catchment. The data-color classification method (into five groups, as described in the legends) is the Quantile technique, in which each class contains the same number of polygons. Thus, polygons in the highest class would fall into the top fifth of all polygons

geometric considerations of surface and subsurface water flow. Topography does indeed cause large variations in soil characteristics, as well as in soil processes and water flow, and yet—as Horton himself identified, and as clearly is expressed in the traditional clorpt model—topography is not the only factor affecting water erosion, and other factors may also affect variations in soil properties in agricultural catchments. Two of the more influential clorpt components on water erosion, and the possible data aggregation and basic units they give rise to—namely, administrative field boundaries and image pixel—are described in the next two sections.

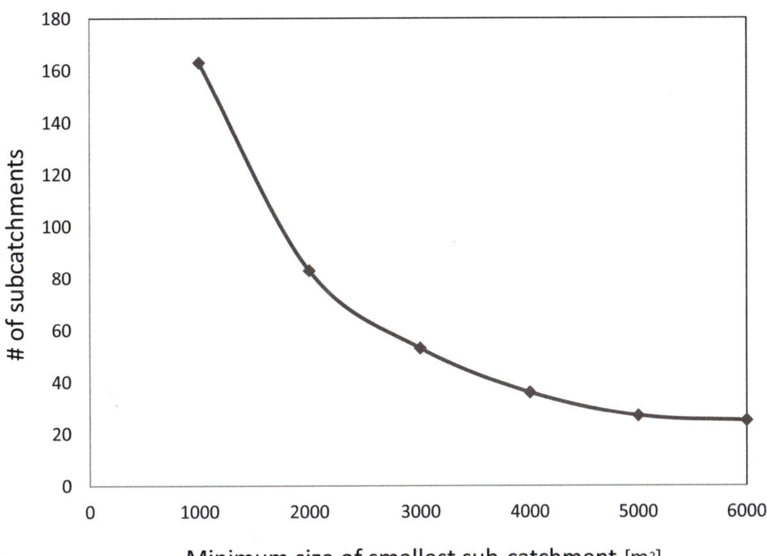

Fig. 4.8 The number of subcatchments is a function of the size of the smallest subcatchment chosen by the user. The function extracted from the curve in Fig. 4.8 can be used to quantify the implications of the decision about the size of the minimum subcatchment chosen, on the number of subcatchment entities in the output

4.1.2 The Field-Based Approach

An important clorpt component that may cause transformation in soil properties and in erosivity —especially in agricultural catchments—is human activity. Many erosion studies of agricultural catchments recognize the importance of agricultural land uses (Renschler 2003) as key factors of soil erosion alongside the effect of topography. A study by Svoray et al. (2015b), for example, leaves little doubt about the scale of impact of agricultural cultivation activities on soil erosivity. Such a dramatic effect calls for a further data model that can display that impact alongside the effects of hillslope and catchment hydrology. Such a data model would naturally be based on agricultural field or plot boundaries and land use maps, rather than on morphometric characteristics.

As a general term, land use refers to the actual use of the physical environment by humans—including settlements, road network, croplands and rangelands, forests, and nature reserves. Rainfall infiltration into the soil, overland flow development, and flow and consequent soil erosion and loss may differ substantially between roads, rangelands, and built-up areas (Svoray and Markovitch 2009). However, in agricultural catchments, the effect of land use is most evident in the cultivation method and cropping system that are applied in the fields (Svoray et al. 2015b).

Soil cultivation causes variation in penetration depth and soil aggregate stability (Svoray et al. 2015a)—since mechanical tools cause modifications to the soil structure, pore size, and other soil properties. These changes in soil properties affect rainfall-runoff ratio, and consequent erosion processes. Soil organic matter, clay, and nitrogen content have been found to decrease significantly with intensive cultivation, while soil erodibility (K factor of USLE, see Chap. 1) increases (Ferreira et al. 2015). Cropping systems—including changes in fertilization and irrigation—may also affect local soil properties (see Chap. 2). Changes in cultivation method and cropping systems can therefore yield spatial patterns in soil properties in the catchment, depending on the location of agricultural fields. This field-dependent spatial pattern is an outcome of agricultural practices, which are usually determined by the farmer or group of farmers involved, based on various cultural, societal, and/or personal considerations. Thus, if a farmer uses a conservation tillage method, and his neighbor uses a traditional cultivation method, the two fields may differ in soil properties, and consequently in the erosion risk to the field. Accordingly, the spatial data model of agricultural practices can be efficiently represented based on agricultural plots as basic units.

For example, Fig. 4.9 shows the spatial variation in the cultivation method and cropping system in the Harod catchment. The agricultural land use layer is polygonal, representing fields of different crops or orchards. The polygons are also tagged by cultivation methods—conventional tillage, reduced tillage, or conservation tillage. The information presented in Fig. 4.8 was collected from farmer's reports that were documented, for administrative and economic purposes in their farm books. To verify the information reported by the farmers, a complementary survey of airphotos and field visits was conducted by our group.

The cultivation method map of the Harod catchment may reflect the following insights that may be relevant to many other catchments: (1) spatial pattern in agricultural land uses arises from the executive decisions of growers, and does not necessarily align with the spatial pattern of other clorpt components; (2) these decisions may be made by a conglomerate of growers—which may result in specific cultivation methods and cropping systems covering large areas and increasing spatial homogeneity in the catchment—but in catchments dominated by fields cultivated by small growers, the heterogeneity of soil properties may be larger; (3) the continuity of the agricultural fields mosaic is, in some locations, disrupted by rangelands, built areas, and natural reserves; and (4) since a relatively large part of the Harod catchment is cultivated by intrusive methods, the scale of the cultivation effect is large (as may well be the case in other agricultural catchments).

he use of agricultural fields as spatial data models helps not only in accounting for erosion patterns in the catchment, better quantification of the rainfall-runoff relationship, and in educated estimations of soil deposition off-site, but also provides new insights into the study of soil response to human activities, and the planning process of soil conservation. The field-based data model is often extended to a related domain—the grid-cell or the image pixel—which is a primary concept of matrix data structure, and commonly used in remotely sensed images. This fairly traditional data model illustrates the rich legacy that has been developed in the history of geoinformatics.

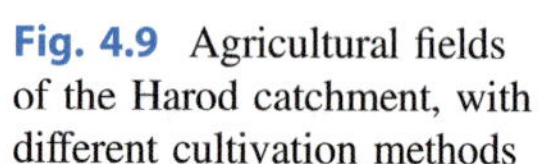
Fig. 4.9 Agricultural fields of the Harod catchment, with different cultivation methods

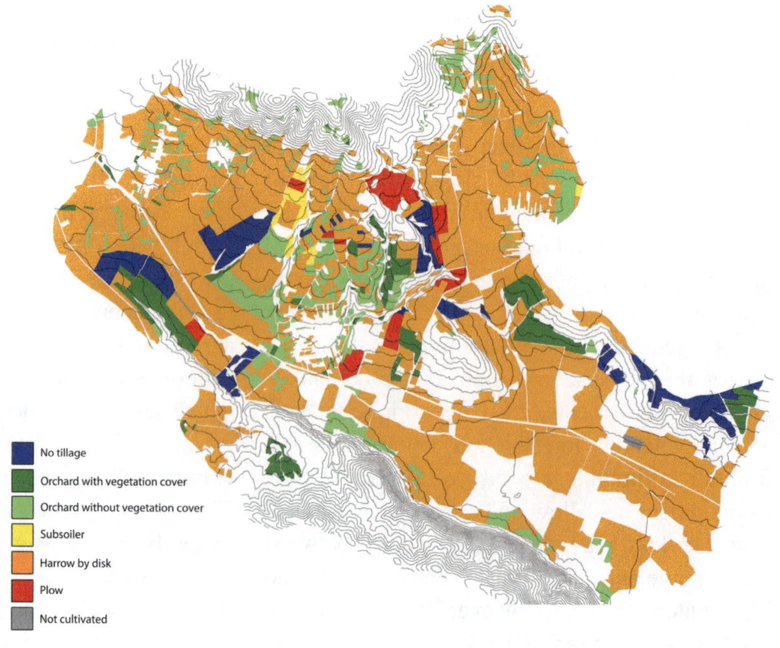

4.1.3 The Image Pixel Approach

The pixel is the smallest picture element of remote sensing images and can only be subdivided by creating repetitive information—i.e., if a pixel is divided, all newly formed smaller pixels will have the same value as the original one (Fisher 1997). The pixel dimensions on the ground are based on the sensor characteristics of the remote sensing platform. The remotely sensed data can be aggregated to the field or topographic units by means of a zonal statistics operator of some sort—but in other studies the data is not aggregated, but kept at the level of the image pixel for further use in a local or focal raster-based GIS analysis. In raster GIS data structures and DEMs, the pixel is often termed as a grid-cell.

A classic example of the use of image pixels as a discrete unit in geoinformatic studies of water erosion is the mapping of climatic factors. Layers of climatic data frequently show complex, heterogeneous patterns, whose properties and dynamics are a function of atmospheric and surface conditions, and therefore may change at multiple scales, ranging from kilometers to meters (Morin et al. 2003). Spatial variation in air and soil temperature, evaporation rate, and, especially, rainfall characteristics may affect soil erosivity, soil moisture saturation, and consequent ponding, in certain parts of the catchment more than others.

Among rainfall characteristics, rainfall intensity, in particular, is a most influential factor on water erosion—especially on splash detachment, and transport capacity (Young and Wiersma 1973). Rainfall intensity has shown high variability at the catchment scale, while the effect of this variability is central to runoff flow and discharge estimations (Arnaud et al. 2002). As a result, the impact of climatic factors on soil erosion, and the probable feedback of water balance change on agricultural soil degradation, requires a different data model from that of topography or administrative field plots. Consider, for example, rainfall intensity estimations based on meteorological radar data and thermal images to estimate soil temperature, visual and IR data to estimate evapotranspiration, and SAR-based data to identify water ponding and soil moisture content. Such data analyses can be exemplified in the Harod catchment, whose clay-rich soils have been observed to seal up during high-intensity rainfall events, thereby preventing deep penetration of rainfall water—which in turn resulted in saturation, ponding, and reduced conditions for crop productivity in the topsoil (Hillel 1998). For this reason, previous studies on water erosion in the Harod catchment, and in many other catchments, have sought to extract spatially, and if possible, also temporally, explicit information on rainfall intensity, in order to produce erosion risk assessment maps (Svoray and Ben-Said 2010).

In recent years, the most commonly used means of producing rainfall intensity maps is the meteorological radar, whose signal is strongly associated with rainfall measurements—in particular, rainfall intensity—at a spatial resolution necessary for catchment scale analyses of water erosion. Radar data has been used for weather estimations for over 60 years, but it is only in the past two decades that it has been used operationally in hydrologic applications (Krajewski and Smith 2002). The radar-based rainfall intensities are computed from the observed radar reflectivity (Morin et al. 2003). The estimated data is usually calibrated with bulk gauge data of relatively high certainty (Morin and Gabella 2007). The scale of observation of the meteorological radar may vary between 10 km^2 and 1 km^2, depending on the measurement platform.

One example of the use of such meteorological radar-based rainfall intensity raster layers, with pixels as basic units, for a long rainfall episode ($\sim$90 mm) occurred during the entire month of March 2010, in the soil erosion risk estimation system at the Harod catchment, as described below. Mean values of rainfall intensity were analyzed at 12-min intervals and processed using Ordinary Kriging (OK) interpolation to a layer calculated based on the maximum rainfall intensities recorded in each of the grid-cells in the studied site at the time of the rainfall event. Figure 4.10 shows three histograms of three distinct short rainfall events during the event: (1) a typical event with low rainfall intensity; (2) a medium-

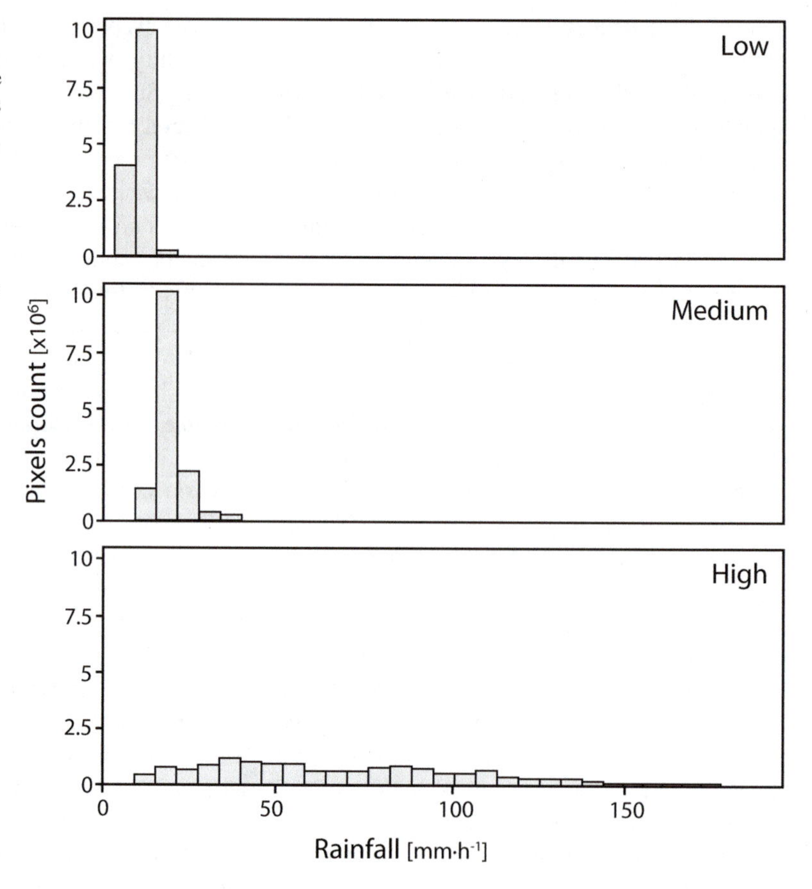

Fig. 4.10 Frequencies of rainfall intensity from meteorological radar in three different short rainfall events in the Harod Catchment, in March 2010. Note how the low- and medium-intensity images have most of the pixels in the left part of the graph while the third image has pixels with much higher rainfall intensity

intensity event; and (3) a high-intensity event. Table 4.2 summarizes the descriptive moments that statistically define this rainfall distribution.

The histograms in Fig. 4.10, and the statistical moments in Table 4.2, show that the spatial variability of rainfall intensity in the Harod catchment was very high in the third event but lower in the first and second events. The difference between the mean value and the median in the low and medium rainfall events is relatively small, and similarly the standard deviation is relatively small in these two cases. The skewness in both events (low and med.) >1—meaning they are significantly right-tailed, or positively skewed (peak on the left part, followed by normal distribution). Kurtosis of the two distributions is >3—i.e., peak distribution. The case of the high rainfall distribution is different, however median and mean show a larger difference than in the case of lower intensity distributions, and the standard deviation is very high. Skewness is however <1—meaning closer to symmetric

Table 4.2 Histogram moments of spatial rainfall intensity [$mm{\cdot}h^{-1}$] data, in the three rainfall events, in the Harod catchment, based on pixel-based meteorological radar (SD = Standard Deviation)

Variable	Mean	Median	SD	Kurtosis	Skewness
Low intensity	10.20	9.87	1.63	10.48	2.14
Med. intensity	18.97	17.48	4.33	6.13	1.71
High intensity	68.06	61.97	36.07	2.36	0.46

distribution, and the kurtosis is <3—meaning closer to flat distribution. Rainfall intensity is therefore, according to Table 4.2, not uniform in the Harod supersite. This is not unique to the Harod catchment and attention must be given to considering spatial rainfall organization in catchment scale analyses of water erosion, particularly in the case of high-intensity events. This insight is also reflected in the rainfall intensity maps in Fig. 4.11, which illustrate the actual interpolated maps representing the spatially explicit pattern in the cases of high, medium, and low rainfall events in the Harod catchment.

Both the descriptive statistics results and the rainfall intensity maps highlight the importance of using a spatially explicit approach when mapping rainfall effect in water erosion studies. Thus, for example, we can see high-intensity values in the western part of the catchment, but the northeastern section—which is characterized by the very steep hillslope of Givat Hamoreh—receives much less rainfall. This may also enhance the importance of a spatially explicit geoinformatics analysis: ignoring this variation by assuming homogenous rainfall intensity may cause overestimation of risk in this part of the catchment. We can clearly see that the spatial pattern of rainfall intensity in the catchment does not align with the spatial pattern of topography of subcatchments and although the data can be aggregated into such units, the original representation needs to be raster based to avoid loss of information.

Maps such as those presented in Fig. 4.11 can indeed be useful for catchment scale analysis of water erosion—but one must remember that rainfall intensity may vary temporally a great deal. That is because the atmosphere varies substantially with time, and because the effect of rainfall intensity of erosion also depends on surface characteristics of the clorpt model. It is therefore crucial to integrate such rainfall intensity maps with the soil and environmental characteristics at the hillslope or catchment scale. The use of pixel-based meteorological radar to map rainfall intensity is being further advanced in recent works. For example, Silver et al. (2020) have shown a 30% improvement in estimations of rainfall intensity, by merging data on elevation, hillslope gradient, hillslope aspect, and distance from the sea, with reflectivity data from a meteorological radar.

To summarize Sect. 4.1, certain discrete landscape units can be used to represent spatially soil, environmental, human-induced, and climatic data, based on class similarities. Hillslope catenary units are based on hydro-pedological processes and subcatchment units are based on catchment topology, while agricultural fields are uniform units in terms of soil cultivation. Another discrete unit is the image pixel, which can be used to represent topography in a DEM, or surface components coverage using optical remotely sensed data, or rainfall intensity from a meteorological sensor, such as converted meteorological radar data.

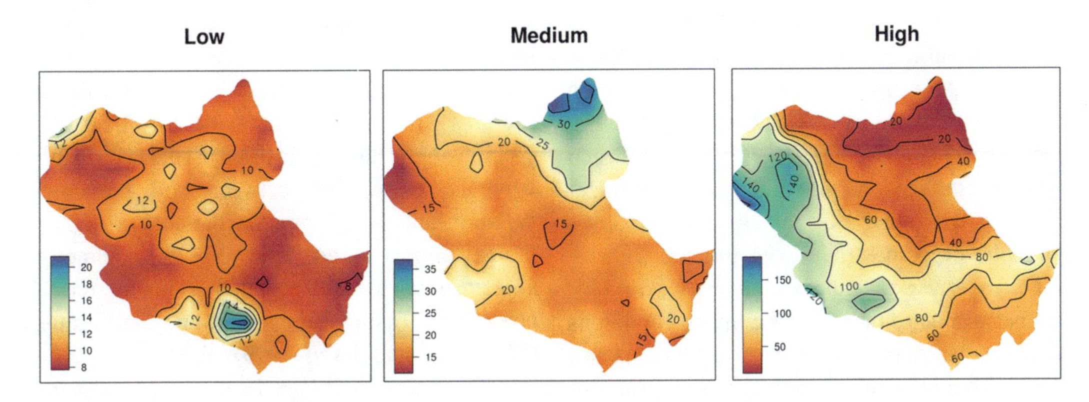

Fig. 4.11 Layers of the rainfall intensity [mm·h^{-1}], in the three rainfall events, in the Harod Catchment. Data from the Radar Meteorological Station from March 2010

4.2 A Suite of Continuous Variables

In Sect. 4.1, we showed how discrete spatial units may be used as spatial data models to divide agricultural catchments into hillslope and catchment units, field parcels, and image pixels. The sharp boundaries of these discrete units are determined by topographic conditions, or based on administrative (in the case of field plots) or logistical considerations (the image pixel boundaries are designed by the remote sensing platform engineering specifications). Soil properties are then aggregated into these units, assuming the variation within the unit is lower than the variation between units.

An alternative method of mapping spatial variation in soil properties is the use of point measurements and their mathematical interpolation in space. Such an interpolation approach may be required when field measurements are used to replace remote sensing in extracting data from deeper soil layers or when laboratory analyses are needed to quantify more complex soil characteristics.

Spatial interpolation techniques rely on existing of patterns of spatial association between the point measurements, which in the case of soil properties, can be the outcome of various clorpt factors or intrinsic soil factors. These complex interactions can be illustrated through the case of spatial variation in soil moisture. Previous studies provided empirical evidence that soil moisture variability in agricultural catchments increases with area size (Famiglietti et al. 2008). The variation in soil moisture depends on the soil's hydraulic properties (Huang et al. 2016), while the pore size and consequent infiltration are largely affected by soil cultivation processes, due to mechanical compaction of the soil, and chemical interactions brought on by fertilizer activities. The effect of soil tilling on the soil's hydraulic properties may even interact with topography at various hillslope catenary positions, and between sites of different climatic conditions (in particular, rainfall intensity and dominant soil types). To add to this complexity, time can also play a role in the patterns of spatial variability. Some tillage practices have been found to have dramatic and instantaneous impacts on soil hydraulic properties, but these have been found to be reducible over time. For example, long-term effects (of a decade or more) appeared less pronounced (Strudley et al. 2008).

For this reason, one must quantify soil variability using well-established mathematical techniques before applying the interpolation technique. Many studies have done so and found differences in the pattern of spatial variation between various soil properties: some properties have been observed to vary substantially in space, while others have not. For example, Denton et al. (2017) measured high variability in available phosphorus and potassium in an equatorial climate, Nigeria ($C_v = >35\%$), while nitrogen, CEC, OC, SAR, and ESP have shown moderate variation (C_v = 15-34%), and pH exhibited low variation ($C_v = <15\%$) (C_v is a standardized measure of dispersion often used in the soil sciences, see for details in Sect. 7.2.1.). They also found negative linear associations between mean and variation values of soil moisture. The differences were observed mainly in relation to maximum porosity and wilting point. Similarly—and depending on the predominant land use and the season—vegetation has been found to cause changes in spatial variability of soil properties (Korres et al. 2015).

Quantification of spatial variation in soil properties is especially called for when applying an upscaling process for matching remotely sensed data with the high-resolution soil profile data measured in the field. Likewise, a downscaling approach used to extract detailed figures from coarse scale satellite data requires a translation to smaller but inherently highly variable sub-areas if they are to be made useful for site-specific applications (Lin 2003). This is probably why the association between soil properties variation and clorpt components such as topography, climate, and land use is commonly examined when seeking spatial estimation of soil properties using interpolation techniques (McKenzie and Ryan 1999). For example, the incorporation of auxiliary Landsat ETM data has markedly improved predictability of spatial variation in soil organic matter—suggesting a degree of spatial correlation due to intrinsic topographic factors that control water flow and storage, soil temperature, and microbiological activity in the soil strata (Mirzaee et al. 2016). Over an even wider region, a wide scale analysis of soil texture and hydraulic conductivity—from data points—was applied to solve the spatial variation in these two properties over the entire landmass. Analyses of the field measurements revealed that approximately half of the spatial variation was observed in areas smaller than 10 km^2, and one-third of the variation in areas smaller than 1 km^2 (Paterson et al. 2018). Spatial variability of soil properties can also vary greatly, due to surface processes such as runoff flow (Costa et al. 2015). The variation of these physical and biological properties has been widely studied, but they are so numerous and complex in their interactions with each other that the outcome variation of the soil properties can be very difficult to estimate. This lack of understanding hampers both estimations and quantification of soil variance at this scale.

The soil, as a complex system comprising a large number of properties, affected by extrinsic and intrinsic factors, and operated by many processes, requires a rigorous and effective tool to unravel this variability, and its effect on water erosion. Geostatistics—a collection of statistical procedures used to study phenomena that vary in space—has been successfully applied in recent decades in the study of variation in soil properties, and in the making of soil maps. The increase in available computational power has fostered the use of numerical methods of geostatistics over large datasets. These developments are useful when studying erosion risk maps, and the state and health of remnant soil in the aftermath of an erosion process. The field of geostatistics is rather developed and previous publications offer very detailed guides on the use of these techniques—such as Oliver and Webster (2014). We will limit ourselves, therefore, to describe the main equations in the field that relate to map soil properties at various levels, depending on soil management practices and the parent material.

This section is divided into three subsections, in accordance with the procedure of analyzing spatially continuous soil properties data from point measurements: (1) we begin with point sampling—including assessing the necessary sample size and the samples' geographical locations; (2) we follow this with autocorrelation analyses using variograms (Curran 1988) and other indices to allow for spatial interpolation as a measure of similarity between nearby observations; and (3) finally we quantitatively describe spatial estimations of values at unknown points, using mainly weighting-based interpolation techniques.

4.2.1 Spatial Sampling

4.2.1.1 Sample Size

Spatial sampling is a controlled procedure of measurements of (soil) properties in any given study area (Delmelle 2014). The sampling procedure involves the computation of two measures: (1) The sample size—namely the number of observations needed to represent the study area population in the statistical sample. This measure may, in practice, be limited by logistical constraints—such as manpower, available time, and budget. (2) The spatial location (x, y) of the sample measurements in the field. Those two quantitative measures are crucially important to the representation of spatial variation in soil properties in the studied area and affect the autocorrelation analysis and subsequent interpolation procedure.

Sample size is the size of the part of the population chosen to represent the entire population in the catchment. Due to the aforementioned logistic considerations, the user will want the minimum number of measurements needed for an acceptable level of error, which always exists. Cline (1944) studied this task and summarized previous research on the use of traditional statistics to study variation in soil properties. Based on this review, he proposed Eq. 4.15 to determine the sample size n and to achieve the desired estimated variance s^2:

$$n = \frac{t_{\alpha}^{2} \cdot s^{2}}{(x-\mu)^{2}}, \quad (4.15)$$

where t is the value of Student's t test at the chosen level of probability α and x is the distance that the analyst is willing to tolerate from μ, which is the estimated mean value of the entire population to be determined from an external source.

Equation 4.15 has been used successfully in many studies, but is based on traditional statistical sampling theory, which does not account for the spatial autocorrelation characteristics of the studied soil property. This is a limitation, because based on Tobler's law (Tobler 1970), spatial variation of measured (soil) properties is negatively biased toward distance. Namely, the disparity in values of soil properties between two point measurements depends on the distance between the two. Adjacent point locations are expected to be similar while distant points are expected to be dissimilar. This bias may affect the sampling scheme and the sample size. That is because if some locations show higher variability in space, they should be represented with more points than more spatially homogenous regions.

To quantify Tobler's principle in spatial sampling, McBratney and Webster (1983) developed a method for determining sample size, based on the semivariogram characteristics (see Sect. 4.2.2), and by considering the neighborhood effect—resulting in a reduction of the sampling effort—sometimes substantially. The McBratney and Webster procedure works as follows:

Step 1: Compute a semivariogram for the estimated variable.

Step 2: Group variances together to compute the global variance and then the standard error.

Step 3: Minimize the grouped value for a given sample size.

McBratney and Webster (1983) founded their spatially explicit method on two well-established principles: (1) random sampling precision is improved by using a systematic sampling such as data stratification, because the data values of points are related to the distances between them and (2) based on Tobler's Law, soil property values at proximate locations should exhibit less variance than more distantly separated points.

In random sampling, some observations can be in close proximity. According to Tobler, they represent duplicate information because they are similar. Therefore, considering neighboring sampling points through quantification can minimize repetition. Equations 4.16 through 4.20 make it possible to compute the McBratney and Webster (1983) method based on the two above-mentioned principles:

$$\hat{Z}_R = \frac{\sum_{i=1}^{n} V_i \cdot \widehat{Z}_i}{\sum_{i=1}^{n} V_i}, \quad (4.16)$$

where $\widehat{Z}_R$ is the global estimate for any given region R at the studied site, i is the neighborhood with area V_i, and $\widehat{Z}_i$ is the local estimate for neighborhood i.V_i is taken as the grid point x_i of a regular grid. The summation operation is computed for all points n within the boundaries of the region. The error in the global mean is equal to the sum of errors in the local estimates (Eq. 4.17):

$$\hat{Z}_R - Z_R = \sum_{i=1}^{n} V_i \cdot \left(\hat{Z}_i - Z_i\right). \quad (4.17)$$

The variance of a region R, $\sigma_R^2 = E \cdot [\{\widehat{Z}_R - Z_R\}^2]$, cannot be estimated ($E$) by using a simple summation operator, since the estimates at the neighboring points are interdependent. The solution McBratney and Webster (1983) suggests to consider the error that results from using the value at a given observation point to estimate the average value over the portion of the region that is nearer to it than to any other region. This can be done using the boundaries of a Thiessen polygon S. Since the data is usually processed using raster data models, the Thiessen polygon, S, is a square with the observation point at its center and sides equal to the sampling interval. The variance of estimating its average value is (Eq. 4.18)

$$\sigma_S^2 = 2 \cdot \overline{\gamma}(\boldsymbol{x}, S) - \overline{\gamma}(S, S), \quad (4.18)$$

where $\overline{\gamma}(x, S)$ is the mean semivariance between a given point, x, and all other points in the square, and $\overline{\gamma}$ (S, S) is the within-square variance. Thus, if the estimated values for these squares are $\widehat{Z}_{Si}$, the average for the region $\widehat{Z}_{SR}$ is approximately (Eq. 4.19):

$$\hat{Z}_{SR} = \frac{1}{n}\sum_{i=1}^{n} \hat{Z}_{Si}, \quad (4.19)$$

with the error approximately $Z_R - Z_{SR}$, the corresponding global estimation variance is (Eq. 4.20)

$$E\left[\{Z_R - Z_{SR}\}^2\right] \cong \frac{1}{n^2}\sum_{i=1}^{n} E\left[\{Z_{Si} - Z(\boldsymbol{x}_i)\}^2\right] = \frac{1}{n} \cdot \sigma_S^2, \quad (4.20)$$

where x_i is the grid point and the approximation improves as n increases. The error of the global estimate depends on the estimation variance of a small part of the region—and this is likely to be smaller than the variance over the region as a whole.

Equations 4.18–4.20 provide the tools whereby the grid spacing (and hence the sample size) is determined, to attain an explicit level of precision in estimation. Given the semivariograms, the equations are solved for a range of sizes of squared grid, or Thiessen polygons, and the variance is plotted against the sample size, n.

4.2.1.2 The Location of Points

In geoinformatics studies, it is crucial to decide where to locate the sampling points in the studied site, as well as how many points to sample. Although the McBratney and Webster method does provide information on the required distance between points, based on the autocorrelation characteristics of the studied soil properties, several other methods to locate the samples in space do exist, and are based on ancillary information and spatial statistics (Sherpa et al. 2016). Of these, three relatively simple methods are usually used:

Random sampling

In random sampling—a common method in many disciplines—the location of each sample point is chosen at random—namely, in an unbiased fashion, while any point location in the population has the same probability to be chosen to the sample as any other point location. In the past, random numbers of point location coordinates were chosen using tables, in which numbers are prearranged so that each digit has no correlation to the other digits. Nowadays, almost every statistical (such as R) or GIS (such as ARCGIS) package provides random number generators that generate automatically random locations within a predefined study area. The number of points is predetermined using Eq. 4.15 for example. The main shortcoming of random sampling is that due to the high spatial variability in agricultural catchments, one may need a large number of random points to represent the irregularities in soil properties in the site.

The stratified random technique

Stratified random sampling is a closely related alternative to the random sampling approach. With it, one divides the study area into layers of similar conditions—such as areas of similar hillslope gradient or areas of similar parent material formations. One then samples a predetermined number of points per area (stratum). Within the strata (geological formations for example), the location of the point samples is randomly selected. For example, one might divide the Harod catchment into five categories of hillslope gradient [°], as follows: 0–3, 3.1–6, 6.1–9, 9.1–11, >11, with all the polygons in each category denoted as a stratum. The procedure works as follows:

Step 1: Compute the total number of points that you need to sample in the entire area in question (based on Eq. 4.15), that is, the sampling effort.

Step 2: Compute the number of samples per stratum—either equally or weighted proportionally to the area covered by each stratum or any other proportion one chooses.

Step 3: Apply random sampling within the strata.

Stratified random sampling thus ensures that all landscape units of interest are (proportionally) represented within the sample. The random and stratified random strategies are demonstrated in Fig. 4.12 using R.

Matrix sampling technique

With matrix sampling, a nominal net is overlaid the study area within its confines, and a sample point is placed at every intersection of the matrix. The procedure works as follows:

Step 1: Delineate the boundaries of the study area;

Step 2: Create an artificial fishnet, with a grid-cell resolution based on the number of points the sampling effort can bear;

Step 3: Label every intersection of this fishnet matrix;

Step 4: Cover the entire matrix, if resources allow—or use random sampling of intersections if only a limited number of points are available.

The main difference between the stratified random and matrix sampling is that while the former allocates the points by a thematic unit (the stratum), the latter attempts to provide areal cover more effectively. One notable drawback of the matrix-based method is that the point location is predetermined by the previous points, due to the rigid structure of the matrix, i.e., it is necessarily located by the distance and angle dictated by the matrix structure. If the matrix is dense enough, this limitation may be minor, but a high-resolution net requires more sampling points.

Mapping soil variability with point measurements requires spatial sampling strategies, and an educated design of sample size and point location, to avoid misrepresentation of the heterogeneity of the soil property in question: an insufficient number of sample points, or incorrect location of the points, may yield unreliable maps. Statistical and geostatistical equations can be used to determine the number of measurements required for catchment scale mapping of water erosion properties. The equations reviewed in this section can be extended to various catchments, based on the environmental and human-based maps described in Chap. 7. The various strata may be chosen based on the site characteristics and available budget.

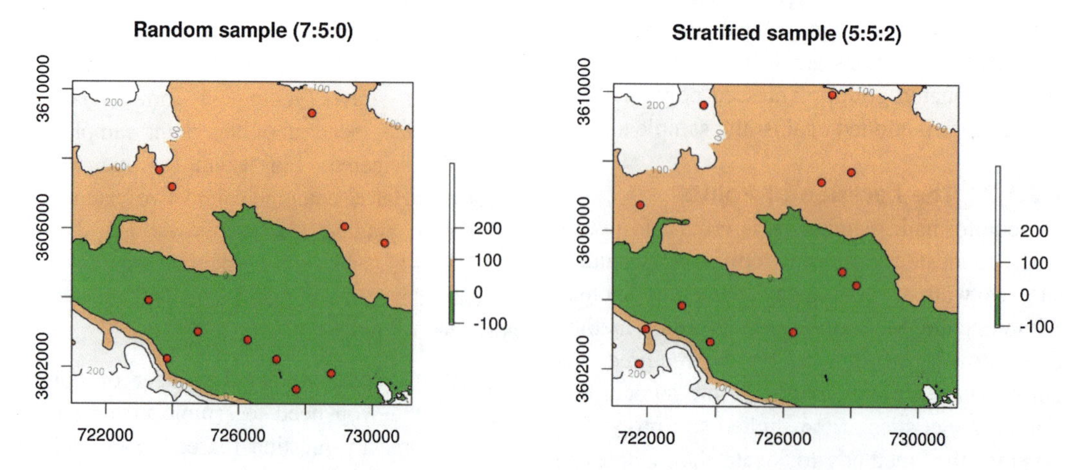

Fig. 4.12 Random (left panel) and stratified random (right panel) sampling techniques applied in the Harod catchment. The red points are the sampling points, and the backgrounds are the independent elevation contour representation from another source of information. The sampling method affects the number of points that are sampled in each elevation group—the titles show the difference between the amounts of points-per-polygon due to the sampling method

4.2.2 Autocorrelation in Soil Properties

Quantification of spatial autocorrelation characteristics of soil properties in point measurements allows to determine their continuity in space and makes the basis for spatial interpolation techniques. On the other hand, if the values of a measurement point's sample maintain spatial autocorrelation, the sample points are not independent, i.e., such a correlation violates the basic statistical assumption of independence between data points. We therefore need quantitative measures to evaluate spatial autocorrelation between point measurements at the site in question.

Three methods of testing autocorrelation in space are described here: the variogram plot, which relies on the association between the distance and semivariance of the point measurements and the nugget–sill (N:S) ratio; Moran's I measure, which is based on an elementary correlation coefficient that allows to quantify the spatial autocorrelation between two values of point measurements at different locations; and the variogram envelope technique, which uses permutation theory to determine the significance of the theoretical variogram. The three measures surveyed in this section have been used in past studies and can be applied to many agricultural catchments.

The variogram

Spatial autocorrelation refers to the mathematical and statistical quantification of how well the value of a (soil) property in a given location $Z(x)$ is associated with this (soil) property values at nearby locations $Z(x + h)$. This quantification is applied within the boundaries of a geographic area—such as a catchment, a hillslope, or an agricultural field. Temporal autocorrelation is a similar concept, referring to measurements' proximity in time, rather than in space. Temporal autocorrelation features are widely described in the literature for the purpose of time series analysis, e.g., Zwieback et al. (2013), but in this book we will focus on spatial variability only.

The study of spatial autocorrelation in soil properties is not a simple task, because soils are complex systems—the products of a large number of interacting physical, chemical, and biological processes (see Chap. 1). Because of this complexity, the spatial variation of soil properties in many studies is treated by scientists as random (Goovaerts 1999). A random variable is a measure that may be assigned with any value within a given set of specified relative frequencies. Consequently, random variables have been, and are being, used to describe a population, or to express an unknown value that a soil property may have at any given space-time location.

Matheron (1965) formulated the mathematical framework for geostatistics based on several assumptions and we will focus here on two of them: (1) soil processes are considered random processes and (2) soil processes are considered stationary, and therefore the same degree of variation occurs across similar distances. The latter assumption is required to bound the solution for a given realization.

In accordance with these two assumptions, $Z(x)$ represents a random process, and the random values of the soil properties in the point measurements x (marked with x, y coordinates), in a region computed as in Eq. 4.21 (Webster and Oliver 2007):

$$Z(\boldsymbol{x}) = \mu + \varepsilon(\boldsymbol{x}), \tag{4.21}$$

where μ is the estimated mean value of the outcome of the soil process and $\varepsilon(x)$ is the error which is usually random. The covariance $C(h)$ is then computed using Eq. 4.22, with h as a vector of values (the lag), E denotes the expected value and signifying the separation between two x-, y-locations in both distance and direction:

$$C(\boldsymbol{h}) = E[\varepsilon(\boldsymbol{x})\varepsilon(\boldsymbol{x}+\boldsymbol{h})]. \tag{4.22}$$

The covariance between the spatial locations can also be defined as in Eq. 4.23:

$$\begin{aligned} C(\boldsymbol{h}) &= E[\{Z(\boldsymbol{x}) - \mu\}\{Z(\boldsymbol{x}+\boldsymbol{h}) - \mu\}] \\ &= E\left[Z(\boldsymbol{x}) \cdot Z \cdot (\boldsymbol{x}+\boldsymbol{h}) - \mu^2\right], \end{aligned} \tag{4.23}$$

where E denotes the expected value, $Z(\boldsymbol{x}$ is the value of the Z soil property at location $\boldsymbol{x}$, μ is the estimated mean value, and $Z(\boldsymbol{x} + \boldsymbol{h})$ is the value at location $\boldsymbol{x} + \boldsymbol{h}$. If μ is not equal to a constant, the covariance cannot exist, and based on the intrinsic stationarity theory of Matheron (1965), the expected subtraction of the two values is equal to zero—$E[Z(\boldsymbol{x}) - Z(\boldsymbol{x} + \boldsymbol{h})] = 0$, and the covariance is therefore changed to 0.5 of the variance of the subtraction of $Z(\boldsymbol{X} + \boldsymbol{h})$ from $Z(\boldsymbol{x})$, and the result is denoted as the semivariance:

$$\gamma(\boldsymbol{h}) = \frac{1}{2} \cdot var[Z(\boldsymbol{x}) - Z(\boldsymbol{x}+\boldsymbol{h})] = \frac{1}{2} \cdot E\left[\{Z(\boldsymbol{x}) - Z(\boldsymbol{x}+\boldsymbol{h})\}^2\right]. \quad (4.24)$$

This is why we usually use semivariance, rather than the covariance. As we see in Eq. 4.24, and as was identified by Tobler, the semivariance $\gamma(\boldsymbol{h})$ is associated with the distance $\boldsymbol{h}$ between the points. The so-called "variogram"—the graph of $\gamma(\boldsymbol{h})$ plotted against $\boldsymbol{h}$—is therefore more representative than the covariance plot and has become the primary tool of geostatistics. The semivariogram is a graph that allows the spatial autocorrelation of the measured soil characteristics at the sample points to be explored. The semivariogram graph has been widely used in the soil sciences literature since the comprehensive work by McBratney and Webster (1986) on choosing the functions for fitting a theoretical function to the empirical measurements of soil properties.

Figure 4.13 shows a model fitted (the fitted line or theoretical semivariogram) to a sample semivariogram of soil health index data from point measurements (the empirical semivariogram) in the Harod catchment. Note that the points in the variogram represent the semivariance values between pairs of point measurements. The point distribution around the theoretical variogram may give an idea about the autocorrelation of the property, but this usually involves a quantitative study of the variogram characteristics detailed below—namely, the nugget, the sill, and the range.

The nugget is the distance of the semivariance from the origin along the Y-axis. Specifically, it represents the semivariance between two measured points that are practically at the same ($\boldsymbol{x}$, $\boldsymbol{y}$) location. In theory, at a zero-separation distance, the semivariance value should be nil, because the property values are measured at the same ($\boldsymbol{x}$, $\boldsymbol{y}$) location. But at an infinitesimally small distance, the semivariogram may often exhibit some nugget effect, which in some cases may be larger than 0. The nugget effect can therefore be indicative of measurement errors inherent in the measurement device being used, or spatial sources of variation at distances smaller than the sampling interval—or both (Oliver and Webster 2014).

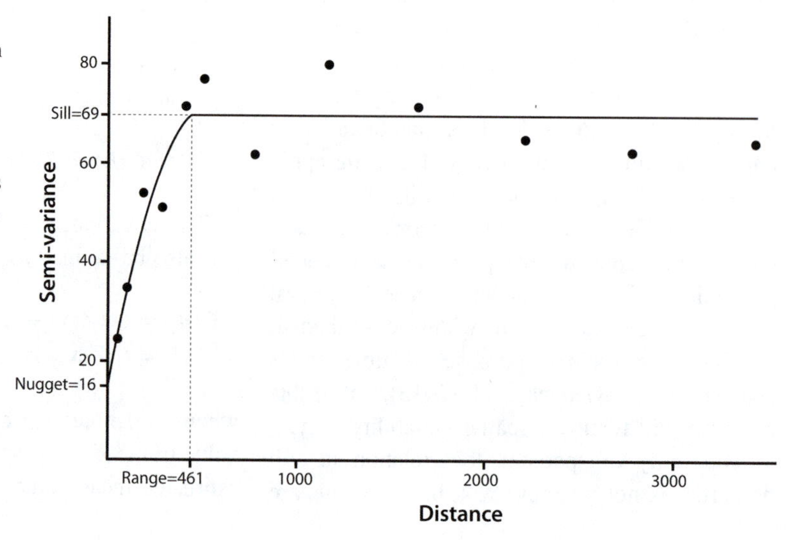

Fig. 4.13 Empirical points and theoretical line variogram of soil health index estimations in the Harod catchment computed with R. Note the nugget at the intercept of Y-axis, the sill as the Y-value at the asymptote, and the range as the minimal X-axis value of the sill

The sill is the semivariance value that limits the variogram asymptote. Points around the curve up to the sill value represent the association between the distance between the pairs of points and the semivariance—while above the sill value, autocorrelation is not maintained, so the semivariance does not change as a function of the distance. The nugget:sill (N:S) ratio is a measure often used in studies of autocorrelation to represent the proportion of variation that cannot be explained by distance to the measurement error of the observed variable. In other words, if the sill value is relatively small compared with the nugget value, then the user should not use the semivariogram, and seek other alternatives or proceed to estimation with caution.

The range is defined as the length along the X-axis (i.e., the distance) of the point from which the differences in semivariance become negligible. The range value in the semivariogram diagram shows the specific distance where the theoretical semivariogram reaches an asymptote —i.e., the distance where the model shows no further increase in the Y-axis versus X-values. It is a highly important measure, because point measurements at distances shorter than the range are expected to spatially autocorrelate, while points at greater distances from other points are not expected to autocorrelate. For example, Karnieli et al. (2008) identified areas prone to degradation/rehabilitation of rangelands in the deserts of Central Asia as a function of distance of the semivarigram range from watering points.

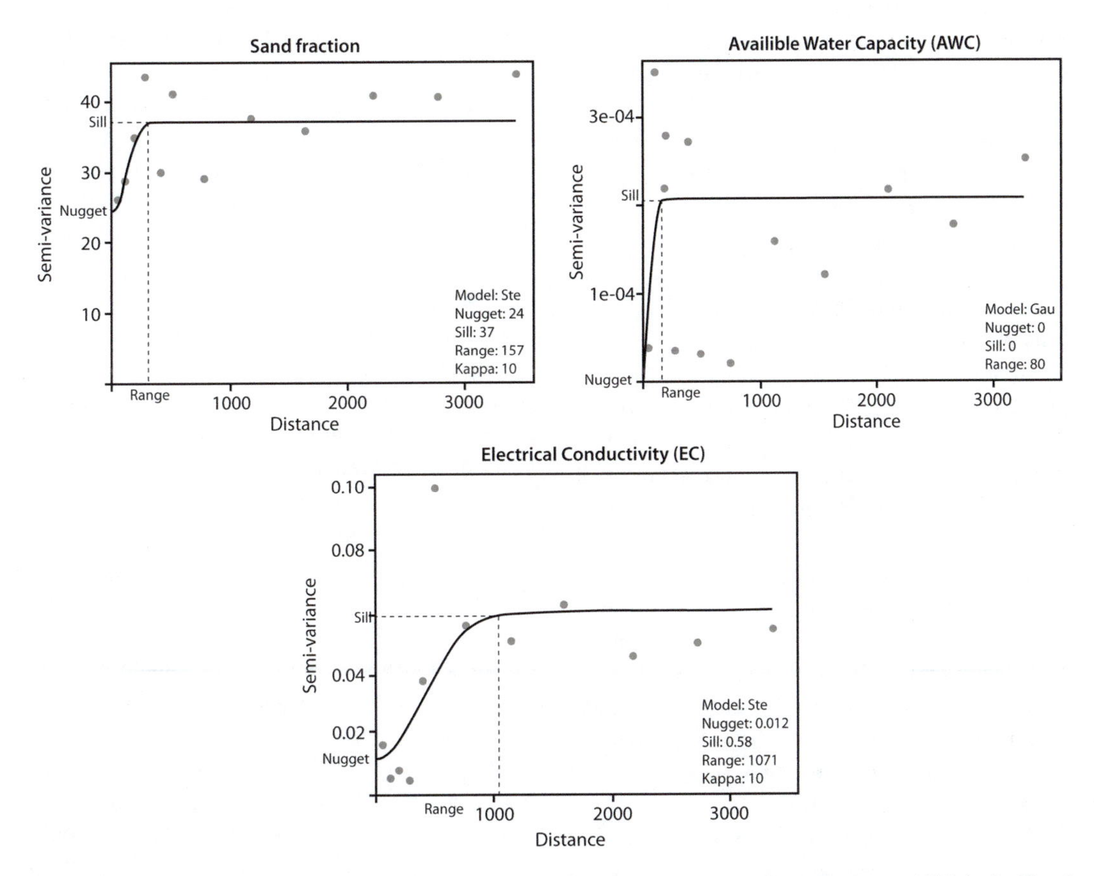

Fig. 4.14 Changes in nugget–sill ratio (N:S) of Available Water Capacity (AWC), sand fraction, and EC, in the Harod catchment, computed with R. Note the very short range in AWC and the small sill and large nugget in the sand fraction. Also note the large dispersion of the observations in the AWC. In the other two properties, the pairs are much closer to the line of the theoretical semivariogram

Table 4.3 Changes in nugget–sill ratio (N:S) of sand fraction content, EC, and Available Water Capacity (AWC), in the Harod catchment. The N:S is higher in the sand fraction, but much lower in EC; n = 130. In AWC, the sill is so low that the semivariogram cannot practically be used for quantification of autocorrelation

Soil property	Nugget	Sill	Range	N:S
Sand fraction	24	37	157	0.65
EC	0.01	0.06	571	0.17
AWC	0	0	80	0

Figure 4.14 and Table 4.3 show the changes in semivariograms and in nugget–sill ratio of Available Water Capacity (AWC), sand fraction, and EC, from point measurements in the Harod catchment.

These three semivariogram characteristics—Nugget, Sill, and Range—are drawn from the theoretical semivariogram, which is a function fitted to the empirical measurements of the empirical semivariogram. The function used for the theoretical semivariogram should support three elements: (1) a small intercept on the Y-axis; (2) a relatively large range; and (3) a relatively large sill. The function that is used to serve as the theoretical semivariogram can be spherical, Gaussian, exponential, or other.

The fitting process of a function as a theoretical semivariogram to the empirical semivariogram is usually conducted using weighted least-squares approximation, with weights proportional to the expected semivariance. This data processing approach is preferable to the computationally more demanding and statistically more generalized least-squares and maximum likelihood techniques (McBratney and Webster 1986). One of the user's main tasks is to choose the most suitable function, among several plausible ones available, to describe the observed variation in the soil properties. In their 1986 paper, McBratney and Webster suggested to select the most suitable model using the *AIC*—the Akaike Information Criterion (Akaike 1973)—as a measure that considers two characteristics: goodness of fit and parsimony.

The *AIC* is widely used in the ecological sciences and was developed as a statistical tool in model selection. In the statistical literature, it is used as an estimator of the relative quality of statistical models that are applied to a prearranged database. Thus, using Eq. 4.25, AIC estimates the quality of each function relative to each of the other functions. The less information the model generates, the higher its quality (see below).

$$AIC = -2 \cdot ln \cdot L + 2 \cdot \mathrm{p}, \tag{4.25}$$

where *L* denotes the maximized likelihood which can be returned by a regression model. *L* is therefore used to assess goodness of fit. Thus, looking at the left expression, if *L* is large (high goodness of fit) the *-2ln* decreases it and vice versa. The p is the number of model parameters that represents a penalty that discourages overfitting, because large number of parameters may improve the goodness of fit but will increase *AIC* due to increased model complexity. So the criterion is a balance between parsimony and goodness of fit while among the various models tested, the model with the lowest *AIC* score is the best one for estimation.

To apply *AIC* in practice, one begins with a set of candidate models (in our case the fitting functions of the empirical semivariogram), then she finds the models' corresponding *AIC* values, with the aim of minimizing the information loss from the most satisfactory fitting function to be used as the theoretical variogram (Webster and McBratney 1989).

However, although the theoretical semivariogram is very extensively used, its three measures of autocorrelation are not the only statistical means of quantifying spatial variation between point measurements of soil properties.

Moran's I

Moran's I (Moran 1950) has been used as an index that quantifies spatial associations in environmental studies, and has been found efficient in examining the autocorrelation of various spatial entities and their properties (Tiefelsdorf and Boots 1995). In practical terms, Moran's I is a correlation coefficient that measures the correlation between two values of the same (soil) property of point measurements at different geographical locations. According to Moran (1950), Moran's I—I(**d**)—is mathematically defined in Eq. 4.26:

$$\mathrm{I}(\mathbf{d}) = \left[\frac{1}{\boldsymbol{W}(\boldsymbol{d})}\right] \cdot \frac{\sum_{\substack{i=1\\ i\neq j}}^{n} \sum_{\substack{j=1\\ j\neq i}}^{n} w_{ij}(d)\cdot(x_i-\overline{x})\cdot(x_j-\overline{x})}{\frac{1}{n}\sqrt{\sum_{i=1}^{n}(x_i-\overline{x})^2}}, \quad (4.26)$$

where ***Wij(d)*** is a matrix of distances that is used to determine if a pair of point measurements is within the distance band d; x_i and x_j are the values of the soil property x at point measurements i and j; and ***W(d)*** is the sum of ***Wij(d)***—the number of pairs of point measurements per distance band (Fortin and Dale 2005). I(**d**) is restricted to the range (−1, + 1), with values of 1 and −1 representing strong relationship (Cliff and Ord 1981). Note that the denominator is simply the standard deviation.

Figure 4.15 illustrates the application of multi-distance Moran's I to the Organic Matter (OM) and Cation Exchange Capacity (CEC) of 130-point measurements of the Harod catchment. The results in the illustration clearly show significant spatial autocorrelation of these two soil properties in the near range (~2.5 km).

As will be discussed in more detail in Chap. 7, quantifying spatial autocorrelation in soil properties—including those affecting water erosion processes—is not always a straightforward task and we therefore need various tools to quantify spatial autocorrelation. Moran's I analysis, as is described in Fig. 4.15, can be a useful addition to quantification of autocorrelation of soil properties with traditional geostatistics, as was done in the case of mapping heavy metal in the soils of Beijing by Huo et al. (2012).

However, despite the successful history of the use of Moran's I in geographical applications—especially in studies related to social geography (Poon and Granger 2003)—it is used less often in the soil sciences. The variogram envelope is a modern alternative that has become a commonly used method of exploring spatial autocorrelation in soils (Annabi et al. 2017)—possibly due to its technical capabilities of simulating random processes.

Variogram envelope

The variogram envelope can be applied using the R function variog.mc.env, which is available in the geoR package. In this function, the envelopes (i.e., the upper and lower boundaries of the possible theoretical semivariogram in the studied site) are

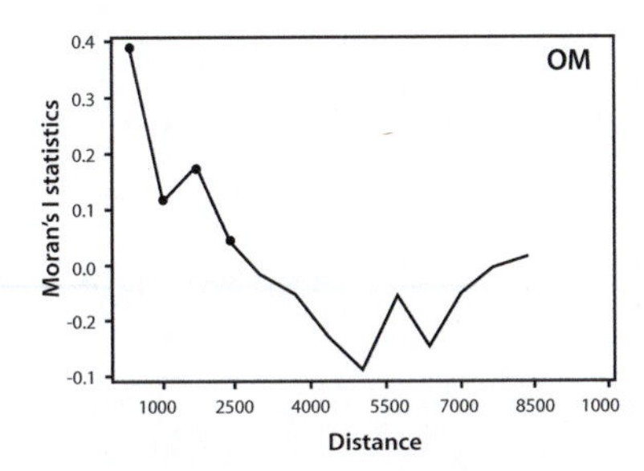

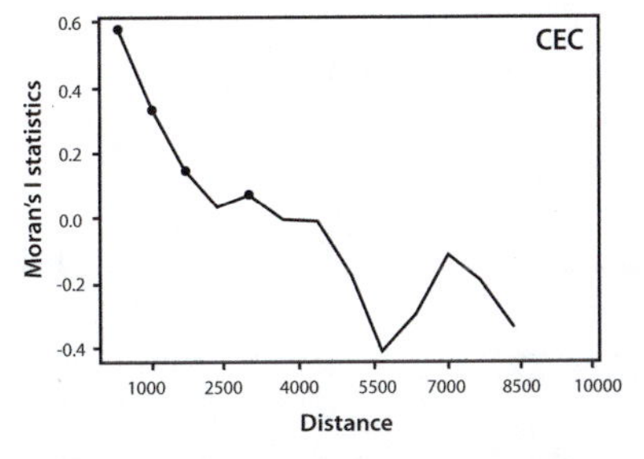

Fig. 4.15 Moran's I correlograms for Organic Matter (OM) and Cation Exchange Capacity (CEC) in the Harod catchment (n = 130). The X-axes represent the space between the point measurements and the Y-axes are Moran's I score. The points indicate spatial autocorrelation. Based on data from: Svoray et al. (2015a), Mapping soil health over large agriculturally important areas. Soil Science Society of America Journal 79(5):1420–1434. (Copyright © by the Soil Science Society of America, Inc.) With permission from Wiley

obtained by permutations, namely, an organized process of changing the order of values in an ordered set of points (Mielke and Berry 2007).

In each permutation of the semivariogram envelope, the data value of each observation point is randomly allocated to an alternative observation point—with a different spatial location—in the site in question. The empirical semivariograms are computed for each permutation, with the same configuration of lag distances and number of lags as was applied for the original—of the real data—semivariogram. The upper and the lower boundaries—or the variogram envelope—are delineated by the maximum and minimum semivariograms at each lag of the simulated data (Gallardo and Paramá 2007). That way, the variogram envelope procedure makes it possible to compute a range of possible semivariograms by randomizing the values of the point measurements.

This process is repeated many times—as a rule of thumb, 100 times. In theory, the total number of repetitions in a permutated process should be *n!*—but since, even in our example of the modest-sized Harod catchment, 130! is an inordinately large number, and the rule of thumb number (100 permutations) should suffice. Each permutation is a round of reassignments of the data point values to the point locations, followed by a computation of the semivariogram (Diggle and Ribeiro 2007). Thus, an envelope of empirical semivariograms is created from permutations of the soil properties data from the point observations, while the location of the point measurements remains constant along the entire process (Fig. 4.16).

After the permutation run ends, the "true' semivariogram is inserted into the envelope between the upper and lower boundaries. If the points at the short distances maintain a large sill and cross the lower boundary of the envelope, the original semivariogram of the real data can be assumed as of significant autocorrelation.

Figure 4.17 illustrates a variogram envelope of a soil health index computed to 130 points distributed in the Harod catchment.

The data in Fig. 4.17 provide evidence of spatial autocorrelation, as the semivariance

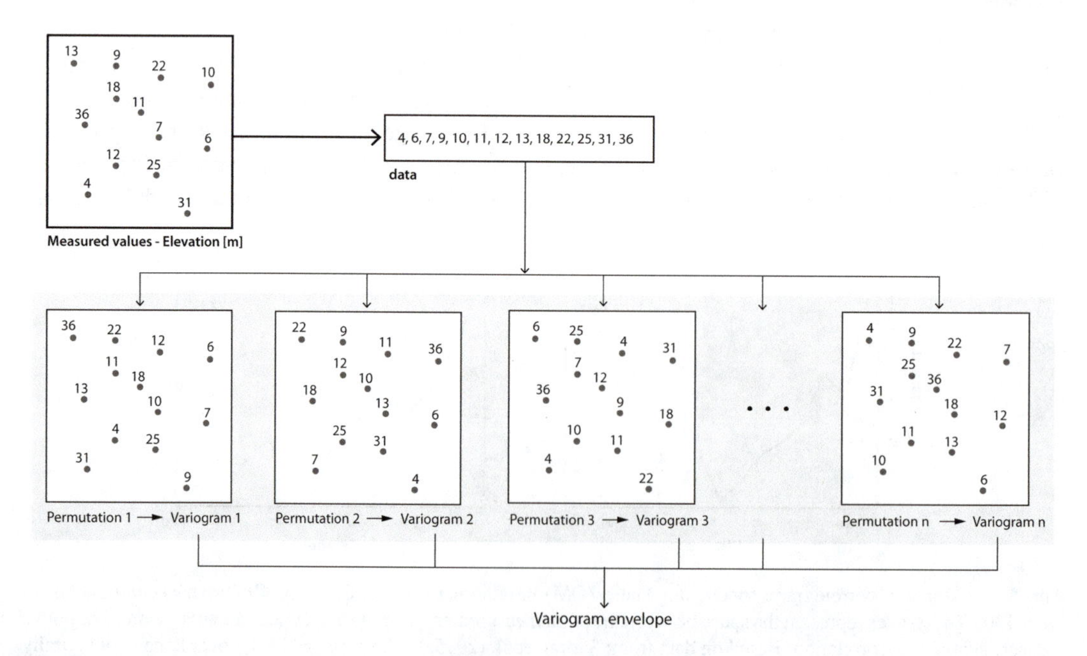

Fig. 4.16 A diagram of permutation process (reassignment of the elevation values to the point locations). The original elevation values—as measured on-site—are presented in the upper left panel. The elevation data values are gathered in a vector data model, and permutated in the various locations in the following permutations, between 1 and n. Often, as a rule of thumb, the permutation process is repeated 100 times

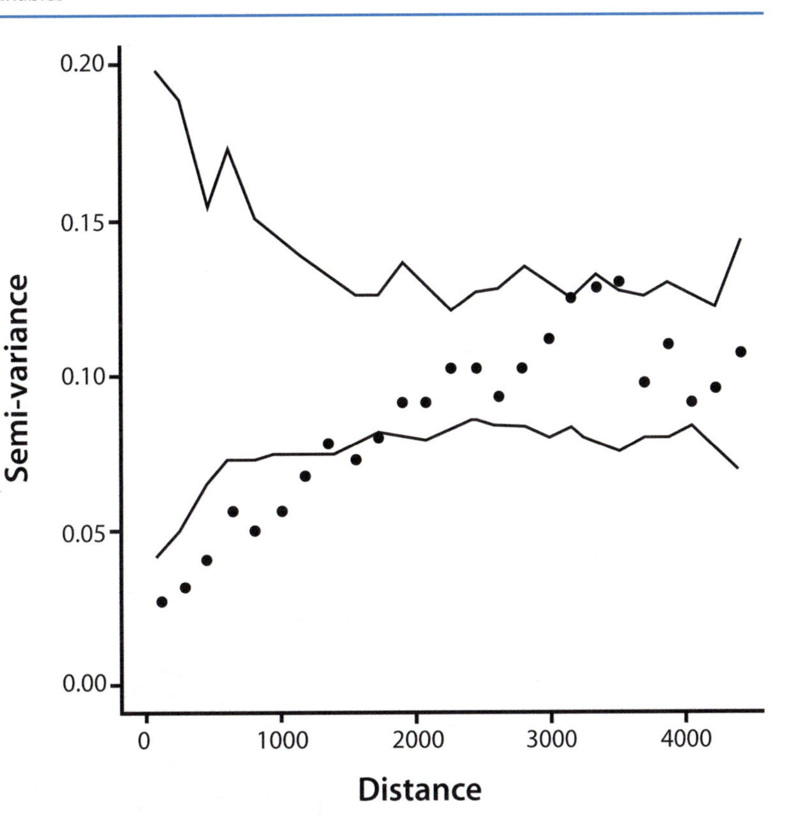

Fig. 4.17 Variogram envelope of soil health index values in the Harod catchment (n = 130), based on data from: Svoray et al. (2015a), Mapping soil health over large agriculturally important areas. Soil Science Society of America Journal 79(5):1420–1434. (Copyright © by the Soil Science Society of America, Inc.) With permission from Wiley. Note that nine pairs cross the lower boundary of the variogram envelope and therefore the data can be assumed to maintain significant autocorrelation

estimates indicate a clear departure from spatial randomness, allowing for an active sill at close distances.

4.2.3 Interpolation of Soil Properties

4.2.3.1 Triangulation and Thiessen Polygons

After collecting data at a sufficient number of points that are spatially distributed in a representative manner, and studying the autocorrelation characteristics of the soil properties, the next step is to interpolate the data from point measurements to estimated spatial surfaces. This step is aimed at the creation of maps of soil properties such as soil texture, soil moisture, or any other soil property that may help in the study of water erosion consequences.

Spatial interpolation techniques are mathematical methods used to estimate the values of surface properties in unknown points, typically forming a continuous surface, based on information from a limited set of measured points (Webster and Oliver 2007). For example, spatial interpolation may be used to estimate soil moisture values in unknown points based on measurements from several TDRs (time domain reflectors) buried in the ground. The data values of the known points are usually stored in point measurements as vector data models and the estimated surface is stored in raster data model with each grid-cell serving as an unknown point and is to be assigned with an estimated value.

In addition to the raster data model, two other geometrical tessellations were used for storing the output surface of spatial interpolation; we will begin this section with a brief description of those two early methods.

The Triangulated Irregular Network (TIN) connects the sample points ("known") to form a network of triangular facets (Peucke et al. 1979). Since its development, TINs have been used in climatological, soil, and hydrological studies as a computationally efficient data model for: representing the rainfall spatial distribution; characterizing geomorphometry of hillslopes and catchments; simulating surface processes; extracting channel networks in detail; and identifying morphological features.

The formation of a TIN layer from points is applied as follows: each of the irregular sampling points is linked to its immediate neighbors by lines to create triangles, such that no triangle contains any other point within its interior area. Because the number of points can be reduced with downscaling, the possible output layers offer multiple resolutions, which translate into computational savings (Ivanov et al. 2004).

Another method, closely related to TIN and extensively used in rainfall studies, is the Thiessen polygons method, based on the Dirichlet tessellation theory (Halls et al. 2001). A Thiessen polygon for point X describes the area where X is the nearest neighbor, i.e., all points in that area are closer to point X than to any other point. The polygon boundaries therefore represent the lines of points that are equidistant, geometrically speaking, to the center points. The process involves creating a TIN layer and drawing a bisector line perpendicular to each of the sides of each triangle. These bisectors form the Thiessen polygons, whose vertices are the intersections of the bisectors (Haggett et al. (1977)—see in Fig. 4.18).

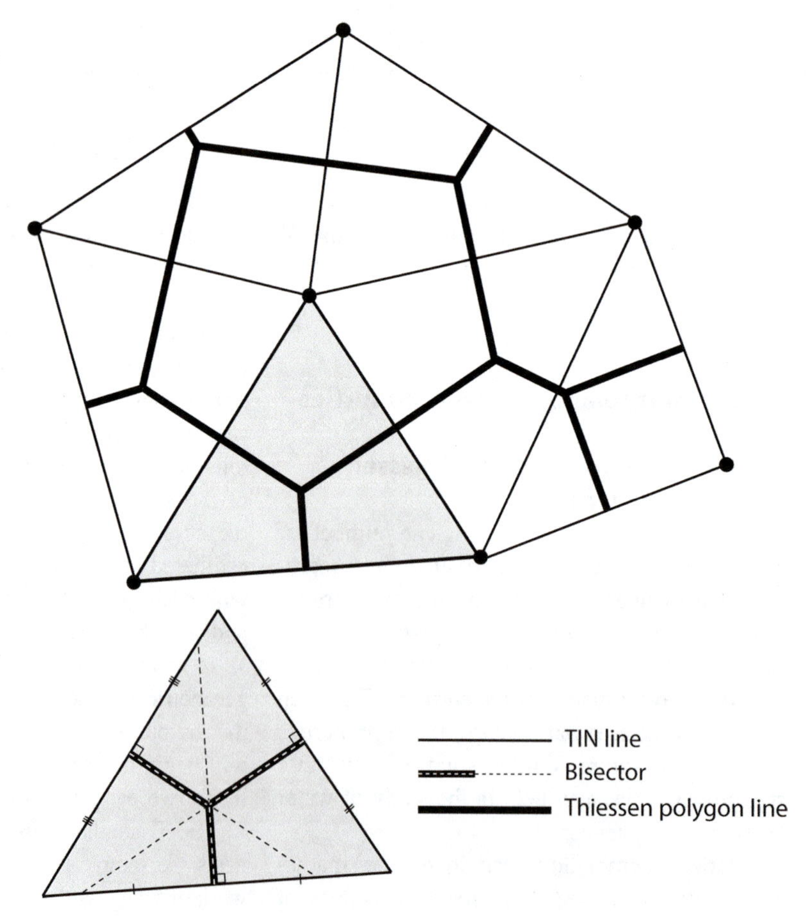

Fig. 4.18 Schematic representation of TIN edges (thin lines) and Thiessen polygons (thick lines). The bisectors of the TIN sides form the Thiessen polygons

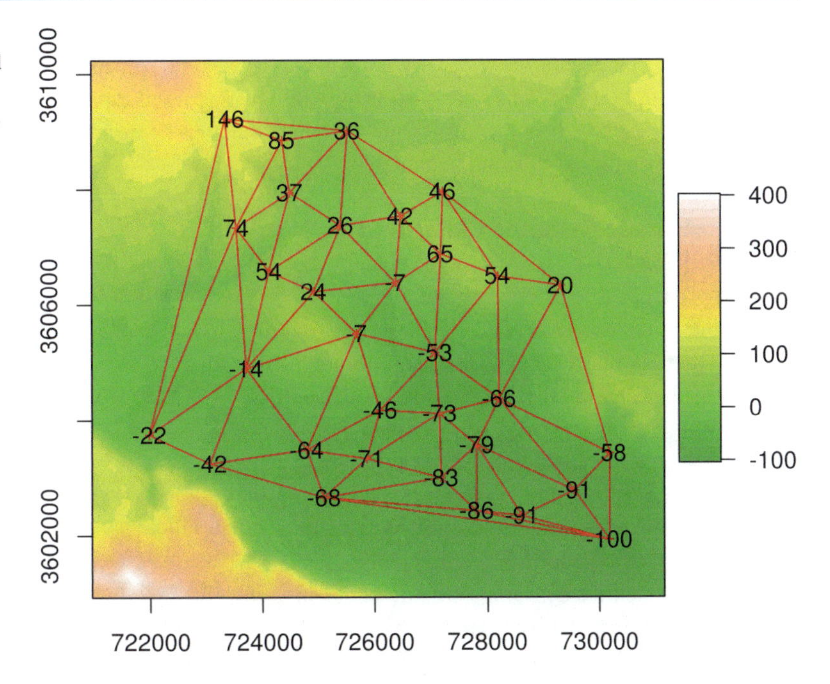

Fig. 4.19 TIN of soil measurements in the Harod catchment: triangles are formed from all the points, such that no point resides within any triangle

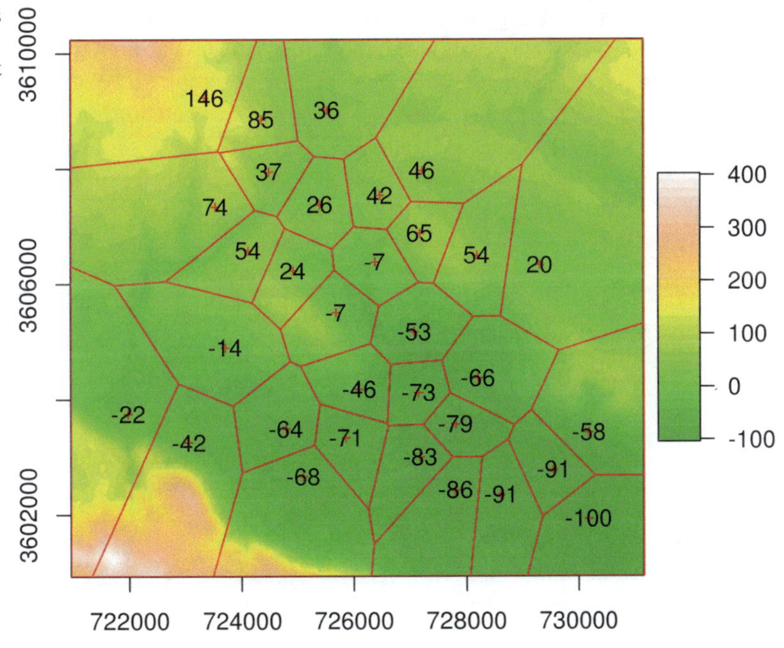

Fig. 4.20 Thiessen polygons of soil measurements in the Harod catchment: the cross at the heart of each Thiessen polygon represents a point measurement

In the context of spatial interpolation, the entire Thiessen polygon is assigned with uniform values based on the particular point the polygon contains, regardless of neighboring points' values. The borders of the polygons are delineated based purely on geometrical considerations, independently of the nature of the spatial continuity characteristics of the (soil) property being measured. Both the TIN and the Thiessen polygons software tools are available in many GIS and statistical software packages. Figures 4.19 and 4.20 represent TIN and Thiessen polygons, respectively, as computed for the Harod catchment data, using the procedure described by Chamberlain and Hollister (2017). This is applied using the R Package 'sf': Simple Features for R (https://cran.r-project.org/web/packages/sf/index.html).

4.2.3.2 Inverse-Distance Weighting (IDW)

Other approaches to converting spatial data from points into facets are offered by a range of methods that take into account the neighborhood effect based on gradually decreasing distance-related weights (as opposed to Thiessen polygons, where the nearest point has a weight of 1, while all other points have a weight of 0). The weights are calculated either deterministically, or based on an examination of the spatial data continuity to identify a spatial structure that, in the following stage, is formulated to extract the weights, in a bid to estimate the soil property values at unknown points.

Inverse-Distance Weighting (IDW) is a deterministic spatial interpolation method that translates into algebraic expression the Tobler concept using a straightforward and easy-to-use computation. The IDW is programmed and available to users in most of the commercial spatial analysis software packages (Lu and Wong 2008). In practical terms, IDW is a weighted average and deterministic multivariate technique that can be computed using Eq. 4.27:

$$\hat{z}(x_0) = \frac{\sum_{i=1}^{n} \lambda_i \cdot z(x_i)}{\sum_{i=1}^{n} \lambda_i}, \tag{4.27}$$

where $\hat{z}(x_0)$ is the estimated value of point x_0, x_i are the measured points $i = 1,2,\ldots n$, and λ_i are the respective weights of the measured points usually assigned as $\lambda_i = d{-}2$, where d is the distance between points x_0 and x_i. The participating $x_i s$ are limited by a search radius, to be determined by the user, to increase computational efficiency. Thus, the weights are not based on the characteristics of spatial variation of a specific soil property as reflected by the autocorrelation depicted in the semivariogram, but are fixed functions of the distance between the points. While this inflexibility may appear to be a drawback of the IDW, when there is a lack of data or limitations in data distribution, for example, IDW can, at times, be a more accurate interpolation method than the kriging family toolkit described below, due to difficulties in correctly estimating autocorrelation

Z	Dis.	W_i	Z×W_i
5	4.3	0.05	0.27
9	3	0.11	1.00
6	2	0.25	1.50
7	3	0.11	0.78
10	1	1.00	10.00
11	1.5	0.44	4.89
7	3	0.11	0.78
7	1	1.00	7.00
4	3	0.11	0.44
5	4.2	0.06	0.28
5	3.5	0.08	0.41
S		3.33	27.3
$\hat{z}(x_0)$ of *A*			**8.21**

Fig. 4.21 Computation—using IDW—of estimated z-values for grid-cells in the output matrix, based on the data from the known points. The search radius of 10 m determines that all points are used in the computation process, and the d^{-2} weight determines that remote points are less influential than nearby points. In this example, based on the attached table, the estimated value for point A is 8.21 m. Accordingly, with the same search radius, B = 6.83 m and C = 6.17 m. This process is repeated for all grid-cells in the raster layer as a whole

characteristics (Erdogan 2009). Figure 4.21 illustrates the application of IDW to three given grid-cells in a raster layer. The search radius is assigned as 10 m which leads to the inclusion of all 11 known points. The distances between all points and the grid-cell A are measured and the values are plugged into Eq. 4.27.

4.2.3.3 Kriging

An alternative to the deterministic and general approach of IDW is the well-known family of interpolation techniques known as kriging. Kriging is an interpolation method which, similarly to IDW, involves a weighted average of the point measurements. However, unlike in IDW, kriging weights are assigned based on the theoretical semivariogram function. Kriging was developed in 1951 by Daniel Krige in his master's thesis, with the aim of offering practical assistance to the mining industry, based on engineering geology principles (Oliver and Webster 2014). Kriging was subsequently formalized by the mathematician Georges Matheron (Matheron 1965) and has been used by many authors in the soil sciences to estimate spatial variation in soil propertiesfrom Webster and Burgess (1983) onward. Kriging is a family of interpolation techniques that can be divided roughly into two main groups.

The first group includes methods that do not use ancillary information, including, first and foremost, the most commonly used simple and robust Ordinary Kriging (OK). OK estimates the values of a given variable (soil property, in our case) in unmeasured points, based on the assumption of a constant but unknown arithmetic mean of the values. Thus, OK can be a suitable method to estimate soil properties when their values are not characterized by any particular spatially varying trend that can be estimated using environmental covariates (see below). Simple Kriging (SK) resembles OK but assumes a known constant estimated mean value which needs to be specified by the researcher. Two more techniques are commonly used in this group: Indicator Kriging (IK), which categorizes the estimated data to binomial coding (1 or 0) based on a threshold value, and Block Kriging (BK) where average estimated values are calculated for polygonal "blocks" rather than for points.

The second group includes interpolation techniques that use secondary variables or ancillary data to enhance estimation certainty by providing a basis for estimations regarding information about the governing processes. One example of the benefit of using ancillary data is the estimation of soil moisture, which can be improved with the use of flow accumulation matrix as a secondary variable due to the large effect of topography on soil moisture content. Universal Kriging (UK) is such a technique, assuming a non-constant trend which is a (linear) function of covariates. UK requires information from a continuous surface covariate, which is available for the entire area, e.g., the X- or Y-coordinates of the layer, to calculate a non-constant trend. Kriging with an external drift (KED) is similar to UK but is based on external data as the trend. Another member of this group is cokriging, which uses information of another variable measured in the same locations as the focal variable, but unavailable for the entire area (e.g., K in soil), to calculate cross variograms. The last example to this group is Regression Kriging (RK), in which the trend is estimated by ordinary least squares, which is unbiased, but does not yield estimates of minimum variance unless the sampling sites have been selected independently at random. Also estimates of the semivariances obtained from residuals from the trend are biased. This is because they depend in a non-linear way on the trend parameters, which are themselves estimated with error. However, it is important to note that previous authors (Hengl et al. 2003) have found that UK, KED, and RK may represent a similar approach that is expected to provide similar estimates with the same input data and parameterization.

A comprehensive discussion of the various kriging techniques could fill a book in its own rightso for more information on the subtleties of the various kriging techniques, the reader is referred to review books that are specialized in geostatistics such as Isaaks and Srivastava (1989). For the purposes of the present discussion, however, we shall briefly review three methods of the commonly used interpolation techniques: OK, cokriging, and Kriging with an External Drift.

OK is the most widely used of the various kriging techniquesboth in general, and in the soil sciences in particular. Its leading assumption is of an unknown constant trend: $\mu(x) = \mu$ unlike simple kriging, where we assume a known constant as a trend, and universal kriging, where we assume a known deterministic function as the trend.

OK is a group of several linear equations (Eqs. 4.28–4.32) which contains semivariances drawn from a fitted semivariogram function. To be implemented, it requires the data extracted from the semivariogram function (Oliver and Webster 2014).

$$\hat{z}(\boldsymbol{x}_0) = \sum_{i=1}^{n} \lambda_i \cdot z(\boldsymbol{x}_i). \quad (4.28)$$

The estimated $\hat{z}$ value at a location $\boldsymbol{x}_0$ is the weighted sum of the measurements in a given surrounding area $\boldsymbol{x}_i$. n denotes the number of known observation points contributing to the estimation of $\hat{z}(\mathbf{x}_0)$. The weights λ_i sum to 1, and the application of OK procedure assumes normal distribution of the data points, and z is the value of the soil property in the location $\boldsymbol{x}_i$. Thus, as an alternative to weighting known points by the square of their inverted distance (as in IDW, OK relies on the spatial distribution of the soil property to determine the weights. This is a more empirically based approach to spatial modeling, because correlation between data points determines the estimated value at an unmeasured point $\boldsymbol{x}_0$.

Another feature of OK that appears in the geostatistical toolkit is the minimized estimation variancealso known as the Kriging Variance (KV)—which was used in the past as a certainty indicator of the Kriging estimations (Isaaks and Srivastava 1989).

Kriging variance $\sigma^2(\mathbf{x}_0)$ for the target point $\boldsymbol{x}_0$ is computed using Eq. 4.29 (Webster and Oliver 2007) where λ_i is the kriging weight; $\gamma(\mathbf{x}_i, \mathbf{x}_0)$ is the semivariance between the ith measurement point and the target point n; and ψ is a Lagrange multiplier.

$$\sigma^2(\boldsymbol{x}_0) = \sum_{i=1}^{n} \lambda_i \cdot \gamma(\boldsymbol{x}_i, \boldsymbol{x}_0) + \psi(\boldsymbol{x}_0). \quad (4.29)$$

The kriging variance was found to increase with increasing nugget effect (Isaaks and Srivastava 1989, p. 305). However, kriging variance is linked to the global semivariogram, which cannot measure reliably the dispersion of local data—only the nature of the soil property that may vary between sites. Thus, once the same data configuration is used, the kriging variance yields the same value, irrespective of the local data values. This is because OK weights and variance are independent of the data values effect (Yamamoto 2000). Similarly, Goovaerts (1997, p. 179) has shown that the variation of data processes by OK is dependent on the formation of the data points and is independent of the data values.

As an alternative to OK, cokriging technique improves estimations by using ancillary data. Cokriging is therefore a multivariate form of OK. Thus, cokriging uses information contained in secondary variables (such as the DEM or its derivatives), to estimate a primary variable such as soil moisture. The estimated $\hat{z}$ value at a location $\boldsymbol{x}_0$ is the weighted sum of the measurements in a given surrounding area. $\lambda_1, \ldots, \lambda_n$ are the calculated weights, with 1 as the known point at spatial location, and a set of n observed data points $z(\boldsymbol{x}_1), z(\boldsymbol{x}_2), \ldots, z(\boldsymbol{x}_n)$, at locations $\boldsymbol{x}_1, \boldsymbol{x}_2, \ldots, \boldsymbol{x}_n$. With $y(s_1'), y(s_2'), \ldots, y(s_m')$, the secondary variables measured at m observations at locations $s_1', s_2', \ldots, s_m'$ where $k_1, \ldots, k_1$ are the calculated weights, such that $z(\boldsymbol{x}_0)$ is the best linear unbiased estimation (Eq. 4.30):

$$\hat{z}(\boldsymbol{x}_0) = \sum_{i=1}^{n} \lambda_i \cdot z(\boldsymbol{x}_i) + \sum_{i=1}^{m} k_i \cdot y(s_i). \quad (4.30)$$

Another common tool for interpolating data with covariates is Kriging with an External Drift (KED). In KED, a covariate is used to model a "drift"—or trend of the dependent variable—in the kriging estimation. In cells without sampling of the dependent variable, the covariate is used to estimate by means of a (linear) function that relates the covariate to the dependent variable. The covariate must maintain autocorrelation in space and must be known at all estimation locations. Kriging with an external drift estimates the value of the soil property ($\widehat{z_{KED}}$) in an unknown point, based on the known points, using a covariate as an additional source of information. KED can add to the conventional kriging procedure a single or several covariates that are usually mapped over wide regions using remotely sensed data. Examples for a drift can be the DEM and its derivatives or a classified image. In many cases, the covariate q is selected so that it has a linear relationship with the estimated soil property in a given point of observation. KED is basically an OK interpolation of the differences between the trend, which is based on the co-variates, and the measured values (i.e., of the trend function residuals), added to the trend to obtain the final estimations (Eqs. 4.31–4.32):

$$\hat{z}_{KED}(\boldsymbol{x}_0) = \sum_{i=1}^{n} \lambda_{KED}(\boldsymbol{x}_0) \cdot z(\boldsymbol{x}_i). \qquad (4.31)$$

For

$$\sum_{i=1}^{n} \lambda_{KED}(\boldsymbol{x}_0) \cdot q_k(\boldsymbol{x}_i) = q_k(\boldsymbol{x}_0); k = 1, \ldots, p, \quad (4.32)$$

where Z is the estimated soil property, the q_k's are the estimator variables, and p is the number of estimators. To apply KED, one must know the value of the mean secondary variable value of the block centered at each experimental location, as well as the mean value of the block to be estimated—while cokriging requires the semivariograms of the two variables as the crossvariogram. KED therefore has the advantage of requiring a less demanding semivariogram analysis than cokriging (Pardo-Igúzquiza 1998). Figure 4.22 is a map of electrical conductivity in the Harod catchment using three interpolation techniques. The two kriging techniqes are very similar but the all three methods show the same northwest–southeast general trend.

To outline the interpolation section, we can assert that mapping soil properties across the entire site in question requires estimations at unmeasured locations (customarily, grid-cells in the raster data model are used for this purpose). Interpolation can be applied using IDWan interpolation technique that, for practical purposes, is the weighted sum of the known points, with the $1{\cdot}d^{-2}$ used as the weight. The weight is then directly and negatively a function of the square of the distance. A more empirical method way of assigning the weights can be applied using the semivariogram that depicts the autocorrelation of the specific soil property. The ordinary kriging interpolation technique takes advantage of the semivariogram to assign the estimation weights. More sophisticated versions of kriging make use of ancillary data, to estimate the soil characteristics with cokriging or universal kriging. One natural candidate for such ancillary data is the DEMwith the effect of topography on soil characteristics due to water-driven

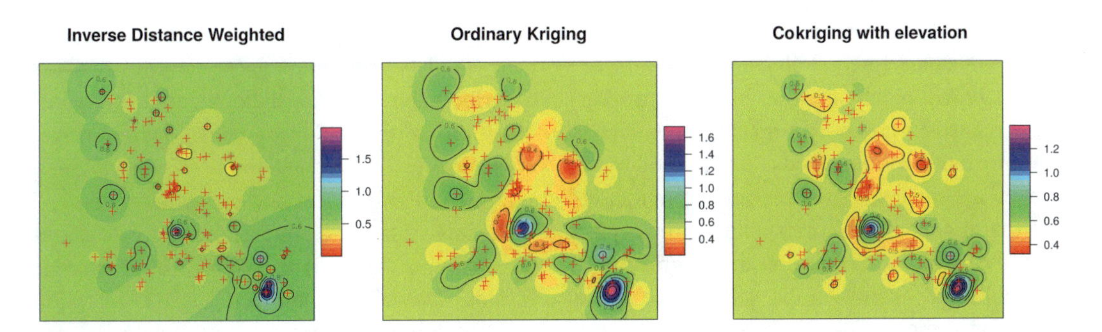

Fig. 4.22 Interpolation of EC using OK, cokriging, based on elevation as a secondary variable, and IDW

propertieshowever, other ancillary data sources such as satellite images and fields parcel data can also be used. Satellite images can serve as a proxy for estimated soil minerals, whereas parcel data may bias soil estimations based on soil cultivation processes.

4.3 Summary

The application of geoinformatics algorithms in the study of water erosion processes in agricultural fields involves accounting for soil variability. We can do this either by dividing the space into discrete units and assume that variation in soil properties within the units is smaller than between the units, or, by modeling the variation within the units, using geostatistical tools.

Four types of discrete units have been identified in the literature and were discussed in this chapter. The first is the pedogeomorphological unit of the hillslope catena that can be delineated using DEM data, based on the procedure proposed by Park et al. (2001) The pedogeomorphological unit leads to the direct analysis and representation of water and sediment flow along the hillslope catena—which is a basic unit of importance in water erosion analyses of catchments. The second is the subcatchment unit that has been thoroughly studied since the days of Robert Horton. The subcatchment unit can also be clearly delineated using a DEM and a pattern-recognition technique. These representations offer the ability to conduct hillslope and catchment scale analyses, based on erosion/deposition processes and empirical evidence. The third level of representation is the parcel (the agricultural field), with data aggregation based on administrative unitswhich are not always uniform from a process-based point of view, but may be very much related to policy and management scenarioswhere cultivation methods that are used by the farmers are mapped based on fields' boundaries. The fourth data model corresponds to the capture of surface conditions (factors) in the grid-cell or pixel unit. Depending on the data source spatial resolution, the grid-cell size ranges from centimeters to tens of meters or even greater—although coarse resolutions are less relevant to catchment scale analyses. The grid-cell/pixel is a very important unit and will be further discussed in Chap. 5 in relation to remotely sensed data, including sub-pixel analysis. It is important to note that the discrete unit approach has been criticized on the grounds that spatial soil variation is often more accurately described through continuous rather than categorical surfaces, and that in most cases the boundaries between discrete soil units are indeterminate.

As an alternative to the four discrete unit types, a set of continuous data analysis tools can be applied to map soil and atmospheric data variation in space, using geostatistical equations. The literature provides us with the necessary procedures to make educated decisions in determining the sample size, locating the sampling points, and testing the autocorrelation of the different properties in space. The arsenal of tools —from random, stratified random, or Webster and McBratney sampling procedures; via variogram envelopes and Moran's I analyses; to the various kriging interpolation techniques (with or without ancillary data)—allows us to map all the spatial variations in the catchment characteristics needed for the water erosion data analyses, with relatively high certainty, given sufficient sample size. The geostatistical models are advantageous for several reasons: they can improve sampling strategy, introduce and quantify the Tobler law in an effort to understand patterns of variation in space, and offer an account of the roles of various climatic and environmental factors in water erosion processes. Conversely, they may suffer from the bull's-eye phenomena (see Chap. 7) and not all variables are spatially continuous. The four data models and the geostatistical analyses described above enable us to account for spatial variation of the soil properties, environmental and climatic factors, and human-induced ones, such as soil cultivation and cropping system.

In this chapter, we presented the procedures for applying these data models in a discrete fashion, as well as the equations necessary for soil sampling, autocorrelation analysis, and interpolation. These data models will serve us in the applications described in Chaps. 5–8, where we also provide further considerations to help the readers decide which of the methods is most suited to their purposes.

Review Questions

1. What are the three different spatial sampling procedures described in this book? Make a list of advantages and disadvantages of each method.
2. Use a DEM to sample 40 points, based on random, stratified random, and matrix techniques. Then interpolate the data using *IDW* and compare the three output layers with the original DEM.
3. What is the *eight flow direction (D8)* method? What is the difference between *D8* and TauDEM by Tarboton (1997)?
4. Compute the steepest hillslope gradient downslope for the center grid-cell in these two matrices of topographic elevation [m].

10	7	11	18	30	21
9	11	10	22	30	29
6	8	10	22	28	25

5. Compute *TWI* and *TCI* to the below database and compare between the two outcomes.

Contributing area a_i					Hillslope gradient S_i				
10	12	12	16	22	6	6	7	8	9
9	13	15	18	21	5	7	6	7	8
11	12	16	18	23	6	6	7	8	7
10	15	16	19	25	7	5	5	7	8
11	14	18	20	24	6	6	6	9	9
10	13	16	21	24	5	7	7	8	10

6. What is the Park et al. model? What geomorphometric principle is it based on? What hillslope units does it divide the landscape into, and what topographic factors is the division based on?
7. Which of the clorpt model components can be easily mapped in catchment sub-units, and represent subcatchments at risk to your catchment? Propose clorpt variables and map them within the subcatchment, using zonal statistics.
8. What is meant by spatial autocorrelation? Why is it so important in point data interpolation? What kind of data can be interpolated based on autocorrelation tests?
9. What is the theoretical difference between *IDW* and kriging? How is it expressed in the interpolation equations? Which of the two is more valuable in mapping the hillslope topography at your site?
10. What is the role of the drift in KED? Why can it be useful in estimating the data values at the unknown points? Give an example from your own real world.
11. How can you use in conjunction the three measures (N:S, *Moran's I* and variogram envelope) to analyze autocorrelation in space in your site? Compute the three measures and explain your results.
12. What is the *AIC* model? How is it being applied? And for what purpose is it being used in the studies of spatial autocorrelation?
13. What is a TIN, and how can it be created from point data? How do you create a Thiessen polygon layer from a TIN? Exemplify from a point data.

Strategies to address the Review Questions

1. Section 4.2.1.2 details the three spatial sampling procedures:
    - *Random sampling*: The location of each sample point is chosen at random and unbiased fashion. Any point location in the population has the same probability to be chosen. Pro: Unbiased; Con: Requires many random points to represent the irregularities in soil properties.
    - *Stratified random*: The study area is divided into groups of similar conditions while in each group the population is randomly sampled. Pro: All landscape units of interest are proportionally represented; Con: The methods suffer some bias (Fig. 4.12).

- *Matrix sampling*: A "net" is overlaid the study area and each intersection of the net is sampled. Pro: Efficient areal cover; Con: Each point is predetermined by the previous points due to the rigid matrix structure.

2. You have to sample the DEM with point data using the three methods. Then you use *IDW* to predict three DEMs from the three different point data layers. Application of *IDW* by solving Eq. 4.27 is described in Sect. 4.2.3.2. Once you come up with three predicted DEMs, you can compare them with the original DEM using RMSE index.
3. The *eight flow direction matrix (D8)* is a method to compute characteristics of drainage structure (Sect. 4.1.1.1). *D8* assumes that water flows from a given grid-cell to one of eight neighbors in the raster data model, based on the steepest hillslope gradient premise. Tarboton (1997), see Eq. 4.2, advanced the D8 approach by accounting the inherent premise that water flows from one grid-cell to another at an angle of 45°. He did so by computing contributing areas and flow direction by directing the flow downslope along the steepest hillslope gradient on the right three-cornered planes located around the center of each grid-cell.
4. Computing the steepest hillslope gradient for the center grid-cell S_{D8} is done using Eq. 4.2. z_8 is the center grid-cell elevation = 11 in the left matrix and 30 in the right matrix. z_i are the elevation of the surrounding grid-cells, x_i is the distance between a given grid-cell and the central grid-cell. We compute the gradient eight times and identify the maximum gradient.

$$\text{Left matrix}: S_{D8} = max_{i=1..8}\frac{z_8 - z_i}{x_i} = \frac{11-7}{1} = 4m$$

$$\text{Right matrix}: S_{D8} = max_{i=1..8}\frac{z_8 - z_i}{x_i} = \frac{30-18}{\sqrt{2}} = 8.485m$$

5. To compute *TWI* we use Eq. 4.10 and to compute *TCI* we need to use Eq. 4.9. The matrices provide data on contributing area a_i and hillslope gradient S_i. The surface curvature C_s can be computed using Eq. 4.8 or by using readily available function in GIS software.
6. Park model is described in Sect. 4.1.1.1. In a nutshell, the model is based on the geomorphometric principle that each point on the landscape can be assigned to a unit that has distinct soil properties. Park et al. reduced the nine-unit soil-landscape model (Conacher and Dalrymple 1977) down to six units: interfluve, shoulder (seepage slope and convex creep slope), backslope (free face and transportational midslope), footslope (colluvial footslope), toeslope (alluvial toeslope), and channel (channel wall and channel bed). That is because three of the original units are too specific while comparing the hillslope processes, and also they are too similar to be separable using the most currently available DEM. The topographic factors that the model is based on are the contributing area, the hillslope gradient, and the hillslope curvature.
7. The clorpt model components are described in Sect. 1.1.2.1 and in Eq. 1.2. They include rainfall characteristics, organisms, relief, parent material, and time. As described in Sect. 4.1, relief is a good example of a clorpt component that can be mapped in subcatchments to estimate soil loss risk. Also, rainfall characteristics from meteorological radar and vegetation by *NDVI* can be good examples. Mapping can be done based on variables at your discretion using readily available GIS software.
8. Spatial autocorrelation is used in point data interpolation to determine the level of continuity in space of the soil property studied. Section 4.2.2 provides the definition of the mathematical and statistical quantification of how well the value of a (soil) property in a given location $Z(\boldsymbol{x})$ is associated with the (soil) property values at nearby locations Z

(x + h). This quantification is applied within the boundaries of a geographic area, such as a catchment, a hillslope, or an agricultural field. Continuous data, based on autocorrelation tests, can be interpolated in space to predict the soil characteristics in unknown points.

9. *IDW* and kriging are discussed in Sects. 4.2.3.2–4.2.3.3. The weights of *IDW* are based solely on the distance between the points (with $\lambda_i = \frac{1}{d^x}$). In Kriging, however, the weights are assigned based on characteristics of the spatial variation of a specific soil property as reflected by the autocorrelation depicted in the semivariogram. Whether *IDW* or Kriging are more certain estimators depends on the data analyzed. For example, when there is a lack of data, *IDW* can be more accurate than the kriging family toolkit due to difficulties in correctly estimating the autocorrelation characteristics that assign the weights.
10. The drift in KED is the trend of the dependent variable (soil property) in the kriging estimation (Sect. 4.2.3.2, Eq. 4.30). KED allows to predict the value of the soil property ($\widehat{z}_{KED}$) in an unknown point, based on the known points, using a covariate (the drift) as an additional source of information for better prediction. A common example for a drift is a DEM and its derivatives. Remotely sensed images can also be used as a drift.
11. To decide which of these three measures is the most suitable for our site, we need to apply the three measures to our data. This requires to sample a minimum number of point measurements of the soil property of interest and the application of the three methods as described in Sect. 4.2.2. Then we can explore the outcomes and determine the level of the soil property continuity based on the agreement between the three methods.
12. The *AIC* is a statistical tool for model selection (Sect. 4.2.2, Eq. 4.25). It is used as an estimator of the relative quality of statistical models that were applied to a prearranged database. To apply *AIC*, we begin with a set of candidate models, then find the model's corresponding *AIC* values, with the aim of minimizing the information loss from the true model. *AIC* was demonstrated in this book for studies of fitting a theoretical semivariogram to an empirical semivariogram.
13. *Triangulated Irregular Network (TIN)* (Sect. 4.2.3.1) is a method that connects the sample known points to form a network of triangular facets. To create a TIN from point data, each of the irregular sampling points is linked with its immediate neighbors by lines that create triangles that do not contain any other point. The irregular output domain offers multiple resolutions, which translate into computational savings, as the number of points can be reduced with downscaling. Creating a Thiessen polygon from a TIN is done by drawing a bisector line perpendicular to each of the sides of each triangle. These bisectors form the Thiessen polygons, whose vertices are the intersections of the bisectors (Fig. 4.18).

References

Akaike H (1973) Introduction to Akaike: information theory and an extension of the maximum likelihood principle, 267–281. Springer, New York, NY. https://doi.org/10.1007/978-1-4612-0919-5_37

Annabi M, Raclot D, Bahri H et al (2017) Spatial variability of soil aggregate stability at the scale of an agricultural region in Tunisia. Catena 153:157–167. https://doi.org/10.1016/j.catena2017.02.010

Arnaud P, Bouvier C, Cisneros L et al (2002) Influence of rainfall spatial variability on flood prediction. J Hydrol 260(1):216–230. https://doi.org/10.1016/S0022-1694(01)00611-4

Barling R, Moore I, Grayson R (1994) A quasi-dynamic wetness index for characterizing the spatial distribution of zones of surface saturation and soil water content. Water Resour Res 30(4)

Boots B, Csillag F (2006) Categorical maps, comparisons, and confidence. J Geogr Syst 8:109–118. https://doi.org/10.1007/s10109-006-0018-9

Burrough PA, Macmillan RA, Deursen W (1992) Fuzzy classification methods for determining land suitability

from soil profile observations and topography. J Soil Sci 43(2):193–210. https://doi.org/10.1111/j.1365-2389.1992.tb00129.x
Campbell JB (1979) Spatial variability of soils. Ann Assoc Am Geogr 69(4):544–556. https://doi.org/10.1111/j.1467-8306.1979.tb01281.x
Carson MA, Kirkby MJ (1972) Hillslope form and process. Cambridge University Press, New York
Chamberlain S, Hollister J (2017) lawn: Client for 'Turfjs' for 'Geospatial' Analysis. https://CRAN.R-project.org/package=lawn. R package version 0.4.2 2017
Cliff AD, Ord K (1981) Spatial processes: models & applications. Taylor & Francis, London
Cline MG (1944) Principles of soil sampling. Soil Sci 58 (4):275–288. https://doi.org/10.1097/00010694-194410000-00003
Conacher AJ, Dalrymple JB (1977) The nine unit landsurface model and pedogeomorphic research. Geoderma 18(1):127–144. https://doi.org/10.1016/0016-7061(77)90087-8
Costa C, Papatheodorou EM, Monokrousos N et al (2015) Spatial variability of soil organic C, inorganic N and extractable P in a Mediterranean grazed area. Land Degrad Dev 26(2):103–109. https://doi.org/10.1002/ldr.2188
Curran PJ (1988) The semivariogram in remote sensing: an introduction. Remote Sens Environ 24(3):493–507. https://doi.org/10.1016/0034-4257(88)90021-1
Delmelle EM (2014) Spatial sampling in handbook of regional science, pp 1385–1399
Denton OA, Aduramigba-Modupe VO, Ojo AO et al (2017) Assessment of spatial variability and mapping of soil properties for sustainable agricultural production using geographic information system techniques (GIS). Cogent Food Agric 3(1):1–12. https://doi.org/10.1080/23311932.2017.1279366
Diggle P, Ribeiro PJ (2007) Model-based geostatistics. Springer, New York
Downs PW, Gregory KJ, Brookes A (1991) How integrated is river basin management? Environ Manage 15(3):299–309. https://doi.org/10.1007/BF02393876
Erdogan S (2009) A comparision of interpolation methods for producing digital elevation models at the field scale. Earth Surf Proc Land 34(3):366–376. https://doi.org/10.1002/esp.1731
Famiglietti JS, Ryu D, Berg AA et al (2008) Reply to comment by H. Vereecken et al. on Field observations of soil moisture variability across scales. Water Resour Res 44(12):1–16. https://doi.org/10.1029/2008WR007323
Fan L, Lehmann P, Or D (2016) Effects of soil spatial variability at the hillslope and catchment scales on characteristics of rainfall-induced landslides. Water Resour Res 52(3):1781–1799. https://doi.org/10.1002/2015WR017758
Ferreira V, Panagopoulos T, Andrade R et al (2015) Spatial variability of soil properties and soil erodibility in the Alqueva reservoir watershed. Solid Earth 6 (2):383–392
Fisher P (1997) The pixel: a snare and a delusion. Int J Remote Sens 18(3):679–685
Fortin MJ, Dale MRT (2005) Spatial analysis. Cambridge University Press, Cambridge
Gallardo A, Paramá R (2007) Spatial variability of soil elements in two plant communities of NW Spain. Geoderma 139(1):199–208. https://doi.org/10.1016/j.geoderma.2007.01.022
Goovaerts P (1997) Geostatistics for natural resources evaluation. Oxford University Press, New York
Goovaerts P (1999) Geostatistics in soil science: state-of-the-art and perspectives. Geoderma 89(1):1–45. https://doi.org/10.1016/S0016-7061(98)00078-0
Haggett P, Cliff AD, Frey A (1977) Locational analysis in human geography. Arnold, London
Halls PJ, Bulling M, White PCL et al (2001) Dirichlet neighbours: revisiting Dirichlet tessellation for neighbourhood analysis. Comput Environ Urban Syst 25 (1):105–117. https://doi.org/10.1016/S0198-9715(00)00035-1
Hengl T, Heuvelink G, Stein A (2003) Comparison of kriging with external drift and regression-kriging. Technical Note, ITC.
Hengl T, Reuter HI (2009) Geomorphometry. Elsevier, Amsterdam
Hewitt AE (1993) Predictive modelling in soil survey. Soil and Fertilizers 56:305–314
Hillel D (1998) Environmental soil physics. Academic Press, US
Horton RE (1932) Drainage-basin characteristics. Eos Trans AGU 13(1):350–361. https://doi.org/10.1029/TR013i001p00350
Horton RE (1945) Erosional development of streams and their drainage basins: hydrophysical approach to quantitative morphology. GSA Bull 56(3):275. https://doi.org/10.1130/0016-7606(1945)56[275:EDOSAT]2.0.CO;2
Huang M, Zettl JD, Lee Barbour S et al (2016) Characterizing the spatial variability of the hydraulic conductivity of reclamation soils using air permeability. Geoderma 262:285–293. https://doi.org/10.1016/j.geoderma.2015.08.014
Hudson BD (1992) The soil survey as paradigm-based science. Soil Sci Soc Am J 56(3):836–841. https://doi.org/10.2136/sssaj1992.03615995005600030027x
Huo XN, Li H, Sun DF et al (2012) Combining geostatistics with Moran's I analysis for mapping soil heavy metals in Beijing, China. Int J Environ Res Public Health 9(3):995
Hurst MD, Mudd SM, Walcott R et al (2012) Using hilltop curvature to derive the spatial distribution of erosion rates. J Geophys Res Earth Surf 117(2):1–19. https://doi.org/10.1029/2011JF002057
Isaaks EH, Srivastava RM (1989) Applied geostatistics. Oxford Univ. Pr, New York 561
Ivanov VY, Vivoni ER, Bras RL et al (2004) Catchment hydrologic response with a fully distributed triangulated irregular network model. Water Resour Res 40 (11):W11102. https://doi.org/10.1029/2004WR003218

Karnieli A, Gilad U, Ponzet M et al (2008) Assessing land-cover change and degradation in the Central Asian deserts using satellite image processing and geostatistical methods. J Arid Environ 72:2093–2105

Kok K, Kim JC (2019) Identification of vulnerable regions to soil loss under the dynamic saturation process. Sci Total Environ 659:1209–1223. https://doi.org/10.1016/j.scitotenv.2018.12.398

Kokulan V, Akinremi O, Moulin AP et al (2018) Importance of terrain attributes in relation to the spatial distribution of soil properties at the micro scale: a case study. Can J Soil Sci 98(2):285–293

Korres W, Reichenau TG, Fiener P et al (2015) Spatio-temporal soil moisture patterns–a meta-analysis using plot to catchment scale data. J Hydrol 520:383–392. https://doi.org/10.1016/j.jhydrol.2014.11.042

Krajewski WF, Smith JA (2002) Radar hydrology: rainfall estimation. Adv Water Resour 25(8):1387–1394. https://doi.org/10.1016/S0309-1708(02)00062-3

Lin H (2003) Hydropedology: bridging disciplines, scales, and data. Vadose Zone J 2(1):1–11. https://doi.org/10.2113/2.1.1

Lu GY, Wong DW (2008) An adaptive inverse-distance weighting spatial interpolation technique. Comput Geosci 34(9):1044–1055. https://doi.org/10.1016/j.cageo.2007.07.010

Malanson G (1999) Considering complexity. Ann Assoc Am Geogr 89(4):746–753. https://doi.org/10.1111/0004-5608.00174

Malone BP, Odgers NP, Stockmann U et al (2018) Digital mapping of soil classes and continuous soil properties in Pedometrics. Springer, Amsterdam

Martz LW, Garbrecht J (1992) Numerical definition of drainage network and subcatchment areas from digital elevation models. Comput Geosci 18(6):747–761. https://doi.org/10.1016/0098-3004(92)90007-E

Matheron G (1965) Les variables régionalisées et leur estimation: Une application de la théorie des fonctions aléatoires aux sciences de la nature

McBratney AB, Mendonça Santos ML, Minasny B (2003) On digital soil mapping. Geoderma 117(1):3–52. https://doi.org/10.1016/S0016-7061(03)00223-4

McBratney AB, Odeh IOA, Bishop TFA et al (2000) An overview of pedometric techniques for use in soil survey. Geoderma 97(3):293–327. https://doi.org/10.1016/S0016-7061(00)00043-4

McBratney AB, Webster R (1983) How many observations are needed for regional estimation of soil properties? Soil Sci 135(3):177–183. https://doi.org/10.1097/00010694-198303000-00007

McBratney AB, Webster R (1986) Choosing functions for semi-variograms of soil properties and fitting them to sampling estimates. J Soil Sci 37(4):617–639. https://doi.org/10.1111/j.1365-2389.1986.tb00392.x

McKenzie NJ, Ryan PJ (1999) Spatial prediction of soil properties using environmental correlation. Geoderma 89(1):67–94. https://doi.org/10.1016/S0016-7061(98)00137-2

Mielke P, Berry K (2007) Permutation methods: a distance function approach. Springer, Amsterdam

Milne G (1936) Normal erosion as a factor in soil profile development. Nature 138(3491):548–549. https://doi.org/10.1038/138548c0

Mirzaee S, Ghorbani-Dashtaki S, Mohammadi J et al (2016) Spatial variability of soil organic matter using remote sensing data. Catena 145:118–127. https://doi.org/10.1016/j.catena.2016.05.023

Moore ID, Gessler PE, Nielsen GA et al (1993) Soil attribute prediction using terrain analysis. Soil Sci Soc Am J 57(2):443–452

Moran PA (1950) Notes on continuous stochastic phenomena. Biometrika 37(1):17–23. https://doi.org/10.2307/21332142.JSTOR2332142

Morin E, Gabella G (2007) Radar-based quantitative precipitation estimation over Mediterranean and dry climate regimes. J Geophysl Res Atmos 112(D20):1–13. https://doi.org/10.1029/2006JD008206

Morin E, Krajewski WF, Goodrich DC et al (2003) Estimating rainfall intensities from weather radar data. J Hydrometeorol 4(5):782–797. https://journals.ametsoc.org/view/journals/hydr/4/5/1525-7541_2003_004_0782_erifwr_2_0_co_2.xml

Mulder VL, Lacoste M, Richer-de-Forges AC et al (2016) GlobalSoilMap France: high-resolution spatial modelling the soils of France up to two meter depth. Sci Total Environ 573:1352–1369. https://doi.org/10.1016/j.scitotenv.2016.07.066

Nettleton WD, Brasher BR, Borst G (1991) The taxadjunct problem. Soil Sci Soc Am J 55(2):421–427. https://doi.org/10.2136/sssaj1991.03615995005500020022x

O'Callaghan JF, Mark DM (1984) The extraction of drainage networks from digital elevation data. Comput Vision, Graphics Image Process 28(3):323–344. https://doi.org/10.1016/S0734-189X(84)80011-0

Oliver MA, Webster R (2014) A tutorial guide to geostatistics: computing and modelling variograms and kriging. Catena 113:56–69. https://doi.org/10.1016/j.catena.2013.09.006

Pardo-Igúzquiza E (1998) Comparison of geostatistical methods for estimating the areal average climatological rainfall mean using data on precipitation and topography. Int J Climatol 18(9):1031–1047. https://doi.org/10.1002/(SICI)1097-0088(199807)18:9<1031::AID-JOC303>3.0.CO;2-U

Park SJ, McSweeney K, Lowery B (2001) Identification of the spatial distribution of soils using a process-based terrain characterization. Geoderma 103(3):249–272. https://doi.org/10.1016/S0016-7061(01)00042-8

Paterson S, Minasny B, McBratney A (2018) Spatial variability of Australian soil texture: a multiscale analysis. Geoderma 309:60–74. https://doi.org/10.1016/j.geoderma.2017.09.005

Peucke T, Fower JR, Little JJ (1979) The triangulated irregular network (Proceeding), pp 199–207. https://doi.org/10.1145/800249

Phillips JD (2017) Soil complexity and pedogenesis. Soil Sci 182(4):117–127. https://doi.org/10.1097/SS.0000000000000204

Poon SH, Granger CWJ (2003) Forecasting volatility in financial markets: a review. J Econ Lit 41(2):478–539. https://doi.org/10.1257/jel.41.2.478
Renschler CS (2003) Designing geo-spatial interfaces to scale process models: the GeoWEPP approach. Hydrol Process 17(5):1005–1017. https://doi.org/10.1002/hyp.1177
Sherpa SR, Wolfe DW, van Es HM (2016) Sampling and data analysis optimization for estimating soil organic carbon stocks in agroecosystems. Soil Sci Soc Am J 80 (5):1377–1392. https://doi.org/10.2136/sssaj2016.04.0113
Silver M, Svoray T, Karnieli A et al (2020) Improving weather radar precipitation maps: a fuzzy logic approach. Atmos Res 234:104710
Strahler AN (1952) Dynamic basis of geomorphology. GSA. Bulletin 63(9):923. https://doi.org/10.1130/0016-7606(1952)63[923:DBOG]2.0.CO;2
Strudley MW, Green TR, Ascough JC (2008) Tillage effects on soil hydraulic properties in space and time: state of the science. Soil Tillage Res 99(1):4–48. https://doi.org/10.1016/j.still.2008.01.007
Svoray T, Ben-Said S (2010) Soil loss, water ponding and sediment deposition variations as a consequence of rainfall intensity and land use: a multi-criteria analysis. Earth Surf Proc Land 35(2):202–216. https://doi.org/10.1002/esp.1901
Svoray T, Hassid I, Atkinson PM et al (2015) Mapping soil health over large agriculturally important areas. Soil Sci Soc Am J 79(5):1420–1434. https://doi.org/10.2136/sssaj2014.09.0371
Svoray T, Karnieli A (2011) Rainfall, topography and primary production relationships in a semiarid ecosystem. Ecohydrology 4(1):56–66. https://doi.org/10.1002/eco.123
Svoray T, Levi R, Zaidenberg R et al (2015) The effect of cultivation method on erosion in agricultural catchments: integrating AHP in GIS environments. Earth Surf Proc Land 40(6):711–725. https://doi.org/10.1002/esp.3661
Svoray T, Markovitch H (2009) Catchment scale analysis of the effect of topography, tillage direction and unpaved roads on ephemeral gully incision. Earth Surf Proc Land 34(14):1970–1984. https://doi.org/10.1002/esp.1873
Tarboton DG (1997) A new method for the determination of flow directions and upslope areas in grid digital elevation models. Water Resour Res 33(2):309–319. https://doi.org/10.1029/96wr03137
Tarboton DG, Bras RL, Rodriguez-Iturbe I (1991) On the extraction of channel networks from digital elevation data. Hydrol Process 5(1):81–100. https://doi.org/10.1002/hyp.3360050107
Tarboton DG, Bras RL, Rodriguez-Iturbe I (1992) A physical basis for drainage density. Geomorphology 5 (1):59–76. https://doi.org/10.1016/0169-555X(92)90058-V
Tiefelsdorf M, Boots B (1995) The exact distribution of Moran's I. Environ Plan A 27(6):985–999. https://doi.org/10.1068/a270985
Tobler W (1970) A computer movie simulating urban growth in the Detroit region. Econ Geogr 46:234–240
Uyan M (2016) Determination of agricultural soil index using geostatistical analysis and GIS on land consolidation projects: A case study in Konya/Turkey. Comput Electron Agric 123:402–409. https://doi.org/10.1016/j.compag.2016.03.019
Webster R, Burgess TM (1983) Spatial variation in soil and the role of kriging. Agric Water Manag 6(2):111–122. https://doi.org/10.1016/0378-3774(83)90003-3
Webster R, McBratney AB (1989) On the Akaike information criterion for choosing models for variograms of soil properties. J Soil Sci 40(3):493–496. https://doi.org/10.1111/j.1365-2389.1989.tb01291.x
Webster R, Oliver MA (2007) Geostatistics for environmental scientists, 2, ed. Wiley, England
Wilson JP, Gallant C (2000) Terrain analysis. Wiley, New York
Yamamoto J (2000) An alternative measure of the reliability of ordinary kriging estimates. Math Geol 32(4):489–509. https://doi.org/1007577916868
Young R, Wiersma L (1973) The role of rainfall impact in soil detachment and transport. Water Resour Res 9(6): 1629–1636
Zhu TX, Cai QG, Zeng BQ (1997) Runoff generation on a semi-arid agricultural catchment: field and experimental studies. J Hydrol 196(1):99–118. https://doi.org/10.1016/S0022-1694(96)03310-0
Zwieback S, Dorigo W, Wagner W (2013) Estimation of the temporal autocorrelation structure by the collocation technique with an emphasis on soil moisture studies. Hydrol Sci J 58(8):1729–1747. https://doi.org/10.1080/02626667.2013.839876

5 Earth Observations

Abstract

To the soil conservationalist, remote sensing methodologies provide profuse and updated information about soil properties, rock exposure, vegetation attributes, and built infrastructure characteristics. This chapter provides an overview of the main characteristics of remotely sensed images, and their use in mapping spatial properties of agricultural catchments. It offers several spectral indices for quantifying soil and vegetation properties, using visible and near-infrared data, by means of algebraic band-ratio methods. Next, it describes soft and hard classification techniques, including pixel-based and object-based classification. Another useful application suggests the use of classified remotely sensed data as a tool for computing and mapping runoff coefficients, as well as channel network data for guiding drainage determination. Finally, in the past decade, the use of high-resolution drones for land use/land cover classification has become common in soil erosion studies. Extracting DEMs and topographic properties has also become an important part in soil degradation analyses; in particular, change detection of soil loss and soil deposition can be used to estimate sediment budgets. The tools reviewed in this chapter further expand the notion that remotely sensed data can be used to provide evidence of the state and dynamics of a given catchment—thereby providing a general framework for depicting and simulating the mechanisms of water erosion in agricultural catchments.

Keywords

Classification · Drainage structure · Drones · Spectral indices · Remote sensing

"*Remote sensing* is the practice of deriving information about the Earth's land and water surfaces using images acquired from an overhead perspective, using electromagnetic radiation in one or more regions of the electromagnetic spectrum, reflected or emitted from the Earth's surface" (Campbell and Wynne 2011, p. 6). Remote Sensing (RS) methodology is also widely applied to observe extraterrestrial bodies—but such applications are beyond the scope of a book on agricultural catchments.

To the soil conservationalist, RS methodologies provide profuse and updated information about soil properties, rock exposure, vegetation attributes and crops physiognomies, agricultural land uses, built infrastructure characteristics, and more. The RS-derived information covers large areas, at high spatial resolution, allows interpretation of reflected radiation in wavelength that the human eye cannot see, and many RS platforms offer high-frequency updates. In applications such as rainfall intensity mapping, crop-cover and yield estimation, repetitive soil moisture assessments, on-site and off-site erosion

T. Svoray, *A Geoinformatics Approach to Water Erosion*,
https://doi.org/10.1007/978-3-030-91536-0_5

damage mapping, land use classification and many more (Atzberger 2013), RS can replace labor-intensive and expensive field visits.

Information is extracted from data, and the data gathered using RS tools is based on quantification of the *radiance* reflected by a given surface, per unit solid angle, per unit projected area [$W \cdot sr^{-1} \cdot m^{-2}$]. The radiation flux is recorded by sensors mounted on platforms such as satellites, airplanes, drones, or tractors/cranes. The sun, or a man-made apparatus, is the radiation source, while the recorded radiance is affected by scattering, absorption, transmission, and reflection of energy within and from the atmosphere (Campbell and Wynne 2011, p. 39). The radiance recorded from the surface is converted into *reflectance data*—namely, the ratio between the amount of energy hitting the surface and the amount of energy leaving it [dimensionless or %]. The reflectance data is stored perpetually in image (2D array) pixels as a *DN* (Digital Number) value—an 8-bit data integer, in the range of 0–255.

As a scientific discipline, therefore, RS is founded first and foremost on known principles of the interaction between spectral radiation and the intertwined earth surface components, under given technical, atmospheric, and environmental conditions. The effect of surface components on spectral reflectance might be governed by the chemical characteristics of the material the surface component is made of, the roughness of its surface, the orientation of the component toward the radiation source, the time of measurement, the wavelengths the recording device is sensitive to, the amount of aerosols in the air and cloud *coverage* in the atmosphere.

The wavelength of the recorded radiance is a major determinant of the interpretation of RS data by the user. The electromagnetic spectrum is broad, and divided into spectral regions (Fig. 5.1). Because spectral regions differ from each other in their sensitivity to various surface characteristics, it is crucial to choose the appropriate spectral region for the specific problem at hand.

Various RS platforms provide useful information in the microwave range—primarily for soil moisture content estimations, aboveground biomass assessments, and soil roughness analysis (Svoray and Shoshany 2003, 2004). However, many of the readily available applications of RS in erosion studies use the infrared bands—in particular, the *NIR* (near infrared) and *SWIR* (Short Wave Infrared)—as well as the *VIS* (visible light data) from RGB cameras. For example, Price (1993) linked the Landsat TM VIS and *NIR* spectral data with soil loss predictions derived from USLE computations in Utah. In this work, the Landsat VIS and *NIR* data was even found as a better predictor of soil erosion than the data collected in the field. In another example, De Jong et al. (1999) combined successfully multi-temporal Landsat TM data to account for vegetation properties in a spatially explicit water erosion model applied to an agricultural catchment in eastern Sicily. These are only two successful examples while many other papers have used RS data for different purposes in studying agricultural water erosion. The review study of Mulla (2013) summarized the more widespread uses of RS data for precision agriculture applications that include: Monitoring soil *organic matter*, assessments of parent material exposure,

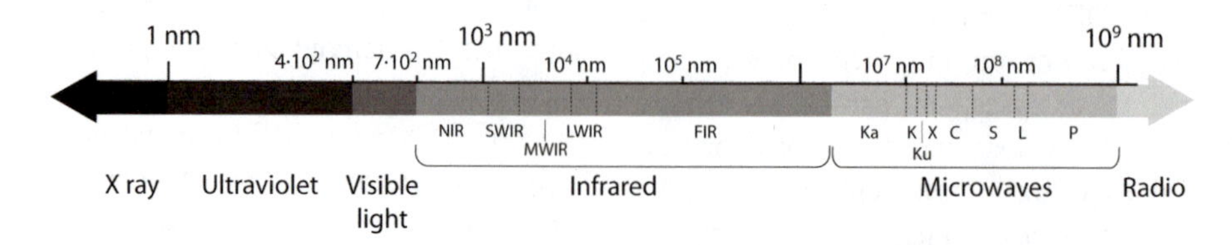

Fig. 5.1 Spectral regions in the electromagnetic spectrum, as divided in commonly used RS jargon. Note that the wavelength units are in nm

assessments of crop stress, mapping crop biophysical or biochemical characteristics, and mapping other soil and crop properties. According to Mulla, a growing interest was observed in time series analysis of RS data, to help in daily, seasonal, and multi-annual soil, crop and pest management.

This chapter focuses on four possible uses of RS data to study water erosion processes. Section 5.1 reviews the use of spectral data—particularly *spectral indices*—to quantify soil and vegetation status and dynamics in agricultural fields. Section 5.2 describes the use of image classification—including clustering, machine learning, and spectral unmixing techniques—to map surface components in the catchment, using spectral, temporal, and other forms of RS data. Section 5.3 exemplifies the fusion of RS data in topographic modeling, to improve the extraction of information on drainage structure. Finally, Sect. 5.4 describes the recently emerged use of drone-derived data to analyze water erosion at high resolution.

Lastly, note that before the RS image DN data is used for *any* of the aforementioned applications, the raw spectral images must undergo a preprocessing stage. This includes:

1. Geometric correction of the image raster data to a known coordinate system, using ground control points and DEM data;
2. Radiometric calibration of each of the image's spectral bands to convert radiance data [$W \cdot sr^{-1} \cdot m^{-2}$] to spectral reflectance [%];
3. Atmospheric correction, using dark objects to reduce atmospheric noise occurring during the image acquisition.

These three preprocessing methods are fundamental to RS data processing and are described in any published RS paper—e.g., Song et al. (2001)—and therefore will not be repeated here.

5.1 Spectral Indices: Spectral Signatures, and Algebraic Expressions

Spectral indices are an umbrella title for a range of simple, useful, and widely used group of RS methods in terrestrial (and water bodies) surface mapping. Mathematically, the spectral indices are algebraic expressions that combine values of spectral reflectance (converted to DN) from two or more spectral regions of the electromagnetic spectrum, to highlight certain surface components over others. This can be done by using additive expressions such as the *Difference Vegetation Index* (*DVI*—Richardson and Wiegand (1977)), as exemplified in Eq. 5.1, with *NIR* as the spectral reflectance values in the near-infrared spectral band, the coefficient a as the slope of the *soil line*, and *RED* as the reflectance values in the red spectral band (see for more details on the soil line Fig. 5.4 in the next pages).

$$DVI = NIR \cdot \mathrm{a} - RED. \tag{5.1}$$

According to Eq. 5.1, the larger the *NIR* DN value and the smaller the *RED* DN value, the larger the *DVI*, and the higher the expected photosynthetic activity in the image pixel (see Fig. 5.2). Nevertheless, most of the spectral indices are based on *band-ratio algebraic manipulation*. Band ratio is used to reduce errors in predictions that are due to the multiplicative effect of constant noise sources in the image pixel—such as topography, illumination angle, and atmospheric aerosols (Chen and Gillieson 2014). An example of such a band-ratio index is the *NDVI*—the *Normalized Difference Vegetation Index* (Eq. 5.2)—that is linked, among other characteristics, to vegetative photosynthetic activity.

$$NDVI = \frac{NIR - RED}{NIR + RED}, \tag{5.2}$$

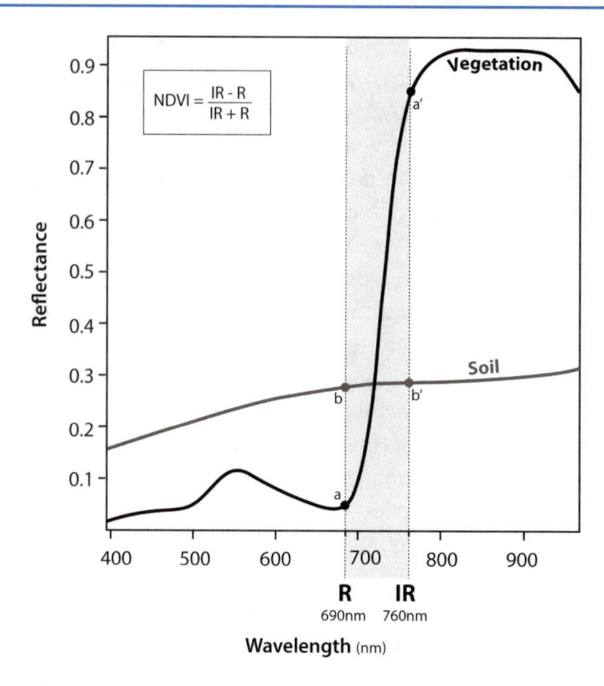

Fig. 5.2 Theoretical spectral signatures of soil and green vegetation in the VIS and *NIR* spectral regions. Note the difference in spectral reflectance values of these two features in the *RED* and *NIR* spectral bands. This difference can be used for a band-ratio manipulation in a vegetation index, such as the *NDVI*. The slope of the curve between a and a' and between b and b' is used in the *NDVI* to dissociate between soil and vegetation. Pixels with small *NIR-RED* differences have low *NDVI* values (soil pixels), and pixels with large *NIR-RED* difference exhibit high *NDVI* values (vegetation pixels). As a reference, typical spectral signatures of water bodies are usually of low, stable, reflectance values, in wavelength between 400 and 900 nm (Decker et al. 1992). That is due to the high absorption of electromagnetic radiation by water

NIR denotes the reflectance in the near-infrared spectral band and *RED* denotes the reflectance in the red band. Both types of spectral indices—additive and band ratio—rely on spectral data to quantify soil properties such as soil pH, moisture, and mineral content (e.g., Ben-Dor et al. 2009), or to assess vegetation *coverage*, productivity, and health (e.g., Xue and Su 2017). However, before we describe some of the available spectral indices to study soil and vegetation properties in agricultural catchments, let us begin with a short portrayal of the theoretical basis for the application of spectral indices. This theory relies on the variation of spectral reflectance values of a given earth surface component (be it soil, or vegetation) with respect to wavelength. This curve represents the *spectral signature* of the surface component, and spectral indices rely mainly on the slope of the spectral signatures in particular spectral regions, and on various absorption features of spectral signature. The difference in these features between soil and vegetation is substantial, and for this reason, spectral indices are often used to map soil and vegetation characteristics (Fig. 5.2).

The unique spectral signature of green vegetation occurs due to the absorption of radiation by the chlorophyll b pigment in the *RED* spectral region, and the very high reflection in the *NIR* spectral range (Das et al. 2019). The latter occurs due to the cellular structure of the plant's mesophyll, which causes the high reflection because each cell directly reflects the radiation in the *NIR* spectral region. The variance between reflectance values of green vegetation in the two spectral regions is substantial—from $15\% < RED < 20\%$ to $35\% < NIR < 60\%$ (Fig. 5.3). Thus, in some cases, when the vegetation is dry, dormant, or

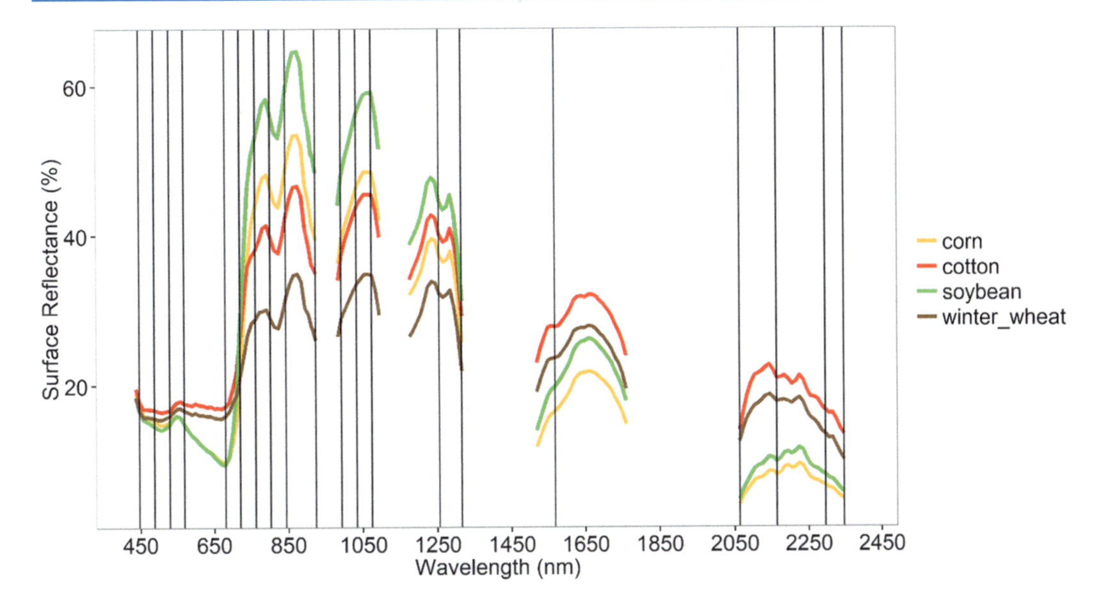

Fig. 5.3 Spectral signatures of four field crops taken in US fields. The absence of data in the 1350–1400, 1800–1950, and 2350–2500 nm is due to water absorption in the atmosphere at these bands. The figure also shows in black vertical lines the corresponding hyperspectral bands for agricultural crops based on Hyperion data (From Aneece and Thenkabail 2018)

sparse, the difference between the reflectance values in the *RED* and *NIR* may not be very different than in the case of the monotonous increase in reflectance with wavelength as is in the case of bare soil (Fig. 5.2). However, the unique spectral trait of the green vegetation has been used since the 1970s in RS applications to develop not only the *NDVI* but more *vegetation indices* based on the *RED-NIR spectral space*. Since the emergence of hyperspectral data use in RS, this field of research has opened up even more possibilities for dissociating between vegetation characteristics for agricultural crop characterization (Thenkabai et al. 2002).

The potential use of the *RED-NIR* spectral space in mapping soil and vegetation properties is illustrated in Fig. 5.4. The black dots in Fig. 5.4 represent the scatter plot of hypothetical image pixels, with each pixel value in the *RED* and *NIR* spectral bands. The closer a pixel is to the upper left-hand corner of the graph, the more the pixel represents a photosynthetically active area. Accordingly, the photosynthetic activity in the pixels is assumed to be lower between the vegetation lines (the dotted lines) toward the soil line at the center of the *RED-NIR* spectral space. The soil line is characterized by image pixels of similar reflected radiance in the *RED* and the *NIR* spectral regions and is the lowest boundary of the point's distribution. We usually do not see points below the soil line, because features in agricultural (and other) catchments usually do not reflect radiance to a greater extent in the red spectral region than in the infrared spectral region. The soil line can be extracted and marked on the *RED-NIR* spectral space, based on sampling of various bare soils image pixels under various clorpt conditions in the catchment in question.

Figure 5.4 shows that the soil line is formed from image pixels (points) with relatively small differences between their reflectance values in the *RED* and *NIR* spectral bands. This observation is in line with the results of the empirical spectral signatures of red and organic soils in Fig. 5.5, as measured under laboratory conditions. Indeed, the spectral reflection from bare soil increases monotonously from the visible to the *NIR* range of the spectrum up to $\sim$1000 nm.

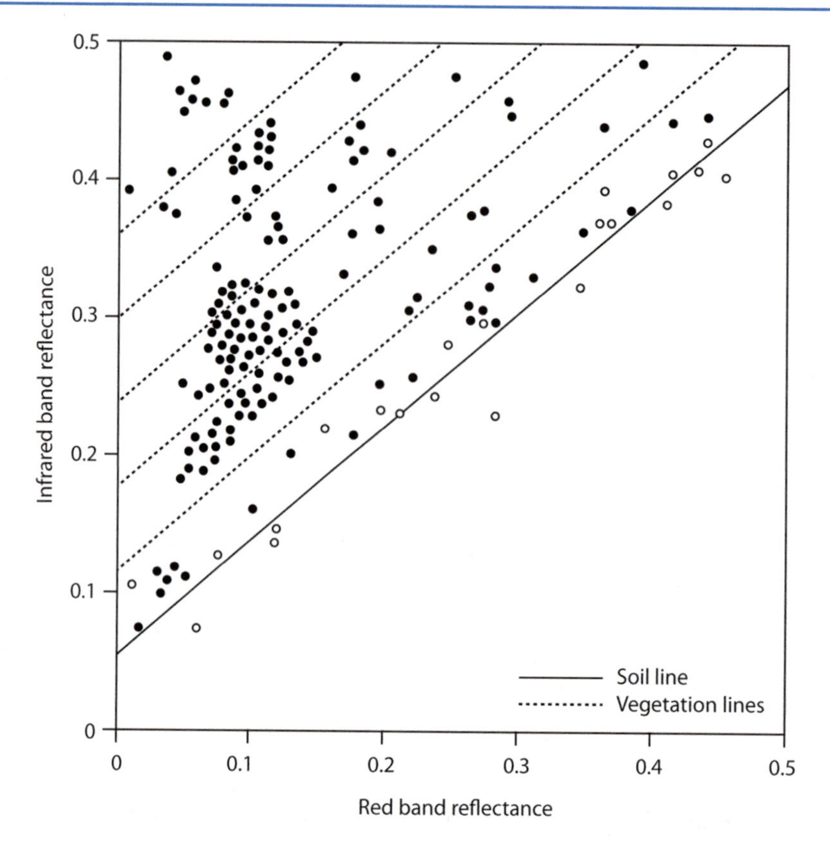

Fig. 5.4 The hypothetical *RED-NIR spectral space*, with black dots representing the image pixels values in the two bands. The estimated *soil line* is extracted as the line of minimum distances from the pixels sampled in bare soils at the site in question, under various clorpt conditions (hollow circles). The parallel dotted lines represent the various zones of vegetation status, with the triangle in the upper left corner representing the zone of highest *NIR* reflectance and lowest *RED* band reflectance values, corresponding to areas with the highest photosynthetic activity

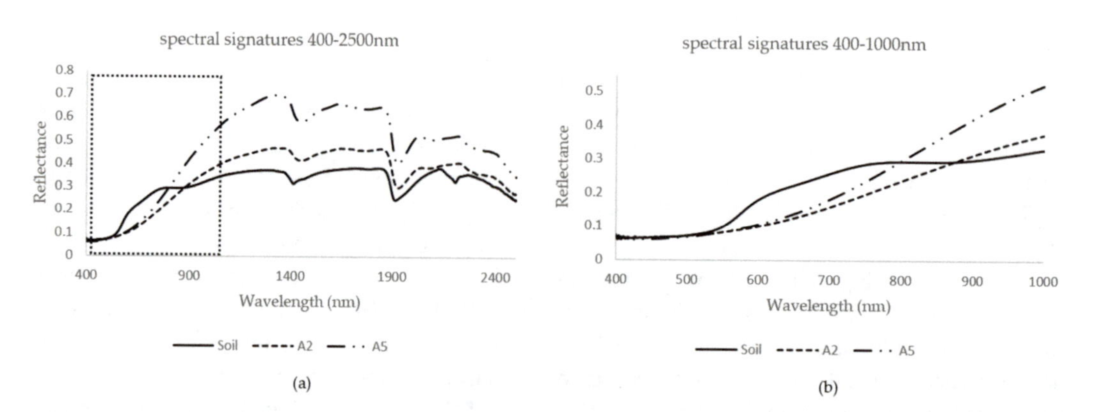

Fig. 5.5 Three typical spectral signatures of soil samples. Note the monotonous increase in reflectance in VIS and *NIR* spectral regions, and the absorption features near 1400 nm and near 1900 nm. The three curves represent reflectance spectra of: (1) Red soil rich in hematite iron oxide (Soil) and (2, 3) Mixed soil and organic matter with compost from two sources (A2 and A5). The left panel represents the entire spectral range across the 400–2500 nm spectral region; and the right panel (**b**) is a zoom-in on the 400–1000 nm spectral region. From Heller Pearlshtien and Ben-Dor (2020)

However, as is observed in the figure, soil properties are characterized by radiation characters based on the chemical characteristic of the origin material, which may vary with environmental conditions.

In a seminal work on the spectral characteristics of soil properties, Ben-Dor (2002) summarized important interactions of electromagnetic radiation with soils, and the spectral signatures of soils, in the *VIS, NIR,* and *SWIR* spectral bands. The soil properties the author examined were *clay minerals*, *carbonates*, organic matter, *soil water content,* and *iron content*.

Ben-Dor (2002)—and other studies—have found three spectral ranges (1300–1400, 1800–1900, and 2200–2500 nm) that are relevant to *clay minerals*, and smectite in particular: The montmorillonite—a common clay mineral, and a member of the smectite group—is characterized by deeper absorption bands in these spectral regions, which make them good candidates for spectral indices. The *carbonates* particles may trigger iron fixation, which affects vegetation productivity. In addition, carbonates increase soil buffering capacity to acidity, with adverse impact on soil processes. The spectral signature of carbonates is found to be mainly active in the *SWIR* spectral region. Ben-Dor and Banin (1990) identified absorption bands as best predictors of carbonate content at: 1800, 2350, and 2360 nm.

Organic Matter (OM) is the cement material of the mineral particles, which reduces erodibility and is therefore a key feature in water erosion risk. OM is made up of the remnants of plant, animal, and fecal material, and decomposes over time. OM is therefore active in the VIS–*NIR* and the *SWIR* spectral range, with a decreasing slope of the spectral signature, as decomposition of OM into the soil progresses. Several absorption bands were found for the OM spectral signature in the *SWIR*, with the deepest absorption band being 1726–2426 nm.

Hydration and *hygroscopic* and *free soil water* were found to be active at the 1400–1900 nm spectral range. Some enhancement of hygroscopic water was found on the 1900 nm band. In addition, as water content increased, the reflectance of the 1400 and 1900 nm diminished.

Monitoring *iron content* in soils from RS data can be applied through exploiting the 900 and 1900 nm absorption features. That is despite emerging complexity through the correlation between iron and other soil properties.

Spectral indices can be a rapid and an easy-to-use methodology for rendering spectral signatures into operative tools, based on the absorption features reviewed earlier. Indeed, some of the spectral indices have become widely used for various ongoing applications around the world. The *NDVI* and its successors have become part of the standard toolkit of agencies such as the USGS to study current and historical changes in vegetation phenology, based on NOAA AVHRR satellite data. The *NDVI* is also used by the Global Inventory Modeling and Mapping Studies (GIMMS) group at NASA Goddard Space Flight Center, to support USDA's Office of Global Analysis (OGA) agricultural monitoring activities, and many others.

However, despite their widespread use and popularity, caution should be used when interpreting the spectral indices output values. That is due to the high surface heterogeneity and the occasionally overlapping features of the various soil and vegetation components in the image pixel signal (Baret and Guyot 1991). Table 5.1 provides a list of some of the spectral indices that may be used for water erosion studies.

Some of the indices in Table 5.1 can be applied using multispectral satellite sensors, while others require the use of narrow-band spectral resolution, such as that provided by hyperspectral sensors. Broadly speaking, this collection of indices can be divided into two types:

1. Vegetation indices aimed at monitoring the crop photosynthetic activity, health, and weed exploration;
2. Soil indices aimed at mapping soil properties that may be useful for soil health testing.

5.1.1 Vegetation Indices

Vegetation Indices (VI) are spectral indices that merge data from different spectral regions to provide qualitative and quantitative expression to

Table 5.1 A collection of soil and vegetation indices

Used by	Aim	Index	Notations
Vegetation Indices			
Seutloali et al. (2017)	Veg. cover assessment for erosion monitoring	$NDVI = \frac{NIR-RED}{NIR+RED}$ $Veg.cover[\%] = \frac{NDVI-NDVI_{min}}{NDVI_{max}-NDVI_{min}}$	*NIR*—near infrared *RED*—red band
Hochschild et al. (2003)	Delineation of erosion classes	$TSAVI = \frac{a(NIR-aR-b)}{[aNIR+RED-ab+X(1+a^2)]}$ $Veg.cover[\%] =$ $1.06+43.5 \cdot TSAVI+97.8 \cdot TSAVI^2$	a, b are the regression coefficients of the soil line. X is a parameter used to minimize soil effects (X = 0.08)
Karnieli et al. (2001)	Aerosol Free Veg. Index	$AFRI_{1600} = \frac{NIR-0.66b_{1600}}{NIR+0.66b_{1600}}$ $AFRI_{2100} = \frac{NIR-0.5b_{2100}}{NIR+0.5b_{2100}}$	*b* refers to the value in the given spectral band
Hunt and Rock (1989)	Water Index	$NDWI = \frac{NIR-SWIR}{NIR+SWIR}$	*NIR*—780–900 nm *SWIR*—1550–1750 nm
Hunt and Rock (1989)	Moisture Stress Index	$MSI = \frac{b_{1599}}{b_{819}}$	*b* refers to the value in the given spectral band
Serrano et al. (2002)	Nitrogen Index	$NDNI = \frac{\log\cdot\left(\frac{1}{b_{1510}}\right)-\log\cdot\left(\frac{1}{b_{1680}}\right)}{\log\cdot\left(\frac{1}{b_{1510}}\right)+\log\cdot\left(\frac{1}{b_{1680}}\right)}$	*b* refers to the value in the given spectral band
Serbin et al. (2009)	Crop residue	$SINDRI = \frac{100\cdot(SWIR6-SWIR7)}{SWIR6+SWIR7}$	*SWIR6*—2185–2225 nm *SWIR7*—2235–2285 nm
Van Deventer et al. (1997)	Crop residue	$NDTI = \frac{b_6-b_7}{b_6+b_7}$	b_6—1570–1650 nm b_7—2110–2290 nm
Sandholt et al. (2002)	Temp. Index	$TVDI = \frac{T_s-T_{min}}{T_{max}+T_{min}} = 1-\frac{\theta_v-\theta_{min}}{\theta_{max}-\theta_{min}}$	*T*—Thermal surface temp θ_v—volumetric surface moisture content
Soil Indices			
Bernstein et al. (2012)	Mud Index	$NDMI = \frac{b_{795}-b_{990}}{b_{795}+b_{990}}$	*b* refers to the value in the given spectral band
Boettinger et al. (2008)	*Calcareous Sediment Index*	$CS = \frac{SWIR-Green}{SWIR+Green}$	*SWIR*—Landsat7 ETM 1550–1750 nm
Castellanos-Quiroz et al. 2017)	*Iron Index*	$FI = \frac{RED}{Blue}$	*Blue*—the blue band *RED*—Red band
Ben-Dor (2002)	*OM Index*	$OM = b_{722}\cdot 0.14+b_{2328}\cdot 0.04-b_{705}$ $\cdot 0.12+b_{1678}\cdot 0.02+0.05$	*b* refers to the value in the given spectral band
Ben-Dor (2002)	*pH Index*	$pH = b_{722}\cdot 0.52+b_{2118}\cdot 0.73+8.04$	*b* refers to the value in the given spectral band
Rockwell (2013)	Clay Mineral Index	$CI = \frac{SWIR}{SWIRL}$	*SWIR*—1550–1750 nm *SWIRL*—2080–2350 nm

vegetation photosynthetic activity, inter-species differences, and variations in canopy structure and composition (Huete et al. 2002). In many cases, crop canopy in agricultural fields does not fully cover the soil surface, so the spectral signature of crops, or weeds, is affected not only by the biotic components, but by the underlying soil brightness, and also by, topographic effects, shadow, soil color and moisture, and by atmospheric conditions (Bannari et al. 1995). Despite these limitations, VIs have been widely used, and have proven effective in various agricultural uses such as crop monitoring (Dong et al. 2019), while the main uses of VIs in erosion studies are (Cyr et al. 1995):

1. To study crop canopy status as an indicator to soil degradation;
2. To quantify canopy cover [in % terms] that acts as a protection layer against raindrop impact and subsequent runoff and erosion processes;
3. To quantify temporal changes in the plant growth and crop phenological development—because from a certain stature and up, the plants stop the runoff and subsequent soil transportation.

Among the most common VIs are the aforementioned *NDVI*, and the *TSAVI* (Transformed Soil Adjusted Vegetation Index). These two indices (see details in Table 5.1) have been used to predict vegetation cover based on empirically based coefficients extracted from regression plots between their values and measured vegetation cover, or LAI, in the field (Carlson and Ripley 1997). The *AFRI* (Aerosol Free Vegetation Index) is another VI that exploits the spectral characteristics of the *SWIR* bands to penetrate the aerosols in the atmosphere and estimate crop canopy status and change even under conditions of smoke, mist, or pollution in the studied site (Karnieli et al. 2001). Another indicator of crop health and, to some extent, of the underlying soil health, especially in dry areas, is the canopy water content. The *NDWI* (Normalized Difference Water Index) can be used as a measure to quantify water content in the leaves using the *NIR* and *SWIR* bands and may indicate water stress in the vegetation strata due to the diminishing soil layer by erosion processes (Jackson 2004). Similarly, the *MSI* (Moisture Stress Index) has been linearly correlated with measurements of leaf water content (Hunt and Rock 1989), thus representing lack of water due to drought stress. Another measure of catchment health is the nutrients content in the crops leaves. The *NDNI* (Normalized Difference Nitrogen Index) is a nitrogen content index in the vegetation canopy that may attest to soil health, since nitrogen concentration is key to the functioning of crops, and especially to aboveground net primary production and decomposition processes (Serrano et al. 2002).

The two crop cultivation indices, *SINDRI* (Shortwave Infrared Normalized Difference Residue Index) and *NDTI* (Normalized Difference Tillage Index) are not classic vegetation indices, but nonetheless may be important in quantifying crop residue as a protection against water erosion. Note that VIs have usually been developed and used to study areas that are prone to erosion under semi-natural conditions (such as rangelands), and the minerals indices have been developed mostly for geological purposes, under more sterile conditions than in agricultural soils. The *TVDI* (Temperature Vegetation Dryness Index) for example, is used to quantify surface moisture status in a mixed area with soil and vegetation, using the empirical, usually linear, relationship between surface temperature and moisture measurements, with $T_{s\text{-max}}$ and $T_{s\text{-min}}$ as the dry and wet edges, respectively.

The aforementioned Table 5.1 describes some of the vegetation indices currently available in the literature. More—over a hundred—vegetation indices can be found in Xue and Su (2017). Figure 5.6 illustrates the computation of three vegetation indices of the Harod catchment, and their aggregation at the field level, using zonal statistics operators.

The VI maps of the Harod catchment show fields of higher and lower vegetation cover (based on *NDVI*), crop residue left in the field (*NDTI*), and *SINDRI*. There is indeed similarity between *NDTI* and *SINDRI* predictions and they are different from the *NDVI* layer that reflects the green vegetation cover in the catchment. Namely, fields of high *NDVI* are those of summer crops.

5.1.2 Soil Indices

Soil spectral indices may provide up-to-date measured data on soil properties. The interpretation of this data is not straightforward as the soil reflectance is the integrated outcome of the reflected radiance from a mixture of minerals, organic matter, and the soil liquid and gaseous phases. Current studies of soil spectral signatures and spectral indices are focused on the amount of

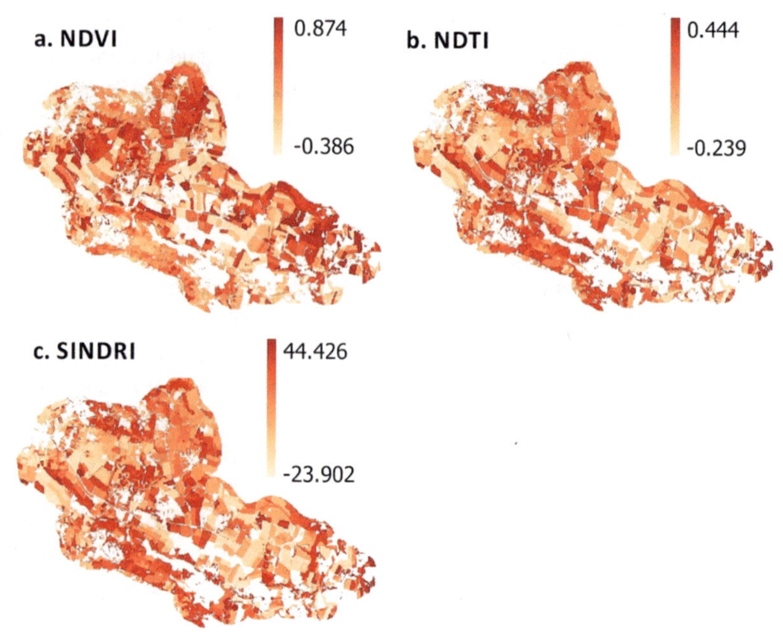

Fig. 5.6 Maps of vegetation indices applied to the Harod catchment. The indices were computed using *VENμS* satellite data of July 2018, and aggregated to the field level using zonal attribute operators

moisture, organic matter and carbon in the soil—but also on the content of nutrients and trace-elements in the soil and on soil mineralogy and texture. Other soil parameters might also be estimated through the use of soil spectral indices (Chabrillat et al. 2013).

Soil moisture indices include the *NDMI* (Normalized Difference Moisture Index) Mud Index that makes it possible to identify water ponding areas in the catchment. This can be done using an empirical threshold. For example, in the work of Bernstein et al. (2012), mud areas due to water ponding have been delineated based on a threshold value of $NDMI > 0.3–0.5$ in agricultural fields at Davis, California. However, to map water ponding, such a threshold value must be determined empirically at each site.

The *CS* Index (Calcareous Sediment Index) was developed to quantify calcareous sediments, using Landsat TM data band 2 (Green) and band 5 (*SWIR*). The normalized difference ratio of the two $\frac{(b5-b2)}{(b5+b2)}$ was found by Boettinger et al. (2008) to be diagnostic for calcareous sedimentary rocks. Using Landsat TM, the *CS* was found to distinguish sedimentary rocks from igneous rocks. The potential for water erosion studies lies in the fact that the *CS* Index can be used to quantify calcareous rock fractions and stones, and their footprint in the soil. This can be used to assess the field's level of degradation. For example, sheet erosion can cause soil loss to the point of exposure of the parent material. Time series *CS* data can be used to monitor the status and change in rock exposure, before the field reaches the point of soil degradation.

The Iron Index (*FI*) has been found useful for the mapping of ferric minerals in the iron-mining industry and is therefore probably more appropriate for identifying large concentrations of iron in the soil. Its use for mapping Iron in agricultural catchments should be therefore used with caution, and must be further validated.

The Organic Matter Index (*OMI*) and the pH Index described in Table 5.1 were successfully validated against field samples and were found effective in mapping both soil properties in the Zevaim height, which is part of the Harod catchment. This was done using data from the DAIS-7915 mission which included a hyper spectral sensor mounted on an airplane (Although note that pH was found to have only indirect spectral assignment). Both OM and pH indices are more difficult to detect, as they require the use of narrow hyperspectral bands—

where the data collection requires higher acquisition and (mainly pre) processing efforts. These efforts are justified, given the ability to isolate the information in narrow absorption features, and the importance of these soil properties—especially that of OM as a cement material for the soil aggregates—to the understanding of water erosion.

With regard to the study of soil texture, a deep drop in spectral reflectance in the *SWIRL* (2080–2350 nm) wavelength compared with the spectral reflectance in the *SWIR* (1550–1750 nm) wavelength of Landsat TM 7 was found as a typical spectral characteristic of the clay minerals group and therefore the ratio between the Landsat TM 5 and 7 spectral bands was used to identify this group of minerals (Rockwell 2013).

Figure 5.7 illustrates the computation of seven soil spectral indices of the Harod catchment, and their aggregation at the field level, using zonal statistics operators. The indices were computed based on a summer image, because large parts of the catchment are exposed at this time of year. These soil properties may be used to compute the soil health indices described in Chap. 7.

In summary, vegetation canopies and soil chromophores have distinct absorption features. These can be used to provide information about the soil and vegetation status and about changes in agricultural catchments—even when using relatively simple algebraic expressions that are usually based on band ratio. Spectral indices provide us with quantitative predictions at the catchment level, based on a solid physical basis that is undisputed and not subject to interpolation error. It must be accepted, however, that the spectral indices provide only ancillary information that should be treated with caution, due to possible errors arising from environmental conditions and the notorious impediment of salt and pepper noise in spectral indices.

With all their advantages and disadvantages, spectral indices alone, do not allow for image pixels to be organized by relevant categories (soil, rock, etc.), so that each image pixel is assigned automatically to a specific group. To do so, we need image pixel classification procedures that are particularly necessary at the modeling stage—as in the case of the C-factor of the USLE (see Chap. 1) mapping in the GeoWEPP for example (see Chap. 3).

5.2 Image Classification Techniques

Classification is the assignment of image pixels or image objects to classes/groups based on similarities in their DN values (Campbell and Wynne 2011, p. 132). Classifiers of RS images can be divided into two main groups (Liu et al. 2011):

1. *Hard classifiers* that directly predict class boundaries, without computing the class probability estimation, and therefore yield crisp estimate (or Boolean true/false membership to the class);
2. *Soft classifiers* that first compute the probability of the candidate to fit the class and then provide assignment of the candidate to a class based on the highest probability.

Outcomes of the two types of classification procedures can be used in water erosion studies to map land covers such as Roads, built areas and utilities, areas of natural vegetation and croplands, crop types for crop-rotation analysis, rock exposure, and degraded areas. A few examples of uses of the two types of classification techniques in water erosion geoinformatics are presented in Table 5.2.

The examples in Table 5.2 may lead to the following three assertions: *First*, new classification techniques are progressively being developed and used in the current years for agricultural purposes; the field is active and evolving. *Second*, as a result, many classification techniques are available now and being further developed. The reader should be able to choose the most appropriate one for the problem in hand, usually based on trial and error procedure, although some generalizations exist, as will be shown later. *Third*, the range of possible applications of image classification to soil erosion

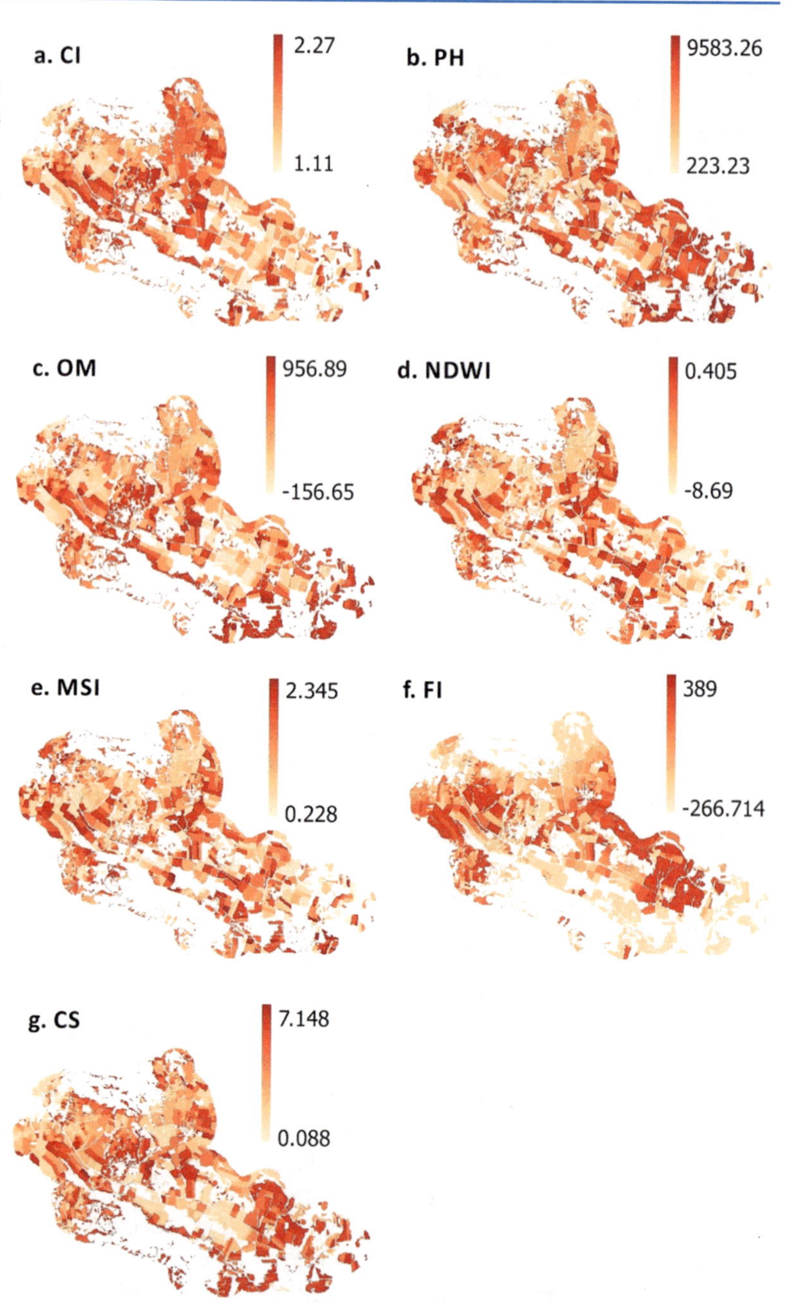

Fig. 5.7 Maps of spectral soil indices, as applied to the Harod catchment. The indices were computed using Landsat 8 data from July 2018, and aggregated to the field level using zonal attribute operators

studies is wide: From soil properties classification to the outlining of flooded areas.

In the next sections we will exemplify several techniques to classify rock, soil, and vegetation groups, however, the reader should note that she can use these methods for many other applications of water erosion studies.

5.2.1 Hard Classification

In the case of hard classification, an image-processing procedure is programmed to group and label image pixels of similar characteristics in a multi-band image, using statistical rules. The input data may be spectral reflectance, or

Table 5.2 A collection of classification applications for water erosion studies. Note that the classification method indicated in the Method column is the most accurate in the corresponding study, while most of these studies applied several classifications techniques to compare their performance

Authors	Aim	Method
South et al. (2004)	Mapping agricultural tillage practices	Spectral angle methods
Cohen and Shoshany (2005)	Crop recognition	Knowledge-Based System
Brungard et al. (2015)	Predicting soil classes	*Random forests* and various machine-learning algorithms
Ganasri and Ramesh (2016)	Mapping C-factor (crop management) for the application of RUSLE	*Maximum Likelihood classification*
Amer et al. (2017)	Distinguishing land from flooded areas	ISODATA of NDWI
Sun et al. (2017)	Soil contaminants (Ni, Zn concentration)	*K-means*
Forkuor et al. (2017)	Mapping sand, silt, clay, cation exchange capacity, soil organic carbon, and nitrogen	Random forests
Wu et al. (2018)	Identifying Soil Texture Classes	Classification trees
Žížala et al. (2018)	Mapping eroded soil via spectral severability of erosion classes	ISODATA
Masoud et al. (2019)	Mapping soil salinity	Spectral mixture analysis

processed data—including, for example, image texture or multi-temporal *NDVI* data (Shoshany and Svoray 2002). The classification procedure is usually executed on a pixel basis, using a local operator—i.e., no information from neighboring pixels is involved in the process.

From a methodological point of view, hard classification techniques are divided into two main groups: *Supervised* and *unsupervised*. The supervised classification group comprises several techniques that use *labeled training sets* (subsets of the image pixels of known labeling) to train the classifier using spectral/other signatures. A training set should be formed for each class and must be large enough to represent various parts of the parameter space of the class to be able to train the classifier effectively. In the unsupervised group, the process of data clustering is automatic, and the groups are formed based on automatic image analysis, so the analyst does not need to provide the classifier with predetermined training sets. The different workflows of the two classification groups are presented in Fig. 5.8.

As will be demonstrated below, supervised classification techniques may be more accurate, but one can choose between the two types, depending on her ability to provide training sets. In the following section, three classification methods among the many available in the scientific literature are described and exemplified.

5.2.1.1 Unsupervised Classification

Unsupervised procedures are used to search for patterns of pixel DN in an image of several spectral bands, without training sets. After robust groups are formed, the user needs to identify the output groups, and label them with real-world land cover, or land use type. Identification of the groups can be applied, for example, by field visits or visual interpretation of RGB data of a high-resolution aerial photo—an information-gathering process that is considerably less demanding than collecting data for training sets. Validation data is still required to test the classification *accuracy* in unsupervised classification, as with all classification techniques.

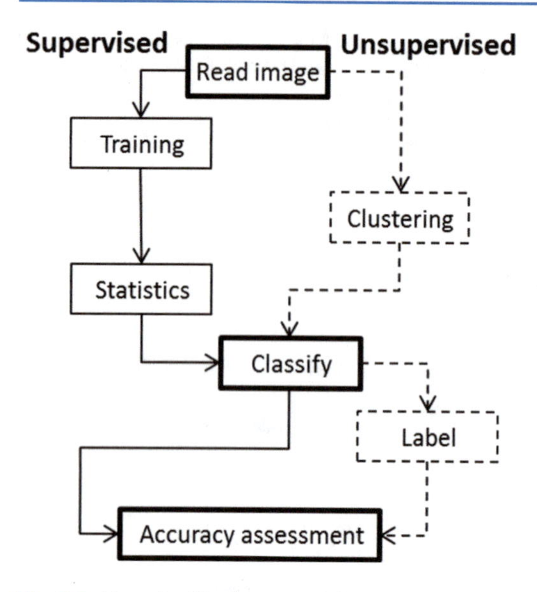

Fig. 5.8 Two classification groups that are distinguished by the different arrow and box types. Solid lines represent the supervised classification workflow; dashed lines the unsupervised one. Bold boxes represent steps shared by the two groups. The supervised approach relies on labeled data provided by the user, while the unsupervised approach applies the learning process, based on data that was not labeled by the user. The RS input data can be spectral, temporal, or other

The literature suggests several unsupervised techniques—including, notably, the archetypal Iterative Self Organizing Data Analysis Technique Algorithm (ISODATA) (Regmi and Rasmussen 2018). ISODATA is based on a set of rules and user-specified values that are incorporated into an iterative procedure. ISODATA splits and merges groups of pixels in the spectral space (e.g., RED-NIR space) based on their DN value distances from the group centroid defined as $\frac{\sum_{i=1}^{n} X_i}{n}, \frac{\sum_{i=1}^{n} Y_i}{n}$, where i is the image pixel and n denotes the number of image pixels in the group. X and Y are the DN values in the input bands. The centroids are placed, and pixels are assigned, based on the shortest Euclidean distance of each pixel DN from the centroid. Groups are split if their standard deviation exceeds the user-defined threshold, and merged if the distance between the groups is below that threshold (Memarsadeghi et al. 2007).

K-means clustering is another unsupervised classification technique, that reduces the within-class variability by minimizing the sums of squares distances (errors) between each image pixel and the group centroid. This principle is similar to that of ISODATA—but while ISODATA does not require the number of groups to be predetermined, in K-means we have to know the number of classes in advance. To explicate the principles of unsupervised classification techniques, we compare the algorithms of two theoretically similar unsupervised classification methods: The K-means, and the *K-means medoids* (Kaufman and Rousseeuw 1990). As we shall see, despite the allegedly small theoretical differences between two algorithms, the empirical results can be very different.

Let us begin with a detailed description of the two algorithms. In K-means:

Step 1: The image spectral domain is divided into sub-spaces, based on predetermined number of groups. (In Fig. 5.9a, there are two groups, partitioned by the bold line, that divides the spectral space into two sub-spaces);

Step 2: A centroid is computed, based on the location of all points in the given sub-space;

Step 3: The partition line is removed, and the points from the entire image domain are assigned to the centroid of the shortest spectral distance;

Step 4: New centroids are computed;

Step 5: The image pixels are reassigned to the centroid of the shortest distance. The process continues until the changes in reassignments are negligible.

In the K-medoids procedure—illustrated in this book using the algorithm *Clara* (Clustering for Large Applications—Kaufman and Rousseeuw (1990)) that is commonly used to classify RS images (Fig. 5.9b):

Step 1: The algorithm creates randomly several subsets from the image of a fixed size;

a. K-means Clustering

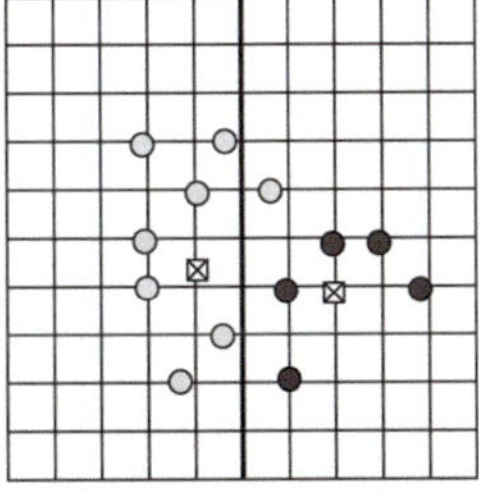

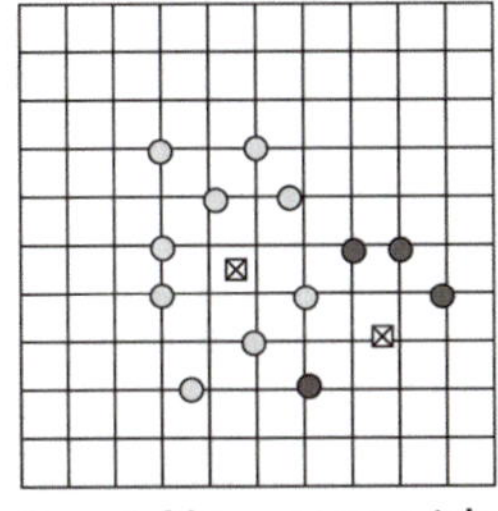

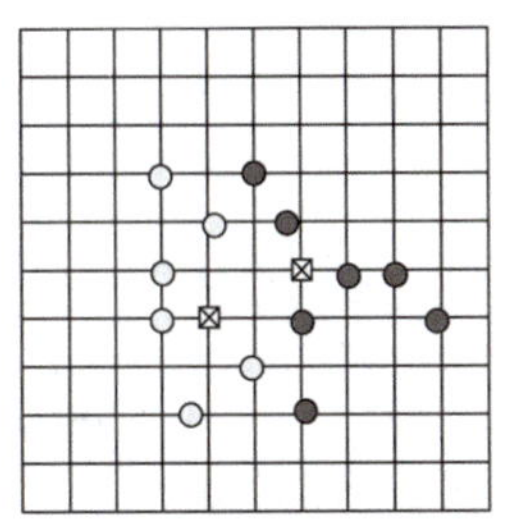

$$\boxtimes = \frac{\Sigma Xi}{n}, \frac{\Sigma Yi}{n}$$

b. K-medoids Clustering

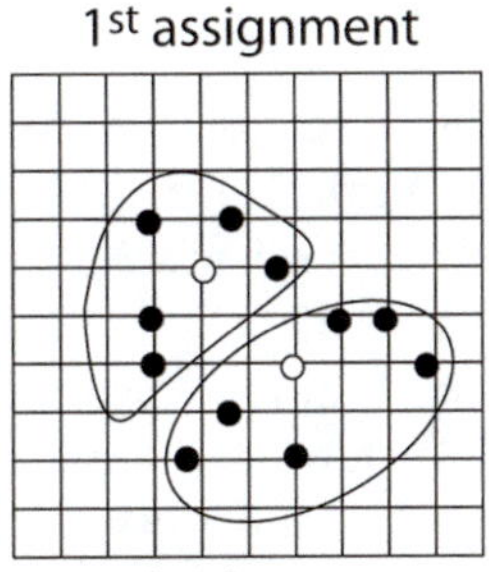

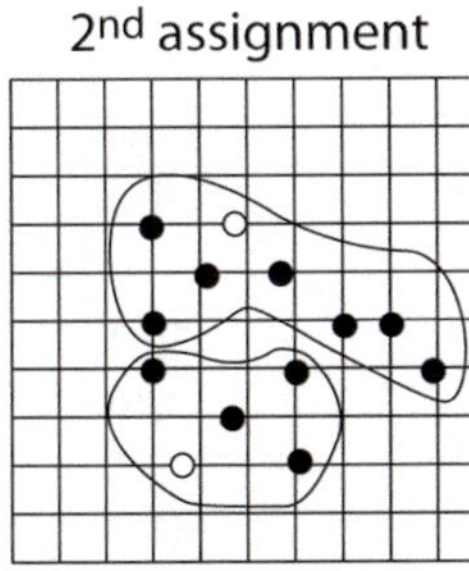

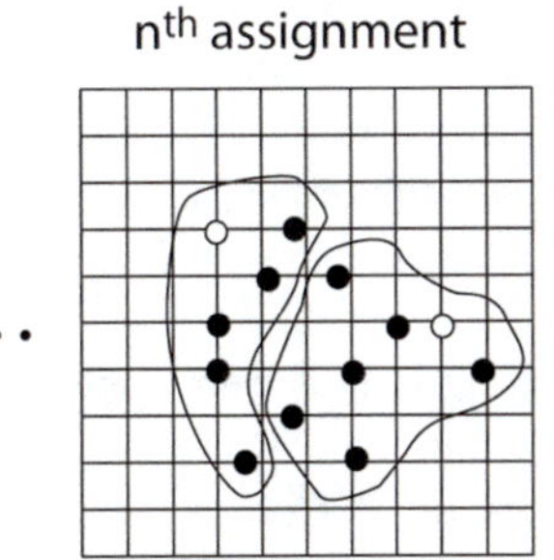

Sub group

D_i - Dissimilarity Index

Fig. 5.9 Unsupervised classification techniques: K-means (**a**) and K-medoids (**b**) clustering (Clara—Clustering Large Applications) illustrated over a background of two spectral bands (the *X* and *Y* axes). In reality we usually use more spectral bands (such as all Landsat TM spectral bands), but for the sake of simplicity only two are illustrated here

Step 2: Then the Partitions Around Medoid (PAM) algorithm is executed on each subset and chooses the corresponding k representative pixels (medoids) as the centroids;

Step 3: The algorithm then assigns all image pixels (N) to the nearest medoid;

Step 4: In each iteration, the pixels used as medoids are randomly replaced with other pixels, and the classification is fully executed;

Step 5: As a measure of the goodness of grouping, the algorithm computes the dissimilarity index for each group, to act as a cost score: If the cost score is lower than the existing score, the new result survives—if not, it is removed;

Step 6: The algorithm retains the sub-dataset where the mean is minimal. Finally, Clara searches for centroids in the spectral space—but unlike the K-means, Clara divides the information into subgroups, instead of sorting the total pixels in the image.

The two algorithms were applied to the Assaf catchment drone data, using the package "stats" for K-means by R Core Team (2020) and the package "cluster" for Clara (Kaufman and Rousseeuw 1990).

After the classification algorithms are executed, the analyst needs to evaluate the classification *accuracy*, after adjusting parameters. The error of the resulting classification is tested against a validation set. *Accuracy* assessment is computed using the *Kappa coefficient* (Cohen 1960), which is extracted from a *confusion matrix*—a matrix that provides a count of the number of observations in each intersection between all the measured and predicted classes (see more detail and explanation in Fig. 5.10).

The Kappa coefficient (Eq. 5.3) that measures the degree of agreement between rating methods is computed using the confusion matrix data (Cohen 1960):

$$\text{Kappa} = \frac{N \cdot \sum_{i=1}^{r} X_{ii} - \sum_{i=1}^{r} (x_{i+} \cdot x_{+i})}{N^2 - \sum_{i=1}^{r} (x_{i+} \cdot x_{+i})}, \tag{5.3}$$

where N is the total number of observations, r is the number of rows in the confusion matrix (in the case of the confusion matrix in Fig. 5.10 = 3); x_{ii} is the number of observations in row i and column i—namely, the diagonal entries ($\Sigma = 210$); x_{i+} and x_{+i} are the marginal totals of row i and column i, respectively, (multiple row with column and $\Sigma = 24{,}310$). Kappa value (in this case 0.67) ranges between −1 and +1 with—according to Landis and Koch (1977)—"Kappa < 0 indicating no

Measured values

		Soil	Rock	Veg.	Row total
Predicted values	Soil	50	20	10	80
	Rock	15	70	5	90
	Vegetation	0	8	90	98
	Column total	65	98	105	268

Fig. 5.10 A confusion matrix (also known as an *error matrix*). The matrix rows present the observations in the predicted classes and the values in the columns present the observations identified by the measurements. The diagonal cells present the number of observations where the prediction and measurement agree. For example, in 50 observations, both the classifier and the measurement in the field agreed that these observations are soils. The other cells present the disagreements. Thus, for example, 15 cases that were observed in the field as soil class were predicted mistakenly by the classifier as rock and zero observations of observed soil were predicted as vegetation. The row and column total signify the number of observations predicted and measured respectively. The data here is synthetic, for illustration purposes

agreement; 0–0.20 representing slight, 0.21–0.40 fair, 0.41–0.60 moderate, 0.61–0.80 substantial, and 0.81–1 nearly perfect agreement." The simple percent agreement would be the sum of the diagonal cells divided by the total observations, 0.78 or 78% in the example in Fig. 5.10.

5.2.1.2 Supervised Classification

The RS literature offers several supervised classification techniques, including the well-known and widely used Maximum Likelihood Classification (MLC—e.g., Awotwi et al. (2018)). Figure 5.11 illustrates the concept of a supervised classification technique, with the MLC algorithm as an example—again, for simplicity, with two spectral bands as input data. The black dots represent members of the soil and vegetation groups of the training data collected by the user ahead of the classification process. The gray point A is the candidate image pixel to be assigned to one of the two classes.

The candidate pixel A is assigned to a class using a maximum-likelihood function. The training set in each class is assumed to have a normal distribution and is represented statistically by the mean and covariance values of both spectral bands. The probability for assignment to each class is computed for each pixel DN in the image. In Fig. 5.11, one can see that the candidate pixel A is more likely to belong to the soil class than to the vegetation one.

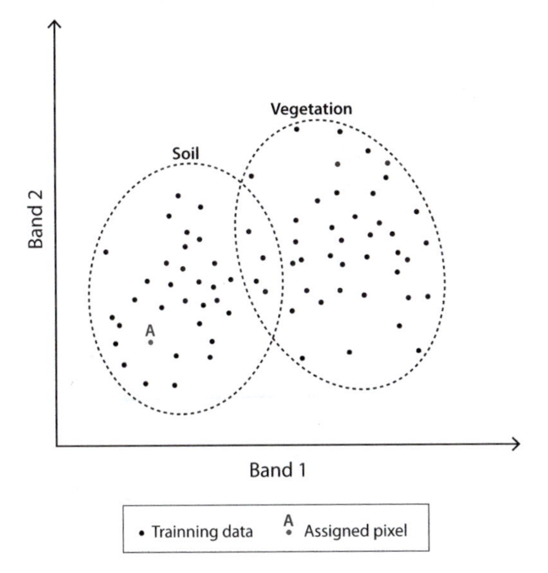

Fig. 5.11 The concept of supervised classification, using the common maximum-likelihood classification. The ellipses are groups of soil and vegetation training sets collected by the user. The gray point A is to be assigned to one of the groups

Random Forests is a more advanced and efficient supervised classification technique. The Random Forests algorithm provides a variable with an importance ranking that can be effectively used to classify the image pixels. It originates in the world of machine learning, and has been used widely in RS for mapping vegetation properties (Cheng et al. 2018), for example. To execute the algorithm, the user must designate dozens of pixels for each group as the training sets. The algorithm then builds random decision trees, from a random division of the subgroups. These forests are then used to predict which group each pixel P_i in the new image belongs to. Figure 5.12 illustrates a dataset (an RS image), and n decision rules applied to the assignment of each pixel P_i, to a group based on the *NIR*, *RED*, and Green data. The final assignment is based, for example, on the majority rules between the assignments of the different trees.

Based on these principles, the Random Forests algorithm was applied in this book using the R package *RandomForest,* based on Liaw and Wiener (2002) as follows:

Step 1: Determine the input layers to be used —namely, the spectral bands, or any other source, such as multi-temporal soil, or vegetation indices data;

Step 2: Delineate representative training sets, based on visual interpretation of high-resolution aerial photos, or field visits. These must represent various phases (illumination angle; mineral content; plant age, etc.…) of the predicted objects (such as soil and vegetation) —but not too many, as it might increase the noise-to-signal ratio;

Step 3: Choose the learning function, and determine the control parameters;

Step 4: Execute an algorithm to learn the associations in the training sets;

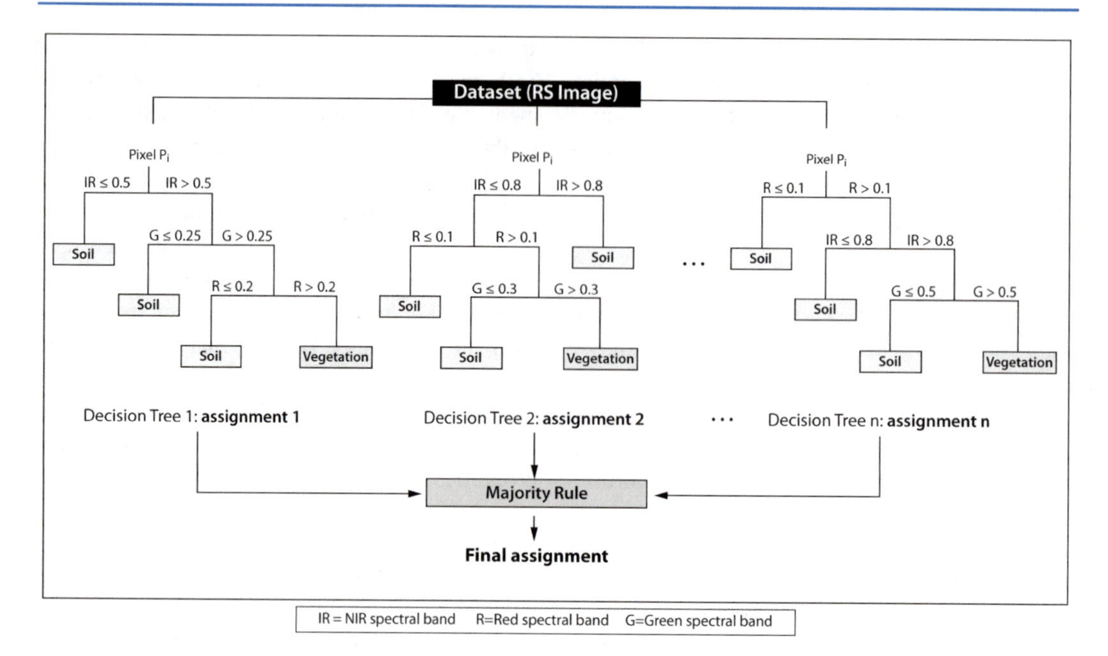

Fig. 5.12 The Random Forests supervised classification technique. The procedure is based on an ensemble of decision trees, where each individual tree in the random forest predicts a class based on the decision rules applied to the variables (in this case, the *NIR*, *RED* and Green spectral bands) and the class with the most votes based, for example, on the majority rule, is the class the pixel P_i is assigned to

Step 5: Evaluate classification *accuracy* (confusion matrix, Kappa) against a validation set which is, of course, different from the training set.

All three classification techniques (K-means, Clara, and Random Forests) were applied to a three-dimensional VIS spectral space—including the Red, Green, and Blue bands (RGB) of sub-meter spatial resolution image of the Assaf catchment in the northern Negev region in Israel (Fig. 5.13).

The three classification groups that the image pixels were assigned to, were bare soil, dry vegetation, and green vegetation (Fig. 5.14). These classification techniques can be applied to any data from multispectral or hyperspectral sensors. A bird-eye visual interpretation of the maps in Fig. 5.14 suggests that the classification results are similar. However, closer inspection reveals changes in pixel assignment between the classification techniques along the paved road, within the channel, and on the north-facing hill-slope in the southern part of the channel. These differences are stark—particularly when comparing the Random Forests supervised classification technique with the two other classification techniques, which are unsupervised.

The relatively small variations that appeared in the visual interpretation of the three maps are even more striking when one compares the confusion matrices (n = 1088) and the Kappa coefficients between the three classification techniques and validation data.

The following three confusion matrices are extracted directly from the report of the R program code for the *accuracy* assessments of the three classification techniques:

Fig. 5.13 An RGB image taken from drone data of the Assaf subcatchment in the northern Negev, Israel

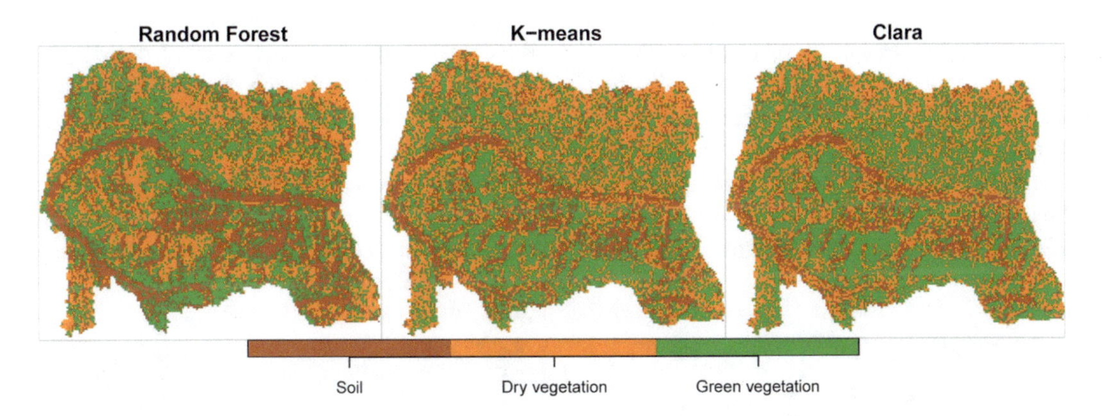

Fig. 5.14 Classification of drone data, using the three classification methods. Note the difference between the supervised random forest and the two K-means which look more similar especially in the center of the image

Random Forests
pred_rf

class	1	2	3
1	567	12	1
2	4	253	0
3	0	0	251

Cohen Kappa and Weighted Kappa correlation coefficients and confidence boundaries

	lower	estimate	upper
unweighted kappa	0.96	0.97	0.99
weighted kappa	0.98	0.99	0.99

```
Number of subjects = 1088
K-means
pred_km
class  1    2    3
  1   517   48   15
  2   70   176   11
  3   11    16  224

Cohen Kappa and Weighted Kappa correlation coefficients
and confidence boundaries
lower estimate upper
unweighted kappa   0.70   0.74 0.77
weighted kappa     0.81   0.83 0.85

Number of subjects = 1088
Clara
pred_cl
class  1    2    3
  1   501   54   25
  2   49   187   21
  3    4    85  162

Cohen Kappa and Weighted Kappa correlation coefficients
and confidence boundaries
lower estimate upper
unweighted kappa   0.61   0.64 0.68
weighted kappa     0.74   0.77 0.79

Number of subjects = 1088
```

Significantly, the Random Forests technique yielded better results (Kappa = 0.97) than K-means (Kappa = 0.74), and substantially better than Clara (Kappa = 0.64). These are very large differences considering the relatively small theoretical difference between K-means and K-Medoids. Not to mention the vastly outperforming of the supervised classification.

The very high classification *accuracy* of the Random Forests procedure is more clearly evident when the classification is applied to a high-resolution RGB ground photograph (Fig. 5.15).

These three classification techniques provide an efficient and easy-to-use tool for identifying land cover and mapping soil and vegetation. Classification techniques can also be used to map

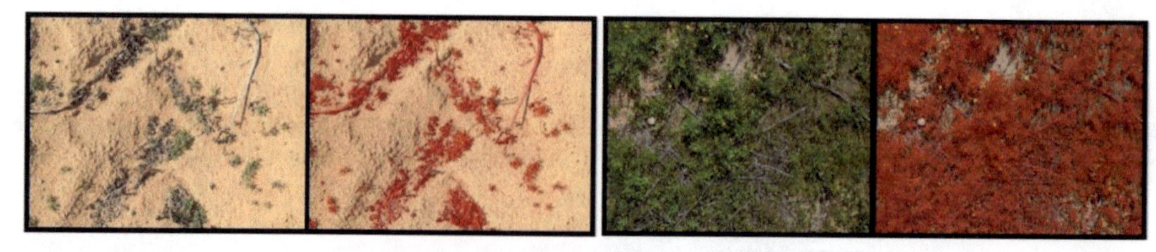

Fig. 5.15 Random Forest classification of RGB close-range field camera images in areas of high and low vegetation cover (by Arthur Khozin). The left panel shows the identification of green vegetation (dwarf shrubs) from a sand dune background under sparse canopy conditions and the right panel is a case of dense canopy conditions. Both cases show the clear distinction between the groups especially in the soil–vegetation boundary

land uses (Ruiz Hernandez and Shi 2018) and agricultural land uses (Vogels et al. 2017)—such as cultivation methods (South et al. 2004). Specifically, crop classification is highly sought-after in water erosion studies, because crops play a major role in reducing soil loss and overland flow. Crops prevent the development of soil crust (Assouline and Mualem 2006), as well as the dispersion of soil particles due to runoff flow. The root system restores organic matter and fosters soil microbial populations around the root system materials—which encourage soil particle cohesion and aggregation. All these increase soil aeration and draining infiltrations, enlarge the soil macropores, and reduce upper and lower runoff.

Crops mapping therefore can be used for:

1. Monitoring the catchment status and change of erosivity;
2. Identifying areas under erosion risk;
3. Recommendations for conservational management.

However, when crops are mapped over large regions, surface heterogeneity as well as technical issues can cause the well-known salt-and-pepper noise that is caused by sharp differences in adjacent image pixels values. One of the solutions to this noise is the *object-based classification* technique.

5.2.1.3 Object-Based Classification

The classification methods described in Sects. 5.2.1.1 and 5.2.1.2 are pixel-based—namely, the algorithms classify the individual pixels based on their own DN values, without any neighborhood effect of the adjacent pixels. The object-based classification method expands this procedure to include the effect of surrounding pixels (Su et al. 2020). This is usually carried out in two steps (Liu and Xia 2010):

Step 1: Aggregation of the image pixels into homogenous spatial objects;

Step 2: Extraction of a classifier to classify the objects into groups.

In Step 1, The analyst aggregates the image pixels into homogenous spatial objects, using existing GIS data from external sources such as the national GIS, or an image segmentation algorithm of some sort (Hossain and Chen 2019). In the segmentation process, several image pixels are grouped into objects that should have similar spectral characteristics because they belong to the same object. Any within-object variation is therefore assumed to be noise—due, for example, to topographic angle, shade, and mixed pixels. The segment boundaries should therefore correspond to real-world entities, such as buildings—or, in agricultural catchments, to the field's boundaries.

In Step 2, a classifier is executed to classify the extracted objects into groups. That is to say, the image classification procedure is applied for *each object and not for each pixel*. For example, in the right-hand panel of Fig. 5.16 we see a Random Forests classification of the Yehezkel catchment, that feeds the classifier with mean values of all image pixels per-field. In this example, the assignment of the field to a particular soil or vegetation class is done based on the mean spectral signatures, of the 12 VIS–*NIR* bands of the *VENμS* satellite, in each object. *VENμS* is a new French-Israeli satellite, with a spatial resolution of 5.3 m and an acquisition cycle of 2 days, specifically oriented to the red-edge spectral bands that are relevant to agricultural applications (https://directory.eoportal.org/web/eoportal/satellite-missions/v-w-x-y-z/venus).

Figure 5.16 demonstrates that an object-based classification does indeed make it possible to eliminate the salt-and-pepper effect—see mainly the southern and most northern agricultural fields. However, in some cases (albeit more rarely), object-based classification may lead to an erroneous result, due to the generalization of spectral signatures at the field level. For example, the third field from the bottom right-hand area of the catchment is actually covered by crops and not a bare soil field as was assigned by the object-based classification. The pixel-based classification clearly shows that most of the

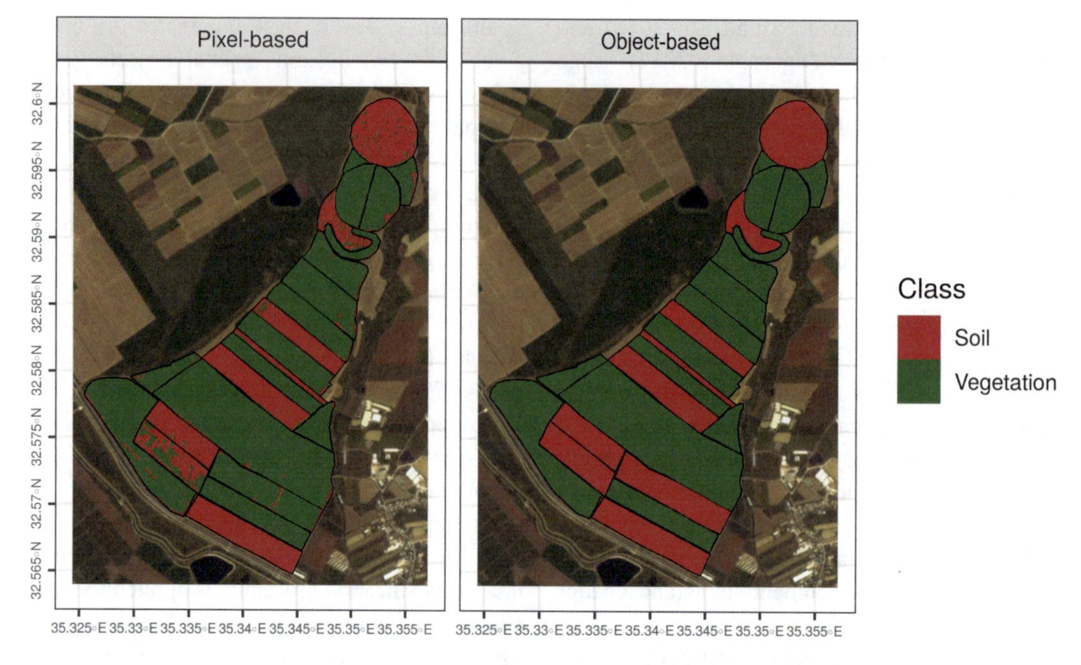

Fig. 5.16 Pixel-based classification (left panel) vs. Object-based classification (right panel) of soil and vegetation in the Yehezkel catchment using a *VENμS* image with 12 spectral bands from 19 April 2020. The objects are polygons of the fields boundaries manually digitized from an orthophoto

field is covered with vegetation, but the object-based classification resulted in its assignment to a soil class, due to its use of a mean spectral signature. Despite this type of possible error, however, object-based classification methods are usually accurate, and are a good planning and soil conservation policy tool, because soil conservation policy is usually designed on a per-grower basis. The object-based classification also provides an efficient tool for *change detection* analysis of cropland mapping, on a field basis (Vogels et al. 2017). However, it should be noted that in more detailed drainage engineering tasks, the user may be interested in mapping parts of the field, to identify crop-cover effects within it, so she will require sub-plot resolution.

5.2.2 Soft Classification—Spectral Mixture Modeling

As was aforementioned, soft classification provides the user with a membership score (based on probability assessment) of the object—the image pixel in the case of RS—to a given class, e.g., rock, soil, vegetation, rather than a Boolean assignment of yes or no. Spectral unmixing modeling allows computing the fraction of each of the image pixel components within the pixel boundaries and therefore can enable to compute a 0–1 membership score of the pixel to each of the mapped classes. There are other examples of soft classifiers and they will be described in Chap. 6.

5.2.2.1 Theoretical Basis

The image pixel DN values epitomize the spectral radiance measured by the sensors on the platform. Due to the high variability of the earth surface in general, and in agricultural catchments in particular, the recorded radiance comes from mixed pixels. The pixel DN is therefore an integrated sum of the radiances of several components (such as soil, rock, and vegetation) within the sensor's field of view, rather than the radiance reflected from a single homogenous component (Quintano et al. 2012). As a result, using the hard classification techniques described in Sect. 5.2.1, the mixed-pixels spectra may cause inaccurate classification results, because mixed pixels may cause intertwined spectral signatures. As an alternative, spectral mixture techniques are applied to the sub-pixel domain, and provide more detailed information on the proportional *coverage* of each surface component within the pixel area.

Spectral Mixture Modeling (SMM) is a mathematical method used to determine the relative contribution of each surface component to the total reflected radiance recorded in a given pixel. The relative spectral contribution is assumed to be proportional to the surface *coverage* of that component within the pixel. Thanks to its strong theoretical basis (e.g., Adams et al. (1986)) and wide empirical validation (e.g., Zhu et al. (2016), Svoray et al. (2004)), SMM is thought to be a reliable classification technique, and is usually applied to multispectral or, preferably, hyperspectral images. SMM has been primarily applied in geophysical studies to map surface mineralogy and has provided efficient information about composition and mineral abundance. SMM was developed and tested in its early stage for relatively sterile laboratory data and was later applied to in situ soil data using sensors mounted on airborne platforms (Ben-Dor and Kruse 1995). SMM has been also used to map eroded areas, based on quantification of the proportions between rock/soil/vegetation components—on the assumption that areas with higher contribution of rock fractions, and lower soil and crop fractions, are more degraded (Li et al. 2015).

Three principles form the foundation for using SMM for studying water erosion, and more specifically, for mapping eroded areas (Metternicht and Fermont 1998):

1. The within-pixel mixture of the spectral signatures of the surface components is assumed to be linear. That is to say, it is assumed that the sun-sensor radiation beam interacts with only one surface component. This assumption may not always be true in heterogeneous areas, where multiple scattering may occur between the earth surface components;
2. The identification of eroded features includes an attempt to unmix between soil and parent material (the underlying rock) of exposed areas—i.e., the spectral signature of these two surface components must be significantly different;
3. The procedure requires a "pure" state of a sufficient number of pixels that can be properly identified and spatially located, for end-members to be used as training sets.

When these three conditions are met, SMM can be applied using readily available multispectral satellite images, to cover large agricultural catchments.

As was shown in Chaps. 2 and 3, water erosion processes remove the topsoil layer, resulting in a decrease of organic matter and other chemicals such as iron oxides content. Over time, more parent material is exposed, and the originally homogenous field becomes patchier, with soil, eroded soil, and rock alterations. The user can identify the eroded areas by dissociating between the soil and the rock spectral mixture. In the case of calcium carbonate rocks, this is relatively easy, as there is a large spectral contrast between the white rocks and the brown soil. Nevertheless, Svoray et al. (2004) have shown that in basaltic soils, too, there is sufficient spectral difference to apply SMM as a soft classification technique to unmix the soil from the exposed rocks—even if the rocks are of a dark hue.

Spectral distinction between vegetation types using unmixing methods—to dissociate between different kinds of crops and weeds, for example —is more difficult because, due to the complex radiation interaction between the leaves within the vegetation canopy, the spectral mixture is non-linear. Non-linear unmixing process requires complex equations to treat the non-linear effects of shadowing, leaf orientation, leaf size, etc. (Ichoku and Karnieli 1996). In the case of vegetation strata, therefore, other approaches—such as *multi-temporal unmixing* methodology—may be needed (Shoshany and Svoray 2002). Such a multi-temporal approach makes it possible to dissociate between the vegetation types based on their phenological phases, rather than by their spectral signatures. In another example, De Jong et al. (1999) used SMM applied to Landsat TM images, to assess vegetation abundance in Sicily, independent of the soil background. Due to the complexity involved in unmixing vegetation types and vegetation with other components, many researchers use linear unmixing models mainly to dissociate between soil and rock fractions over exposed soils when the crop strata are removed.

Since SMM has been applied successfully on various RS platforms—such as Viking (Adams et al. 1986) and Landsat TM (Svoray et al. 2004)—it may be a useful addition to the toolkit of soil conservationalists when seeking to identify exposed areas under degradation risks. SMM is also important for mapping the status of the remaining soil, and variations in soil health (see Sect. 7.1). In such instances, SMMs are used to map exposed rocky areas, as well as the increase in stoniness. The following sections describe a procedure for mapping eroded soil by unmixing soil from rock endmembers.

5.2.2.2 Procedure and Results

This section describes the governing equations and necessary steps for the use of SMM to map eroded areas in agricultural catchments. Such a procedure can be applied using multispectral and hyperspectral images and readily available image-processing software tools including, for example, ENVI (https://www.harrisgeospatial.com/Software-Technology/ENVI) or ERDAS Imagine (https://www.hexagongeospatial.com/products/power-portfolio/erdas-imagine). The procedure described here was applied using Landsat 8 data of the Harod catchment, with R—RStoolbox: Tools for Remote Sensing Data Analysis. R package version 0.2.6. https://CRAN.R-project.org/package=RStoolbox developed by Benjamin Leutner, Ned Horning and Jakob Schwalb-Willmann in 2019.

SMM aims at solving the mixed spectral reflectance from the RS image pixel, using pure endmember spectra (Eq. 5.4 from Adams et al. (1986)):

$$Ref_{bn} = f_{n1}EM_{1b} + f_{n2}EM_{2b} + e_b, \qquad (5.4)$$

Ref_{bn} denotes the reflectance data in a spectral band $\boldsymbol{b}$ for a given pixel n; f_{n1} is the relative contribution (fraction) of the first endmember; EM_{1b} is the typical reflectance of the first endmember; and e_b represents the residuals and noise component.

An *endmember* is a surface cover that lies at the extreme end of surface cover phases, in terms of pureness of the spectral signature. Surface covers such as rocks often appear in nature in various mixtures with other components, such as soil patches or vegetation cover; in these instances, the endmember class used for training the classifier must be pure as possible. The term endmember is derived from mineralogy—meaning a mineral that is pure, unadulterated by other minerals.

Equation 5.4 is applied to a RS image, inverted and solved using a least-squares regression, under the constrains that $\sum_{i-1}^{n} f = 1$. For mathematical reasons, the number of surface components cannot exceed the number of spectral bands, plus one. The procedure is illustrated in Fig. 5.17.

The application of the SMM in Eq. 5.4 to an agricultural catchment is done as follows (see also Fig. 5.17):

Step 1: Image data is selected and preprocessed: Geometric, radiometric, and atmospheric corrections are applied;

Step 2: A sufficient number of plots of endmembers, e.g., pure soil and pure rock plots, are delineated on the images, based on orthophoto analysis and/or field visits, using GPS, for example. Unmixing of soil/rock *coverage* must be applied in the dry season, when the vegetation cover is low, and the soil and rock can be observed by the sensors. Previous studies (e.g., Svoray et al. (2004)) have used 20 to 30 sites—but fewer replications can be used if there are administrative constraints. Naturally, the more pure sites one has, the more one can express the surface heterogeneity, thereby enhancing final *accuracy*;

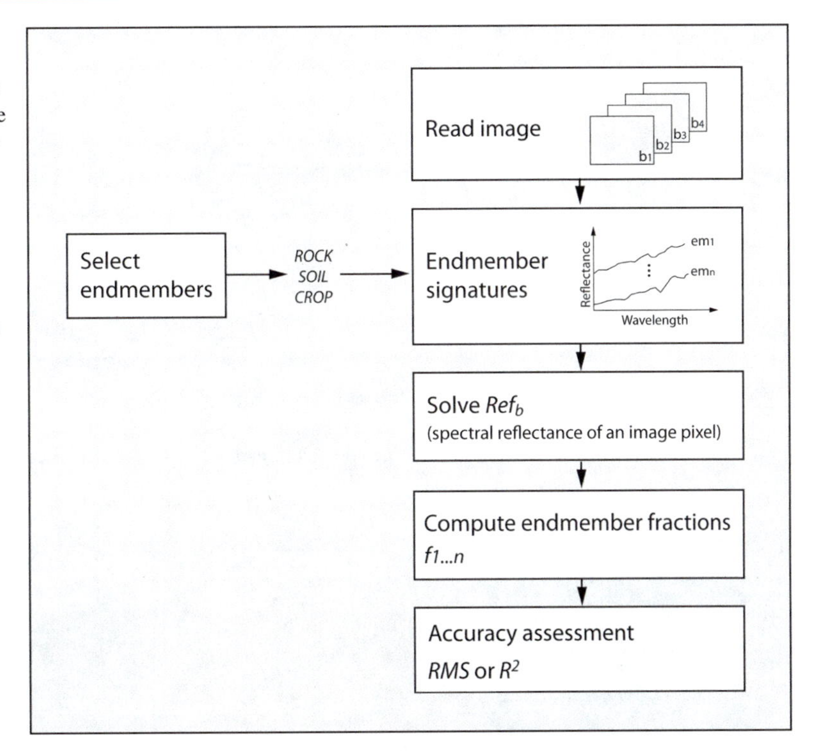

Fig. 5.17 The spectral unmixing procedure. The multi/hyperspectral image is read and preprocessed, while endmember components are identified and sampled to quantify spectral signatures. The unmixing solution is applied, the within-pixel fractions are computed, and *accuracy* assessment is applied, against a validation set measured in the field

Step 3: Mean values of spectral signatures of the pure endmembers—in our example rock and soil—of the study site are computed from the image data;

Step 4: Equation 5.4 is applied and inverted, to produce the soil and rock cover fractions in each pixel;

Step 5: Because pure endmembers may vary between catchments—and, with it, the *accuracy*—it is useful to validate the results against field measurements of rock cover [%].

Such a validation process can be applied using ground photography. The percentage cover of the two endmembers is quantified for the entire site as the mean value from all the photographs. Next, the rock *coverage* from the SMM and that measured in the field, are regressed and a significant linear correlation is expected when the SMM is a good predictor. Figure 5.18 shows the endmember spectra of soil and rock in the Harod catchment, based on a Landsat 8 image in the summertime.

As evident in Fig. 5.18, there is a clear distinction between the vegetation, the rock, and the soil spectra. This difference has been already discussed in the previous section. The rock and soil spectra are also substantially different, as more energy is reflected from the rocks than from the soil components. While the soil comprises minerals from the parent material, it has undergone various pedogenetic processes, so its spectral characteristics differ from those of the original rock. In this and in other studies, this difference has proven to yield high-*accuracy* SMM-based maps.

Figure 5.19 confirms that the SMM maps of two parts of the Harod catchment do indeed differ from one another: The fields with more yellowish hues in the rock fractions indicate fields undergoing more intensive erosion processes. The fields with high vegetation cover mask the dissociation between rock and soil fractions, but in other cells the soil/rock ratio can indicate the degradation level. This is definitely an added value to hard classification, as it makes it possible to quantify the ratio within the pixel level.

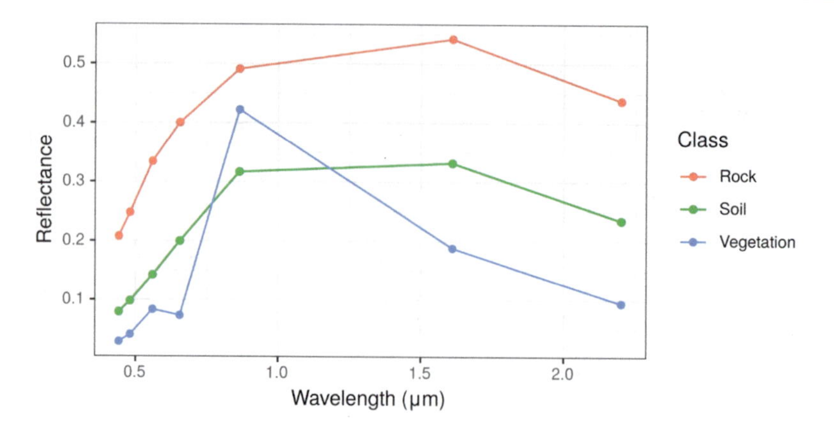

Fig. 5.18 Endmembers of pure pixels of vegetation, soil, and rocks, in the Harod catchment

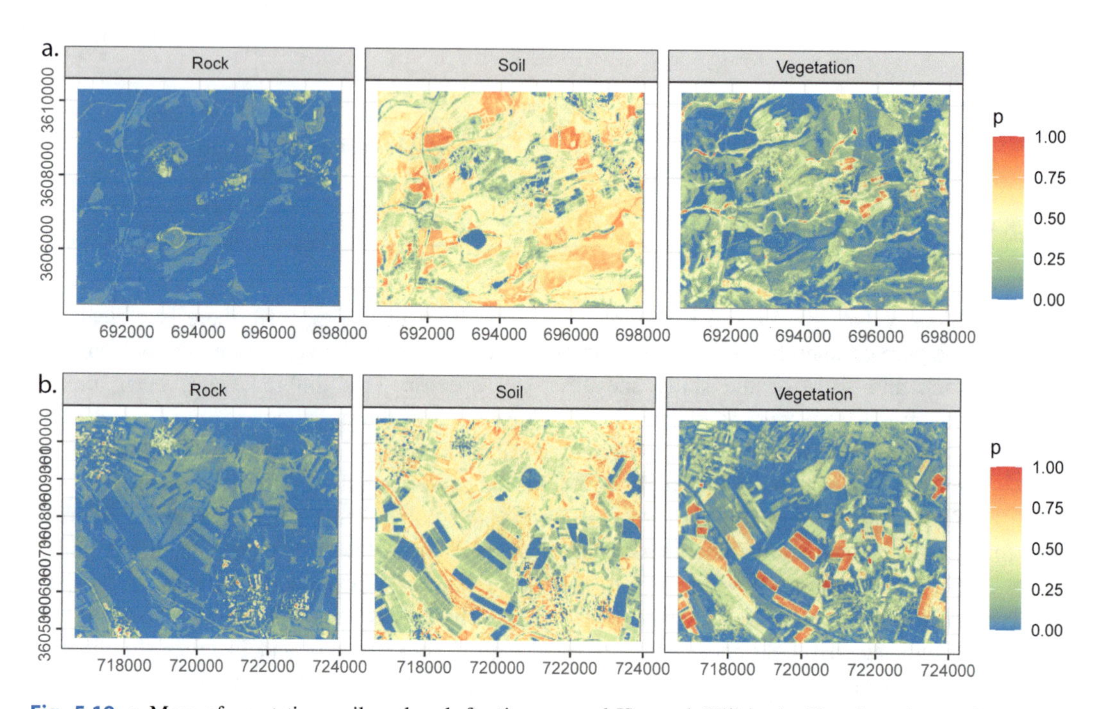

Fig. 5.19 **a** Maps of vegetation, soil, and rock fractions around Hamoreh Hill in the Harod catchment; **b** Maps of vegetation, soil, and rock fractions in the Yehezkel catchment. The p-value represents the percentage of cover of each of the endmembers in the image pixel *n*

As with all RS analyses, in SMM processing the user strives to ensure sufficient prediction *accuracy*. Whether the extracted empirical endmember signatures are measured in the field, or derived from laboratory spectra available in the literature, they limit the error of the predicted cover fractions *f*. The user can increase representativeness by increasing sample size, or by selecting the representative endmember sites as closely as possible. Further improvements in prediction can be achieved by using more spectral bands in hyperspectral platforms. Reducing the within-class variance can be done by selecting the spectral bands that contain the greatest difference between the endmembers—such as the red edge (Cui and Kerekes 2018) between soil and vegetation, and various absorbing bands between soil and rock. In practice, this may require careful exploration of the endmembers' spectra, as described above. Alternatively, a systematic trial and error problem-solving process may be applied against data measurements —but this may be more laborious, and may require specific classification for each site. The effectiveness of SMM for mapping surface components therefore depends on:

1. The quality of the unmixing algorithm;
2. The selection of archetypal endmembers with good representativeness of the field parameter space of environmental conditions (hillslope gradient and aspect, shade, etc.…);
3. Low noise-to-signal ratio in the image.

5.3 Synergy of RS Data in Catchment Models

5.3.1 Background

This section illustrates the use of classified RS data embedded in a DEM-based model, to create a more realistic and updated description of catchments drainage structure. The modeling effort includes determining flowdirections, estimating runoff contributing areas, and extracting channel networks. The methods described in this section were proposed by Svoray (2004), and developed further by Ackermann et al. (2008), Bruins et al. (2019).

Determining drainage structure is a computational task, whose characteristics and dynamic properties depend on intertwined links between rainfall and surface properties. Rainfall exerts control on water balance and is responsible for many infiltration-runoff feedbacks—while topography, surface roughness, and soil/vegetation/rock cover, affect runoff development and flow routes above- and belowground. Quantifying drainage structure in space and time provides basic tools for hydrological modeling and erosion-risk assessment. The subjects involved lie at the heart of erosion studies: Quantifying channel dimensions, and—at larger temporal scales—simulating drainage network development and landscape evolution models (Rodriguez-Iturbe et al. 1994). Quantifying drainage structure also provides a baseline for modeling pathways of mass and energy transport through the catchment in hillslope and channeling processes (Pelletier 2008).

DEM-based methodologies were advanced to automate drainage structure determination using mainly the *eight-flowdirection matrix* method (D8, O'Callaghan and Mark (1984)) which is implemented in two steps:

Step 1: Computing a flowdirection matrix from a DEM;

Step 2: Computing the number of grid-cells that contributes to each of the DEM grid-cells, while using a threshold value to delineate the channel network.

For more details on the procedure, see Sect. 5.3.2. The D8 procedure initially assumed flowdirection as though the water were flowing toward a single line, using the common angle between grid-cells in matrices of 45°. However, it has since been further developed to account for a more realistic water flow route. This modification was done by using the steepest hillslope gradient downslope on the eight triangular facets centered on each DEM point, and computing the contributing area by proportioning flow between two downslope DEM grid-cells, based on how close this flowdirection is to the direct angle to the downslope DEM grid-cell (Tarboton 1997). The modified D8 method is encapsulated in the TauDEM (Terrain Analysis Using Digital Elevation Models) software application (https://hydrology.usu.edu/taudem/taudem5/index.html).

Apart from limitations of the DEM-based algorithms, the *accuracy* of the predicted drainage structure is also affected by the quality of the DEM used for the drainage structure computations. Each DEM contains both arbitrary and deterministic inaccuracies, because of imprecisions in the measurement tools used to create it, but also due to systematic error propagation in the DEM processing procedure (Wise 2000). In particular, contour-based and low-resolution DEMs from satellite data, which are still widely used, suffer low vertical *accuracy*. These sources of error can cause defects in the output drainage layers, such as artifact depressions—especially in plateaus and fan deltas that are characterized by small *elevation* differences, discontinuities in the streams, and coarse flowdirections, such as the known error of right-angle curves in channels which are extremely rare in reality.

Several studies (e.g., Tribe (1992), McCormack et al. (1993)) suggest algorithms to cope with the aforementioned challenges—however, these important studies could not provide a full solution, and their algorithms may be subject to other types of error.

A logical solution to these DEM errors is to use ancillary data, which may be more accurate, detailed, and/or certain than the source DEM (Hutchinson 1989). A pivotal example of the use of ancillary data is provided by Turcotte et al. (2001), who used a Digital River and Lake Network (DRLN) data established from digital maps to improve the modeled watercourse network. Turcotte et al.'s method is based on the idea that for DEM matrix grid-cells that are overlapped with the ancillary channel network, flowdirections are biased toward the channel network using the DRLN nodes.

However, classified RS data can be used as a modern substitute to the digital maps used by Turcotte et al. RS platforms provide readily available, repetitive, independent, and detailed information about the earth's surface. For example, hyperspectral data acquired by the AISA sensor has been used to monitor effects of channel morphology on depth retrieval (Legleiter and Roberts 2005). In another work, identification of water bodies from space has been made possible using MODIS ocean color and SMOS salinity data (Matsuoka et al. 2016). More recently, the Surface Water and Ocean Topography (SWOT) airborne data has been used to measure temporal variations in river water surface *elevation* (Altenau et al. 2019).

The use of RS as an ancillary data source in DEM-based modeling of drainage structure (Fig. 5.20) is based on the following four assumptions:

1. RS images are usually of higher spatial resolution than DEMs, which are frequently extracted from RS images using digital photogrammetry;
2. Channels are clearly identified in RS images, because they absorb light, due to the shadowing effect of channel banks, and therefore have significantly different DN values than the exposed hillslopes;
3. RS images provide information on surface entities—such as soil, rock, and vegetation—so they can be exploited to assign runoff coefficients;
4. RS data can also introduce errors into the DEM analysis, due to errors encountered, for example, with classification techniques (Livne and Svoray 2011), or cloud cover and the salt-and-pepper effect.

These assumptions illustrate the potential and possible shortcomings of the integrated approach, and the reason for cautious use of image-processing tools, while still preserving the most important topographic features and hydrological processes.

While RS data has its own limitations, with the increased availability of RS platforms from drones to satellites, its benefits should grow over time. The following sections provide a description of the synergy of RS data in a DEM-based model for drainage structure quantification. The RS is used in this procedure for two purposes:

1. To improve the channel network extraction process;
2. To assign runoff coefficients to the flowaccumulation grid-cells.

5.3.2 The Procedure

Embedding the RS data in the D8 algorithm is executed as described in Fig. 5.21 with the image-processing technique for the identification of the channels from an RS source—in this case a *VENμS* satellite image. The center of Fig. 5.21 describes the DEM analysis based on D8 and the

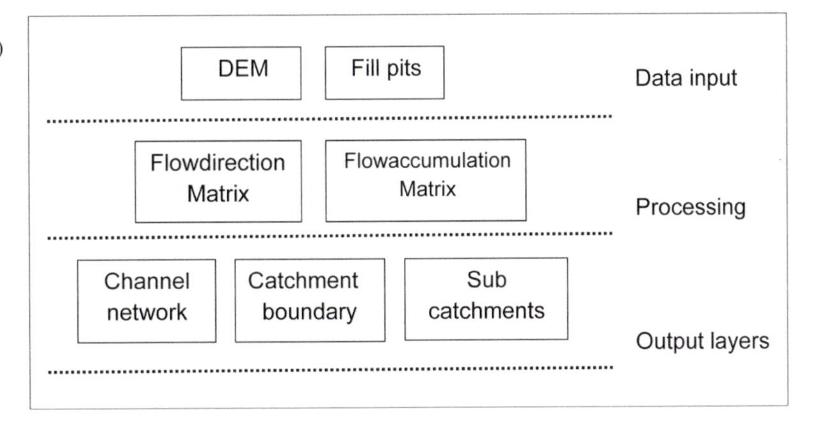

Fig. 5.20 The eight flow direction matrix method (8d) for drainage structure modeling using DEM as a data source

right part of the figure describes the validation process. The overall procedure steps go as follows:

Step 1: Apply image-processing algorithms, to identify the channel network from RS data;

Step 2: Practice the "stream-burning" approach to bias computations of flowdirections from the DEM toward the channel network—identified in the RS data—and thus improve channel network prediction. In other words, the second step is to "Burn" the channel network identified in Step 1 into the DEM;

Step 3: Apply the D8 algorithm, using the "burned" DEM, to bias flowdirection computations toward the grid-cells overlapping the channel network identified in the image;

Step 4: Compute flowaccumulation using runoff coefficients from RS data and set a threshold value to dissociate hillslope grid-cells from channel grid-cells;

Step 5: Test the predicted channel network against validation data, using *accuracy* assessment indices.

5.3.2.1 Data

A *VENμS* image acquired at July 15, 2018 was geometrically, radiometrically, and atmospherically corrected (Fig. 5.22). Although the Harod channels are not easily identified visually, the image does provide important information on these lineaments, which can be identified through advanced image-processing techniques. In this example, the most efficient *VENμS* spectral band for channel detection was band 5 with a center wavelength of 620 nm. The photogrammetric DEM with horizontal resolution of 30 m was extracted from SRTM (Shuttle Radar Topography Mission). SRTM data can be freely downloaded from https://www2.jpl.nasa.gov/srtm.

5.3.2.2 Image Processing Procedure

The image-processing procedure used here was applied in four steps (Svoray 2004). Specific configurations of each procedure and the threshold values must be determined empirically for every site.

Step 1: Filtering elongated features

The first step is to enhance the channels in relation to the surrounding hillslope pixels. This can be done due to the shadowing effect of the channel's banks that causes light absorption, which decreases the spectral reflectance and

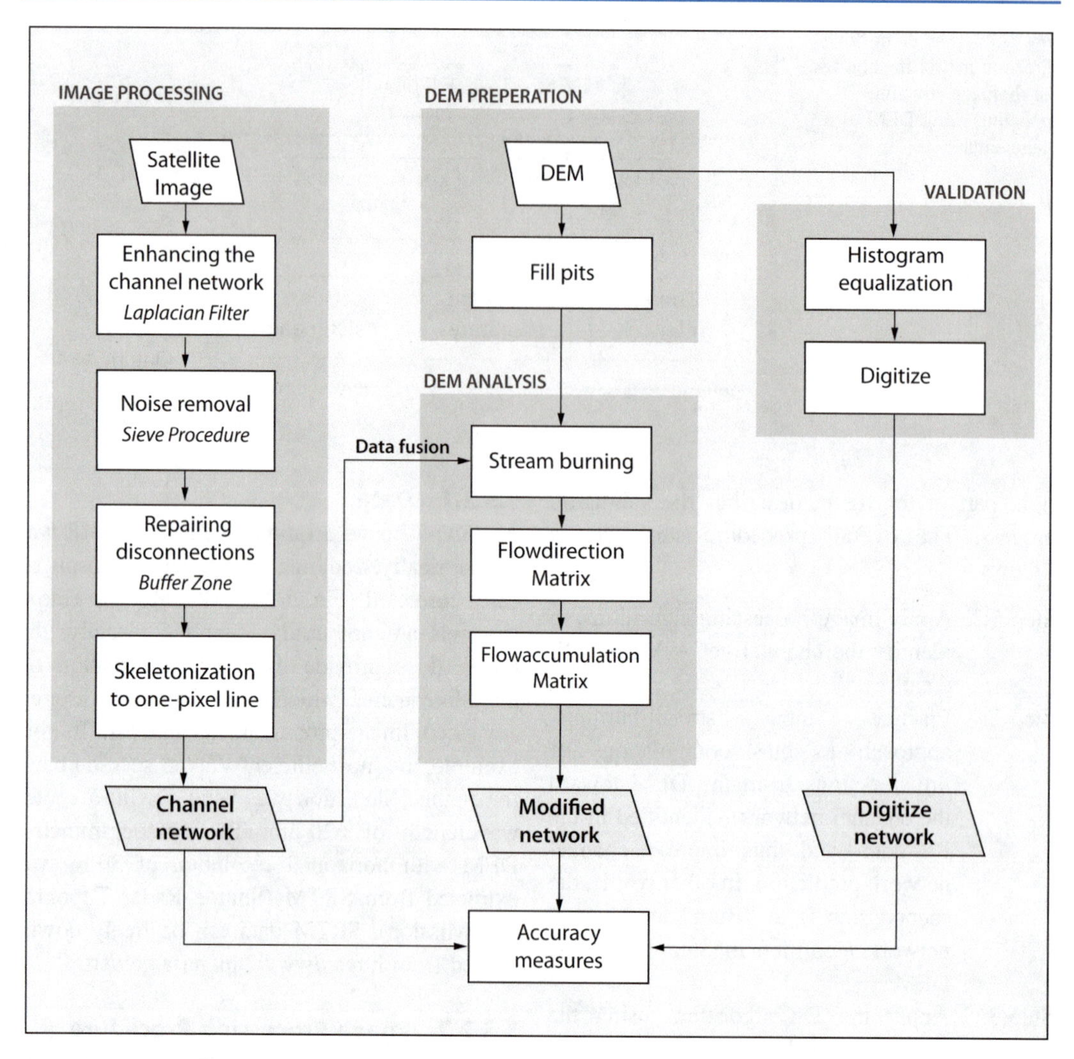

Fig. 5.21 A flowchart of the channel network procedure, using image data embedded in the DEM-based procedure. The left part of the figure includes the four steps applied for identification of a channel network in a satellite image (or any other RS source); the center part presents the data fusion of the RS-based network in the DEM and the DEM analysis D8 and the right part of the image is the validation process

assigns them with low values compared with the exposed hillslope pixels. Even if this effect cannot be seen by the naked eye, it provides sufficient contrast for identification of the lineaments. A 3 · 3 *Laplacian filter* is used to detect the edge between hillslope and channel pixels. This was executed using the following filter configuration: −20 for the center pixel; 4 for horizontally and vertically nearest neighbors; and 1 for nearest neighbors on the diagonal. Laplacian filter identifies image pixels with sharp color difference because it is the outcome of the spatial derivative in *X* and *Y* directions. The outcome of the Laplacian filter execution has yielded an inverted enhancement of the channel pixels over the hillslope pixels with channel pixels as edges becoming bright in the output and hillslope pixels becoming darker.

Step 2: Noise removal

In Step 2, we eliminate the hillslope pixels, to remove background effects which are basically noise to the channel network. To do so, we apply two procedures. In the first procedure we set a

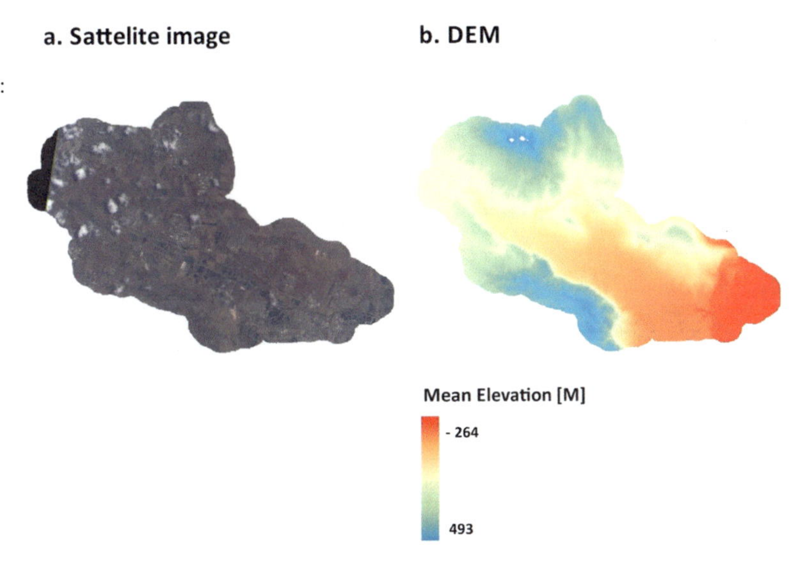

Fig. 5.22 The input data needed for the channel network extraction algorithm: **a** A *VENμS* image with 5 m spatial resolution, and **b** A SRTM DEM with 30 m spatial resolution

threshold value (DN = 5 in our example here—see Fig. 5.23) to the image DNs in the output layer of Step 1. Image pixels with DN values below this threshold are set to 0. Because hillslope pixels are characterized by low-DN values, due to the Laplacian filter manipulation, they are assigned 0 and the channel pixels remain 1.

In the subsequent procedure, clumps—collections of contiguous image pixels with DN = 1—of a size >3 pixels, that were not erased in the previous step, were set to 0 using the *raster-GIS Sieve procedure based on their small size*. The sieving procedure is efficient, because the channel pixels (DN = 1) create patches >3 pixels, and therefore remain as channel pixel during this procedure run, while a large number of small clumps that generate salt-and-pepper noise in the process are omitted in the outcome of Step 2.

Step 3: Buffering

The output of Step 2 is a layer with enhanced channels and is relatively free of salt-and-pepper noise of 1–3 pixel size clumps of hillslope. In Step 3, to minimize the amount of disconnections in the channels, it is extended by implementing a 1-pixel buffer zone. This buffering step is aimed at repairing disconnections <3 pixels in size, subject to their configuration in space. Indeed, from the empirical point of view, performing Step 3 was found to substantially reduce the number of disconnections.

Step 4: Skeletonization

The buffered channel network is wider than in reality, and must be *skeletonized* to one-pixel line to direct the biased flowdirection to the center of the channel. The skeletonization procedure deletes superfluous cells, based on their configuration in space, to create a single grid-cell line network. One of the most commonly used skeletonization procedures is the Riazanoff et al. (1990) code. This was coded in Svoray (2004) and applied to create streams one pixel in width.

Figure 5.23 shows how the channel image pixels are enhanced after the computation of the Laplacian filter; see the pixel within the black rectangle. This allows a better distinction between the two landscape features and a more certain determination of the cutoff point (see Step 2). This 0 (hillslope pixels) and 1 (channel pixels) layer was clumped into clumps (a collection of touching pixels) of different size using the clump operation and is being sieved to remove the small "channel" clumps. (Clump number 3 with 2 pixels and clump number 4 with 1 pixel.) Note especially in the next step in Fig. 5.23 how the disconnections in the channel network are being repaired by the buffer operator. That is because the B pixels are assigned as part of the channel network. Finally, the skeletonization process in Fig. 5.23 shows how redundant touching pixels of the main channel network are

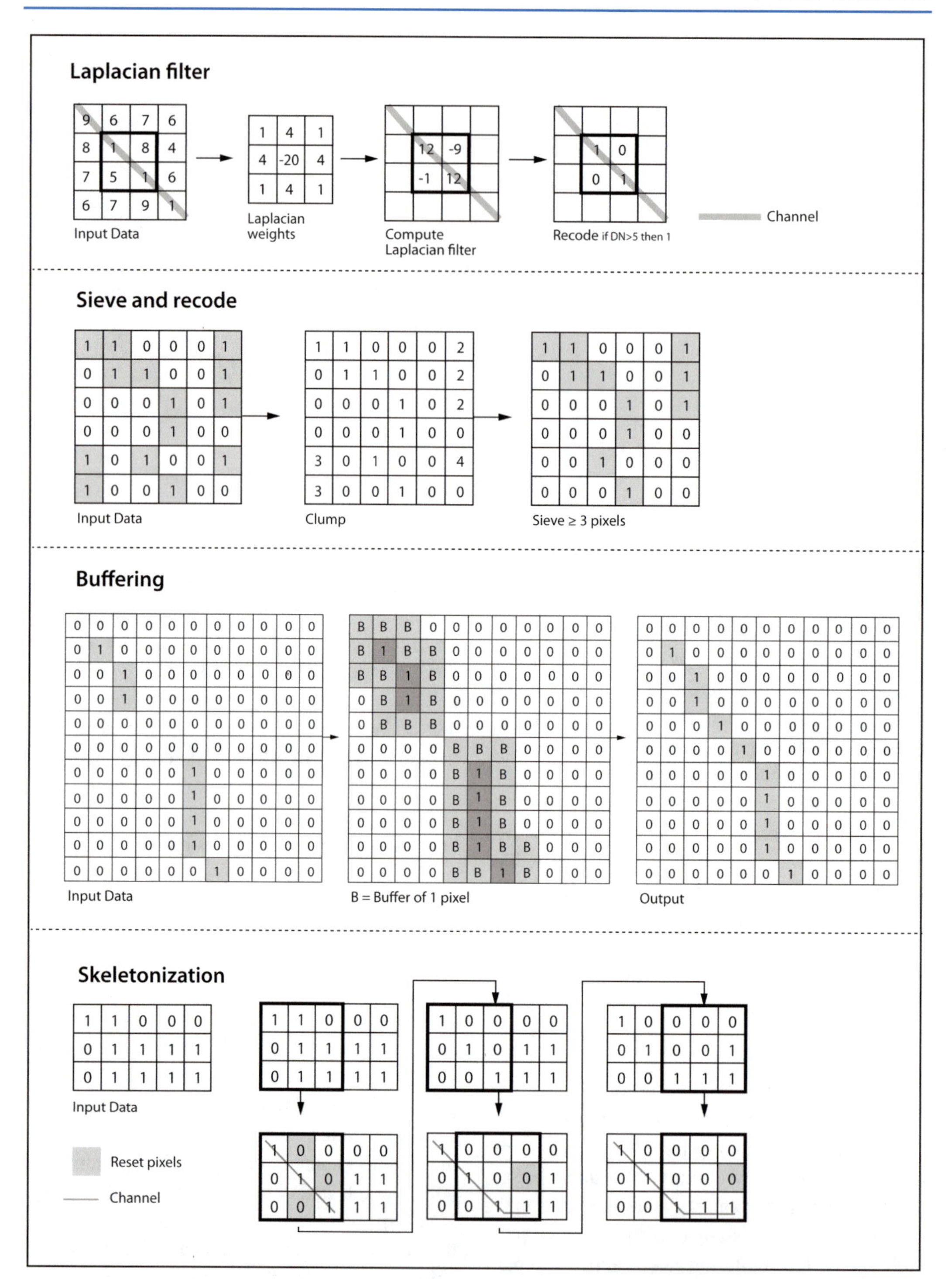

Fig. 5.23 Computation of raster data to extract the channel network from an RS image data. The Laplacian filter shows how the specific weights configuration enhances the channel data compared with the hillslope pixels. The *Clump and Sieve* step shows how that data is recoded based on contingency and the clumps <3 pixels are erased. Then the *buffer* step illustrates how disconnections are connected with the buffer. Finally, the skeletonization step shows how pixels that are > from one-pixel line are erased

being removed to create a one-pixel line channel network.

Figure 5.24 shows the outcomes of the execution of the four steps described above to *VENμS* data of the Harod catchment. One can clearly see the enhancement of the channel vs. the hillslope pixels, the removal of the small clumps, and the conversion of the channel network into a one-pixel line layer.

These four steps presented here were executed to various satellite images (Landsat, SPOT) from the summer of 2018. The use of *VENμS* was found to yield the highest *accuracy* compared with field data from the Survey of Israel.

5.3.2.3 The DEM-Based Algorithm and Data Fusion

This step can be carried either with D8 or Tau-DEM—both readily available GIS software tools. The drainage structure determination algorithm involves:

Step 1: DEM depressions—DEM grid-cells that all their neighboring grid-cells are higher than them—are filled, by increasing their *elevation* to a level that water can flow downslope from the depression to the lowest adjacent grid-cell. That is done to avoid confusion when computing the critical hillslope gradient in the next step;

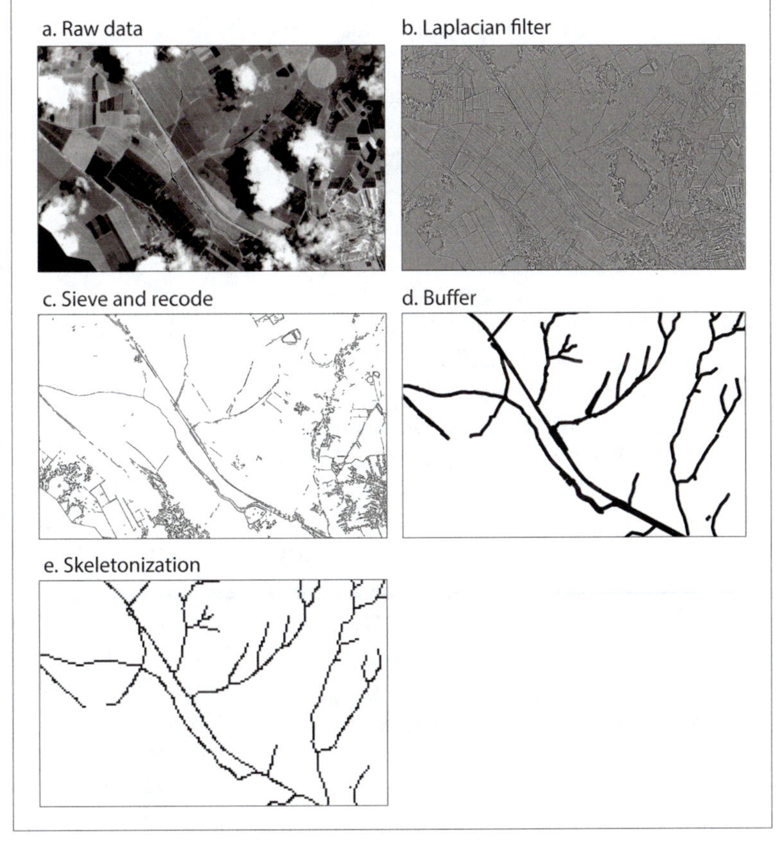

Fig. 5.24 An illustration of the procedure to extract a channel network from an RS image. **a** Shows the raw data of one of the spectral bands of *VENμS* also showing the effect of other linear features, such as roads, that may introduce errors. These can be eliminated using information such as a road map. **b** The output of the Laplacian filter, that enhances the small channels, as evident from the image. **c** Shows patches of hillslope pixels that were marked by the sieve procedure. **d** Shows the buffer processing that enhances the channels width **e** The skeletonization procedure once again thinning the patches to the center of the channels

Step 2: Computing flowdirections to each raster grid-cell based on the steepest descent approach. Namely, the direction of the given pixel n to the grid-cell that has the steepest hillslope gradient with it, is assigned as its flowdirection;

Step 3: Flowaccumulation matrix is computed, by assigning a counter to each grid-cell that counts the accumulated number of grid-cells that flow into it, as determined by the flowdirection matrix. The grid-cell contribution can be normalized to the runoff coefficient of the contributing grid-cells using RS data (see Sect. 5.3.2.4);

Step 4: The grid-cell is assigned a channel if its flowaccumulation value exceeds a user-defined threshold value. Namely, Step 4 includes extracting the channel network, using a conditional operator based on a threshold value to be determined empirically for each catchment;

Step 5: Catchment and subcatchment boundaries are delineated, based on the flowdirection toward the catchment outlet or network branching point.

The DEM can be extracted from LiDAR (Light Detection and Ranging) data or stereoscopic aerial photos using photogrammetry—but often it is contour-based DEM or SRTM data, which offers spatial resolution of tens of meters. One precondition of the data-fusion procedure is to adjust the RS data, and in this case *VENμS* image and the DEM data, to the same coordinate system grid. Geometric correction must be done at below grid-cell resolution, so that all DEM grid-cell locations fit to their spatially corresponding image pixels. Since the DEM spatial resolution is usually coarser, the image-based channel network pixels need to be resampled to DEM grid-cell size, using the nearest-neighbor algorithm.

Once geometric correction is applied, the RS information is merged with the DEM data, by means of *stream burning*—i.e., lessening the *elevation* in the DEM grid-cells that overlap the RS-based channel network and their surrounding grid-cells in a gradual manner (Lindsay 2016). This can be materialized using Eqs. 5.5–5.7:

$$E_{\mathrm{MOD}(i,j)} = E_{(i,j)} - P_{(i,j)}, \tag{5.5}$$

where $E_{\mathrm{MOD}(i,j)}$ is the new *elevation* layer with grid-cells located each in the *i, j* location; $E_{(i,j)}$ is the original *elevation* value from the original DEM; and $\mathrm{P}_{(i,j)}$ is the value to lower the original DEM value to bias the flowdirection toward the channel network identified in the RS data.

The $\mathrm{P}_{(i,j)}$ value was assigned by Turcotte et al. (2001) as a coefficient of the located grid-cell *i,j* that gradually increases toward the image-based channel network. As a result $E_{MOD(i,j)}$ is lowered gradually as the grid-cell *i,j* is closer to the image-based channel network. This was done by using an inverse power-law function (Eq. 5.6)—where R_m is the maximum distance increment (for example with a maximum radius of 3 grid-cells from the image-based channel network); $R_{(i,j)}$ is the distance between each grid-cell in the DEM layer and the nearby image-based channel network [# of grid-cells]; and α is the inverse exponential coefficient.

$$\mathrm{P}_{(\mathrm{i,j})} = \frac{1}{2} \cdot \left(\frac{R_m}{R_{(i,j)}}\right)^{\frac{1}{\mathrm{a}}}. \tag{5.6}$$

Thus:

$$\alpha = \frac{\ln(R_m) - \ln(R_m - 1)}{\ln(3)}, \tag{5.7}$$

α is to be calibrated empirically, using the maximum radial influence to be decided by the user. For example, if the user decides on a radius of 3 grid-cells, $R_m = 3$ yields an α of 0.37 and as a result, $\mathrm{P}_{(i,j)}(3) = 0.5$, $\mathrm{P}_{(i,j)}(2) = 1.5$, $\mathrm{P}_{(i,j)}(1) = 9.8$. Thus, the inverse power-law function forces greater values for grid-cells near the channel network and it tends to reduce substantially away from the channel network. The $E_{(i,j)}$ values in the first neighborhood are subtracted by $\mathrm{P}_{(i,j)} = 9.8$, the next by $\mathrm{P}_{(i,j)} = 1.5$, and the last by $\mathrm{P}_{(i,j)} = 0.5$. Beyond the point and within the channel network $\mathrm{P}_{(\mathrm{i,j})}$ (3) the original DEM values are used. The

Fig. 5.25 The reduction of P (i,j) as a function of distance from the DRLN. Note the outcome of smaller reduction in $E'(i,j)$ with increasing distance from the DRLN. $E(i,j)$ is the DEM, the DRLN is the channel network from ancillary source, $P(i,j)$ is computed from Eq. 5.6 and $E'(i,j)$ is computed from Eq. 5.5

a. **E(i, j)**

21	20	19	25	29	20	23	23
20	20	18	23	28	19	23	22
23	24	16	23	29	19	23	22
23	26	16	23	21	25	27	19
24	23	15	23	25	25	27	15
25	22	12	23	24	25	24	23
20	19	14	24	23	23	24	24
19	18	15	23	23	23	24	24

b. **DRLN path**

21	20	19	25	29	20	23	23
20	20	18	0	28	19	23	22
23	24	16	0	29	19	23	22
23	26	16	0	0	25	27	19
24	23	15	0	0	0	0	15
25	22	0	0	0	0	24	23
20	0	14	24	23	23	24	24
19	18	15	23	23	23	24	24

c. **P(i, j)**

0	0	1	2	1	0	0	0
0	1	2		2	1	0	0
0	1	2		2	1	1	0
0	1	2			2	2	1
0	1	2					2
1	2					2	1
2		2	2	2	2	1	0
1	2	1	1	1	1	0	0

d. **E′(i, j)** = E(i, j) - P(i, j)

21	20	18	23	28	20	23	23
20	19	16		26	18	23	22
23	23	14		27	18	22	22
23	25	14			23	25	18
24	22	13					13
24	20					22	22
18		12	22	21	21	23	24
18	16	14	26	22	22	24	24

functions to determine $P_{(i,j)}$ may vary due to localities.

After the RS stream network is burned into the DEM, we apply the flowdirection algorithm on the output data to determine the flow path of all DEM grid-cells in the modified DEM. Once we have the flow paths we can compute the contributing area with the flowaccumulation algorithm (Fig. 5.25).

5.3.2.4 Weighting Flowaccumulation

When the counter counts the number of grid-cells to assign to the flowaccumulation matrix, it assumes that each grid-cell contributes runoff from its entire area to the immediate downslope grid-cell. In other words, it assumes zero infiltration in the contributing grid-cells. This assumption may be unrealistic, because grid-cells covered with vegetation—or even bare soil grid-cells—contribute less runoff downslope than less-permeable cells that are covered with rocks, or roads. To compensate for this variation in permeability and runoff contribution, RS data is used to classify the grid-cell cover type, and to assign the consequent runoff coefficients to the grid-cell, thereby assigning it a corresponding weight. The runoff coefficient/weight can be assigned based on field measurements (e.g., Table 5.3).

Table 5.3 Runoff coefficients, based on empirical field studies

References	Surface cover	Hillslope length [m]	Runoffcoeff	Rainfall [mm]
Yair and Kossovsky (2002)	Rock	33–63	0.51–1	1–3
Yair and Kossovsky (2002)	Bare soil	4–20	0.015–0.37	4
Lange et al. (2003)	Rock\soil	18	0.16	16
Lange et al. (2003)	Vegetation	18	0.8–0.9	53

The procedure for computing flowaccumulation with weights is operating as follows:

Step 1: The RS source image is classified as belonging to one of three groups: Bare rock, vegetation, or bare soil. More classes with unique runoff coefficients may be added, such as roads and cemented areas. The spectral signatures of vegetation, soil, and rock are very different, and can therefore be easily identified, using one of the classification procedures described in Sect. 5.2;

Step 2: Contributing area is computed, using D8 or TauDEM. The flowaccumulation in each cell is computed using a weighted linear combination, with weights assigned based on the surface cover (Table 5.3).

The weighted flowaccumulation is computed using Eq. 5.9—but first, Eq. 5.8 shows the general formula for computing flowaccumulation:

$$Flowacc_{ij} = \sum_{i=1}^{n} P_{dir}, \quad (5.8)$$

where $Flowacc_{ij}$ is the accumulated runoff in the *ij-th* grid-cell [# of grid-cells], n is the number of pixels in the flow path and P_{dir} represents all the cells directed toward the given *ij-th* grid-cell. Equation 5.8 is simply a counter that counts the grid-cells directed toward each grid-cell in the flowaccumulation matrix. To express the effect of reduction in contributing area due to water infiltration in each grid-cell, based on runoff coefficients, using the weighted linear combination in Eq. 5.9, use the formula:

$$\mathrm{Flowacc}_{\mathrm{wei}(i,j)} = R \cdot W_R + S \cdot W_s + V \cdot W_v, \quad (5.9)$$

where $\mathrm{Flowacc}_{\mathrm{wei}(i,j)}$ is the accumulated runoff in the *ij* cell [# of cells]; *R is* the number of cells covered by rocks and W_R is the weight (runoff coefficient) of the rocks; *S* is the number of cells covered by soil, and W_S, is the weight (runoff coefficient) of the soil; *V* is the number of cells covered by vegetation, and W_V is the weight (runoff coefficient) of the vegetation (Fig. 5.26).

Once weighted contributing areas are assigned, we can compute the annual water depth for each grid-cell, *AWP* [mm] using Eq. 5.10:

$$AWP = ADR + ACR, \quad (5.10)$$

where *ADR* is annual rainfall amount, and *ACR* is the annual contributed runoff, both measured in [mm]. *ACR* is the sum of contributing runoff $\mathrm{Flowacc}_{\mathrm{wei}(i,j)}$, as computed in Eq. 5.9. Due to the differences in runoff coefficients under saturated and unsaturated conditions (Lange et al. 2003), the computation of runoff water depth is divided into two phases as shown in Eq. 5.11: Unsaturated conditions *unsat*, and saturated conditions *sat*.

$$ACR = \left(\sum_{i=1}^{n} DR_i \cdot \frac{CA}{FA}\right)_{unsat} + \left(\sum_{i=1}^{n} DR_i \cdot \frac{CA}{FA}\right)_{sat}, \quad (5.11)$$

where n denotes the quantity of rainfall events in a hydrological year and i is the grid-cell; *DR* denotes rainfall depth [mm]; *CA* is the weighted flowaccumulation and *FA* is the agricultural field area—both measured in [m^2]. DR_{unsat} is the cumulative rainfall amount that is computed for rainfall events of 3–53 mm rainfall depth; similarly, DR_{sat} is the accumulated rainfall amount in every rain event of 53 mm depth. Below 3 mm rainfall, it is assumed that no runoff occurs. Based on Ackermann et al. (2008), in the unsaturated phase *unsat*, the runoff coefficient for areas covered by rock outcrops (W_R in Eq. 5.9) was assigned 0.51, and bare soil W_S and vegetated W_V grid-cells 0.015. In the saturated phase *sat*, the runoff coefficient for areas covered by rock outcrops was raised to 0.9, and for soil and vegetated surface to 0.07. *CA/FA* is the ratio between the computed weighted flowaccumulation that feeds the field and the field area. The act of $\mathrm{DR_i} \cdot \frac{\mathrm{CA}}{\mathrm{FA}}$ represents runoff subsidy, in physical units equivalent to rainfall depth. This can be true if we assume that runoff distributes equally in space in the field.

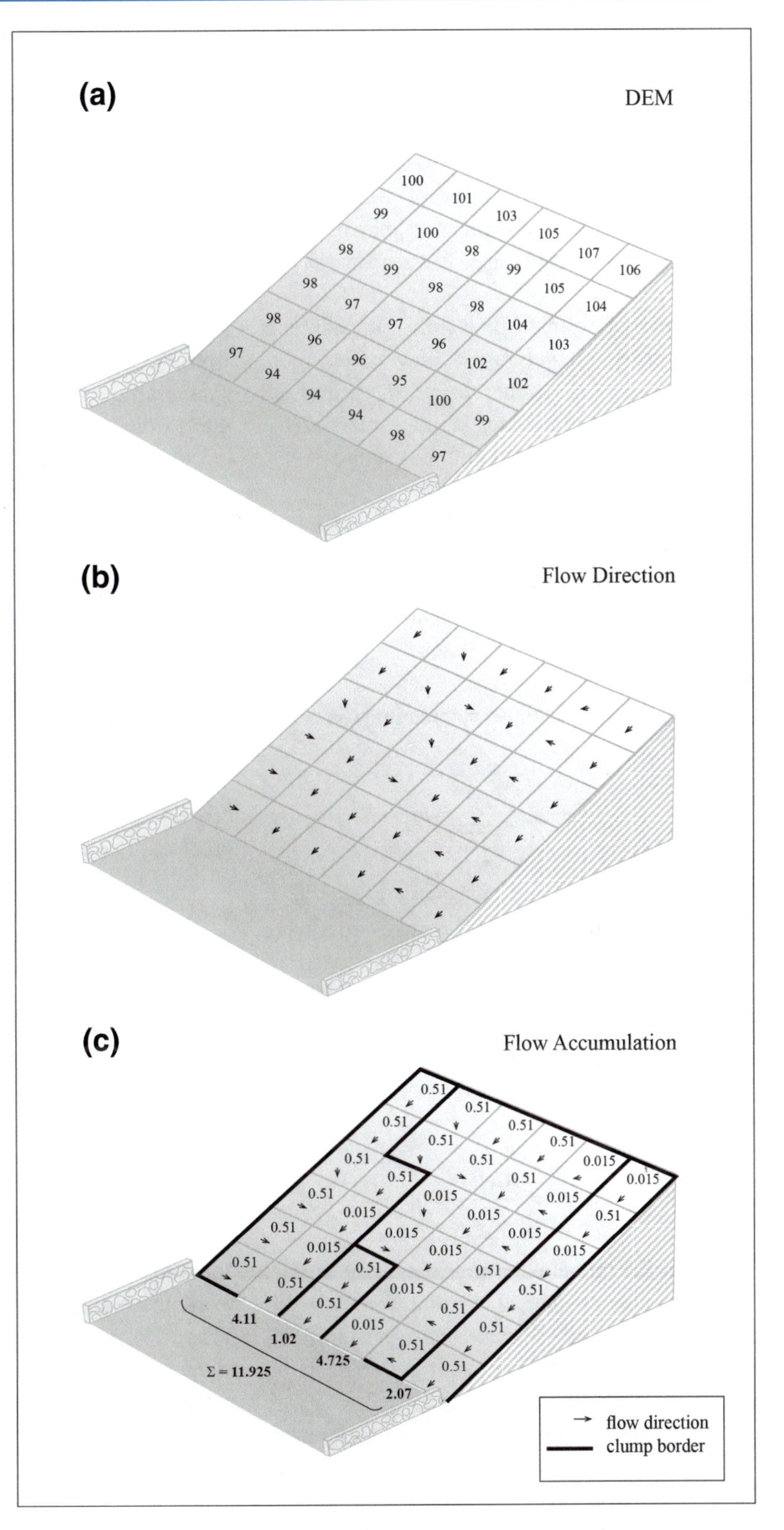

Fig. 5.26 **a** *Elevation* data [m]; **b** Determination of Flowdirection to each *i*th, *j*th grid-cell in the DEM, as the steepest hillslope gradient between the grid-cell and the eight adjacent grid-cells; **c** Accumulating contributing area to each grid-cell by counting the number of grid-cells in the flow path to it (as determined by Eq. 5.9). In principal, the contributing area assumes that complete grid-cells contribute runoff along the flow path (each grid-cell weight = 1). Here, the grid-cells weight considers the land cover along the flow path, whereby in this example rock weight = 0.51 and the weight of soil and vegetation surface = 0.015. The contributing area within the clump border is the weighted sum of all grid-cells within the clump

Rainfall intensity, a variable that is so important to runoff yield in semiarid catchments, is not expressed in the method. Rainfall duration that can be important in humid areas is also not expressed. These drawbacks, however, do not directly affect the assignment of weights. For more conservative estimation, maximal runoff coefficients for all landscape components may be used.

The use of weights in the D8 procedure is crucial for simulating runoff flow under more realistic conditions. In that regard, the use of RS in the process contributes another tier to the drainage structure computation procedure.

5.3.2.5 Error Assessment

Error assessment of the predicted channel network can be applied against reference data, using the following four measures: *Coverage, Purity, Connectivity, Accuracy* (Ichoku and Karnieli 1996).

Coverage [%] is the underestimation error (Eq. 5.12):

$$Coverage = \frac{OL}{TR} \cdot 100, \quad (5.12)$$

where *OL* (overlapping) is the number of grid-cells of the reference channel network that were assigned as channels in the overlapping locations in the predicted network, and *TR* (total reference) is the total number of grid-cells in the reference data, assigned as channels.

Purity [%] is the overestimation error (Eq. 5.13):

$$Purity = \frac{ART}{TR} \cdot 100, \quad (5.13)$$

where *ART* is the number of grid-cells of artifact channels in the predicted layer that were detected as "no channel" in the reference layer, and *TR*, once again, is the total number of grid-cells in the reference data assigned as channels. *Purity* may exceed 100% because due to classification error, the number of artificial grid-cells can be even larger than the number of the channel grid-cells in the reference network.

Connectivity is the number of cases of disconnections along the predicted channel network. The source of this error is usually the image-processing part of the procedure. This step was not implemented automatically but by applying visual interpretation.

Finally, *accuracy* is the number of grid-cells identified as geographically distorted channels. Namely, grid-cells that are assigned as channels but are distorted geographically with a small shift from the location that they were supposed to be. This kind of distortion may occur, for example, due to the 45° angle limitation of flowdirection calculations in D8. To minimize the effect of location error; a buffer of a single grid-cell is established for the reference data. Within this buffer, wherever the predicted grid-cell is assigned as a channel grid-cell it is counted as true prediction.

These four *accuracy* indices can be materialized as follows, using simple map algebra operators of the raster data of the reference and predicted layers (Fig. 5.27).

Coverage is computed based on the number of "channel" grid-cells 1 that do overlap in the predicted and reference channel networks. Thus, if the grid-cell value is 1 in the predicted network, and 1 in the reference one, the result is a single certain grid-cell c_i, and the summation of all c_is is *OL*.

Purity can be computed by identifying grid-cells that are assigned as channels 1 in the predicted network and 0 in the reference network. Thus, if a grid-cell in the predicted layer is equal to 1 and in the reference layer is equal to 0, this grid-cell is assigned as an error e_i and the summation of all e_is is the *ART*.

The *connectivity* error can be traced by counting the number of grid-cells along the network that cause disconnections to water flow. Thus, if a grid-cell along the stream network is assigned 1, and 0 in the predicted layer, it is counted as a single *connectivity* error. See the couple *A* grid-cells in Sect. 5.3 of Fig. 5.27. These are two *connectivity* errors, $\sum e_i = 2$.

In the case of the *accuracy* measure, the user uses a buffer zone operator to allow a tolerance

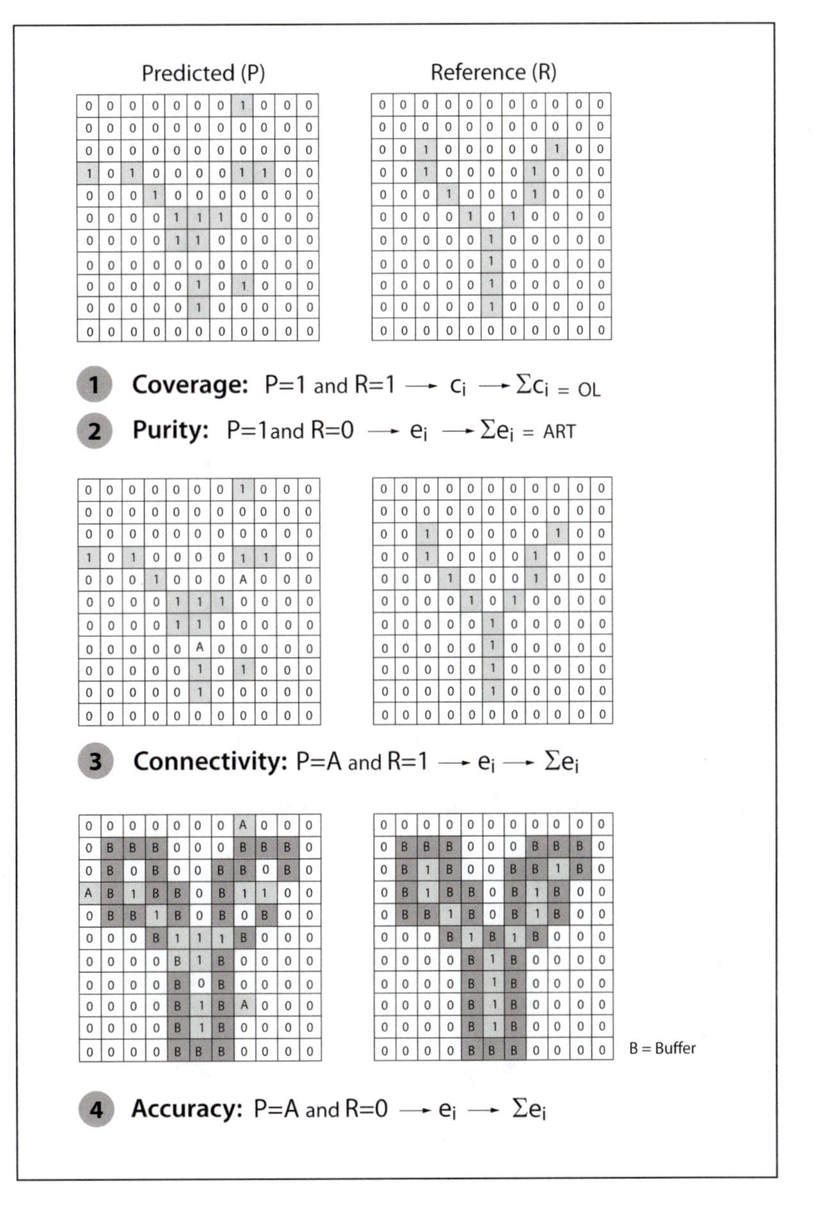

Fig. 5.27 Illustration of the four error measures used to assess the model functionality. See explanation in text

of a one-grid-cell and then, if an identified predicted grid-cell falls within a one-grid-cell buffer zone, as illustrated in Sect. 5.4 of Fig. 5.27 it is counted as a validated predicted channel grid-cell. See the *A* grid-cells beyond the buffer zone, in this case *accuracy* error $\sum e_i = 3$. The light gray grid-cells that fall within the boundaries of the buffer zone are counted as part of the predicted channel network.

These four error assessment measures are complementary. An integration of these four different aspects to describe the network may advance the entire procedure.

5.3.3 Results

The procedure described in Sect. 5.3.2 was applied to the Harod catchment. The output network is continuous—representing subtle changes in the stream network—and suffers none of the unrealistic flowline problems (Fig. 5.28).

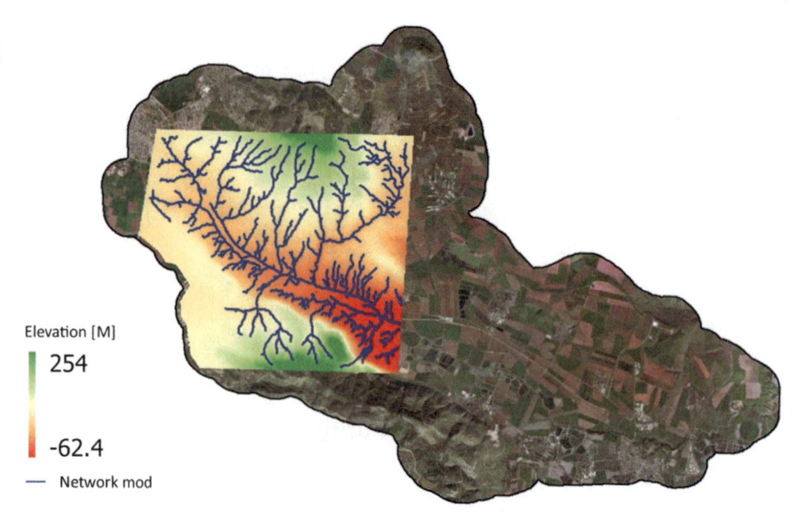

Fig. 5.28 The channel network in part of the Harod catchment, created by the modified DEM procedure

Two meters resolution channel network from the Survey of Israel, was used as a reference layer. The reference data was verified by certified surveyors. The error test of the four indices shown the following results: *Coverage* of 92% because the method did not fail in locating the channel grid-cells, *connectivity* level was perfect with zero disconnections and *purity* was observed to have 27 artefact grid-cells. Finally, *accuracy* had shown 45 distorted grid-cells. Svoray (2004) had shown in a previous study in a different location that the modified DEM methodology outperformed the "regular" D8 procedure in all four error assessment measures.

The synergy of updated data in the DEM-based modeling offers the benefits of a complementary source but requires the use of robust image-processing techniques. Such a framework can be used for highlighting areas at risk of water erosion, by computing accurate flowdirection and flowaccumulation matrices, as well as channel network extraction, which allows the user to dissociate between hydrological and geomorphological sink and source areas. The source areas may be subject to soil loss, and the sink areas to soil deposition and water ponding (Assouline et al. 2007). Computation of flowdirection and flowaccumulation matrices can be improved by using higher resolution *RS* data from drones, planes, or satellites, depending on the scale and area of interest. Due to the fact that the extraction of fine scale DEMs is not feasible in every unit of soil conservation, the use of coarse scale DEMs is widely used and requires the synergy with ancillary data to improve predictions (Li et al. 2017).

The methodology suggested here could extend the use of geoinformatics in hydrology, by enriching the DEM information. However, for future use, the reader is advised to note that:

1. More field-based information is needed to determine runoff coefficients of surface covers, that can be mapped using RS data—such as roads, rock, vegetation, built area covers, and runoff generation under various tillage methods (Langhans et al. 2019); similarly, more field studies are needed to determine the bias of the network system places on flowdirection (Lindsay 2016);
2. Higher resolution DEMs and RGB images are now more readily available to identify microtopography, and small-scale barriers. The goal is to enable programming of more detailed monitoring systems of catchment morphology. These are relatively attainable, and continued development of field studies in this direction may meet these requirements in the near future.

5.4 Drone Remote Sensing

Drones—a.k.a. Unmanned Aerial Vehicles (UAV)—are emerging, efficient, and easy-to-use RS platforms that have recently been used for a wide range of agricultural and ecological applications. These include, inter alia: Rangeland monitoring, geomorphological and hydrological studies, precision agriculture applications, and wildlife tracking (Singh and Frazier 2018). Because drones fly at low altitude, they provide high spatial resolution and can offer water erosion studies with RS data at the sub-meter resolution, which is highly sought-after for accurate planning of soil conservation and drainage activities (d'Oleire-Oltmanns et al. 2012). Drones are operated by a ground-based human operator, who uses a communication protocol to control the flight route and data acquisition. The operator can either fully control the drone's path, or program it prior to the flight, so that the drone is autonomous in the field, and operates according to computer code instructions. Drones can be easily reused, which facilitates the production of high temporal-resolution time series. They also provide multi-angular data, so can be used to extract high-resolution DEMs and topographic information.

For all these reasons, drones are an important complementary resource to aircraft and satellite RS. Their use is increasing rapidly, and is expected to increase even more in the next decade (Der Merwe et al. 2020). The literature roughly categorizes four types of sensors that are commonly used in drone RS for agricultural studies (Maes and Steppe 2019). The *first* type is the high-resolution and relatively low-cost RGB cameras, that are readily available and used for land use/land cover classification, as well as for generating high-resolution DEMs and output layers of vegetation stature. The *second* group is of multispectral cameras, that consist of sensors of different lenses, each sensor sensitive to a single spectral region of the electromagnetic spectrum (mainly V*IS or NIR*), which can be very useful for computations of soil and vegetation indices (see Sects. 5.1.1 and 5.1.2). The *third* group is of the more advanced (and expensive) hyperspectral cameras that cover large spectral regions of the electromagnetic spectrum. This range is divided into many narrow spectral bands of high spectral resolution, each of which can be <10 nm (Adão et al. 2017). Finally, the *fourth* group is of thermal cameras, that are used to extract canopy and soil temperature, but at a lower resolution, and with only one band sensitive in the longwave infrared (7–12 mm) region. In practice, these four groups of sensors sample most of the spectral information currently used to study water erosion in agricultural catchments (see Sect. 5.1 for details).

In recent years, drones have increasingly become a most important tool in water erosion studies, and are being used in various applications in this field. This section describes the following three possible applications: DEM extraction; the application of the Topographic Threshold (TT) to identify areas at risk of channel initiation; and the application of a *change detection* procedure, to identify areas under soil loss and deposition conditions. The use of RGB data for land cover classification using drone data was demonstrated in Sect. 5.2 and is therefore not repeated here.

5.4.1 DEM Extraction

DEM is one of the most important data sources, for a wide range of water erosion applications in geoinformatics. High-resolution DEMs are sought-after for tasks such as quantification of erosion rates, mapping hillslope units (see Sect. 3.1), identifying initial points of channeled erosion, computation of geomorphometric measures, such as hillslope gradient and orientation, mapping wetness index and the topographic threshold, and many other applications.

Readily available contour-based DEMs—traditionally produced manually, using topographic maps—offer low spatial resolution of tens of meters (Mizukoshi and Aniya 2002). As an alternative, many of the photogrammetric DEMs are extracted from airborne data, at the spatial

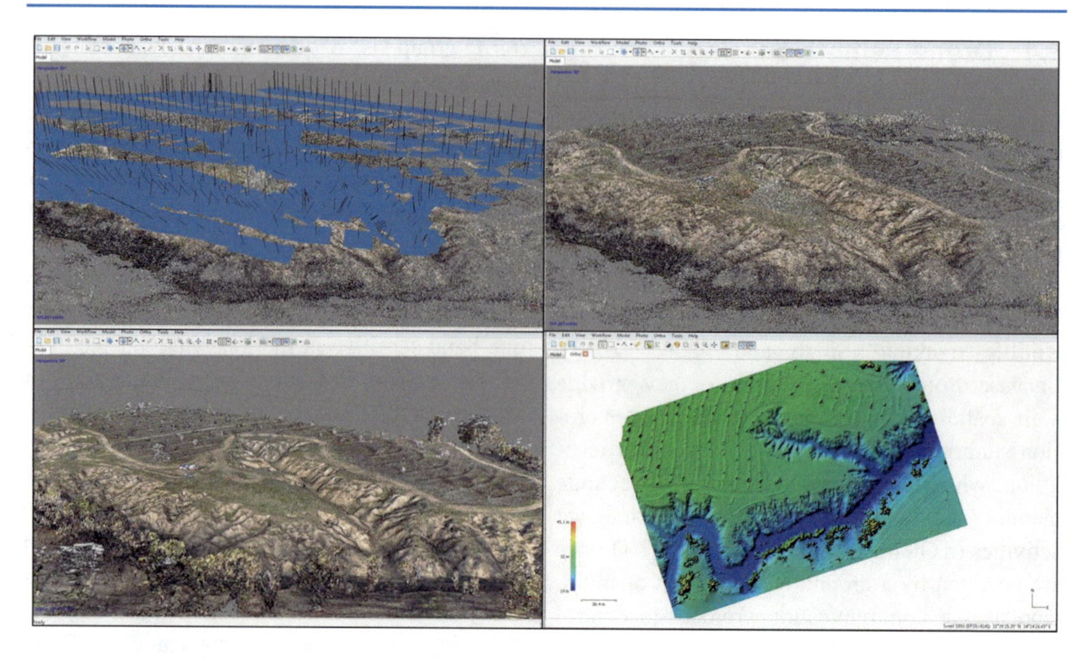

Fig. 5.29 DEM extraction, from Agisoft Metashape software. The upper left panel shows the reconstructed camera orientation; the upper right panel and the lower left panel are the high- and low-density point clouds; and the lower right panel is the extracted DEM

resolution of a few meters. Another option for spatial *elevation* data is airborne LiDAR (Light Detection and Ranging) data, which is very accurate, with even higher horizontal resolution than photogrammetric DEMs (Li et al. 2019; Neugirg et al. 2016); However, LiDAR data is very expensive, and the data processing procedure is not straightforward. This limits the use of LiDAR for several applications, for example, for time series analysis of soil loss and sediment deposition during the wet season or between years.

Compared with these three DEM sources, drones data can provide a cost-effective solution for the extraction of high-resolution and multi-temporal DEMs. To demonstrate the potential of drone-based DEMs, this chapter describes a procedure of DEM extraction from multi-angular drone data. In this illustration, a data scan from a UAV-type DJI Phantom 4 Pro, carrying a 12MP RGB camera, was processed to extract a DEM from stereo pairs of drone images, using the Agisoft Metashape software (https://www.agisoft.com/). The software activation procedure is described step-by-step in the software's manual, but we shall very briefly review the principles here (Fig. 5.29).

Step 1: The Agisoft algorithm searches for common points on the series of images taken by the drone and matches their location. The algorithm also finds the camera's position for each image and refines camera calibration parameters to locate the same points in two or more overlapping images. Then, a sparse point-cloud model is simulated and the algorithm reads the geometrical characteristics of the camera for the different images. The sparse point-cloud represents the orientation of images and it is required to extract the DEM in the next steps;

Step 2: To increase the vertical accuracy of the DEM product—twenty Ground Control Points (GCPs) are sampled in representative locations the field, using high-accuracy RTK DGPS, with the GCPs used as reference points;

Step 3: To generate a dense point-cloud based on the camera positions and the drone images. The dense point-cloud is editable, and classified prior to the surface generation which includes a 3D representation of the object surface;

Step 4: After the *elevation* surface is constructed, an orthomosaic is created and projected onto a DEM.

Most importantly, the DEM construction process described here is relatively simple, easy-to-use, and can be followed from the aforementioned Agisoft Metashape software manual. Note that, as a precondition, one must program the drone camera to photograph the study area with stereo pairs.

The DEM produced here for the Assaf catchment was tested for vertical *accuracy* against a validation set of field measurements, using RTK DGPS with sub-centimeter horizontal *accuracy*. A high agreement was observed between the values of the extracted DEM and field measurements at overlapping points (RMSE = 0.16 cm; n = 16). Such a high-*accuracy* and high-resolution DEM can be used for various applications. Figure 5.30 shows the DEM extracted in this process from the Assaf catchment, and the hillslope and topographic wetness index layer computed from this DEM layer.

The raw DEM data is indeed very detailed and provides identification of the narrow channel, and even the smallest channels that branch off it. Moreover, the *elevation* values appear realistic, with relatively small differences in the plateau. The hillslope gradient layer provides further information—indicating the sharp hillslope gradient is adjacent to the channel. The topographic wetness index (see Eq. 4.10) enhances the dissecting flow lines in the plateau and the middle of the channel, at a higher resolution. Note that the

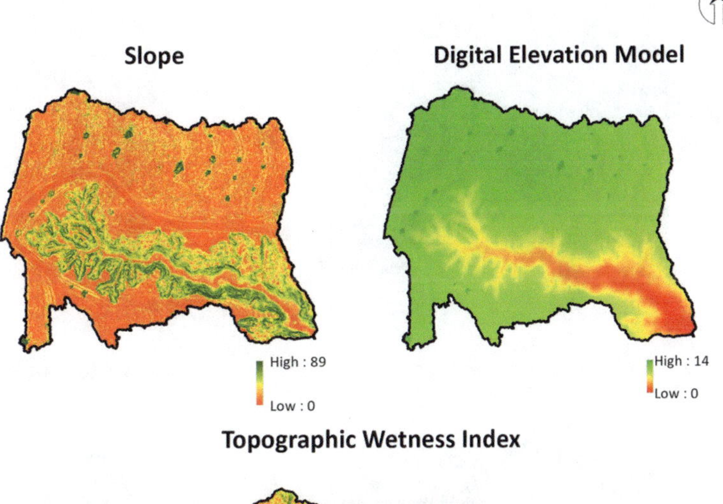

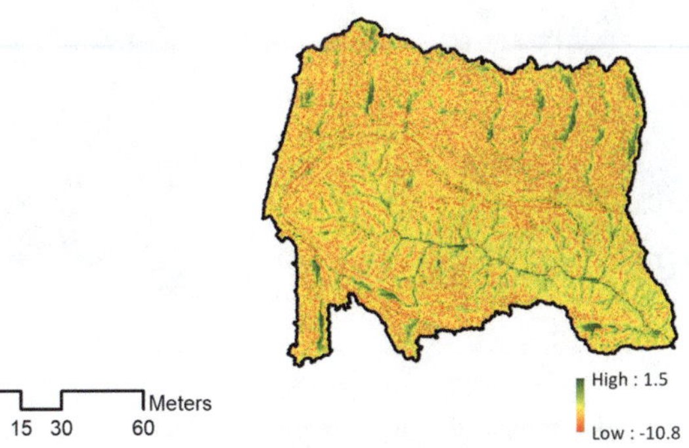

Fig. 5.30 Layers of DEM and its derivatives, as extracted from drone data. Horizontal *accuracy* of the DEM is 16 mm, and spatial resolution is 2.5 cm

very infrequent negative wetness index values are caused by the very high spatial resolution of the DEM, which results in low-flow accumulation values in some locations (DEM grid-cells). In these grid-cells, the numerator is smaller than the denominator in the wetness index equation, and the Ln of a fraction is a negative number.

5.4.1.1 High-Resolution Topographic Threshold

The Topographic Threshold (TT—see Chap. 2) is a commonly used measure to test the topographic limits of water erosion, based on the hillslope gradient *S*—which contributes area relationships A_s. The TT has been found useful in various environments as a proxy for the more detailed critical shear stress threshold (Montgomery and Dietrich 1994). The computation of TT is described in Eq. 2.10, and the entire procedure of how to apply it using DEM data is described in Chap. 6, Sect. 6.1. Figure 5.31 shows the use of the DEM drone data of the Assaf subcatchment, to compute TT under various configurations of slope and exponent coefficients of the TT. The four pairs of coefficients were extracted from other studies to exemplify the outcome of the use of various coefficients. The larger difference between the four layers in Fig. 5.31 is in the exponent value (b coefficient), while the multiplier coefficients were modified to a lesser degree between the four equations. The grid-cells that exceeded the TT were designated at-risk (*Yes*), and those below the critical TT were designated not at risk (*No*). The application of TT to drone-derived DEM is especially interesting, as it can predict failure points at very high resolution, which can be used as indicators to explore initial points of channeled erosion.

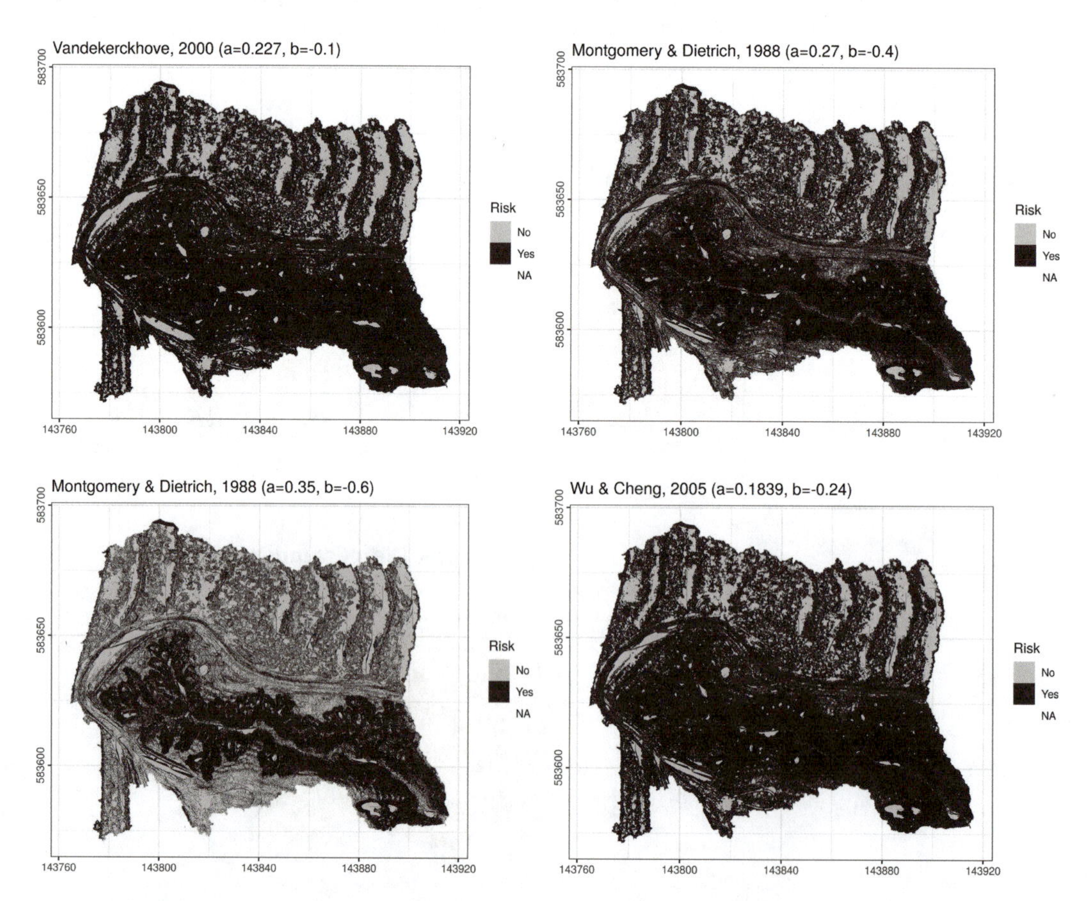

Fig. 5.31 Layers of erosion risks, computed using four configurations of TT coefficients, based on a literature review from Svoray and Markovitch (2009). See also Table 6.1

The four TT output layers in Fig. 5.31 reveal a difference between the four simulations in the distribution of at-risk areas. The largest area at risk (9% of the area) was evaluated based on the Vandekerckhove et al. (2000) coefficients (a = 0.23, b = −0.1), from the Cerro Tonosa, Spain. The other three runs yielded <1% risk area: Wu and Cheng (2005), from the Loess Plateau of North China (a = 0.18, b = −0.24); Montgomery and Dietrich (1988) (a = 0.27, b = −0.4), from California, US; and the Montgomery and Dietrich (1988) coefficients from the Sierra Navada, US (a = 0.35, b = −0.6). It is highly important, therefore, always to extract the appropriate coefficients of the area in question.

5.4.1.2 Change Detection

In a *change detection* procedure, we seek to identify locations where the data value of a given property changes between two dates. When studying water erosion, detection of changes in soil depth or *elevation* can be very useful, and provide a quantitative tool for estimating soil loss and deposition processes. For example, we can ask if the *elevation* value of a given grid-cell in an early DEM at time t_1 is different from the *elevation* value of the same grid-cell at a later date DEM t_2 as described in Eq. 5.14:

$$\Delta\ elevation = elevation_{t_2} - elevation_{t_1}. \quad (5.14)$$

In other words, we search for areas of soil loss and deposition by comparing DEMs of different time steps. Two consecutive dates of interest might be, for example, before and after a rainfall event. In another example, changes within the same hydrological year were computed to assess the dynamics of soil loss/deposition during the wet and dry seasons. Figure 5.32 illustrates a change detection process between two DEMs extracted from the drone-derived DEMs of the Assaf subcatchment, between February 2018 t_1 and October that year t_2.

The change detection output in Fig. 5.32 shows an overall change in *elevation*, ranging from +4.85 cm to −5.25 cm. This is a relatively large range, with negative values (soil loss) along the channel, and positive values (deposition) near the channel. In view of these results, one might ask how does one determine statistically what is considered a no-change class (i.e.,—would it be only absolutely zero change?), and what is the

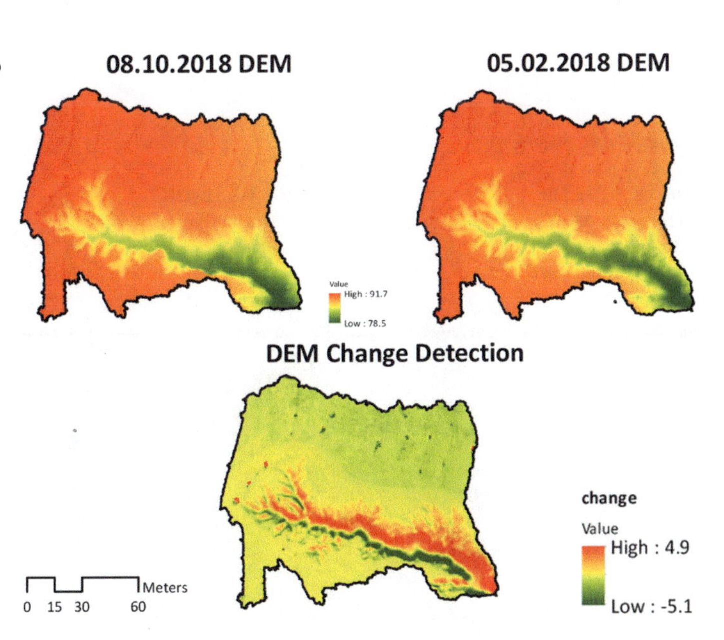

Fig. 5.32 A change detection map of the Assaf catchment in the northern Negev, as computed from two drone-extracted DEMs

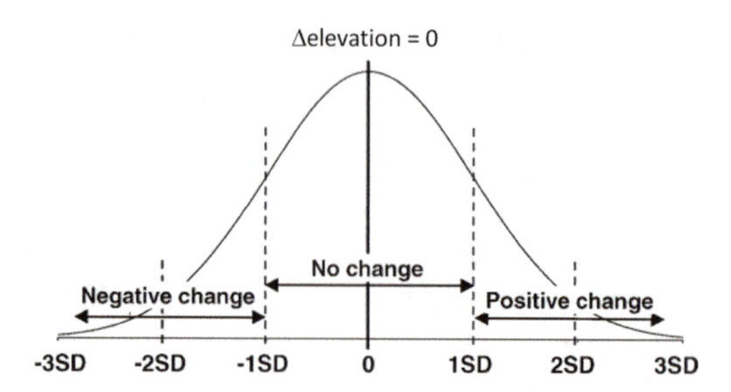

Fig. 5.33 The use of standard deviation levels as thresholds, to determine *elevation* change statistically and the direction of change. Δ*elevation* values between −1 and +1 std are designated as *No change*. Lower and higher values are designated *negative* (deposition) and *positive* change (loss), respectively. Reprinted from: Volcani, Karnieli & Svoray, The use of remote sensing and GIS for spatio-temporal analysis of the physiological state of a semiarid forest with respect to drought years. Forest Ecology and Management 215(1):239–250, (2005) with permission from Elsevier

difference between the positive and negative changes in terms of degree. The answers to these questions can come from descriptive statistics.

One common way to statistically validate changes is based on a determination of standard deviation levels as thresholds from the mean Δ*elevation* values (Fig. 5.33). Thus, changed and unchanged grid-cells can be distinguished, and negative (soil deposition) and positive (soil loss) changes can be labeled (Volcani et al. 2005).

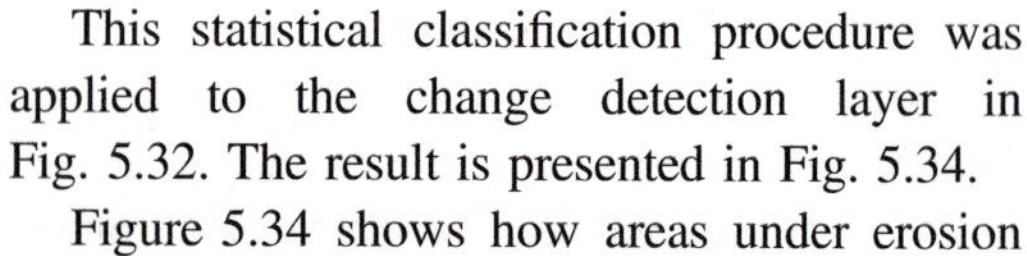
This statistical classification procedure was applied to the change detection layer in Fig. 5.32. The result is presented in Fig. 5.34.

Figure 5.34 shows how areas under erosion and definition processes can be identified based on direct observations and classified with statistical validation process.

5.5 Summary

Remote sensing is a powerful tool for collecting repetitive data over a wide area, with relatively small effort. Such a geoinformation source may provide a possible solution to the challenge that expensive and labor-intensive field measurements pose to the soil conservationalist. Over four decades, researchers have developed RS methods to monitor patterns and processes within the earth surface (Cracknell 2018). Much progress has been achieved in RS of the environment, and many tools are now available for water erosion studies. Nonetheless, some gaps in the field still exist with regard to several attributes that cannot be measured using RS data, despite accelerating progress in technology. Nonetheless, modern RS platforms provide high *coverage* of the electromagnetic spectrum, and high accessibility in time and space, while various facets of this prospect offer several well-shaped

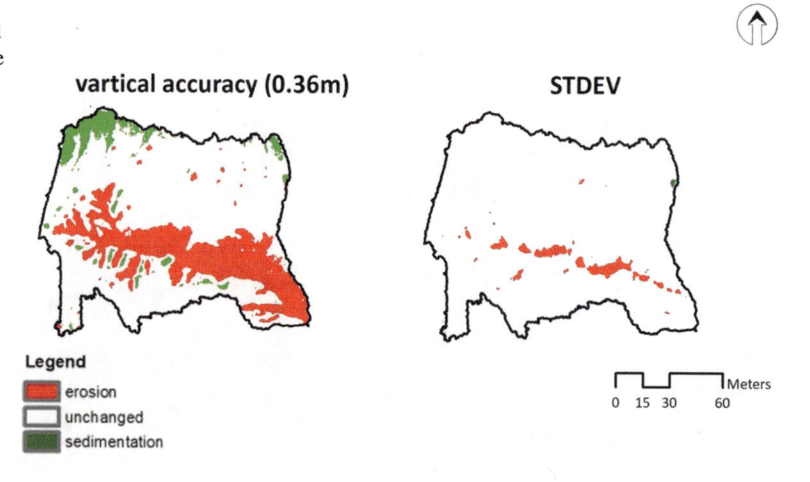

Fig. 5.34 A classified change detection layer, based on the classification procedure presented in Fig. 5.31. The result shows classes of soil erosion, sedimentation, and no change between the two dates

implementations, that share the following five aspects of traditional RS:

1. **Soil spectral indices** provide information on soil properties—including soil moisture content, texture, and nutrients content. This is an important addition to the soil data extracted using field measurements;
2. **Vegetation indices**, and especially the *NDVI*, provide valuable information on crop productivity, field *coverage* of vegetation canopies, and crop health, by monitoring crop and weeds yields;
3. **Image classification techniques** provide accurate and high-resolution data on surface cover—including, in the main, rock/soil/vegetation *coverage*, as well as crop classification and cultivation history. Object-based classification techniques make it possible to aggregate classified data at the field level, while soft classification techniques allow the extraction of data at the sub-pixel level;
4. **Image processing tools**, and the aforementioned classification techniques, provide further information about land cover, for catchment analysis of drainage structure, flowdirection, and flowaccumulation algorithms;
5. **Close range measurements** from drones provide extremely high-resolution RS and *elevation* data (with a vertical *accuracy* of mm).

Recently, several novel RS approaches have been developed, using machine-learning techniques and data-mining classification methodologies. These enable the identification of crop type, gulling initial points, piping, and even minerals in the soil, and are promising tools for soil conservation applications. Water erosion risks can be identified, and soil quality state and change can be quantified and used as a source of data for conservation planning and policy. This provides new opportunities for soil conservation over large areas, by providing evidence of the actual damage and the mechanisms at play in water erosion processes. Most generally, recent interpretations of agricultural RS data consider the catchments as complex systems that consist of many simple rock–soil–vegetation elements, each reacting differently to the system sensors, from satellites to drones. This understanding merges classic frameworks (such as soil and vegetation indices) with machine learning, unmixing and statistical classification tools, pattern recognition, and DEM-based algorithms.

The tools reviewed and provided here further expand the idea that RS data can be used to provide evidence of the state and dynamics of a catchment, thereby providing a general framework for depicting and simulating the mechanisms of water erosion in agricultural catchments.

Review Questions

1. List and describe the possible uses of RS data in water erosion studies, and describe the advantages of RS data over other data sources for these purposes.
2. The following Table represents spectral signatures of four surface components, using image DN values. Which component applies in each of the four cases?

Surface component	Blue	Green	Red	*NIR*
1	16	20	27	32
2	15	17	13	11
3	21	40	29	66
4	26	35	25	47

What might be the difference between surface components 3 and 4?

3. What is the *RED-NIR spectral space*—and where is the soil line located within it? Why are the vegetation lines expected to increase in vegetation cover toward the upper left corner? Use readily available Landsat 8 data to create the *soil line* in your agricultural catchment.
4. Compute *NDVI* of the two spectral signatures in Fig. 5.2. Between the two signatures, which one is expected to have higher *NDVI* values and why?

5. Use Landsat 8 images, or a closer range RGB aerial photo, to classify three land covers: Soil, rock, and vegetation in your agricultural catchment (*supervised*, and *unsupervised*). Then, using visual interpretation, identify 100 validation points for each class, and compute *Kappa coefficient* based on a *confusion matrix*. Which of the two classification methods was more accurate?
6. Use Landsat 8 to compute the spectral indices in Table 5.1 in your catchment.
7. Compute *D8* in your catchment, with and without weighted contributing areas. In the weighted contributing area, assign crop-cover fields with vegetation weights from Table 5.3, and uncultivated fields with bare soil weights from the same Table. What is the difference between the two output layers?
8. Use field boundaries layer of your catchment to compare object-based classification with pixel-based classification. Do you observe a large degree of salt-and-pepper effect in your catchment? What are the reasons for the different outcomes?
9. Compute the four error measures—*Coverage*, *purity*, *connectivity*, and *accuracy*—for the two-channel networks below.
10. Use a high-resolution aerial photo to classify hard and soft classification of soil, rock, and vegetation groups in your catchment. Then assign the soft classification groups into crisp groups, and compare the results.
11. Identify potential areas for water ponding in your agricultural catchment, using the Mud Index in Table 5.1. The threshold for water ponding can be the upper 0.25 std of the WI distribution in your catchment—or larger (0.5, or even 1 std) if you wish to accentuate. Then compare the result with a computed DEM-based WI of the same classification categories.
12. Classify your area into five groups of erosion/deposition, using two measurements of *elevation* (DEMs): (1) Compute change detection; (2) Classify the output layer into five groups: *No change* ($\mu + -0.5 \cdot$ std); *Eroded areas* ($\mu + 0.5$–$1 \cdot$ std); *Severe erosion* ($\mu + 1.5 \cdot$ std); *Deposited areas* ($\mu - 0.5$–$1 \cdot$ std); *Severe deposition* ($\mu - 1.5 \cdot$ std).

Strategies to address the Review Questions

1. Uses of RS data in water erosion studies are described in various sections:
   - Section 5.1 reviews the use of spectral data and algebraic expressions to highlight surface components over others. Section 5.2 describes classification procedures designed to group and label pixels of similar characteristics. Section 5.3 exemplifies the fusion of RS data in topographic modeling. Section 5.4 describes the use of drone-derived data to study small-scale processes.
   - Applications include, e.g., rainfall intensity mapping, crop-cover and yield estimation, repetitive soil moisture assessments, on-site and off-site erosion damage mapping, and land use classification.
   - The advantages of RS data over other data sources: replace labor-intensive and expensive field visits; large areas coverage at high spatial resolution; provides information on reflected radiation in wavelength that the human eye cannot see; offers high-frequency updates.

2. See Fig. 5.2 that illustrates spectral signatures of soil and vegetation. In accordance:

   - Component 1 shows a monotonous increase in reflectance with wavelength, which is typical to bare soil spectral characteristics.
   - Component 2 shows stable low reflectance along the wavelength axis, which is a typical spectral signature to water bodies (See caption).
   - Component 3 is typical to vegetation spectral signature but with relatively high value in *NIR*.
   - Component 4 is typical to vegetation spectral signature but with relatively lower value in *NIR*.

- The lower *NIR* value in the 4th component may occur due lower vegetation cover, longstanding vegetation cover, or unhealthy one.

3. The *RED-NIR spectral space* (Sect. 5.1) is a plot of the reflectance values of different surface components in these spectral bands. The plot is especially useful to dissociate between soil and vegetation properties.

   - The *soil line* (Fig. 5.4) represents the spectral reflectance of different soils in the NIR and RED spectral bands.
   - The closer a pixel is to the upper left-hand corner in the *RED-NIR* spectral space, the more the pixel is of higher *NIR* and lower *RED* values and therefore more photosynthetically active.
   - The soil line can be extracted and marked on the *RED-NIR* spectral space, based on sampling of a sample of image pixels of bare soils under various clorpt conditions in the studied catchment.

4. Using the values in Fig. 5.2 we can calculate the *NDVI* of a and b:

$$NDVI_a = \frac{0.85 - 0.05}{0.85 + 0.05} = 0.888$$

$$NDVI_b = \frac{0.3 - 0.28}{0.3 + 0.28} = 0.034$$

Higher *NDVI* values are clearly expected in a because the difference between *NIR-RED* reflectance is much higher between a and a' than between b and b'.

5. This task requires data from your catchment of interest. Section 5.2.1 describes both *supervised* and *unsupervised* classification procedures. After classification is applied, you can use a *confusion matrix* (Fig. 5.10) to compute the *Kappa coefficient* (Eq. 5.3). The latter is used to compare measured (visual interpretation) and predicted (classified images both supervised and unsupervised) data to decide which of the two classification methods is more accurate.
6. This is a task that needs to be practiced using your own data by applying the relevant equations in Table 5.1.
7. This is a task that needs to be practiced using your own data by applying the steps of the drainage structure determination algorithm discussed in Sect. 5.3.2.3 and computing flowaccumulation with weights as described in Sect. 5.3.2.4. Using contributing areas without weights is based on the assumption of zero infiltration in the contributing grid-cells. Using weighted contributing areas allows to compensate for the variation in permeability and runoff contribution. For weighting the contributing area grid-cells, RS data is used to classify the grid-cell cover type, and to assign the consequent runoff coefficients to the grid-cell, thereby assigning it a corresponding weight.
8. Pixel-based classification (Sects. 5.2.1.1 and 5.2.1.2) and object-based classification (Sect. 5.2.1.3) mainly differ in the generalization level. Figure 5.16 shows the difference between pixel-based and object-based classifications for the Yehezkel catchment. Due to the generalization of spectral signatures at the field level we expect the object-based classification to reduce the salt-and-pepper effect. However, note that generalization may lead to erroneous results.
9. The four error measures can be computed using the procedures in Sect. 5.3.2.5:

   - *Coverage:* The underestimation error, computed using Eq. 5.12:

$$Coverage = \frac{13}{16} \cdot 100 = 81.25\%$$

   - *Purity:* The overestimation error is computed using Eq. 5.13:

$$Purity = \frac{6}{16} \cdot 100 = 37.5\%$$

   - *Connectivity*: the number of cases of disconnections along the predicted

channel network, which is the number of sections of "1" we have in our predicted network. In our networks the Connectivity is 2.
- *Accuracy*: the number of grid-cells that are assigned as channels but are distorted geographically with a small shift from the location that they were supposed to be. In our networks we see 1 grid-cells as such (in the left branch).

10. Again, "your own data practice." The practice of hard classification is described in Sects. 5.2.1.1 and 5.2.1.2 and that of soft classification in Sect. 5.2.2.2. Assignment to crisp groups can be done using a majority operator. Namely, the groups with the highest fraction become the crisp group of the grid-cell.
11. The procedure is basically described in the question: First, run the indices—*NDMI* from Table 5.1 and WI from Eq. 4.10 and assign a threshold to yes/no ponding. Then, compare the output ponding grid-cells (layers assigned with 0 for no ponding and 1 to ponding grid-cells, based on the threshold values) between the two indices. The comparison can be executed using a change detection procedure (Sect. 5.4.1.2, Eq. 5.14) between the output ponding layers.
12. Change detection is described in Sect. 5.4.1.2 and the computation of Δ*elevation* between the two dates can be applied based on Eq. 5.14. Figure 5.32 shows the use of std to classify the change detection output layer. The monitoring of change in elevation or soil thickness after erosion events is an important tool for soil degradation studies that was not studied in full in the literature.

References

Ackermann O, Svoray T, Haiman M (2008) Nari (calcrete) outcrop contribution to ancient agricultural terraces in the southern Shephelah, Israel: insights from digital terrain analysis and a geoarchaeological field survey. J Archaeol Sci 35(4):930–941. https://doi.org/10.1016/j.jas.2007.06.022

Adams JB, Smith M, Johnson P (1986) Spectral mixture modeling: a new analysis of rock and soil types at the Viking Lander 1 Site. J Geophys Res 91(B8):8098–8112. https://doi.org/10.1029/jb091ib08p08098

Adão T, Hruška J, Pádua L et al (2017) Hyperspectral imaging: a review on UAV-based sensors, data processing and applications for agriculture and forestry. Remote Sens 9(11):1110. https://doi.org/10.3390/rs9111110

Altenau EH, Pavelsky TM, Moller D et al (2019) Temporal variations in river water surface elevation and slope captured by AirSWOT. Remote Sens Environ 224:304–316. https://doi.org/10.1016/j.rse.2019.02.002

Amer R, Kolker AS, Muscietta A (2017) Propensity for erosion and deposition in a deltaic wetland complex: implications for river management and coastal restoration. Remote Sens Environ 199:39–50. https://doi.org/10.1016/j.rse.2017.06.030

Aneece I, Thenkabail P (2018) Accuracies achieved in classifying five leading world crop types and their growth stages using optimal earth observing-1 Hyperion hyperspectral Narrowbands on google earth engine. Remote Sens 10(12):2027

Assouline S, Mualem Y (2006) Runoff from heterogeneous small bare catchments during soil surface sealing. Water Resour Res 42(12):W12405. https://doi.org/10.1029/2005WR004592

Assouline S, Selker JS, Parlange JY (2007) A simple accurate method to predict time of ponding under variable intensity rainfall. Water Resour Res 43(3): W03426. https://doi.org/10.1029/2006WR005138

Atzberger C (2013) Advances in remote sensing of agriculture: context description, existing operational monitoring systems and major information needs. Remote Sens 5(2):949–981. https://doi.org/10.3390/rs5020949

Awotwi A, Anornu GK, Quaye-Ballard JA et al (2018) Monitoring land-use and land cover changes due to extensive gold mining, urban expansion, and agriculture in the Pra River Basin of Ghana, 1986–2025. Land Degrad Dev 29(10):3331–3343. https://doi.org/10.1002/ldr.3093

Bannari A, Morin D, Bonn F et al (1995) A review of vegetation indices. Remote Sens Rev 13(1–2):95–120. https://doi.org/10.1080/02757259509532298

Baret F, Guyot G (1991) Potential and limitations of vegetation indices for LAI and APAR assessment. Remote Sens Environ 35(2–3):161–173

Ben-Dor E (2002) Quantitative remote sensing of soil properties. In: Anonymous advances in agronomy, vol 75. Elsevier Science & Technology, p 173–243

Ben-Dor E, Banin A (1990) Near infrared reflectance analysis of carbonate concentration in soils. Appl Spectrosc 44:1064–1069

Ben-Dor E, Chabrillat S, Demattê JAM et al (2009) Using Imaging Spectroscopy to study soil properties. Remote Sens Environ 113:S38–S55. https://doi.org/10.1016/j.rse.2008.09.019

Ben-Dor E, Kruse FA (1995) Surface mineral mapping of Makhtesh Ramon Negev, Israel using GER 63 channel scanner data. Int J Remote Sens 16(18):3529–3553. https://doi.org/10.1080/01431169508954644

Bernstein LS, Jin X, Gregor B et al (2012) Quick atmospheric correction code: algorithm description and recent upgrades. Opt Eng 51(11):111719. https://doi.org/10.1117/1.OE.51.11.111719

Boettinger JL, Ramsey RD, Bodily JM et al (2008) Landsat spectral data for digital soil mapping in digital soil mapping with limited data. Springer, Netherlands, Dordrecht

Bruins HJ, Bithan-Guedj H, Svoray T (2019) GIS-based hydrological modelling to assess runoff yields in ancient-agricultural terraced wadi fields (central Negev desert). J Arid Environ 166:91–107. https://doi.org/10.1016/j.jaridenv.2019.02.010

Brungard CW, Boettinger JL, Duniway MC et al (2015) Machine learning for predicting soil classes in three semi-arid landscapes. Geoderma 239–240:68–83. https://doi.org/10.1016/j.geoderma.2014.09.019

Campbell J, Wynne R (2011) Introduction to remote sensing, 5th edn. The Guilford Press, New York

Carlson TN, Ripley DA (1997) On the relation between NDVI, fractional vegetation cover, and leaf area index. Remote Sens Environ 62(3):241–252. https://doi.org/10.1016/s0034-4257(97)00104-1

Castellanos-Quiroz HOA, Ramírez-Daza HM, Ivanova Y (2017) Detection of open-pit mining zones by implementing spectral indices and image fusion techniques. DYNA 84(201):42–49. https://doi.org/10.15446/dyna.v84n.60368

Chabrillat S, Ben-Dor E, Viscarra Rossel RA et al (2013) Quantitative soil spectroscopy. Appl Environ Soil Sci 2013:1–3. https://doi.org/10.1155/2013/616578

Chen Y, Gillieson D (2014) Evaluation of landsat TM vegetation indices for estimating vegetation cover on semi-arid rangelands: a case study from Australia. Can J Remote Sens 35(5):435–446. https://doi.org/10.5589/m09-037

Cheng Z, Lu D, Li G et al (2018) A random forest-based approach to map soil erosion risk distribution in Hickory Plantations in western Zhejiang Province. China. Remote Sensing 10(12):1899. https://doi.org/10.3390/rs10121899

Cohen J (1960) A coefficient of agreement for nominal scales. Educ Psychol Measur 20(1):37–46. https://doi.org/10.1177/001316446002000104

Cohen Y, Shoshany M (2005) Analysis of convergent evidence in an evidential reasoning knowledge-based classification. Remote Sens Environ 96(3):518–528. https://doi.org/10.1016/j.rse.2005.04.009

Cracknell AP (2018) The development of remote sensing in the last 40 years. Int J Remote Sens 39(23):8387–8427. https://doi.org/10.1080/01431161.2018.1550919

Cui Z, Kerekes JP (2018) Potential of red edge spectral bands in future landsat satellites on agroecosystem canopy green leaf area index retrieval. Remote Sens (Basel, Switzerland) 10(9):1458. https://doi.org/10.3390/rs10091458

Cyr L, Bonn F, Pesant A (1995) Vegetation indices derived from remote sensing for an estimation of soil protection against water erosion. Ecol Model 79(1–3):277–285. https://doi.org/10.1016/0304-3800(94)00182-h

Das B, Sahoo RN, Pargal S et al (2019) Comparative analysis of index and chemometric techniques-based assessment of leaf area index (LAI) in wheat through field spectroradiometer, Landsat-8, Sentinel-2 and Hyperion bands. Geocarto Int 35(13):1415–1432. https://doi.org/10.1080/10106049.2019.1581271

De Jong SM, Paracchini ML, Bertolo F et al (1999) Regional assessment of soil erosion using the distributed model SEMMED and remotely sensed data. Catena 37(3):291–308. https://doi.org/10.1016/S0341-8162(99)00038-7

Decker AG, Malthus TJ, Wijnen MM et al (1992) The effect of spectral bandwidth and positioning on the spectral signature analysis of inland waters. Remote Sens Environ 41(2):211–225. https://doi.org/10.1016/0034-4257(92)90079-Y

Der Merwe D, Burchfield D, Witt T (2020) Drone in agriculture. In: Sparks D (ed) Advances in agronomy Elsevier, UK

d'Oleire-Oltmanns S, Marzolff I, Peter K et al (2012) Unmanned Aerial Vehicle (UAV) for monitoring soil erosion in Morocco. Remote Sens 4(11):3390–3416. https://doi.org/10.3390/rs4113390

Dong T, Liu J, Shang J et al (2019) Assessment of red-edge vegetation indices for crop leaf area index estimation. Remote Sens Environ 222:133–143. https://doi.org/10.1016/j.rse.2018.12.032

Forkuor G, Hounkpatin OKL, Welp G et al (2017) High resolution mapping of soil properties using remote sensing variables in south-western Burkina Faso: a comparison of machine learning and multiple linear regression models. PLoS One 12(1):s. https://doi.org/10.1371/journal.pone.0170478

Ganasri BP, Ramesh H (2016) Assessment of soil erosion by RUSLE model using remote sensing and GIS-a case study of Nethravathi Basin. Di Xue Qian Yuan 7 (6):953–961. https://doi.org/10.1016/j.gsf.2015.10.007

Heller Pearlshtien D, Ben-Dor E (2020) Effect of organic matter content on the spectral signature of iron oxides across the VIS–NIR spectral region in artificial mixtures: an example from a red soil from Israel. Remote Sens 12(12):1960

Hochschild V, Märker M, Rodolfi G et al (2003) Delineation of erosion classes in semi-arid southern African grasslands using vegetation indices from optical remote sensing data. Hydrol Process 17 (5):917–928. https://doi.org/10.1002/hyp.1170

Hossain MD, Chen D (2019) Segmentation for Object-Based Image Analysis (OBIA): a review of algorithms and challenges from remote sensing perspective. ISPRS J Photogramm Remote Sens 150:115–134. https://doi.org/10.1016/j.isprsjprs.2019.02.009

Huete A, Didan K, Miura T et al (2002) Overview of the radiometric and biophysical performance of the MODIS vegetation indices. Remote Sens Environ 83(1–2):195–213. https://doi.org/10.1016/s0034-4257(02)00096-2

Hunt ER, Rock BN (1989) Detection of changes in leaf water content using near- and middle-Infrared reflectances. Remote Sens Environ 30(1):43–54. https://doi.org/10.1016/0034-4257(89)90046-1

Hutchinson MF (1989) A new procedure for gridding elevation and streamline data with automatic removal of spurious pits. J Hydrol 106:211–232

Ichoku C, Karnieli A (1996) A review of mixture modeling techniques for sub-pixel land cover estimation. Remote Sens Rev 13(3–4):161–186. https://doi.org/10.1080/02757259609532303

Jackson T (2004) Vegetation water content mapping using Landsat data derived normalized difference water index for corn and soybeans. Remote Sens Environ 92(4):475–482. https://doi.org/10.1016/j.rse.2003.10.021

Karnieli A, Kaufman YJ, Remer L et al (2001) AFRI—aerosol free vegetation index. Remote Sens Environ 77(1):10–21. https://doi.org/10.1016/s0034-4257(01)00190-0

Kaufman L, Rousseeuw PJ (1990) Finding groups in data: an introduction to cluster analysis. Wiley, New York

Landis JR, Koch GG (1977) The measurement of observer agreement for categorical data. Biometrics 33(1):159–174. https://doi.org/10.2307/2529310

Lange J, Greenbaum N, Husary S et al (2003) Runoff generation from successive simulated rainfalls on a rocky, semi-arid, Mediterranean Hillslope. Hydrol Process 17(2):279–296. https://doi.org/10.1002/hyp.1124

Langhans C, Diels J, Clymans W et al (2019) Scale effects of runoff generation under reduced and conventional tillage. Catena (Giessen) 176:1–13. https://doi.org/10.1016/j.catena.2018.12.031

Legleiter CJ, Roberts DA (2005) Effects of channel morphology and sensor spatial resolution on image-derived depth estimates. Remote Sens Environ 95 (2):231–247. https://doi.org/10.1016/j.rse.2004.12.013

Li L, Nearing MA, Nichols MH et al (2019) Using terrestrial LiDAR to measure water erosion on stony plots under simulated rainfall. Earth Surf Proc Land 45 (2):484–495. https://doi.org/10.1002/esp.4749

Li X, Li J, Shen H et al (2017) DEM generation from contours and a low-resolution DEM. ISPRS J Photogramm Remote Sens 134:135–147. https://doi.org/10.1016/j.isprsjprs.2017.09.014

Li Y, Wang H, Li XB (2015) Fractional vegetation cover estimation based on an improved selective endmember spectral mixture model. PLoS One 10(4):e0124608. https://doi.org/10.1371/journal.pone.0124608

Liaw A, Wiener M (2002) Classification and regression by randomforest. R News 2(3):18–22

Lindsay JB (2016) The practice of DEM stream burning revisited. Earth Surf Proc Land 41(5):658–668. https://doi.org/10.1002/esp.3888

Liu D, Xia F (2010) Assessing object-based classification: advantages and limitations. Remote Sens Lett 1(4):187–194. https://doi.org/10.1080/01431161003743173

Liu Y, Zhang HH, Wu Y (2011) Hard or soft classification? Large-margin unified machines. J Am Stat Assoc 106(493):166–177. https://doi.org/10.1198/jasa.2011.tm10319

Livne E, Svoray T (2011) Components of uncertainty in primary production model: the study of DEM, classification and location error. Int J Geogr Inf Sci 25 (3):473–488. https://doi.org/10.1080/13658816.2010.517752

Maes WH, Steppe K (2019) Perspectives for remote sensing with unmanned aerial vehicles in precision agriculture. Trends Plant Sci 24(2):152–164. https://doi.org/10.1016/j.tplants.2018.11.007

Masoud AA, Koike K, Atwia MG et al (2019) Mapping soil salinity using spectral mixture analysis of landsat 8 OLI images to identify factors influencing salinization in an arid region. ITC J 83:101944. https://doi.org/10.1016/j.jag.2019.101944

Matsuoka A, Babin M, Devred EC (2016) A new algorithm for discriminating water sources from space: a case study for the southern Beaufort Sea using MODIS ocean color and SMOS salinity data. Remote Sens Environ 184:124–138. https://doi.org/10.1016/j.rse.2016.05.006

McCormack E, Gahegan MN, Roberts SA et al (1993) Hoyle feature-based derivation of drainage networks. Int J Geogr Inf Syst 7(3):263–279

Memarsadeghi N, Mount D, Netanyahu N et al (2007) A fast implementation of the isodata clustering algorithm*. Int J Comput Geom Appl 17(1):71–103

Metternicht GI, Fermont A (1998) Estimating erosion surface features by linear mixture modeling. Remote Sens Environ 64(3):254–265. https://doi.org/10.1016/S0034-4257(97)00172-7

Mizukoshi H, Aniya M (2002) Use of contour-based DEMs for deriving and mapping topographic attributes. Photogramm Eng Remote Sens 68(1):83–93

Montgomery DR, Dietrich W (1994) A physically based model for the topographic control on shallow landsliding. Water Resour Res 30(4):1153–1171

Montgomery DR, Dietrich WE (1988) Where do channels begin? Nature (London) 336(6196):232–234. https://doi.org/10.1038/336232a0

Mulla DJ (2013) Twenty five years of remote sensing in precision agriculture: key advances and remaining knowledge gaps. Biosys Eng 114(4):358–371. https://doi.org/10.1016/j.biosystemseng.2012.08.009

Neugirg F, Stark M, Kaiser A et al (2016) Erosion processes in calanchi in the Upper Orcia Valley, Southern Tuscany, Italy based on multitemporal high-resolution terrestrial LiDAR and UAV surveys. Geomorphology 269:8–22

O'Callaghan JF, Mark DM (1984) The extraction of drainage networks from digital elevation data. Comput Vis Gr Image Process 28(3):323–344. https://doi.org/10.1016/S0734-189X(84)80011-0

Pelletier JD (2008) Quantitative modeling of earth surface processes. Cambridge University Press M.U.A

Price KP (1993) Detection of soil erosion within pinyon-juniper woodlands using Thematic Mapper (TM) data. Remote Sens Environ 45(3):233–248. https://doi.org/10.1016/0034-4257(93)90107-9

Quintano C, Fernández-Manso A, Shimabukuro YE et al (2012) Spectral unmixing. Int J Remote Sens 33(17): 5307–5340. https://doi.org/10.1080/01431161.2012.661095

Regmi NR, Rasmussen C (2018) Predictive mapping of soil-landscape relationships in the arid southwest United States. Catena 165:473–486. https://doi.org/10.1016/j.catena.2018.02.031

Riazanoff S, Cervelle B, Chorowicz J (1990) Parametrisable skeletonization of binary and multi-level images. Pattern Recogn Lett 11(1):25–33. https://doi.org/10.1016/0167-8655(90)90052-4

Richardson AJ, Wiegand CL (1977) Distinguishing vegetation from soil background information. Eng Remote Sens 43:1541–1552

Rockwell BW (2013) Automated mapping of mineral groups and green vegetation from landsat thematic mapper imagery with an example from the San Juan Mountains, Colorado. https://search.datacite.org/works/10.13140/rg.2.1.2507.7925.

Rodriguez-Iturbe I, Marani M, Rigon R et al (1994) Self-organized river basin landscapes: fractal and multifractal characteristics. Water Resour Res 30(12):3531–3539. https://doi.org/10.1029/94WR01493

Ruiz Hernandez IE, Shi W (2018) A Random Forests classification method for urban land-use mapping integrating spatial metrics and texture analysis. Int J Remote Sens 39(4):1175–1198. https://doi.org/10.1080/01431161.2017.1395968

Sandholt I, Rasmussen K, Andersen J (2002) A simple interpretation of the surface temperature/vegetation index space for assessment of surface moisture status. Remote Sens Environ 79(2–3):213–224. https://doi.org/10.1016/s0034-4257(01)00274-7

Serbin G, Hunt ER, Daughtry CS et al (2009) An improved ASTER index for remote sensing of crop residue. Remote Sensing (Basel, Switzerland) 1 (4):971–991. https://doi.org/10.3390/rs1040971

Serrano L, Peñuelas J, Ustin SL (2002) Remote sensing of nitrogen and lignin in Mediterranean vegetation from AVIRIS data: decomposing biochemical from structural signals. Remote Sens Environ 81(2):355–364. https://doi.org/10.1016/S0034-4257(02)00011-1

Seutloali KE, Dube T, Mutanga O (2017) Assessing and mapping the severity of soil erosion using the 30-m Landsat multispectral satellite data in the former South African homelands of Transkei. Phys Chem Earth Parts A/B/C 100:296–304. https://doi.org/10.1016/j.pce.2016.10.001

Shoshany M, Svoray T (2002) Multidate adaptive unmixing and its application to analysis of ecosystem transitions along a climatic gradient. Remote Sens Environ 82 (1):5–20. https://doi.org/10.1016/S0034-4257(01)00346-7

Singh KK, Frazier AE (2018) A meta-analysis and review of unmanned aircraft system (UAS) imagery for terrestrial applications. Int J Remote Sens 39(15–16):5078–5098. https://doi.org/10.1080/01431161.2017.1420941

Song C, Woodcock CE, Seto KC et al (2001) Classification and change detection using landsat TM data: when and how to correct atmospheric effects? Remote Sens Environ 75(2):230–244. https://doi.org/10.1016/S0034-4257(00)00169-3

South S, Qi J, Lusch DP (2004) Optimal classification methods for mapping agricultural tillage practices. Remote Sens Environ 91(1):90–97. https://doi.org/10.1016/j.rse.2004.03.001

Su T, Zhang S, Liu T (2020) Multi-spectral image classification based on an object-based active learning approach. Remote Sens 12(3):504. https://doi.org/10.3390/rs12030504

Sun W, Zhang X, Zou B et al (2017) Exploring the potential of spectral classification in estimation of soil contaminant elements. Remote Sens (Basel, Switzerland) 9(6):632. https://doi.org/10.3390/rs9060632

Svoray T (2004) Integrating automatically processed SPOT HRV Pan imagery in a DEM-based procedure for channel network extraction. Int J Remote Sens 25 (17):3541–3547. https://doi.org/10.1080/01431160410001684992

Svoray T, Gancharski SBY, Henkin Z et al (2004) Assessment of herbaceous plant habitats in water-constrained environments: predicting indirect effects with fuzzy logic. Ecol Model 180(4):537–556. https://doi.org/10.1016/j.ecolmodel.2004.06.037

Svoray T, Markovitch H (2009) Catchment scale analysis of the effect of topography, tillage direction and unpaved roads on ephemeral gully incision. Earth Surf Proc Land 34(14):1970–1984. https://doi.org/10.1002/esp.1873

Svoray T, Shoshany M (2003) Herbaceous biomass retrieval in habitats of complex composition: a model merging SAR images with unmixed landsat TM data. TGRS 41(7):1592–1601. https://doi.org/10.1109/TGRS.2003.813351

Svoray T, Shoshany M (2004) Multi-scale analysis of intrinsic soil factors from SAR-based mapping of drying rates. Remote Sens Environ 92(2):233–246. https://doi.org/10.1016/j.rse.2004.06.011

Tarboton DG (1997) A new method for the determination of flow directions and upslope areas in grid digital

elevation models. Water Resour Res 33(2):309–319. https://doi.org/10.1029/96wr03137
Thenkabai P, Smith R, Pauw E (2002) Evaluation of narrowband and broadband vegetation indices for determining optimal hyperspectral wavebands determining optimal hyperspectral wavebands for agricultural crop characterization. Photogramm Eng Remote Sens 68(6):607–621
Tribe A (1992) Automated recognition of valley lines and drainage networks from grid digital elevation models: a review and a new method. J Hydrol 139(1):263–293. https://doi.org/10.1016/0022-1694(92)90206-B
Turcotte R, Fortin JP, Rousseau AN et al (2001) Determination of the drainage structure of a watershed using a digital elevation model and a digital river and lake network. J Hydrol 240(3):225–242. https://doi.org/10.1016/S0022-1694(00)00342-5
Van Deventer AP, Ward AD, Gowda PH et al (1997) Using thematic mapper data to identify Contrasting soil plains and tillage practices. Photogramm Eng Remote Sens 63(1):87–93
Vandekerckhove L, Poesen J, Oostwoud Wijdenes D et al (2000) Thresholds for gully initiation and sedimentation in Mediterranean Europe. Earth Surf Proc Land 25(11):1201–1220. https://doi.org/10.1002/1096-9837(200010)25:113.3.CO;2-C
Vogels MFA, de Jong SM, Sterk G et al (2017) Agricultural cropland mapping using black-and-white aerial photography, object-based image analysis and random forests. ITC J 54:114–123. https://doi.org/10.1016/j.jag.2016.09.003
Volcani A, Karnieli A, Svoray T (2005) The use of remote sensing and GIS for spatio-temporal analysis of the physiological state of a semi-arid forest with respect to drought years. For Ecol Manage 215(1):239–250. https://doi.org/10.1016/j.foreco.2005.05.063
Wise S (2000) Assessing the quality for hydrological applications of digital elevation models derived from contours. Hydrol Process 14(11–12):1909–1929. https://doi.org/10.1002/1099-1085(20000815/30)14:11/123.0.CO;2-6
Wu S, Chen L, Wang N et al (2018) Modeling rainfall-runoff and soil erosion processes on hillslopes with complex rill network planform. Water Resour Res 54 (12):1–17. https://doi.org/10.1029/2018WR023837
Wu Y, Cheng H (2005) Monitoring of gully erosion on the Loess Plateau of China using a global positioning system. Catena (Giessen) 63(2–3):154–166. https://doi.org/10.1016/j.catena.2005.06.002
Xue J, Su B (2017) Significant remote sensing vegetation indices: a review of developments and applications. J Sens 2017:1–17. https://doi.org/10.1155/2017/1353691
Yair A, Kossovsky A (2002) Climate and surface properties: hydrological response of small arid and semi-arid watersheds. Geomorphology 42(1):43–57. https://doi.org/10.1016/S0169-555X(01)00072-1
Zhu X, Helmer EH, Gao F et al (2016) A flexible spatiotemporal method for fusing satellite images with different resolutions. Remote Sens Environ 172:165–177. https://doi.org/10.1016/j.rse.2015.11.016
Žížala D, Juřicová A, Zádorová T et al (2018) Mapping soil degradation using remote sensing data and ancillary data: South-East Moravia, Czech Republic. Eur J Remote Sens 52(sup1):108–122. https://doi.org/10.1080/22797254.2018.1482524

Assessments of Erosion Risk

6

Abstract

This chapter presents methods for identifying and classifying areas subject to soil degradation risks—such as soil loss, water ponding, and sediment deposition. It begins in the first section with the exploration of the application of the topographic threshold in mapping areas that are, or are not, at risk of soil incision. Then, the second section describes an expert-based system for identifying areas at risk, based on weighting the contributing variables as defined by experts. The third section of the chapter describes data-mining techniques for studying soil degradation risks, using machine-learning methodology and training sets. Finally, in its fourth and final section, a fuzzy rule-based system is used to simulate soil loss and deposition in a spatially and temporally explicit manner, based on subjective probability, as formulated through fuzzy mathematics. Fuzzy rule-based modeling makes it possible to bypass more rigorous computations of exact physically based models. The methods demonstrated in this chapter are simple and easy-to-use—and can be applied to map soil degradation risks using the geoinformatics data stored in GIS layers.

Keywords

AHP · Expert systems · Data mining · Fuzzy logic · Machine learning · Topographic threshold

Assessing erosion risk over wide regions effectively involves putting together an advanced spatial inventory of various soil degradation risks. In such an inventory, each single soil degradation risk—such as soil loss, sediment logging, or water ponding—is identified in a spatially explicit layer that quantifies the level of risk in every basic unit such as the grid-cell, the field, hillslope unit, or a subcatchment. The inventory of identified areas at risk allows the analyst to determine which parts of the catchment can be disregarded, and which parts should be dealt with urgently (see Chap. 8).

In addition to the physically based and empirical modeling tools that can be used to identify risk levels in the spatial realm (Rosas and Gutierrez 2020), various geoinformatics methods have been developed for this purpose. These methods combine spatially and temporally explicit datasets with machine-learning and statistical techniques. Of the plethora of methods available, two very distinct groups stand out. The first group includes risk-mapping systems founded on established procedures for knowledge extraction from human experts and the formalization of this knowledge in a computer language. The second group includes risk-mapping

T. Svoray, *A Geoinformatics Approach to Water Erosion*,
https://doi.org/10.1007/978-3-030-91536-0_6

systems based on empirical data-mining methodologies. The data-mining methods use a collection of field and remotely sensed data, in a bid to discover meaningful risk-mapping patterns in those datasets through machine learning, advanced statistical techniques, and training sets that are labeled before the process and can be used to train the machines to predict areas that are at risk, and those that are not.

The data models used in both groups of methods are usually raster-based, corresponding to their central inputs: DEMs and remotely sensed images. Accordingly, the output risk prediction layers are also usually rasters which can be summarized, through *zonal statistics*, into vector-based hillslope data models of Park et al. (2001), subcatchments, field borders, etc. (see Chap. 4).

The methods in both groups have been used for various applications. Whether we need to further develop more data-driven methodologies, or more experts-based systems, is still an open question. This chapter therefore presents procedures and governing equations of four geoinformatics techniques from *both groups*, for generating output layers of water erosion risks (Table 6.1).

This chapter begins with a description of the procedure for applying the *Topographic Threshold* (TT) technique (Sect. 6.1). In the TT procedure, the areas at risk are identified by their characteristics of contributing area and hillslope gradient. Section 6.2 describes the use of *multi-criteria analysis* (MLC—Malczewski 1999) to predict areas as at risk, using expert knowledge and clorpt factors. The subsequent Sect. 6.3 illustrates the use of *data-mining* (DM) approaches to predict erosion risks, also based on the clorpt factors. Finally, Sect. 6.4 demonstrates the use of *fuzzy logic* (FL) to apply space–time dynamic prediction of erosion risks in a combined process-based and geoinformatics approach, again based on the same clorpt factors (Table 6.2 and 6.3).

The input layers used in this chapter for estimating erosion-risk levels include DEMs and their derivatives, layers of cultivation methods, linear unpaved roads, and meteorological radar data used to simulate space–time variations in rainfall intensities, and their impact on water erosion (see full list of layers in Table 6.4). Furthermore, using simulations, these GIS layers can be used to estimate potential water erosion responses to changes in climatic conditions and soil cultivation practices.

The approaches described in this chapter may also be useful in obtaining insights into fundamental geoinformatics questions—such as: What is the difference in the degree of consistency between using one expert, as opposed to several? Should we rely more on data gathering—or on experts? How can we deal with rare cases in training sets? What happens when we evaluate a fitted model using a different dataset? Do we benefit from more knowledge extracted from the experts—or do we have to "keep it simple"? Answers to such questions may assist the reader in deciding whether to collect more data for data-driven approaches or to mine more knowledge from experts.

Table 6.1 The four methods described in this chapter to estimate erosion risk levels, in an agricultural catchment, in a spatially explicit manner, using various inference methods combined with geoinformatics

Methods of erosion risk estimation described in this chapter				
Section	6.1	6.2	6.3	6.4
Input data	DEM	Clorpt	Clorpt	Clorpt
Geoinformatics technique	Topographic Threshold (TT)	Multicriteria (MLC)	Data Mining (DM)	Fuzzy Logic (FL)
Concept	Topographic	Expert-based	Data-based	Process-based
Dynamic	Static	Semi-dynamic	Static	Dynamic
Output	Binary 0/1	Range 0–1	Range 0–1	Range 0–1

Table 6.2 The TT coefficients used in this section to illustrate the effects of changing the coefficients on the risk-mapping by TT

Source	a	b	Category
Svoray and Markovitch (2009)	0.0048	−0.55	Extracted to the Harod catchment
Govers (1991)	0.0035	−0.40	Minimum a coeff. found in our search
Vandekerckhove et al. (2000)	0.227	−0.10	Maximum a coeff. found in our search

Table 6.3 Predicted critical hillslope gradient S_{TT} [°] under different a (rows) and b (columns) coefficients of the TT, for three representative flow accumulation conditions: Minimal flow accumulation (upper panel, 0.09 ha), median flow accumulation (0.27 ha, middle panel), and maximal flow accumulation (18,820 ha, lower panel)

	B						
a		−0.60	−0.48	−0.35	−0.23	−0.10	Flowacc [ha]
	0.004	0.57	0.57	0.57	0.57	0.00	Min. = 0.09
	0.199	40.00	32.00	25.00	19.00	14.00	
	0.395	59.00	51.00	43.00	34.00	27.00	
	0.591	68.00	62.00	54.00	46.00	37.00	
	0.787	73.00	68.00	61.00	53.00	45.00	
	0.004	0.57	0.57	0.57	0.00	0.00	Median = 0.27
	0.199	24.00	20.00	18.00	15.00	13.00	
	0.395	41.00	37.00	32.00	28.00	24.00	
	0.591	52.00	48.00	43.00	38.00	34.00	
	0.787	60.00	56.00	51.00	47.00	42.00	
	0.004	0.00	0.00	0.00	0.00	0.00	Max. = 18,820
	0.199	0.00	0.00	0.57	1.00	4.00	
	0.395	0.00	0.00	0.57	2.00	9.00	
	0.591	0.00	0.57	1.10	3.00	12.00	
	0.787	0.00	0.57	2.00	5.10	16.00	

6.1 Topographic Threshold

6.1.1 The Approach

TT is a prevalent physically based measure that has been used widely since the 1980s to map risks for soil incision, and even for estimating soil loss rates. Recent TT studies have quantified gully formation in time using high-resolution DEMs and images, extracted from unmanned drones (Gudino-Elizondo et al. 2018). With the development of new, accurate and detailed remote sensing tools, the application of TT—as a spatial measure—may allow to delineate areas that can potentially exceed the threshold for soil surface sheering in the catchment, based on a relatively small field-sampling effort.

The theoretical background of TT—which was established originally in the geological sciences—is described in Chap. 2. Briefly put, TT evolved from the ideas of Patton and Schumm (1975) on channeled erosion as a threshold process, and those, in turn, were inspired by the work of Schields (1936) on critical shear stress. Later, Begin and Schumm (1979) used the hillslope gradient and contributing area as a proxy for the hydraulic radius of the flow and for surface erosivity. Continuing this line of thought, identifying grid-cells with hillslope gradient that exceeds a threshold value for a given contributing area can be used as a proxy for predicting possible incisions, and can therefore be applied as an instrument to predict water erosion risk—in particular, risk of channeled erosion, such as gullies or rills (Torri and Poesen 2014). Section 6.1 accordingly focuses on the procedure of applying TT to map areas at risk in agricultural catchments using DEM data.

The TT equation (Eq. 6.1) assumes that concentrated overland flow produces shear stress in excess of a critical value needed to erode the surface that can be quantified using power law.

$$S = a \cdot A^{-b} \tag{6.1}$$

The analyst must geographically sample observed initiation points of gullies in the studied catchment, and compute the hillslope gradient S and contributing area A for each of these gullies initiation points. Then, the TT line is practically an empirical line in which the multiplier a and exponent b coefficients are estimated by trial and error. The extraction of coefficients for TT line implementation on data values from the Yehezkel catchment (a subcatchment of the Harod, 13 km^2 in size) is illustrated in Fig. 6.1. Note that the physical units of the contributing area are in hectares [ha]; the hillslope gradient units are in meters [$m \cdot m^{-1}$]; and the axes are converted to log–log scale. The nineteen incision points in Fig. 6.1 were measured in three different environments: Natural sites (i.e., rangelands), tilled areas, and (in three instances) the downslope area of an unpaved road (Svoray and Markovitch 2009).

The area at risk of erosiongully incision, for example—comprises all grid-cells whose combination of contributing area and hillslope gradient lie above the lowermost-points of observed erosion events. In those grid-cells, shear stress may exceed the surface resistance. Contrariwise, the area below the observed lowermost points represents safe sites.

A notable limitation of the TT approach is in the generalization of the coefficients. The TT graph represents the empirically estimated conditions in the agricultural catchment in question, which do not necessarily represent a universal form of the contributing area-hillslope gradient relationship. Indeed, as we shall see later in this section, different TT line coefficients have been extracted at various locations around the world. One must therefore sample all incision points in the catchment in question, to properly cover the parameter space of topographic conditions, and to avoid missing conditions of a possible incision. However, due to the effect of various dynamic climatic (e.g., rainfall intensity) and human-induced conditions (e.g., tilling actions), the pattern of gully heads in one year may differ from that of another, resulting in the analyst possibly missing incision points. Thus, gully heads should be sampled, when possible, at more than one season.

The advantage of TT as an estimator of water erosion risks is indeed that it has a sound physical basis. However, one of its disadvantages is that it ignores the clorpt variables other than topographysuch as surface cover (vegetation cover, roads, soil crust) that may have a considerable impact on runoff generation and surface resistance. As shown in Sect. 5.3.2.4, the effect of surface cover can be integrated into DEM-based computations by weighting the runoff coefficient. However, a more comprehensive multivariate analysis can be applied using the methods in the sections below.

6.1.2 Procedure

A TT procedure may be applied to estimate erosion risk through the following five steps:

Step 1: DEM acquisition and preprocessing;
Step 2: Computing the flow accumulation and hillslope gradient rasters (see Sect. 5.3);
Step 3: Extracting local TT coefficients from an empirical graph (a and b coefficients, see Eq. 6.1 and Fig. 6.1);
Step 4: Applying the equation from Step 2 on the two rasters from Step 3;
Step 5: Subjecting the predicted layer to an accuracy test.

For the data-based illustration, in Step 1, a contour-based DEM was used as the input topographic layer with vertical resolution of ~1 m². The DEM was compiled by a certified surveyor and was provided to us by the Israeli Ministry of Agriculture. As a preprocessing stage to the DEM, all artificial pits were removed (see Sect. 5.3.2 and Tarboton et al. (1989) for more information about filling pits). In Step 2, the TT coefficients a and b were extracted using field data on ephemeral gullies initiation. To illustrate the effect of using different coefficients, several configurations of coefficients from other studies were also applied (Table 6.2). In Step 3, hillslope gradient [°] and flow accumulation matrix [m²] were computed for each of the DEM grid-cells, using the TauDEM software by Tarboton (1997), available online at https://hydrology.usu.edu/taudem/taudem5/documentation.html. To compare the Harod TT coefficients with those from other studies, the physical units were converted as follows: Hillslope gradient values were converted from decimal degree units [°] to length ratio units [$m \cdot m^{-1}$], and flow accumulation values were converted from square meters [m²] to hectares [ha].

In Step 3, nineteen gullies were observed in the field, and for each of their gully heads, the mean hillslope gradient and contributing area values were computed for the immediate vicinity, using a 3 · 3 kernel function. Note that in the beginning of the growing season, before the crop seedlings cover the field, it is possible to detect and delineate the ephemeral gullies and the initiation points both from high-resolution remotely sensed data and in field visits. The TT coefficients a and b were estimated by plotting the contributing area-hillslope gradient graph as a log–log relationship (Fig. 6.1). Following Patton and Schumm (1975), a line that connects the lowest dots was assigned as the threshold between possible soil loss and no risk for soil loss. Under a given flow accumulation value A_i in a given cell i, all cells with a hillslope gradient

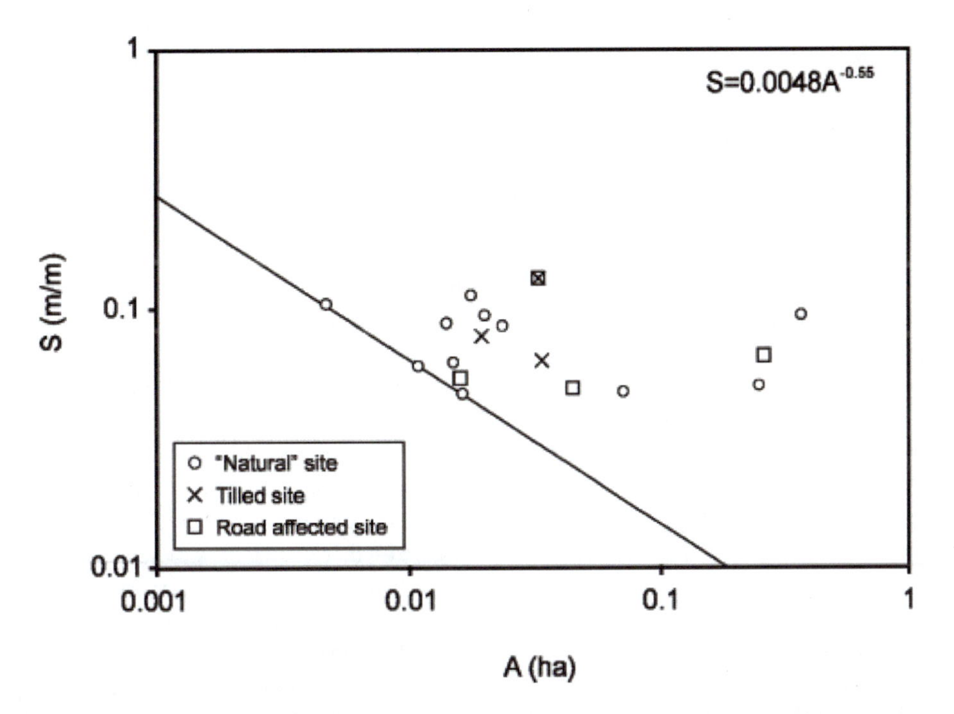

Fig. 6.1 The Topographic Threshold (TT) extracted from the Hillslope gradientS and Contributing area A graph at the Yehezkel catchment. The points represent observed gully heads and the TT is determined as the two lowermost-points line. Note that the axes are logarithmic. the line is used to compute the predicted areas at risk. Namely, the area above the line (i.e., high contributing area and slope) corresponds to high-erosion risk. From: Svoray and Markovitch (2009), Catchment scale analysis of the effect of topography, tillage direction and unpaved roads on ephemeral gully incision. Earth surface processes and landforms 34(14):1970–1984. doi: https://doi.org/10.1002/esp.1873. With permission from Wiley (Copyright © 2009 by John Wiley & Sons, Ltd.)

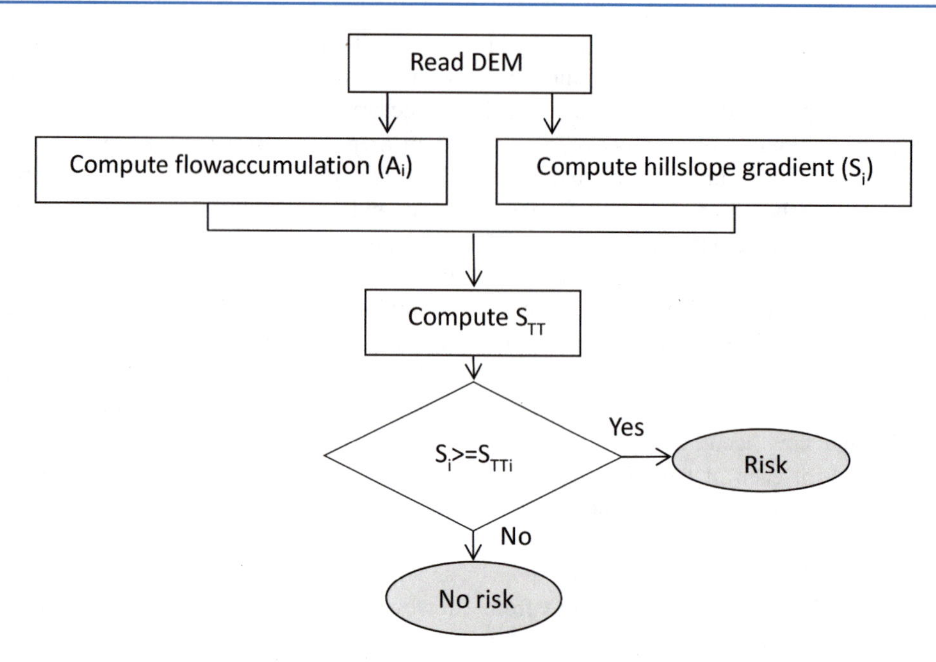

Fig. 6.2 Applying the TT technique to a DEM. Note that S_{TT} is computed using the TT equation as in Fig. 6.1 and Eq. (6.1), based on the chosen coefficients

S_i value exceeding the threshold value S_{TTi} were assigned with risk for soil loss by gully erosion (Fig. 6.2).

In Step 4, the contributing area-hillslope gradient relationship was applied to the Harod catchment, using the coefficients from Table 6.2. Figure 6.3 illustrates the computation on a small 3 · 3 pixel subset of the rasters. The result is a binary raster of *Risk/No risk* values for each grid-cell (bottom-right in Fig. 6.3).

In Step 5, the accuracy assessment for the Risk/No Risk layer is applied using error measures that can be either Kappa coefficient, or ROC and AUC (see Sect. 6.3). This requires a validation dataset of gully heads that were observed in the field and mapped using differential GPS.

6.1.3 Results and Discussion

6.1.3.1 Application to the Yehezkel Catchment

The final output of the TT procedure (Fig. 6.2) is a binary layer of soil erosion risk. The rightmost panel in Fig. 6.4 shows the output of applying the TT procedure to the DEM of the Yehezkel catchment, which is a subcatchment of the Harod. The coefficients used are from Svoray and Markovitch (2009), with $a = 0.0048$ and $b = -0.55$.

The hillslope gradient layer highlights the sharp topography of the northern section of the catchment (Hamore Hill) and along the flow lines in the center. The flow accumulation (i.e., contributing area) raster shows the water accumulation along the flow lines. These characteristics are evident in the TT-based risk map: The northern part of the catchment is mostly at risk, while risk diminishes, in a patchy fashion, toward the south, depending on the contributing area. In the southernmost part—where the Harod River is located—there are still some areas at risk, even though the hillslopes are very gentle (However, this may be also due to an overestimation effect that characterizes the TT predictions due to the use of binary classification and a cutoff point). Based on the result shown in Fig. 6.4, 58% of the Yehezkel catchment is at risk—which means that for the purposes of prioritizing the use of soil conservation techniques in the catchment, this method may be impractical.

S_i [mm-1]

0.04	0.01	0
0.07	0.01	0.1
0.03	0.02	0.01

A_i [ha]

0.01	0.01	0.1
0.05	0.07	0.08
0.01	0.1	0.05

S_{TTi}

0.060	0.060	0.017
0.025	0.021	0.019
0.060	0.017	0.025

$$S_{TTi} = 0.0048 A_i^{-0.55}$$

Risk

0	0	0
1	0	1
0	1	0

if $S_i < S_{TTi}$ then 0
if $S_i > S_{TTi}$ then 1

Fig. 6.3 Computing incision risk for a DEM, using hillslope gradient S_i and contributing area A_i data in the grid-cell i. The S_{TTi} layer in the lower left panel is computed using the coefficients extracted from Svoray and Markovitch (2009), plugged into Eq. (6.1)

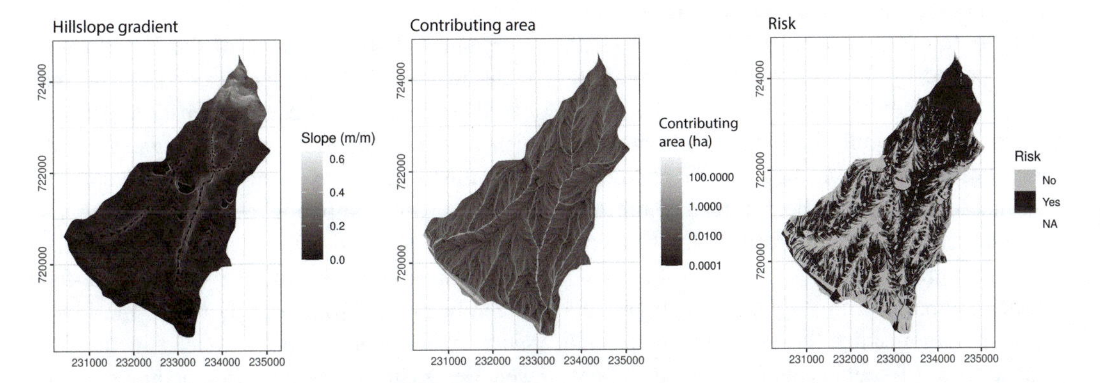

Fig. 6.4 Water erosion risk map in the Yehezkel catchment using the TT method with $a = 0.0048$ and $b = -0.55$. The center panel shows the flow accumulation raster; the left panel shows the hillslope gradient; the right panel shows the predicted erosion risk raster. The area at risk adds up to 58% of the entire catchment

6.1.3.2 Application to a Larger Catchment

The Yehezkel catchment is relatively small, covering an area of just ~ 13 km^2. When the TT is applied to a larger catchment—such as the entire Harod catchment, which is 193 km^2 (Fig. 6.5)—the effect of a much larger parameter space of topographic conditions can be observed on the TT computations.

In the bottom panel of Fig. 6.5, the output map—based on the relatively high coefficients of Vandekerckhove et al. (2000)—predicts almost the entire catchment (94%) as free of incision risk. The areas at risk are at the margins of the catchment—in the southern section, at the foot-slopes of Mt. Gilboa, and in parts of the foot-slopes of Hamoreh Hill in the northern section of the Harod catchment. This result is clearly an underestimation, because gully incision was also observed in other parts of the Harod catchment. At the other extreme, the map obtained from the coefficients by Govers (1991) is also suspect, as it shows almost the entire catchment to be at risk—which is both an unreasonable result, and impractical for soil conservation purposes. The two maps illustrate what happens when the analyst chooses coefficients that are unsuited to the environmental conditions of the catchment, resulting in very large under- or overestimation.

Conversely, the risk map based on the local empirical coefficients by Svoray and Markovitch (2009) shows higher variability. However, even this result—using coefficients which are well suited to the local catchment conditions—still were reported by the authors to suffer overestimation, as it shows large swathes of the catchment as prone to gully incision. While many of the gullies in areas exceeding the TT conditions may indeed incise, the sheer extent of the area classified as at risk is impractical for soil conservation purposes.

The comparison in Fig. 6.5 illustrates the importance of making the effort to choose the most suitable TT coefficients for each catchment. Less suitable coefficients may cause severe under- or overestimation. The selection of suitable coefficients may be achieved by sampling gully heads in representative areas of the parameter space of the specific catchment (see Sect. 4.2.1).

6.1.3.3 Aggregation at the Field Scale

Aggregation of spatial information on grid-cells at risk, at the field level (see Sect. 4.1.2), may be useful for soil conservation activities and planning, because the recommendations can be more easily tailored to the farmer needs. Also, aggregation to the field level can reduce noise due to salt-and-pepper effect and provide a different fragmentation of the catchment than apparent at the grid-cell level, and may allow new patterns of risk to emerge, through visual interpretation.

A well-known geoinformatics tool for applying an aggregation to the field level is the *zonal statistics* tool offered in many GIS software tools—such as the Spatial Analyst extension of ArcGIS; the "Raster Layer Zonal Statistics" tool in QGIS; the "exactextractr" package in R (Baston 2020); and more. In general, zonal statistics functions summarize the values of all grid-cells coinciding with particular "zones" delineated in another dataset—which can be a raster or a vector layer—applying a function such as the mean, median, minimum, and maximum to obtain a single value per "zone." This approach is based on the two layers system (value layer and zone layer) that was developed in the map algebra language in the earliest days of the development of GIS as a field in its own right (Fig. 6.6).

Figure 6.7 shows a map of the Harod catchment in which the raster-based TT risk map based on Svoray and Markovitch (2009) coefficients (see Table 6.2) was aggregated to the agricultural field level using zonal statistics. This was done by computing the mean of the risk values of all grid-cells coinciding with each agricultural field. The mean risk score per-field therefore ranges from 0 to 1, where the lowest score means that all the grid-cells in the field were designated as no risk, and the highest score means that all grid-cells in that field were designated as risk. The result in Fig. 6.7 displays a pattern of higher risk at the catchment margins,

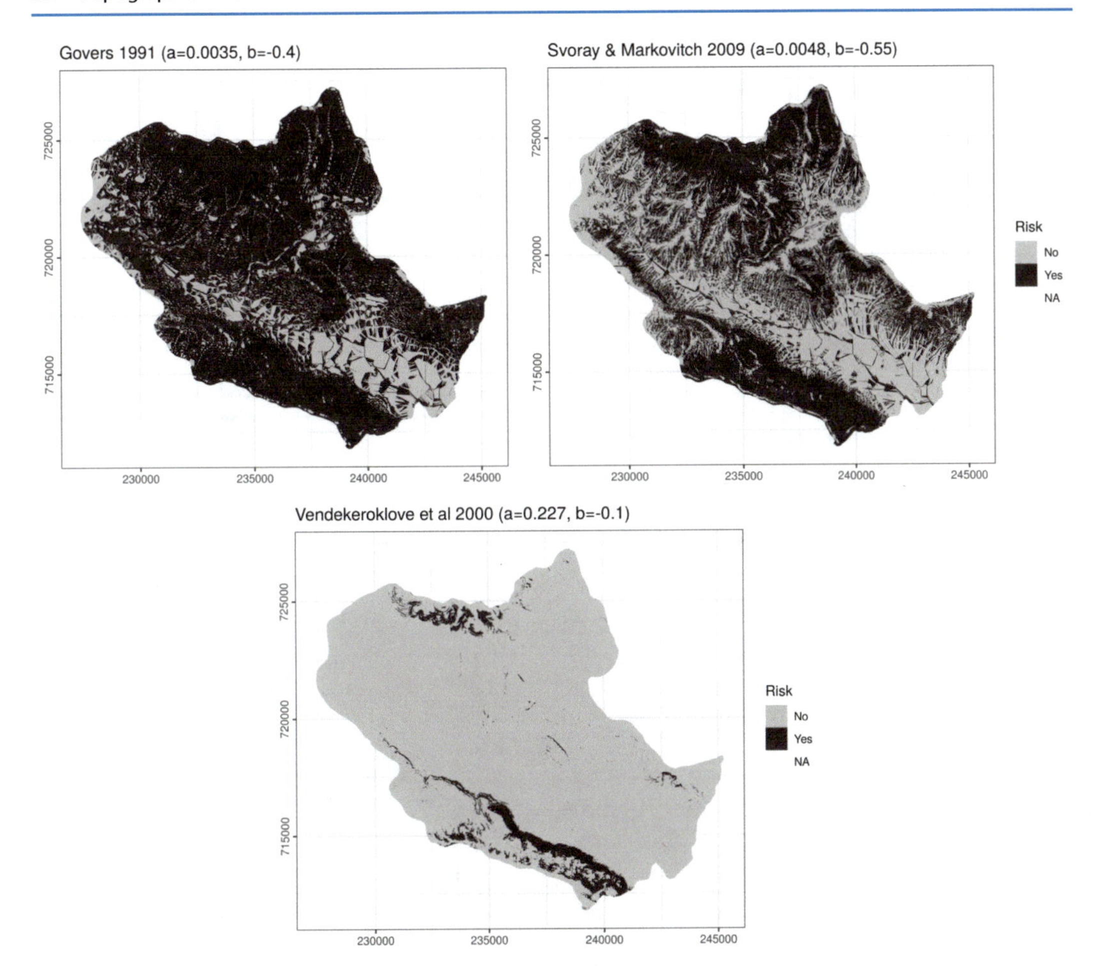

Fig. 6.5 A mapping of areas at risk in the Harod catchment, based on three configurations of TT coefficients. Based on the coefficients of Govers (1991), the predicted area at risk is 78%. Based on the locally estimated coefficients by Svoray and Markovitch (2009)—56%. The coefficients of Vandekerckhove et al. (2000) appear to underestimate the risk, with only 6% of the area predicted to be at risk

and lower risk at the center of the catchment. Additionally, there is small-scale variation between the fields in each part of the catchment.

The obtained field-level risk pattern can be used by the soil conservation planners to prioritize soil conservation activities at the field/farmer level—for example, by assigning a threshold value to the field score, and allocating resources to fields that score above it (see Chap. 8).

6.1.3.4 Sensitivity Analysis

Different TT coefficients were computed in many catchments around the world and were published in various research studies. Many of them were summarized in the seminal review paper on TT in different environments by Torri and Poesen (2014). Similarly, Svoray and Markovitch (2009) assembled 26 studies that applied TT to agricultural catchments in different parts of the world, to study water erosion processes. The coefficient values extracted in these 26 studies varied as follows: (1) The a coefficient varied between 0.0035 and 0.787; and (2) The b coefficient varied between −0.6 and −0.104. The variation in TT coefficients between studies is probably due to differences in local conditions, such as rainfall characteristics, crop plant type, soil properties (including texture, and hydraulic

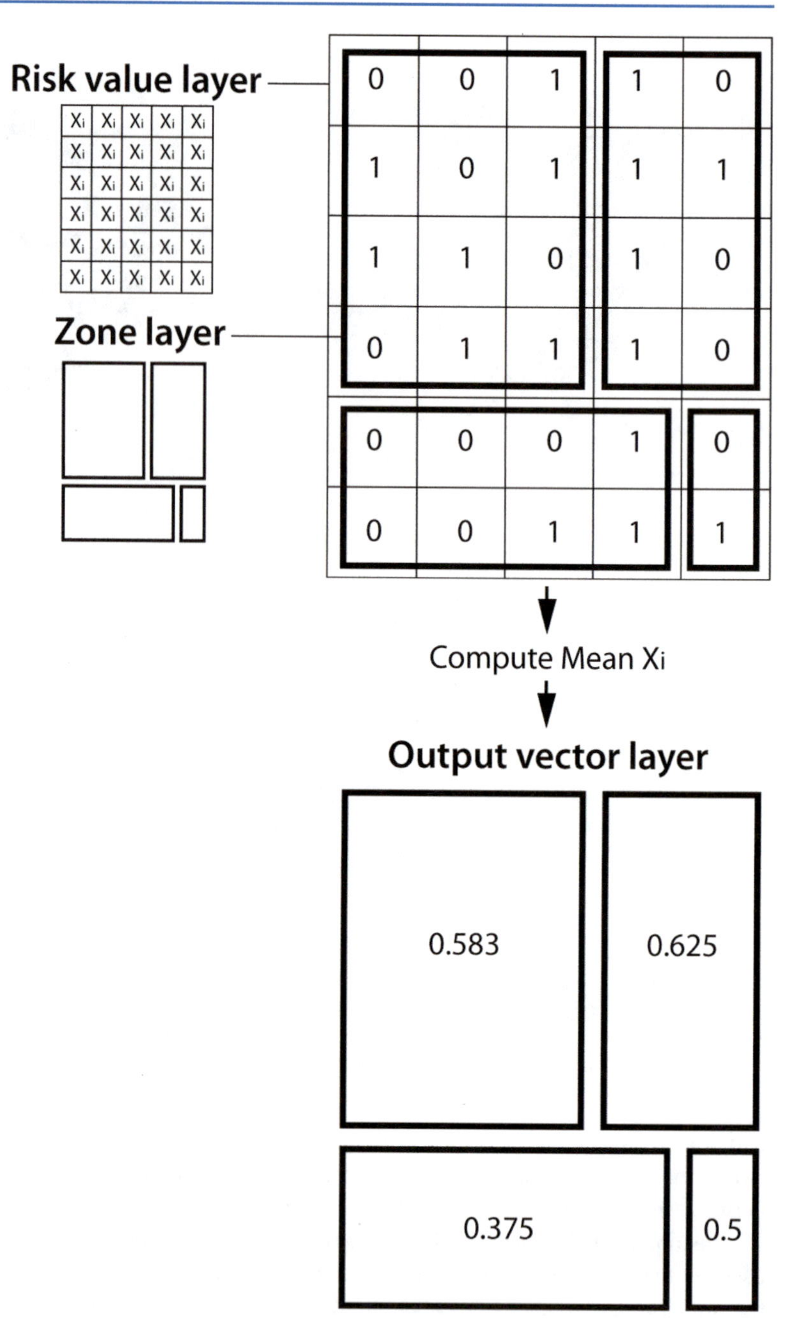

Fig. 6.6 The *zonal statistics* idea is a tool to aggregate raster grid-cells into a polygon layer using statistical operators. This example shows the use of the mean function, but many other functions—such as standard deviation, median, and mode —can be applied instead

conductivity, for example), cultivation method, and method of measurement of the contributing area and hillslope gradient.

Since analysts can sometimes fail to identify initiation points in their studied agricultural fields, for logistical reasons, they may have to apply TT using coefficients derived from the literature. It is therefore imperative to be aware of the catchment variability and to choose the most appropriate coefficients to the studied site. To express the effect of using different coefficients on predicted S_{TT} (depending on A_i), Table 6.3 shows different S_{TT} values computed for the Harod catchment, using different combinations

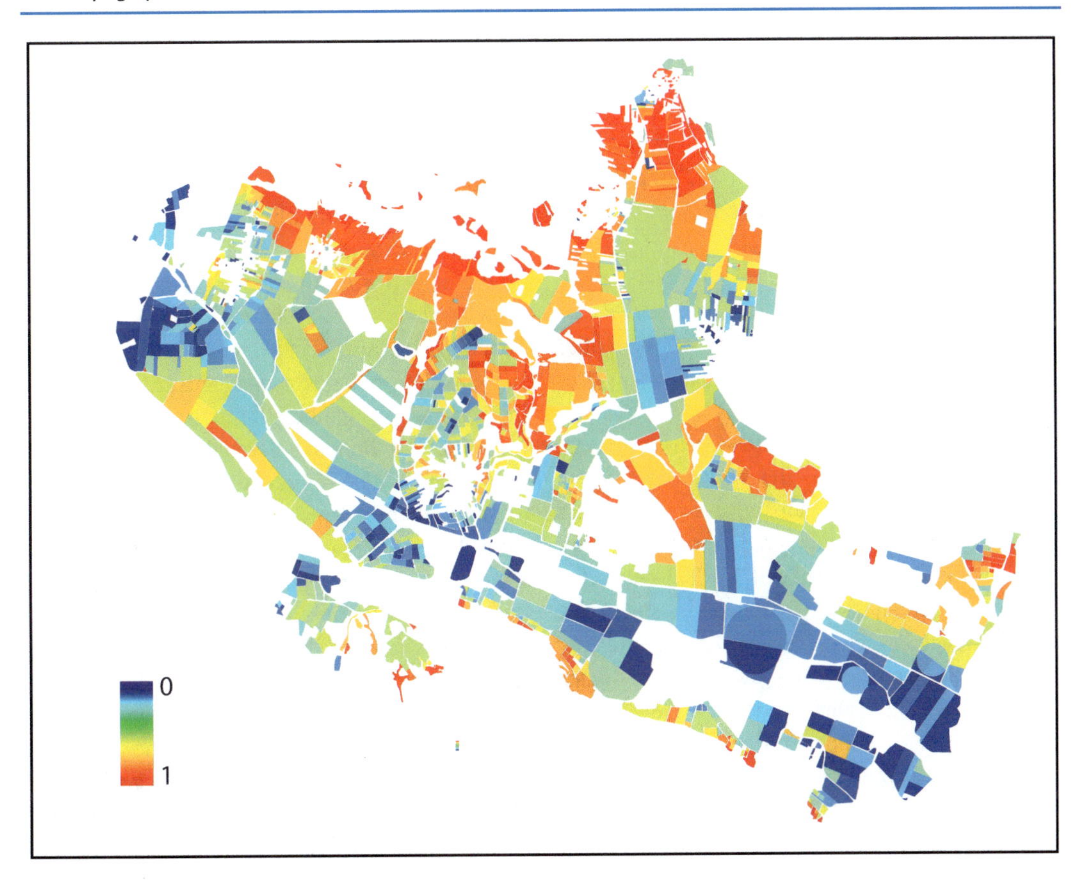

Fig. 6.7 Mean erosion risk of grid-cells per-field in the Harod catchment. Bluish hues are closer to zero risk; reddish hues indicate are closer to maximal risk

of coefficients. The hillslope gradient values of S_{TT} were converted from [m · m^{-1}] to [°]. The *Sensitivity* analysis was applied to compute S_{TT} for three representative area sizes: (1) An area with the minimal upslope contributing area in the Harod catchment (flowacc = 0.09 ha); (2) An area with the median contributing area in the catchment (with flowacc = 0.27 ha); (3) And an area representing the maximal contributing area in the entire Harod catchment (the outlet), with flowacc = 18,820 ha.

Table 6.3 shows that in the cases of the largest upslope contributing area, at the catchment outlet, the change in predicted S_{TT} due to the effect of coefficients is relatively minor, because in almost all cases, the S_{TT} required to initiate incision is very small (with the exception of two or three cases in the very large coefficients). In the areas with median upslope contributing area, the difference between the selected coefficients is more pronounced, and one can see that as the a coefficient increases and b decreases, the S_{TT} substantially increases. This trend is also evident in the case of minimal upslope contributing area, but in a more extreme fashion, since the required S_{TT} to induce erosion risk gets steeper as the contributing area decreases in size.

The findings presented here show that as a method to identify areas at risk of erosion, TT can suffer large overestimation—thus accurately identifying many of the areas that are at risk, but also falsely designating grid-cells as areas at risk when in fact they are not. Hence, TT can be used in agricultural catchments as a first step in identifying safe areas, or for better understanding of the effect of topography, using well-established physical equations. The resulting TT map can help the analyst understand the topographic structure of the catchment.

6.2 Expert-Based Systems

6.2.1 The Approach

From the hillslope scale to that of the entire catchment, the spatial distribution of water erosion phenomena is affected by two fundamental characteristics: The overland flow generation and dynamics—in particular, the water pressure on the soil—and the soil erosivity or resistance. These two characteristics are governed not only by topography—as simulated by TT—but also by other climatic, environmental, soil, and human-induced factors. The TT coefficients express—to some extent—the effect of other clorpt factors on water erosion, but there is a need to explicitly model those factors, with the help of more advanced analytical tools.

Two modern approaches offer sophisticated modeling and empower multivariate analysis of the effect of clorpt variables on water erosion. These are the *expert-based approach*, founded on knowledge extraction from human experts, and described in this section; and the *data-mining approach*, which is a data-driven methodology based on discovering patterns in geoinformatics datasets through machine-learning techniques, described in Sect. 6.3.

An *expert system*, in practical terms, is an information system that produces recommendations in a given problem domain through the reasoning of human experts and is capable of explaining the rationale behind the decision rules (Malczewski 2004). The primary use of such a predictive system is to emulate the professional knowledge and inference mechanism of human experts. The expert system can provide the decision-maker either with a "hard" and automatic answer (e.g., "There is a certain risk for gully incision at this location") and a recommended single course of action (e.g., "Change tilling practice to no tillage")—or with a "soft" and semi-automatic answer that may help the decision-makers in making a more educated decision (e.g., "There is a high/low probability of water erosion—to lessen the probable risk, use one of the following alternatives: Implement an alternating cropping system, or apply reduced tillage or perform hillslope terracing").

A typical expert system comprises four components (Jackson 1998): (1) The *Knowledge Base*, which includes a mechanism that imitates the human reasoning, through some sort of translation of human inference into a computer language; (2) The *Inference Engine* that supplies the computational procedures to make the decisions based on the aforementioned Knowledge Base rules. Inference engines may involve various architectures, such as probabilistic and Bayesian inference, fuzzy logic, Dempster–Shafer theory, and others; (3) The *Explanatory Interface* allows the analyst to explore how the rules were formulated and hence the reasoning mechanism, and how the expert perceives the effect of the input variables on the (erosion) process; and (4) The *User Interface*—namely, the software the analyst uses to communicate with the system.

The literature describes many expert-based systems. Here, we are going to review a simple and an effective method of integrating expert knowledge in a geoinformatics database, to study water erosion risks: *Analytical Hierarchy Processes* (AHP), which is a structured technique that uses the well-known *Multicriteria Analysis* (MCA) method (Malczewski 1999) to extract weights of different criteria toward a decision. MCA allows expert knowledge to be incorporated into geoinformatics tools, using relatively simple mathematical equations and computer programming. The idea of MCA is based on the integration of several factors to predict the phenomena in question—in this case, areas at risk of water erosion. As described in the next section, one of the leading ideas in MCA is to express the effect of each (clorpt) variable on each (water erosion) process by means of weights.

6.2.2 Weighting

Weights are mathematical tools to quantify the magnitude of the effect of each variable on each process—usually, the larger the weight, the larger the impact. Three approaches are commonly used in the literature for weighting: *Ranking*, *Rating*, and the *Pairwise comparison method* (Saaty 1977).

The Ranking approach is based on ordering the input variables by a human expert. The generated order is then used to compute the weights for each variable, based on one of the Eqs. (6.2)–(6.4) (Malczewski 1999):

(1) The *Rank sum* method, can be computed by Eq. (6.2):

$$w_i = \frac{n - r_j + 1}{\sum r_k}, \tag{6.2}$$

where w_i denotes the weight of importance for the ith variable, n is the number of elements in the population of variables, r_j is the rank position of the variable, k is the index of ranking positions running from 1 to n (k = 1,2…n) and $\sum r_k$ in the denominator is the sum of all ranking positions.
For example, if we order hillslope gradient (1, first in importance), contributing area (2, second in importance), and soil texture (3, third in importance) in determining soil loss, the normalized weight according to the Rank sum method (Eq. 6.2) would be 0.17 for soil texture, 0.33 for contributing area, and 0.5 for hillslope gradient, where $\sum w_i = 1$ and $\sum r_k = 6$.

(2) The *Rank Reciprocal weights*, are simply extracted from the normalized reciprocal rank of participating variables by the expert, as in Eq. (6.3):

$$w_i = \frac{\frac{1}{r_j}}{\sum\left(\frac{1}{r_k}\right)}, \tag{6.3}$$

where, in the aforementioned example of the three soil properties, the weights according to Eq. (6.3) are 0.18 for soil texture, 0.27 for contributing area and 0.55 for hillslope gradient. Once again, $\sum w_i = 1$ and $\sum r_k = 6$. However, note that here the sigma is summarizing $\frac{1}{r_k}$.

(3) the *Rank exponent method*, based on Eq. (6.4):

$$w_i = \frac{(n - r_j + 1)^p}{\sum (r_k)^p}, \tag{6.4}$$

where p is the normalizing power coefficient, with p = 0 and w_i = 1—that is, equal weight for the variables and for p = 1 the system yields Rank sum weights. The larger the p, the larger the difference between weights. In the aforementioned example, when p = 2, the weights according to Eq. (6.4) are 0.07 for soil texture, 0.29 for contributing area, and 0.64 for hillslope gradient, while, once again, $\sum w_i = 1$.

In the Rating approach, the analyst determines a scale—for example between 0 and 10. Then, the expert is requested to assign a weight value to each variable between 0 and 10, and these are subsequently summarized. Then, each assigned weight is divided by the sum of all weights to impose $\sum w_i = 1$. The values of the variable's weights can be determined based on multiplications of importance between the weights of the variables. Thus, if the hillslope gradient is twice as important as the contributing area and contributing area is three times more important than texture, the hillslope gradient can be assigned 6, contributing area 3 and soil texture with the value of 1. The normalized weights of the three variables in this case would then be 0.6 and 0.3, and 0.1, respectively.

Although simple and effective, the ranking and rating methods are less commonly used than the pairwise comparison approach (Saaty 1980), which is more structured, can be used to detect errors in the consistency of the experts' assessments, and is cognitively more established. Pairwise comparison is used in many scientific fields—from multiagent AI systems, to social sciences and psychology. In the pairwise comparison method, the expert compares pairs of clorpt variables, one pair at the time, to extract the importance weightings of all variables. This provides the expert with a tool to determine which of the two variables is more important than the other, to quantify how important it is, or to ascertain if the variables are of identical effect. In practical terms, the expert compares pairs of variables at the intersections of the pairwise comparison matrix (Fig. 6.8), and cross-checks all pairs.

The Saaty (1980) scale (Fig. 6.9) ranges from 1 to 9, while 1 denotes equal importance of the two variables, and 9 represents supreme importance of one of the variables over the other.

More specifically, the pairwise comparison matrix is filled as follows. The expert has to assign a value of relative importance between two variables in the table intersection cells. The comparison of importance between variables is conducted between every possible pair of variables of the upper triangle (gray cells) by means of the Saaty Appreciation Scale. The diagonal cells are all designated 1, because in those instances the variable is compared with itself. The cells in the lower triangle (white cells) are computed and filled automatically, because their values are the respective reciprocal (X^{-1}) values of the cells of the upper right triangle.

The intersection cells are filled in, while the expert compares the row variable against the column variable. For example, if the hillslope gradient is moderately more important than the contributing area, then at the intersection between the two where the hillslope gradient is in the row and the contributing area is in the column, the number assigned is 3, and 1/3 in the commensurate cell at the other intersection. Similarly, if the expert estimates that hillslope gradient is strongly or very strongly more important than soil texture, the assigned values are 6 and 1/6, respectively.

Once the matrix is filled, the weight of each variable is extracted as follows (Malczewski 1999):

Step 1: Sum the values of the pairwise comparison matrix in each column;

Step 2: Divide each cell in the pairwise comparison matrix by its column sum;

Step 3: Average the values in each row of the resulting matrix to obtain the weight of each variable (see the example in Fig. 6.10 based on the pairwise comparison matrix from Fig. 6.8).

In order to estimate the consistency of the expert in filling in the pairwise comparison matrix, we have to begin with the following procedure. First, the analyst has to compute the *weighted sum vector* by multiplying the values in the columns of the original pairwise comparison matrix with the weights of the variables.

	Hillslope gradient	Contributing area	Soil texture
Hillslope gradient	**1**	3	6
Contributing area	1/3	**1**	4
Soil texture	1/6	1/4	**1**

Fig. 6.8 The pairwise comparison matrix, in which the expert draws comparisons between every two variables. The relative importance of each variable is assigned according to the Saaty scale (Fig. 6.9). The diagonal cells are designated 1, because the variable is compared with itself, and the upper right (gray) triangle is filled by the expert. The lower left triangle mirrors that of the gray cells, so can be filled automatically. Thus, if hillslope gradient is perceived by the expert to be moderately more important than contributing area, it is assigned the value 3 at the intersection between the two variables

Importance	1	2	3	4	5	6	7	8	9
Definition	Equal importance	Equal to moderate importance	Moderate importance	Moderate to strong importance	Strong importance	Strong to very strong importance	Very strong importance	Very to extremely strong importance	Extreme importance

Fig. 6.9 The Saaty (1980) Appreciation Scale, used to compare two input variables in terms of their importance to the occurrence of a given phenomenon, or its effect on a given process

	A			**B**			**C**
	Hillslope gradient	Contributing area	Soil texture	Hillslope gradient	Contributing area	Soil texture	w_i
Hillslope gradient	**1**	3	6	0.67	0.70	0.55	0.64
Contributing area	0.33	**1**	4	0.22	0.24	0.36	0.27
Soil Texture	0.17	0.25	**1**	0.11	0.06	0.09	0.09
Σ	1.5	4.25	11	1	1	1	1

Fig. 6.10 Extracting the weights from the pairwise comparison matrix. This is done in three steps: (1) Each column of the pairwise comparison matrix is summarized (**A**); (2) Each intersection is divided by the sum of its column (**B**); and (3) The weight of the variable is the average in the row (**C**)

Thus, the column of the first variable (hillslope gradient) in the original pairwise comparison matrix is multiplied by the weight of the hillslope gradient. Same is done for the second column of the contributing area variable which is multiplied by its weight. Similar computation is applied for the third column that is multiplied by the soil texture weight. As a second step, the analyst computes the summation of the new values over the rows. And finally, in the third step, the analyst divides the weighted sum vector by the criterion weights to compute the *consistency vector* (Fig. 6.11).

In the fourth step, using Eq. (6.5) we compute lambada λ—the *average consistency vector*. In this example, $\lambda = 5.567$.

$$\lambda = \frac{\sum_1^n CVEC}{n} = \frac{16.7}{3} = 5.567. \tag{6.5}$$

In computing the *Consistency Index CI*, the analyst needs to assume that the *average consistency vector* λ is greater than the number of variables studied n for positive and reciprocal matrices. The average consistency vector is equal to the number of variables if the pairwise comparison matrix is consistent (Malczewski 1999). *CVEC* is the summation of the consistency vector (Fig. 6.11). Thus, inconsistency is expressed by the normalized distance of n from λ—as is Eq. (6.6):

$$CI = \frac{\lambda - n}{n - 1}. \tag{6.6}$$

In this case *CI* is $\frac{5.567-3}{3-1} = 1.2835$. The consistency relationship *CR* is determined by Eq. (6.7):

$$CR = \frac{CI}{RI}, \tag{6.7}$$

where the *Random Index RI* is a parameter depending on n (Saaty 1977). For example, in a case of a system with 3 variables, the appropriate value of *RI* is 0.58 (Malczewski 1999) and therefore *CR* = 2.21. *CR* values <0.1 indicate a consistent matrix (Malczewski 1999). Therefore, the results presented in the pairwise comparison matrix in Fig. 6.8 are inconsistent.

	Hillslope gradient	Contributing area	Soil texture	Σ (Rows)	Σ (Rows)/ wi (Consistency vector)
Hillslope gradient	0.64	1.92	3.84	6.4	10.00
Contributing area	0.09	0.27	1.09	1.45	5.37
Soil Texture	0.01	0.02	0.09	0.12	1.33
				ΣCVEC	16.7

Fig. 6.11 The consistency matrix. Each intersection is the original pairwise comparison value multiplied by the weight of the corresponding variable

6.2.3 Decision Rules

The weightsdetermined by one of the above three methodscan be combined to perform a multivariate analysis, by a variety of methods of two main types (Jankowski 1995). The first type of methods is the compensatory method, usually based on summation. In decision rules of summation, variables with higher values can compensate for variables with lower values. For example, if the contributing area is large, no matter how small the value of rainfall intensity is, the overall score for risk will be high. In the compensatory methods, the input variables are normalized to a quantitative scale, between 0 and 10, for example, then assigned with weights. The second type—the non-compensatory methods—includes those algebraic or other expressions that prevent the possibility of compensating between the variable's values. The non-compensatory method allows for a specific criterion to force a final score. For example, if the rainfall intensity is low, no matter what the value of all other criteria is, the overall score for water erosion risk will be low. Since the second approach entails, at most, only a serial ranking of the variables, this approach requires less attention from ranking experts. An example of a non-compensatory method is given in Sect. 6.4 as one of the joint membership functions; for now, we shall focus on two compensatory methods.

A very straightforward and commonly used compensatory method of multivariate prediction is the *Weighted Linear Combination* (WLC—Eq. 6.8), which is a weighted summation of the input variables for (risk) estimation:

$$S_j = \sum_{i=1}^{n} W_{ij} \cdot X_i, \tag{6.8}$$

where, for example, the following configuration is relevant to predictions of water erosion phenomena: S_j is the level of risk to a given water erosion phenomenon j, such as soil loss, or water ponding; n is the number of variables; W_{ij} is the weight of a variable i; and X_i is the variable's value within the normalized scale.

Applying WLC involves the following three preliminary steps:

Step 1: Selecting the variables that affect the predicted outcome;

Step 2: Extraction of a weight for each variable;

Step 3: Normalizing the range of the original variables values to a quantitative scale, between 1 and 10, for example, Basnet et al. (2001) or 1–5 as in the example at Table 6.4.

To apply WLC to a catchment, the user must normalize the variables' values in each layer into levels of influence for each water erosion phenomena, based on expert knowledge (see for example Table 6.4).

The normalization procedure solves problems such as, for example, when contributing area becomes more influential in a WLC than rainfall intensity, just because contributing area is measured by hundreds of m^2, and rainfall intensity by tens of mm in a 30-min period. In the normalization process, contributing area can be divided into five levels of risk to soil loss, as follows: 0–100; 101–800; 801–900; 901–1000; >1000, based on expert knowledge extracted from a drainage engineer. Similarly, the rainfall intensity can be divided into five groups: 0–3; 3.1–10; 10.1–25; 25.1–50; >50, based on a climatologist's experience. The raw data transfer into categories allows the X_i values in WLC to be fed with data on a normalized range of 1–5—or any other range the reader may choose. This means that the normalization step requires further knowledge input from the experts.

Another, more advanced, method for combination of multivariate data, is the *Compromise Programming Technique* (CPT). CPT was developed to express the object (grid-cell) conditions, based on points of predetermined extreme reference in the input variables, and computing the "distance" from those reference points (Eq. 6.9—Tkach and Simononic 1997):

$$L_p = \left\{ \sum\nolimits_k w_k^p \cdot \left[\frac{f_{k+}(x) - f_k(x)}{f_{k+}(x) - f_{k-}(x)} \right]^p \right\}^{\frac{1}{p}}, \quad (6.9)$$

L_p denotes the *distance metric* of the total score value of a given grid-cell from two well-defined points: *Ideal* and *anti-ideal*. The *ideal* point refers to the value of the criterion from which the expert expects that there will be definitely no erosion damage. The *anti-ideal* point refers to the value of the criterion from which the expert definitely expects erosion damage. w_k is the weight provided by the pairwise comparison matrices for the given variable *k; p* is a power variable, ranging from unity to infinity; $f_{k+}(x)$ is the ideal point; $f_k(x)$ is the value of the variable *k* in the grid-cell *x*; $f_{k-}(x)$ is the anti-ideal value. The *p* variable controls the effect of the deviation from the ideal point. In the case of $p = 1$, the weights of the deviations are similar; as *p* increases, the weight grows with the deviation.

CPT requires more knowledge from the expert than the WLC does. That is to say, in CPT, the expert needs to set threshold values for the ideal point and the anti-ideal point for each variable. For example, the expert might determine that for the hillslope gradient variable, the ideal point is the value 2°. In the ideal point (and more mild hillslope gradients) the expert assumes that no water erosion is expected to occur. In parallel, the expert may assign 18° as the anti-ideal point. In the anti-ideal point—or in steeper hillslope gradients—the expert assumes that erosion will definitely occur. Examples for ideal and anti-ideal points for four clorpt variables in the Harod catchment, as were determined by experts, are detailed in Table 6.5.

Figure 6.12 simulates the increase in the compromise programming score (L_p from Eq. (6.9) with a single variable, $k = 1$, and $f_k(x)$ are represented in the X-axis) as a function of an increase in hillslope gradient, from the ideal point 2° to the anti-ideal point 18°. The different *p* curves show how the compromise programming score becomes closer to linear with the decreasing *p*. The score at the ideal point converge $L_p = 0$, and at anti-ideal $L_p = 1$.

The use of CPT raises the question: Does supplementary expert information yield a more accurate estimation, or does it reduce the quality of estimation, due to the limited knowledge of human experts, which can result in a magnified error?

Table 6.4 An example of the normalization of the values (X_i from Eq. 6.8) of the variables into five levels of influence for the application of a weighted linear combination

Level of influence	Hillslope gradient [°]	Tillage direction	Contributing area	Rainfall intensity	Crop type	Vegetation cover	Road contribute	Road barrier	Aspect	Stone Cover
1	0–2	0–1	0–100	0–3	Dense woodland	91–100			340–90	80–100
2	2.1–6	2–3	101–800	3.1–10	Pasture	81–90			90–110 320–340	50–80
3	6.1–10	4–5	801–900	10.1–25	Sparse woodland	71–80	3	3	110–130 270–320	40–50
4	10.1–25	6–7	901–1000	25.1–50	Orchard	51–70	2	2	225–270 130–160	20–40
5	25<	>7	>1001	>50	Cropland	21–50	1	1	160–225	0–20

Table 6.5 An example for possible assignment of ideal and anti-ideal points for four clorpt variables

	Soil loss		Sediment deposition		Water ponding	
	Ideal	Anti-ideal	Ideal	Anti-ideal	Ideal	Anti-ideal
Hillslope gradient	2	18	6.5	0	6.5	0
Vegetation cover	100	50	0	90	50	90
Rain intensity	0	50	0	10	0	50
Crop type	Orchards	Field	Field	Pasture	Orchards	Field

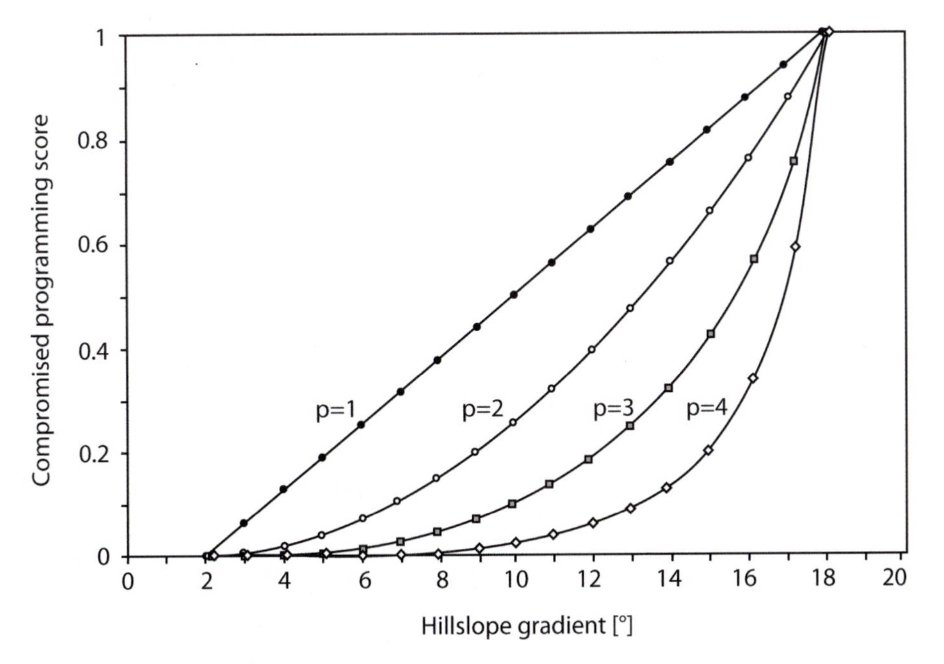

Fig. 6.12 The compromise programming concept, as exemplified in a scenario where 2° is the ideal point, and 18° is the anti-ideal point. Note how the compromise programming score is less affected by the hillslope gradient in mild topography as the *p*-value increases from 1 to 4

6.2.4 Procedure for Estimating Risk Levels

The pairwise comparison and WLC/CPT procedures were implemented to the Harod catchment (Fig. 6.13) in order to (1) Study how knowledge is extracted from experts, and in particular to see how a single best expert performs against a group of experts in predicting three water erosion risks: Soil loss, water ponding, and sediment deposition; (2) Compare the performance of WLC versus CPT, assuming the latter demands more details from the expert; and (3) Apply simulations of the effect of variation in clorpt variables on the prediction of the three risks.

The procedure in Fig. 6.13 allows the experts' knowledge to be explored, in a bid to estimate the risk levels of the aforementioned three soil degradation phenomena for each grid-cell in a raster layer of the catchment. This is done through the following seven steps:

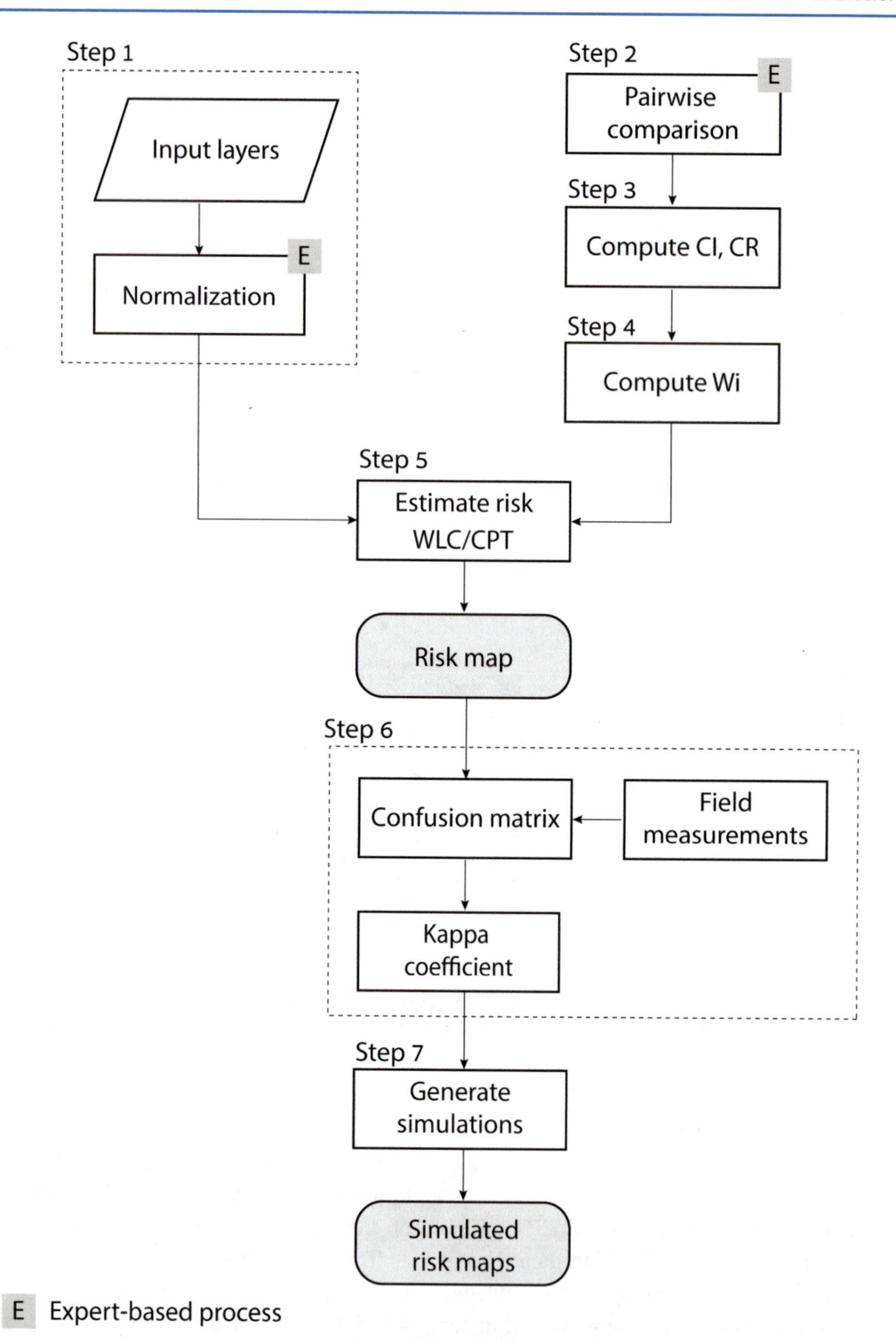

Fig. 6.13 The application of the spatial multicriteria expert-based system to the Harod catchment using GIS layers

Step 1: Design a spatial raster dataset with the clorpt variables that affect the water erosion in your catchment (e.g., hillslope gradient, contributing area, rainfall intensity from meteorological radar, etc.) The values of the variables in the database are then normalized into levels of the same scale;

Step 2: Fill in the pairwise comparison matrices as described in conditions in your catchment and availability (e.g., a soil scientist; a drainage engineer; a farmer etc.);

Step 3: Compute *CI* and *CR* for each expert, and decide on the final matrices to be used to extract the weights;

Step 4: Compute the weights;

Step 5: Estimate the risk, by means of WLC, and by CPT;

Step 6: Test the estimation results using confusion matrices and Kappa coefficients (see Sect. 5.2.1.1);

Step 7: Prepare flowchart (such as in Fig. 6.14), for consistent simulation analysis.

6.2.5 Simulations

A powerful tool for studying water erosion risks using expert systems is to simulate variation in the input variables values to test their effect on the spatial distribution of soil degradation phenomena in the catchment (Fig. 6.14). The harsher the simulated conditions in the given criterion C_n, the greater the area that is expected to be at risk in the output layer O_m. Moreover, an expert

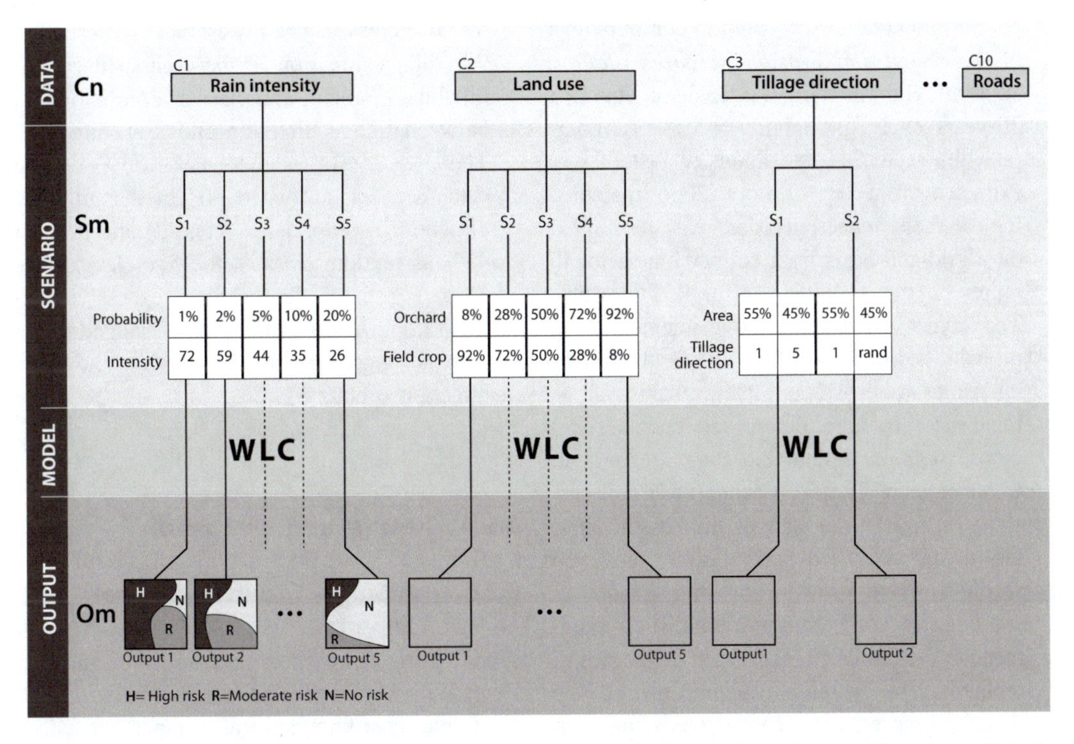

Fig. 6.14 Simulation process of alterations in rainfall intensity, land use, and tillage direction. The criteria C_n in the database represent the input layers of clorpt variables. The scenarios S_m represent the various alterations in the three aforementioned variables, and the output layers O_m represent the changes in the three risk levels as a result of the changing conditions in the different S_ms. Note that the coverage of the low risk group increases as rainfall intensity decreases, for example

system such as that in Fig. 6.13 may make it possible to quantify the rate of increase in the severity of erosion risk in the catchment, and to determine which fields are likely to be damaged in each scenario. Such a strategy can be applied to three of the more dynamic clorpt variables: Rainfall intensity, land use, and tillage direction.

In each simulation run S_n, one of the criteria (rainfall intensity, for example) is altered in methodical fashion, while the other variables are kept constant, with their original assigned values as observed in the catchment. The simulations of changes of the simulated variable input data values were applied here as illustrated in Fig. 6.14, but these can, of course, vary according to the hypotheses tested.

- Rainfall intensity: The baseline for rainfall intensity simulations ($S_1 \ldots S_5$ in C_1 dataset in Fig. 6.14) was the rainfall intensity measured in the large October 2006 event, with 44 mm per 30-min period. According to computations of the *Israeli Soil Erosion Research Station*, an event with this rainfall intensity occurs in a 20-years cycle (probability = 5%). To compute the scenarios, the intensity raster layer extracted from the October 2006 meteorological radar reflectivity data was multiplied on a grid-cell basis by a conversion factor to represent the following rainfall intensities: The layers representing low intensity bore probabilities of 10 and 20%, while the higher intensities were of 1 and 2% probability.
- Land use: In agricultural catchments, field crops are more vulnerable than orchards to water erosion. That is because their soil erosivity is high at the start of the season, after the fields are tilled, and before the crop seedlings have grown. For this reason, in some cases, soil conservationalists recommend to the farmers to shift from field crops to orchards. The simulation of land use change therefore represents a continuous change in the catchment from field crops to orchards. To initiate the simulation, the land use conditions of the Harod during the October 2006 event were used once again as a starting point. At that point in time, the Harod catchment was covered by more than ninety percent of field crops and the rest was covered by orchards. To simulate a gradual shift between field crops and orchards, an interval of 22% was assigned (see the swap between land uses in $S_{1.0.5}$ under C_2, in Fig. 6.14). In practical terms, the simulation process included vegetation cover as a quantitative measure, to highlight the difference between field crops and orchards. The mean value of vegetation cover, extracted from the October 2006 orthophoto, was 60% for orchards and 0% for crop fields. These two mean values were assigned to all fields in each of the simulations using stratified random sampling, to maintain a similar field size in all four simulations.
- Tillage direction was simulated by representing two extreme states of tillage direction in the entire catchment (S_1 and S_2 in C_3 simulation). To do so, as a starting point, the computed tillage directions were classified into five classes. The lowest class representing 90° tillage direction to the computed critical hillslope gradient, and class 5 representing 0° between tillage direction and flow direction. Then, S_1 was assigned with 55% of the catchment as cultivated 90° to the hillslope direction (using classes 1 and 2) and the rest 45% as random class of the five classes (see S_2 in Fig. 6.14). In the more excessive scenario (S_1), the other 45% were changed to the highest degree of risk (0° from the flow direction = class 5).

6.2.6 Results and Discussion

6.2.6.1 A Single Expert or Several Experts?

When using an expert system, the question always arises as to whether we should work with the single most knowledgeable expert, or with a group of several less experienced ones. More specifically, we ask, does a single expert provide more valuable insight than a less consistent group of several experts and can they provide more accurate predictions of soil degradation risks? Svoray and Ben-Said (2010) explored these questions based on knowledge extracted from four experts: A drainage engineer, two soil conservationalists and a professor of soil sciences. All four experts were found to be consistent (*CR* <0.1) when tested as a single expert, with regard to *most* of the risks studied (soil loss, water ponding, and sediment deposition). Nevertheless, the use of the four pairwise comparison matrices together yielded *CR* <0.1 in *all* cases. In other words, the integration of knowledge from several experts provides a more consistent "average" result, compared to the result from each individual expert. This outcome may be due to the fact that completing the forms of the pairwise comparison matrices by experts may be susceptible to subjectivity (Ni and Li 2003). In addition, converting physically based information such as that related to water erosion into the Saaty scale may cause loss of information (Gilliams et al. 2005). Therefore, the use of the Saaty scale by one person may be less consistent or even inaccurate, while increasing the number of experts may compensate for methodological limitations between different schools. An example of the subjectivity of the single expert is a very low weight assigned by a single—most consistent—expert to hillslope gradient and to rainfall intensity. This is despite the fact that many previous studies—both empirical and theoretical—found those two variables are key factors in water erosion processes, especially during heavy storms (see Chap. 3). Such an omission of variables that were found to be influential in many previous studies can clearly be magnified when a single expert is relied upon—no matter how knowledgeable they may be. The idea that a group of experts can decrease the effect of subjectivity has already been observed by Dalkey (1969) and Ni and Li (2003). To some extent, this idea is also in line with the recently emerging practice of crowdsourcing in geoinformatics (Haklay 2010). Although crowdsourcing refers mainly to the replacement of expert knowledge with assessments by non-professionals, rather than those of several experts, our results nonetheless show that the use of additional experts, even if they are less consistent than a single authoritative one, enhances predictive ability, rather than decreases it. This contradicts the intuitive expectation that less knowledgeable experts decrease predictive ability.

Aside from consistency, the predictive ability of group of experts was found to be higher than that of the single, most consistent expert. With a single, most consistent expert, the Kappa coefficient value of the risk prediction, when compared with the validation data from the field, is a mere 0.41—versus 0.72 provided by mixed matrices of all experts. This is a very substantial difference —enough for most users to reject the accuracy achievable by a single expert in favor of that of a group of experts.

6.2.6.2 A Complex or a Simple System?

Based on the *Occam's Razor principle*, one might conjecture that the simpler the model and the less it is founded on unnecessarily complex assumptions, the more accurate its predictions will be. This hypothesis is not a straightforward one—data-wise, at least—because occasionally when we strive to reduce the complexity in input data by omitting expression of an important variable to the process, we may detract from the model's predictive ability. One such example is disregarding the effect of cultivation method. As demonstrated by Svoray et al. (2015), adding the cultivation method variable to the MLC significantly boosts predictive ability from Kappa = 0.72 to Kappa = 0.93. However, the comparison between the WLC and the more demanding CPT has shown that the latter reduces predictive ability, from Kappa = 0.72 with the WLC, to Kappa = 0.62 through CPT (Svoray and Ben-Said 2010). Moreover, CPT predictions for all three soil degradation processes generally overestimated the risk values. Namely, the additional knowledge needed to determine the ideal and anti-ideal values for each variable appeared to be inaccurate because it was the only difference between the two predictions. One possible reason for the overestimation observed in the CPT is the experts' extra caution due to possible confusion when establishing the ideal and anti-ideal points. In other words, additional knowledge from the experts, which may be unfounded, may create noise and decrease prediction accuracy. That said, compromise programming has been found to be successful in various other applications, such as land use planning (Baja et al. 2007), and more recently, for soil conservation practices (Arabameri et al. 2018).

6.2.6.3 Which are the Most Important Clorpt Factors?

The assignment of weights to the clorpt variables using the pairwise comparison matrix can shed light on the perception of the process by experts. We know—from Chap. 3—how the effect of clorpt variables on erosion processes has been quantified by empirical models, and how weights of importance are extracted using data-fitting techniques. These results can be compared with an expert's assessment. In the case of predicting soil loss risk, Svoray et al. (2015) have shown—in line with the findings of the soil loss models described in Chap. 3—that the hillslope gradient has the highest weight among the evaluated variables. The second most important factor in preventing runoff development was vegetation cover, and the third most important contributor to soil loss is presence of roads. All other variables are of lower, uniform, weights. Surprisingly, the contributing area does not have a particularly high weight. Rainfall intensity was also found unexpectedly to have low weight, despite its apparent importance in the studies described in Chap. 3.

In the estimation of water ponding risks, experts rated—in different pairwise comparison matrices than those of the soil loss matrices—hillslope gradient to be the influential factor, followed by contributing area, then all the other variables. This is perhaps because hydrological properties that result in overland flow and water ponding are affected by the interaction between rainfall characteristics and the soil surface (see Chap. 3). In clayish soils, such as the soils of the Harod catchment, raindrops only penetrate the upper layer, so water absorption by the soil profile is limited even during events of large rainfall amounts (Bauder 2005). Sediment deposition occurs when soil particles that are detached from hillslopes in the upper parts of the catchment are transported to and accumulate in lower parts toward the outlet (Hoober et al. 2017). Crop coverage, size of contributing area, and the hillslope gradient are the criteria that mainly dictate the amount of soil deposited downslope (see Chap. 3). The most important of all the variables is the hillslope gradient, due to its contribution to overland flow. The second group of factors—deemed by the experts to be roughly equal in effect, and much less than hillslope gradient—included contributing area, roads, tillage direction, and vegetation cover.

6.2.6.4 Risk Maps

What are the possible agricultural applications of such an expert-based system? Eq. (6.8) shows that WLC with weights from the pairwise comparison matrices can be used to predict erosion risk for each raster grid-cell. The score—which in the example of the Harod catchment is represented by a continuous variable ranging between 0 and 5—can be further categorized into discrete levels of risk, such as no risk, moderate risk, and high risk. Depending on the original values of the variables in each grid-cell and the weight of this variable, the final WLC score may be used to identify areas at risk, and to assess the status of the entire catchment, based (for example) on the spatial distribution of areas at high risk. Figure 6.15 is a map of the Harod catchment, with mean scores of the soil loss risk estimates as predicted for each of the grid-cells within a given agricultural field, averaged using the zonal mean operator (see description in Fig. 6.6). The weights used in this risk map were computed by Svoray et al. (2015). Note that while pixel scores range between 0 and 5, agricultural field have a narrower range of values due to zonal averaging. The spatial distribution of soil loss risk per-field does not follow the topography and therefore it is not the decisive factor in the experts' risk analysis. In fact, we see clusters of fields with mixed soil erosion scores—especially in the central section of the Harod. This is mainly due to the different soil cultivation practices between the fields, but also due to the presence of roads, and other local factors. It should also be noted that there is no clear east–west trend due to variations

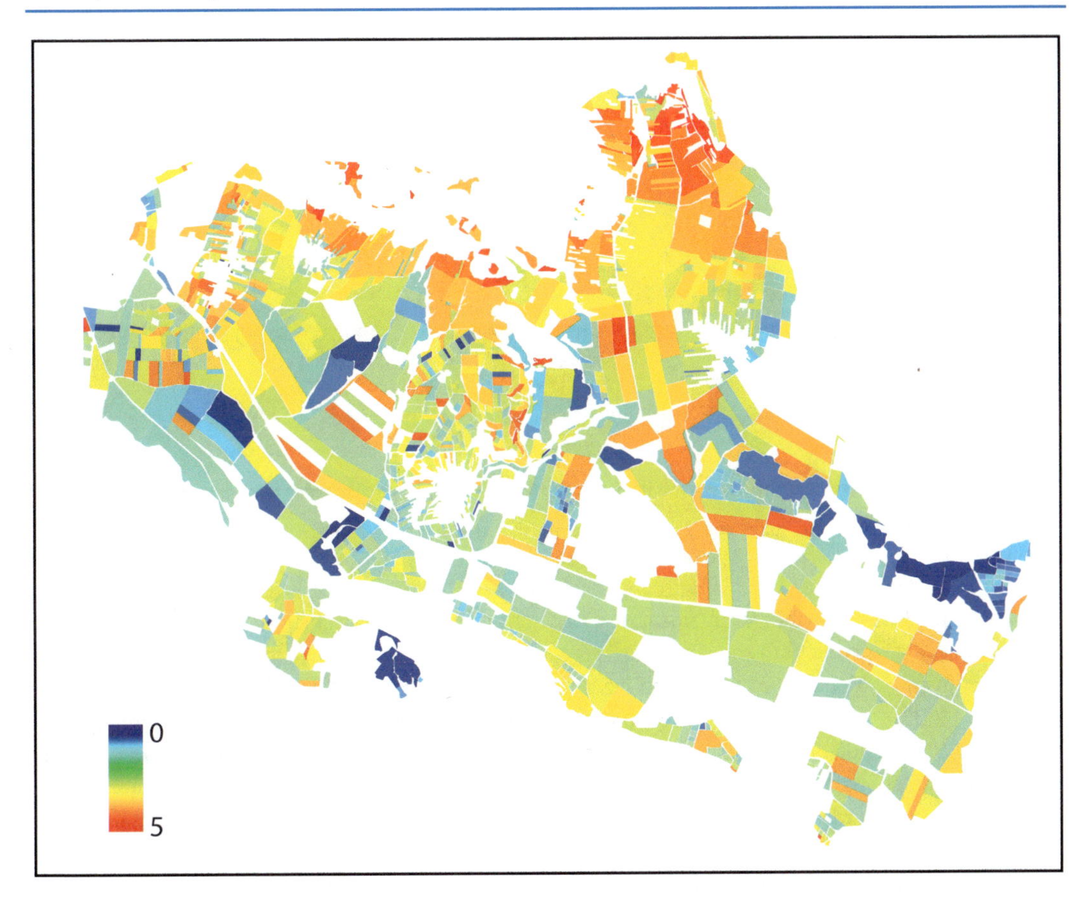

Fig. 6.15 Soil loss risk map in the Harod catchment using WLC with weights extracted from experts using pairwise comparison

of rainfall intensity as is expected in this area (see Sect. 4.11). Some topographic effect can be seen in the northern part of the catchment at the footslopes of Hamoreh Hill—but not in the southern part of the catchment, which also features steep topography.

Using the same methodology as in the soil loss risk map in Fig. 6.15, other water erosion risks can be predicted using their respective weights (Fig. 6.16).

The risk maps of the three soil degradation phenomena (Fig. 6.16) are clearly different. Indeed, some of the soil loss risk mirrors the risk of water ponding and sediment logging, which are more similar but not identical, because there is a difference between overland flow and sediment transfer. Maps such as those in Figs. 6.15 and 6.16 can be used to prioritize soil conservation actions in the catchment, as they can help soil conservationalists identify fields at greater risk, and to allocate resources to the areas that must be dealt with urgently.

6.2.6.5 Scenarios

Rainfall Intensity

We should begin with a methodological note, that rainfall intensity simulations are an example of how remotely sensed data and geoinformatics analysis can replace logistically complex field measurements where, for example, instruments may suffer recurring damage during very intensive rainfall events. More related to the phenomena studied, erosion risk in the fields has been found (as expected in semiarid catchments) to rise with the rarity of the rainfall event and its intensity (Table 6.6). For example, between the

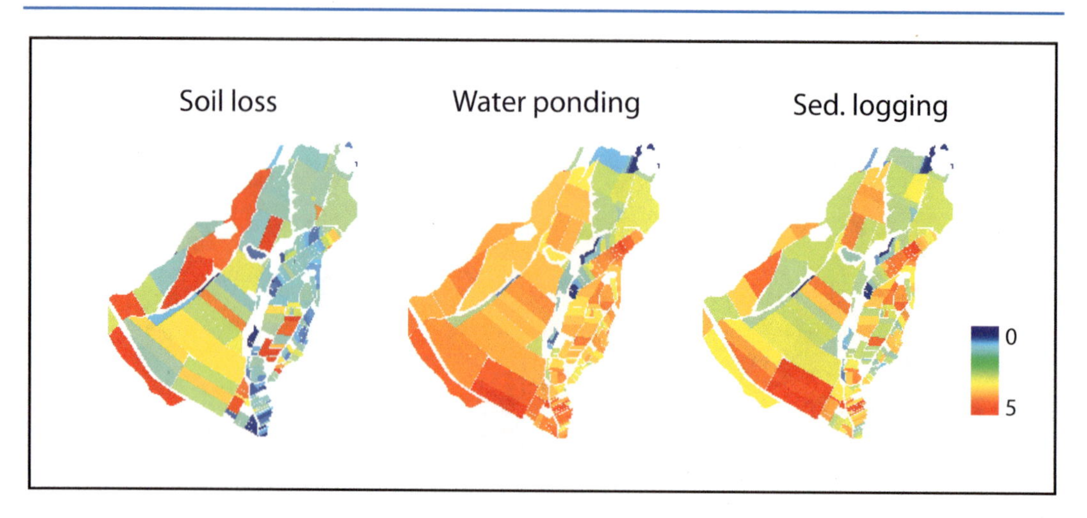

Fig. 6.16 Soil loss, water ponding, and sediment logging risk maps in the Yehezkel catchment, using WLC with weights extracted from experts, using pairwise comparison

scenarios of 20% probability to that of 1% probability, the pixels at the highest risk increased from ~40% coverage of the catchment to ~60%. Such a change may require heightened preparation effort when such an event occurs. Simulated risk maps may show where such risks may increase, due to the extreme conditions. The increase in rainfall intensity is a clear and present danger, because the frequency of extreme rainfall events globally is expected to increase (Woodward 1999) due to global warming, coupled with shorter rainy seasons, a higher number of extreme events per season, and consequent greater overland flows (Wittenberg et al. 2007). Therefore, quantification of catchment response to increased rainfall intensity is highly sought-after.

Land Use

Vegetation in agricultural areas is hypothesized to reduce risks of water erosion, as the canopy is assumed to shield the soil from raindrop impact: Raindrops fall on leaves and then onto the ground, thereby losing energy. In addition, the roots of plants help to hold the soil particles together (Zhou et al. 2008). Therefore, experts assume that bare soils are more susceptible to erosion processes, compared to the areas with higher vegetation cover, e.g., at the beginning of the winter season. However, the extent of the effect of spatio-temporal changes in vegetation cover on erosion risk is still not fully measured, and therefore not fully understood (Knox 2001).

Table 6.6 quantifies the effect of field crops and orchards on soil loss, water ponding, and sediment logging—from 92% coverage of the catchment with crops (with 0% vegetation cover at the start of the winter) and 8% orchards (60% vegetation cover at the start of the winter), to the opposite inverted ratio. Even in this extreme scenario, the expected large improvement in the catchment's resistance to water erosion is not apparent, notwithstanding some reduction in the erosion risks for the three soil degradation phenomena. In other words, although the modeling shows some reduction in risk, even if the Harod were entirely replanted as orchards, it would still be at serious erosion risk. That is probably due to the fact that the orchards' tree canopy would provide only 60% coverage, meaning that various erosion processes may occur in the separation strips and surplus water may flow within the entangled trees roots, reveal them and cause soil degradation.

Tillage Direction

As demonstrated in Chap. 2, tillage-induced roughness can divert overland flow, and consequently, it is hypothesized that it might affect soil loss patterns, as well as water ponding, and sediment logging. Theoretically, fields with tillage lines along the topographic flow directions

Table 6.6 The effect of simulated rainfall, land use and tillage direction variation on the proportion of area at three levels of water erosion risk: High, Moderate, and Low. Note how the area of the High-risk scenario diminishes from the top to bottom within each risk category, due to improved (i.e., less extreme) rainfall amount. In the case of tillage direction scenario 2 means risky tillage orientation (see below text). Based on data from: Svoray and Ben-Said (2010), Soil loss, water ponding and sediment deposition variations as a consequence of rainfall intensity and land use: A multi-criteria analysis. Earth Surface Processes and Landforms, 15. doi: https://doi.org/10.1002/esp.1901. With permission from Wiley (Copyright © 2014 by John Wiley & Sons, Ltd.)

		Rainfall effect					Land use transformation					Tillage direction			
	Prob. 30-min (%)	Mean	Hi risk	Risk	No risk	Field crops (%)	Mean	Hi risk	Risk	No risk	Till Sce	Mean	Hi risk	Risk	No risk
Soil loss	**1**	2.64	61.82	19.75	18.43	**92**	2.26	43.51	27.14	29.35	**1**	2.49	50.57	33.44	16.00
	2	2.57	55.28	24.31	20.41	**71**	2.18	35.22	29.96	34.82	**2**	2.54	49.53	27.85	22.62
	5	2.49	47.55	29.40	23.05	**50**	2.11	26.25	35.03	38.72					
	10	2.44	42.63	32.12	25.25	28	2.04	17.28	40.09	42.62					
	20	2.40	38.64	34.44	26.92	**7**	1.97	10.73	39.53	29.35					
Sed. Dep	**1**	2.64	50.11	24.23	25.67	**92**	3.14	54.00	25.24	20.76	**1**	2.62	37.90	31.54	30.56
	2	2.57	46.08	24.49	29.43	**71**	3.05	46.10	27.53	26.38	**2**	2.63	38.44	31.50	30.06
	5	2.48	41.02	25.23	33.74	**50**	2.96	30.74	32.27	36.99					
	10	2.43	37.94	25.48	36.58	**28**	2.88	28.05	33.81	38.15					
	20	2.38	35.05	25.64	39.31	**7**	2.83	25.39	35.32	39.29					
Wat. pond	**1**	2.87	53.70	23.36	22.94	**92**	2.77	46.89	29.83	23.28	**1**	2.71	45.20	26.94	27.86
	2	2.81	50.64	24.69	24.67	**71**	2.74	43.37	30.84	25.78	**2**	2.80	51.13	24.33	24.54
	5	2.76	46.76	26.33	26.90	**50**	2.70	39.45	32.51	28.04					
	10	2.72	44.24	27.21	28.55	**28**	2.67	36.77	32.95	30.28					
	20	2.69	41.97	28.05	29.98	**7**	2.63	31.63	35.85	32.52					

are expected to experience increased runoff and downslope water ponding, while those with tillage lines perpendicular to the flow lines are expected to decrease runoff risks (see Sect. 2.1). In practical terms, the question is how much does improved tillage direction affect the erosion risk in the catchment. Tabel 6.6 shows that even if the farmers will cultivate the soil along the flow direction lines in large parts of the catchment, the effect on soil loss, sediment transport and soil deposition will be relatively small. Moreover, the response of a 1–2% change in the area between the risk groups can be within the prediction limits of the system. However, as expected, converging tillage lines with flow direction lines can affect runoff more substantially than sediment transport. Thus, the high-risk group for water ponding increased considerably (6% coverage) due to misuse of tillage direction. This suggests that tillage lines of this sort—in our case here, a maximum of ~400 m long lines with widths of tens of cm that were observed in the field—do not create enough water stress for shearing, or do not have enough energy for transport capacity to deposit soil particles downslope.

Summary

Mapping erosion risks can be done using a spatially explicit application of TT to the catchment area, or by means of physically based/empirical models. It can also be done through remote sensing data analysis or, according to traditional methodology, by means of field visits and visual interpretation of aerial photos. However, another mean might be derived from a structured and formal questioning of experts—such as soil conservationists, drainage engineers, scientists, or farmers—and translation of their knowledge into computer language, to automate erosion risk-mapping.

Automatic mapping of erosion risks by means of expert knowledge requires a consistent questioning procedure, and a suitable methodology to convert the human expert's inferences into decision rules and equations. It also requires a rigorous practice, to test the expert's consistency and level of predictive accuracy. While several methodologies exist to translate the expert's knowledge into a consistent procedure, the *Multicriteria Analysis* (MCA) is a simple and commonly used one. It has a well-proven predictive ability and it provides a useful tool for testing expert opinions. Although the method may suffer from subjectivity in completing the forms and limitations of the Saaty Appreciation Scale, it has proven (R^2 = 0.72) to be successful in predicting areas at risk of common soil degradation—such as soil loss, water ponding and sedimentation—in small sediment deltas in the fields (Svoray and Ben-Said 2010).

The following ideas may be summarized for future use of MCA in predicting areas at risk of erosion in other agricultural catchments. First, MCA is, overall, a useful tool for translating expert knowledge into computer language. It can be computed with simple software means and it provides accurate estimations of soil loss, sediment logging, and water accumulation. Second, we need many experts to derive accurate weights. In estimation of the weights, several experts offer greater accuracy than a single "canonical" one. Experts from different fields, in particular, may be more efficient at predicting the risks and appear to compensate for the inconsistencies and possible subjectivity of a single expert. Third, in continuation with the previous idea, merging together pairwise comparison matrices increases the consistency of the single experts, so it is recommended to merge matrices. Fourth—keep it simple: The simpler method weighted linear combination achieved better results than the computerized programming technique. Using the CPT, the experts tended to consistently overestimate the risks. That is probably due to the need to assign the ideal and anti-ideal points which require more effort from the experts. In that case, the more information required from the experts acts against them. Fifth—scenarios of possible changes in the dynamic clorpt variables can help estimate catchment reaction to extreme events without the need of risking expensive devices in the field. Here, for example, the scenarios have shown a much greater catchment reaction to rainfall intensity than to changes in land use, or tillage direction.

In the following section we shall demonstrate the use of a completely different approach for risk-mapping—namely, the data-mining approach which originates from the world of data-driven approaches. We will also compare between the two methods, i.e., data-mining and expert-based systems, to show the pros and cons of using data-versus knowledge-based systems.

6.3 Data Mining (DM)

Knowledge Discovery in Databases (KDD) refers to a set of organized procedures for searching meaningful patterns of simulated real-world phenomena—such as areas at risk of water erosion—in large (geoinformatics) datasets. One example of such a process is the identification of patterns of gulling in agricultural fields, by mapping areas prone to gully incision based on GIS layers that represent clorpt variables—such as contributing area, cultivation method, and rainfall intensity.

The core of KDDs is the Data Mining (DM) procedure, which includes the inferring algorithms that actually explore the data, discover previously unknown patterns, and develop the predictive mathematical formulas. DM algorithms are commonly used to gain a better understanding of processes and forms, to analyze them, and to predict their occurrences in unknown locations (Maimon and Rokach 2010, p. 1).

For example, satellite images offer spectral data that may be analyzed for soil moisture monitoring and for vegetation productivity assessments. The soil moisture and biomass datasets can be then manipulated by DM procedures to retrieve information on locations of areas under threat of water ponding, and to map their spatio-temporal patterns. These patterns can also reveal the underlying processes that caused the observed water ponding phenomena.

A typical DM process consists of three primary steps (Fig. 6.17):

Step 1: Preprocessing of the dataset—including data collection, and data cleaning, as well as a division of the original dataset into training and validation sets, which are both labeled data points that have been measured and verified in the field;

Step 2: Classification—actual application of the DM algorithms or statistical tools (usually referred to as classifiers), to analyze the patterns;

Step 3: Validation—testing the results against field measurements which did not participate in model fitting.

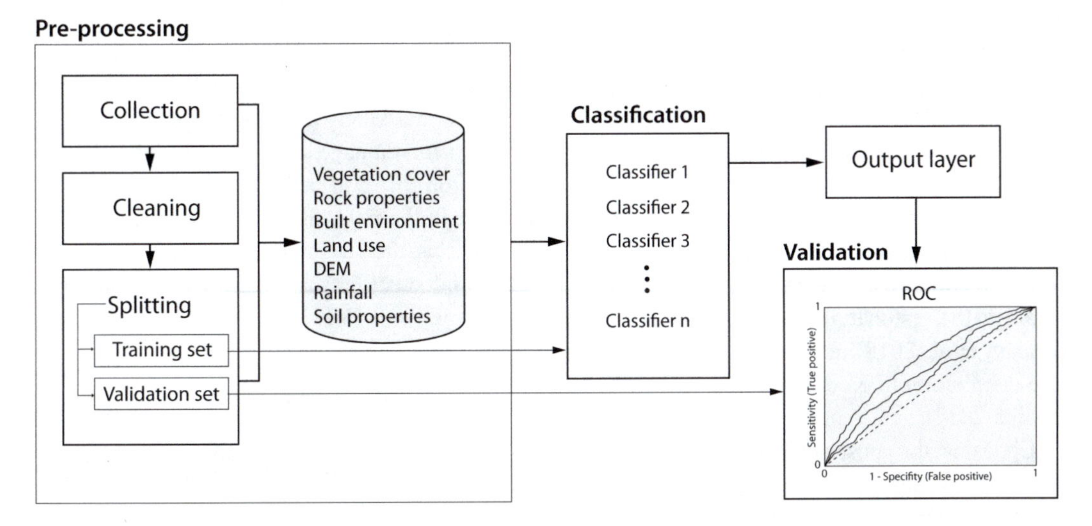

Fig. 6.17 The three steps of the Data Mining (DM) process: (1) Preprocessing including data collection, cleaning, and splitting the dataset into training and validation sets; (2) Classificationapplying classifiers; and (3) Validation (using ROC see Fig. 6.19)

		Measured values	
		Positives (Risk)	Negatives (No risk)
Predicted values	Positives (Risk)	TP—True positive	FP—False positive
	Negatives (No risk)	FN—False negative	TN—True negative

Fig. 6.18 A 2 · 2 confusion matrix that defines four possible combinations of prediction results. *TP* are the observations where both model and validation sets agree that there is risk; *FP* are the cases where the model predicts a risk, but the validation data say no risk; *FN* are the cases where the model predicts no risk, but the validation data finds a risk; and *TN* are the cases where both predict no risk

With the increased availability of spatio-temporal data in water erosion studiesincluding, for example, readily available remotely sensed data, and more frequent digital field sampling of soil and vegetation properties by farmers, using various devices mounted on tractorsopportunities for exploring patterns of soil properties in agricultural catchments are consistently increasing, thereby enhancing the potential of using DM algorithms to study water erosion. Most importantly, the extracted spatial information can be transformed into knowledge for the benefit of farmers, in a bid to reduce soil degradation.

DM tools are designed to be as automated as possible, because the datasets available are too large to browse manually, and visual interpretation may suffer from subjective interpretation. The existing methods include classification techniques, clustering analysis, and statistical tools, such as regression analyses. In recent years, machine-learning classifiers, such as *Random Forests*, *Adaboost* (Adaptive Boosting), *artificial neural nets*, and many others, have been extensively used for: Spatial prediction of soil properties (Dharumarajan et al. 2017), mapping soil formations using legacy data (Hounkpatin et al. 2018), estimating landslide susceptibility (Kadavi et al. 2018), and predicting the particle size distribution of eroded soils (Lagos-Avid and Bonilla 2017).

Choosing the most accurate classifier among the many available is a critical part of the DM procedure. The *Receiver Operating Characteristic* (ROC) is a common method for evaluating classifiers (Swets 1988). ROC combines the information from multiple *confusion matrices* (see Sect. 5.2.1.1) covering the entire range of threshold values that determines, for instance, whether there is a soil loss risk, or not. In other words, the classifier estimates a probability value between 0 and 1 for each grid-cell (0.1, 0.2 …. 0.9). Based on the given threshold, each grid-cell is then classified into the binary categories, such as under soil loss risk (1) or not (0). Once the observed and predicted binary classifications are available for each observation (from the "validation" dataset), a 2 · 2 confusion matrix can be constructed to examine the classifier predictive ability (Fig. 6.18).

The confusion matrix offers two measures of the classifier predictive ability. Sensitivity is defined as the proportion of grid-cells observed in a field to have an event (erosion damage) that were correctly identified by the classifier as having the event (erosion risk). That is, a highly sensitive classifier more fully identifies areas with erosion damage. Specificity is defined as the proportion of grid-cells that have no erosion damage in the field, that were correctly identified by the classifier to have no erosion risk. Sensitivity and specificity are calculated according to Eqs. (6.10) and (6.11):

$$\text{Sensitivity} = \frac{TP}{TP + FN}, \tag{6.10}$$

$$\text{Specifity} = \frac{TN}{TN + FP}, \tag{6.11}$$

where *TP*s are the observations where both the classifier and the measurements agree that there is risk, and *FN* are the observations where the classifier predicts no risk, but the measured data finds a risk; *TN*s are the cases where both predict no risk; *FP* are the cases where the model predicts a risk but the measured data say no risk.

The ROC space is therefore defined as the graph of 1−Specificity against Sensitivity, whereby the points on the curve (Fig. 6.19) represent threshold values in the 0–1 range, with x-values are false positives, and y-values are the true positives. The thresholds are discrimination values—for each single classifier—of an observation to belong to the positive class (e.g., erosion risk). Thus, as the threshold value is assigned higher in the classifier, the higher the number of observations that are assigned to belong to the positive class. The example in Fig. 6.19 shows ROC curves of three different single classifiers, each with different threshold values. Namely, each point on a curve represents a different threshold value.

In all three cases, the evaluated threshold values range from 0 to 1, from left to right (i.e., the first threshold from the left is 0 in all three cases; next is 0.1; etc.) The best threshold to be used in each classifier is the point where the ROC curve is closest to the "optimal point" (i.e., $x = 0$, $y = 1$): In ROC1, it is the threshold of 0.2 (namely, the second point on the ROC1 curve); in ROC2 it is 0.3 (third point on the ROC2 curve); and in ROC3 the most efficient threshold is 0.7, the seventh point on the curve.

Each curve in Fig. 6.19 represents a different classifier—or the same classifier with a different configuration of model parameters. Thus, when comparing the curves, the analyst can identify the best performing classifier. This is done by assuming that the optimal point is located at the upper left corner where there are no false positive observations (false positive = 0) and no false negative observations (true positive = 1). In other words, the analyst strives for true

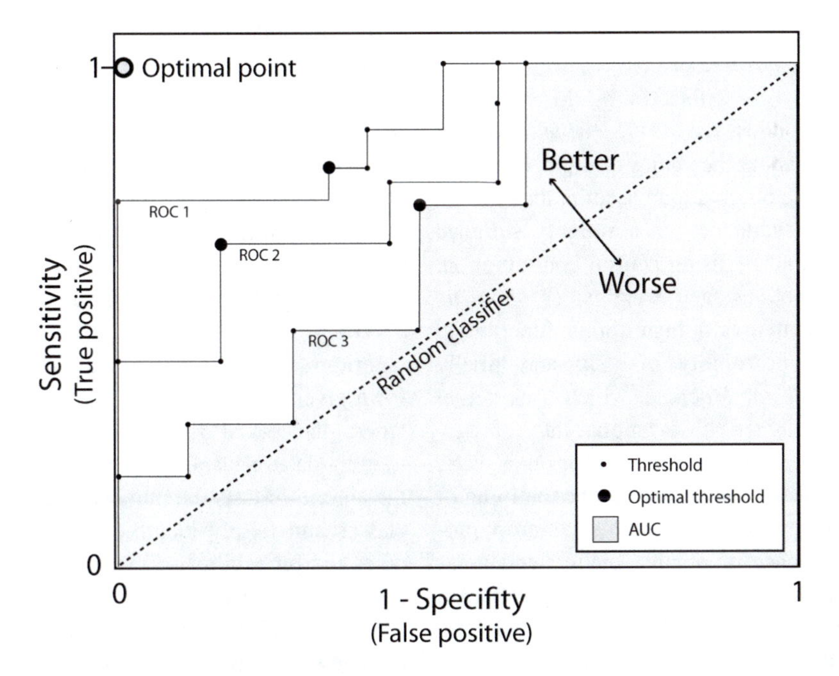

Fig. 6.19 The ROC (Receiver Operating Characteristic) curve: The upper left corner is the optimal point of classification, with the highest score for true positive (1) and the lowest (0) for false positive (1 − Specificity). The closer the ROC is from the upper left corner–at the optimal threshold point–the better the results—and vice versa. The points below the diagonal are worse than random guess. AUC is defined as the Area Under Curve of each ROC (exemplified here for ROC3 with the gray area)

identification (Sensitivity = 1), and low values of false identification (1 − Specificity = 0). A low threshold value assigned to the classifier naturally produces high Sensitivity. That is because many observations will be identified as areas at risk, but also many points that are not at risk will be identified as if they are. Conversely, a high threshold value provides a low false positive rate —because not many grid-cells are assigned as being at risk. This choice, however, can lead to an outcome of less true positive predictions. Thus, when the analyst compares classifiers, or configuration of classifiers, she wishes that the ROC curve will merge with, or be the closest to, the optimal point. The Euclidian distance between a threshold value and the optimal point can be used as a measure and the classifier with the shortest distance from the optimal point could be regarded as best performing.

The *Area Under Curve* (AUC) is a single metric to summarize the ROC curve, in terms of classification precision, for each classifier. The AUC ranges from 0 to 1, with an AUC of 0.5 for the random classifier, and an AUC of 1 for the perfect classifier, offering a true positive of 1, and false positive of 0.

Section 6.3.1 describes four DM classifiers for mapping soil loss risk: (1) *Decision trees,* that ask questions about the values of input variables, and classify the observation based on the answer; (2) Artificial neural nets, that roughly simulate the human mind by using neuron configuration, activation functions, and weights; (3) *Logistic regression*, that uses a logarithmic function to quantify the probability of risk; and finally (4) *Support Vector Machine*, which uses kernel functions to assign observations into classes. These classifiers can also be used for other water erosion analyses, such as areas prone to piping or water ponding; areas of diminished crop productivity; absence of organic matter; and areas suffering from nitrogen shortage. We begin with a description of the theoretical basis behind these four DM algorithms.

6.3.1 Theoretical Basis

While many classifiers have been developed for data-mining purposes (Rokach and Maimon 2008), the following four different strategies have proved to be efficient and can be implemented by readily available tools such as Weka (https://www.cs.waikato.ac.nz/ml/weka/), or R.

6.3.1.1 Decision trees

Decision trees are collections of learning algorithms that can be applied to classify an object (such as a grid-cell in a raster layer) into a desired set of classes (e.g., Risk/No risk), based on the values of the input variables (such as hillslope gradient, cultivation method, etc.…) that characterize that object. Decision trees have been used for decades in various applications and have been found to be efficient when dealing with large datasets. They are straightforward to interpret and particularly well suited to handle interactions between input variables.

A decision tree combines the input variables in a hierarchy, with the most important variable at the root node. Each node in the tree then refers to a variable, and in each leaf, the grid-cell is assigned to a class (Risk/No risk) that represents the most suitable class value. Thus, all grid-cells in the raster—covering the entire study area—are classified by going through the decision tree, from the root node to the specific terminal leaf node, according to the values of independent variables for that particular grid-cell.

When using decision trees as classifiers, the algorithm is trained to compute the probability for a given class, using a group of observations where the actual risk is known. That is the training data. Thus, once the rules are generated, the algorithm tests the number of truly identified classes and falsely identified classes, and computes the probability of the grid-cell to become at-risk based on the training data.

In Fig. 6.20 for example, the ratios come from the known training data. Thus, in the case of the

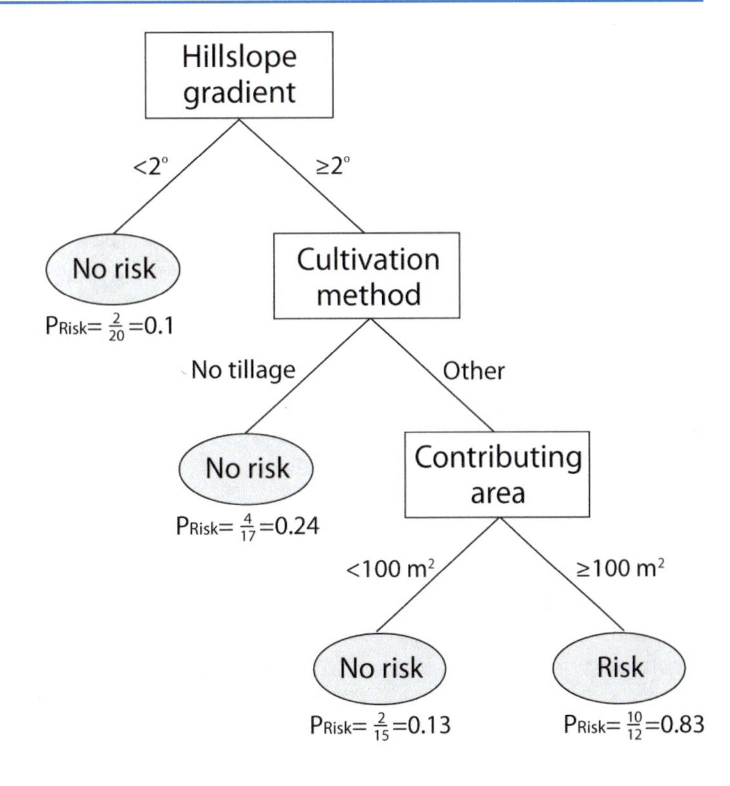

Fig. 6.20 A decision tree, starting from the root node (top) to the leaf nodes (gray ellipses predicted class). The rectangles are the input variables, and the branches (lines) are the decision rules. The nodes are the splitting points at every variable. Thus, if the hillslope gradient is 3°, the cultivation method used in it is conservation tillage, and the contributing area is 200 m^2, the predicted outcome is "risk"

branch with the rule of *hillslope gradient* >2° and *no tillage*, that was predicted as no risk, four of the total 17 cases in the training data (total = cases identified at risk + cases identified with no risk) were found to be correct, and therefore the probability for the grid-cell to be at risk is 0.24 (4/17). Similarly, all other probabilities for risks in the leaves of Fig. 6.20 were computed. These probabilities are used—based on a threshold value—to determine if the grid-cell is, or is not, at risk. Thus, if a given grid-cell is assigned a probability value of $P_{\text{risk}} = 0.13$, and the one next to it is assigned $P_{\text{risk}} = 0.83$, and the user-defined arbitrary threshold is 0.5, the former grid-cell is designated no risk, and the latter as risk, etc. Therefore, as we saw in the ROC analysis, the selection of the threshold has consequences for the predictive ability of the classifier.

Decision trees are usually represented in the form of leaves and branches, as shown in Fig. 6.20. However, if the graphs are too complex or large, they can be replaced by a more graphically condensed collection of textual rules. Thus, each collection of branches and leaves is translated into a simple "if–then" rule. For example:

> **If** the hillslope gradient is greater than 2 and cultivation method is no tillage, **then** the grid-cell is assigned as an area at risk with a probability of 0.24 to have a risk.

We can imagine the training data for the decision tree in Fig. 6.20 as a cube with hillslope gradient, cultivation method, and contributing area as three dimensions, while every observation in the training data is located inside the cube. The observations belonging to the same outcome class are then grouped, and the respective rule to characterize the group can be derived.

Real-world problems are of course more complex than the example presented in Fig. 6.20, so decision trees are often combined to form a "forest" (as described in relation to the Random Forests algorithm, Sect. 5.2.1.2). The assemblage of several trees helps overcome overfitting, i.e., models that do not generalize well to new data. Such a forest of decision trees is implemented in the AdaBoost (Adaptive Booster) algorithm

(Freund and Schapire 1996). Adaboost is a widely used collection of classifiers that, in each iteration, increase the weights associated with the misclassified points and decreases those weights associated with the correctly classified ones (Rokach and Maimon 2008). Such a strategy obviously leads the classifier to enhance the misclassified points. This approach is especially useful in training sets of rare cases (Joshi et al. 2002), as is clearly the case of gully heads, with a few tens of training sets to be identified among millions of grid-cells. However, despite its advantages for the problem in hand, AdaBoost may be also sensitive to noise and outliers.

6.3.1.2 Artificial Neural Networks (ANN)

Artificial Neural Networks (ANN) is a classifier that is inspired (roughly) by the structure and functioning of the human brain. ANN classifiers comprise three components: (1) Input layers representing the independent variables, such as geoinformatics layers, usually stored as rasters, where grid-cells comprise the observations; (2) An output layer, which includes the predicted values, also usually stored as raster data models; and (3) The hidden layers, which include activation functions. The basic ANN components are denoted as neurons that are connected by weights (the links, Fig. 6.21). The weights are adjusted at each case study, using training data. Figure 6.21 illustrates the flow of data values in a typical ANN from the left-hand section of the panel to its right-hand section. On the left, we see the input raster layers that feed the input neuron layer X_is. The hidden layer—which can be multilayer —is composed of neurons that are the activation functions, which in their simplest form can be a weighted linear combination. However, the activation functions can be apportioned with more complex functions (Karlik and Olgac 2011). The value of the result computed by the activation function is normalized to the 0–1 range, using simple algebraic normalization. The final result is compared with the training set data to create the most suitable network configuration, in terms of accuracy when predicting the output characteristic. The various network configurations are achieved by changing the weights, the activation functions, the number of hidden layers, etc. Once a successful configuration is achieved (i.e., good agreement between the predicted and observed data), it is applied to a different dataset, and tested against the validation data.

ANNs have been successfully used in many applications of water erosion studies—for: Generating soil erosion maps (Gholami et al. 2018), predicting particle size distribution of eroded sediment (Lagos-Avid and Bonilla 2017), and for identifying areas at risk, including water erosion risks (Mosavi et al. 2020).

The MultiLayer Perceptron (MLP) is a feed-forward ANN classifier that uses layers of supervised classifiers, named perceptrons, as the mechanism for labeling. The information in MLPs flows only forward along the entire neural net, through the nodes and the weighted connectors, from the input nodes to the predicted layers. The literature distinguishes between two different groups of MLPs: (1) Shallow MLPs with 2 or 1 hidden layers; and (2) Deep MLPs are those more complex systems with >2 hidden layers (Grekousis 2019). MLPs are broadly applied in many disciplines and specifically in the soil sciences (Fernandes et al. 2019).

6.3.1.3 The Logistic regression

The logistic regression model aims at predicting the probability of the occurrence of an (erosion) event. The dependent variable in logistic regression is binary—meaning, for example, that the user can predict only whether the grid-cell is at risk of erosion (1), or not (0). The prediction is based on the logistic regression curve, which gives the probability of an erosion event in a given grid-cell, based on the input variables (the hillslope gradient in Fig. 6.22).

For example, if the grid-cell has a hillslope gradient of 30°, there is a 100% chance that the grid-cell is at risk; if its hillslope gradient is 15°, it has a 50% chance of having an erosion event; and on near-level areas, there is 0% chance of an erosion event. Logistic regression differs from linear regression in that the former models binary outcome variables, while the latter models continuous outcome variables. Also, in linear

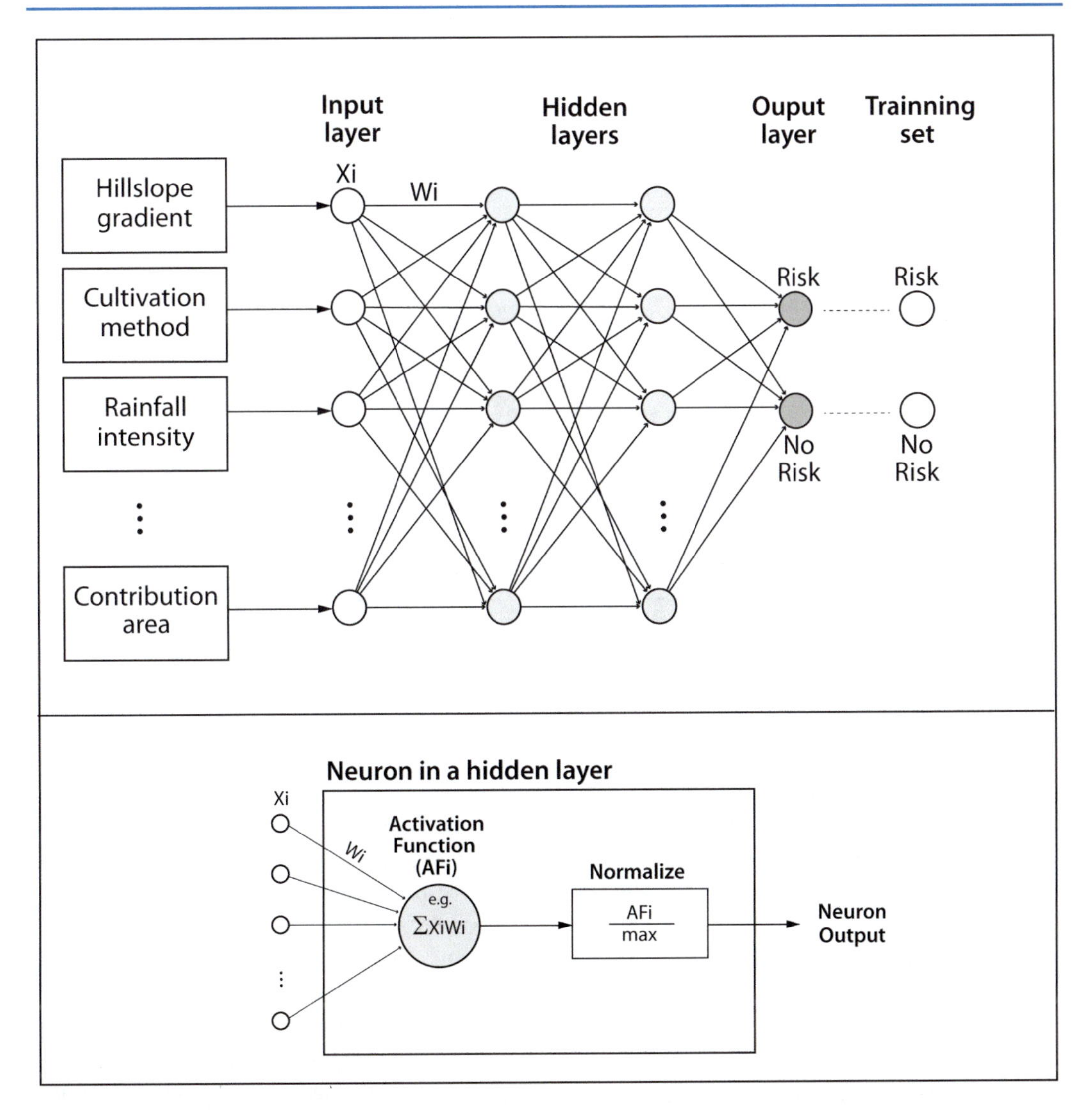

Fig. 6.21 An ANN architecture. The input layer comprises the raster values of the input variables, the hidden layer includes the activation AF_i and normalization functions AF_i/max to feed the output neurons with output values between 0 and 1. These are compared with the training set, to achieve a prediction of the output characteristic

regression, we fit the line using a least-square technique, to minimize the sum of squares of the residuals, while in logistic regression, we use a maximum-likelihood technique to choose the curve that best fits the points among the various curves shifted from the original line.

Logistic regression is used for classification, by using rules. For example, if a grid-cell has a 50% chance of hosting an erosion event, it is classified as an area at risk. Below that threshold value, it is classified as not at risk.

The strategy of the logistic regression is to fit the training dataset to an S-shape logistic function curve using a log-likelihood function. No assumption about the nature of the variable's distribution is required to apply the method. Within this process, optimal weights (i.e., fitted regression coefficients) are determined for each

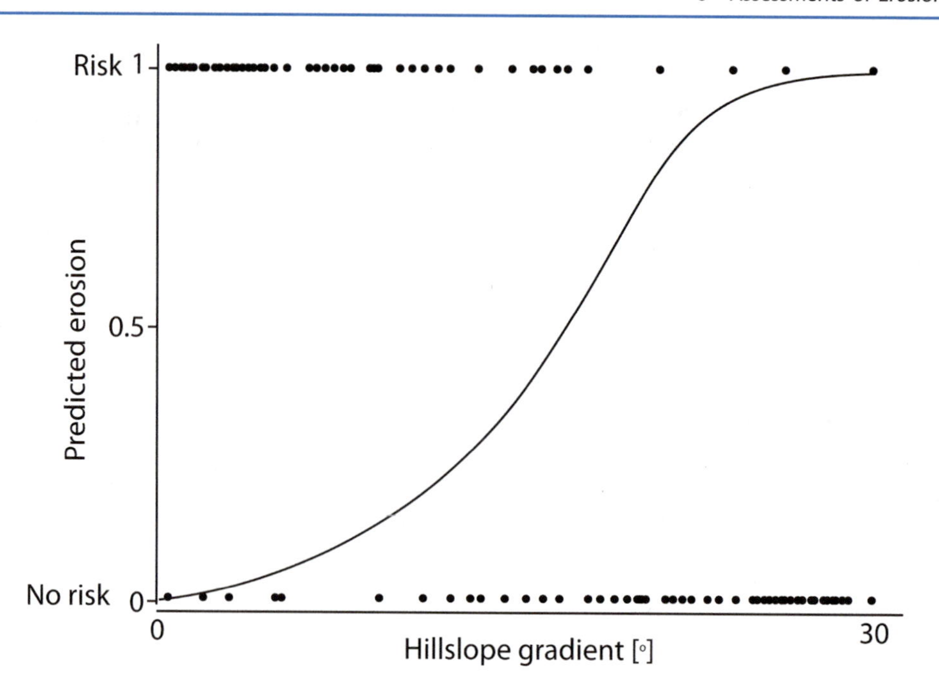

Fig. 6.22 The logistic regression curve—an example of application of water erosion risk prediction. The curve converts any hillslope gradient value to probability in the 0–1 range, with 1 representing risk and 0 representing no risk. The points are observations

of the input variables. An example for the application of logistic regression to predict gully erosion is provided in Eq. (6.12) (Akgün and Türk 2011):

$$\begin{aligned} Risk = {} & 0.008357 \cdot WG + 0.001264 \cdot SPI + 0.001821 \cdot HG \\ & - 0.002798 \cdot DD + 0.951699 \cdot LD \\ & - 0.002235 \cdot LC - 0.000722 \cdot PC - 5.9164, \end{aligned} \tag{6.12}$$

where *Risk* denotes the erosion in the grid-cell, *WG* is the Weathering Grade, *SPI* is the Stream Power Index, *HG* is the Hillslope Gradient, *DD* is the Drainage Density, *LD* is the Liniment Density, *LC* is the Land Cover class and *PC* is the Profile Curvature. The weights were assigned as part of the regression model-fitting process. Logistic regression applied to geoinformatics data is a common method to quantify susceptibility to gulling in an agricultural catchment (Conoscenti et al. 2014).

6.3.1.4 Support Vector Machine (SVM)

Support Vector Machine (SVM) is the fourth and last machine-learning algorithm discussed in this section. SVM is a classifier that assigns observations to classes, based on a threshold value that is the boundary between the edge observations of the classified groups in the training data. This threshold is located at the midpoint between the edge observations and is known as the *optimal hyperplane*. The margins on either side of the optimal hyperplane are the shortest perpendicular distances between the edge-observation points, and the optimal hyperplane.

Observation points (grid-cells) with relatively short Euclidian distance from the optimal hyperplane are denoted *supporting vectors*—hence the name support vector machine. The assignment of the closest points as supporting vectors is done using a straight line that acts as a threshold (Fig. 6.23). In cases when observation points cannot be assigned by using a simple line,

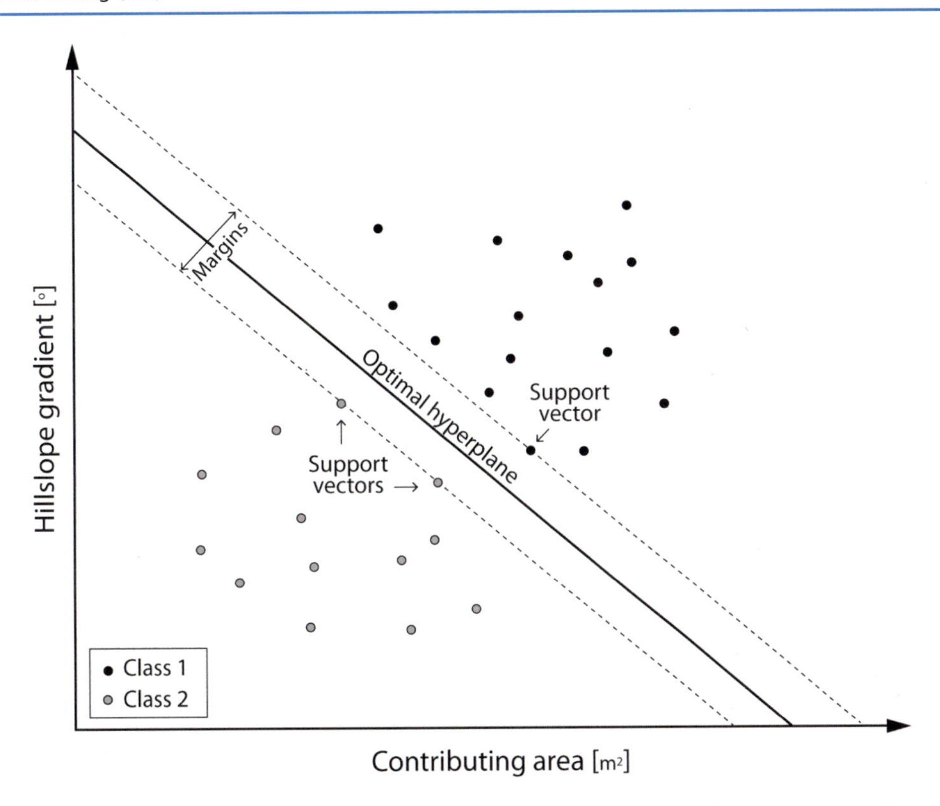

Fig. 6.23 A two-dimensional Support Vector Machine (SVM) model. The gray dots, in this example, represent observations without erosion risk. The black dots represent the observations with erosion risk. The optimal hyperplane is the midpoint between the edge observations denoted as the support vectors. The margins determine the separation zone between the two groups

a kernel function is utilized. Because the margins and the optimal hyperplane may be sensitive to outliers that can distort them both, soft margins are sometimes used to improve classification accuracy. In soft margins, we replace the closest outliers with neighboring observations to determine the margin's buffer zone. Figure 6.23 illustrates how the SVM is used to assign the observation with a tag to a risk area (black dots) or no risk area (gray dots). The procedure goes as follows: The support vector observations are identified, the margins are delineated, and the midpoint is marked as the optimal hyperplane.

SVM is a machine-learning tool used frequently in the soil sciences. In a recent work, for example, Choubin et al. (2019) used SVM to map four levels of flood susceptibility in Iran, based on topographic, soil, and lithological input variables. In a similar work, piping risk was successfully predicted using SVM and geoinformatics data on the aforementioned variables, and additional UAV and soil analyses laboratory data (Hosseinalizadeh et al. 2019).

6.3.2 Procedures

The four aforementioned classifiers were applied to the Yehezkel catchment to map areas at risk—i.e., to classify sections of the area into Risk/No Risk classes. Classification was done based on the ten variables described in Table 6.4—namely: Hillslope gradient, tillage direction, upslope contributing area, rainfall intensity, crop type, vegetation cover, roads as contributors to overland flow, roads as barriers, hillslope aspect, and stone coverage. These are the same input layers that had been used to map erosion risk in the catchment using expert-based knowledge (Sect. 6.2). The training data for the four classifiers was based on: (1) 52 physically measured gully initiation points observed in an airphoto and verified in the field; and (2) 5200 points that were verified in the orthophoto which are not gully heads or located within a gully.

The decision-trees were applied using the AdaBoost.M1 algorithm, with the adabag package (Alfaro et al. 2013) in R. The concept of a collection of decision trees rather than a single tree has been widely examined in the past (Rokach and Maimon 2008), while Adaboost, specifically, has already been used in the soil sciences (Shu and Burn 2004).

To apply the standard logistic regression classifier to the Yehezkel catchment, the training dataset of the observed gully heads was fitted to a logistic function curve. This was done by plotting measured points of gully-heads with classified data, using the log-likelihood function (Dai and Lee 2003). For each input layer of the ten layers, the algorithm determined the weights for all predicting variables. To do so, the logistic regression—R package stats (part of base R) was applied.

The Support Vector Machine (SVM) was applied using R package e1071 (Meyer et al. 2019)—for an overview, see Bennett and Campbell (2000).

The ANN applied to the Yehezkel catchment in this study was induced within the R package neuralnet (Fritsch et al. 2019).

6.3.3 Results and Discussion

Maps of the four classifiers applied to the Yehezkel catchment are presented in Fig. 6.24. There are differences in the predictions of risks for the various grid-cells by the four classifiers, but there are also similarities. In the northern part, for example, all four classifiers identified risk in the footslope of Hamoreh Hill with mild topography, and lower risk values in the eastern part of the catchment.

An aggregation of the predictions of the classifiers to the field level (Fig. 6.25) allows a more generalized comparison between the classifiers. Here we see the differences more clearly. For example, the northern part of the catchment is, at least partially, in agreement between the classifiers, but the central section and the southern section, in particular, shows large differences between classifiers (e.g., the difference between SVM and the logistic regression in the southern section).

The similarities and differences between the classifiers can also be seen in Table 6.7, which presents the four moments of descriptive statistics for the predictions of the four classifiers. The results show that the spatial variability of erosion risk in the Yehezkel catchment was high. The difference between the mean values of the classifiers predictions is relatively large, so the range 0–1 is different between the classifiers and should be considered when analyzing the dataset by visual interpretation.

Histograms all show skewness >1—meaning they are right-tailed, or positively skewed (peak on the left part, followed by a normal distribution). Kurtosis of all layers is >3—i.e., peak distribution. The predicted values of ANN and logistic regression were characterized by exceptionally high values of skewness and kurtosis.

ROC curves of the four classifiers (Fig. 6.26) are fairly similar at some threshold values, with all areas under the curve larger than that of the random classifier. The Adaboost curve was closest to the optimal (0, 1) point, with Euclidean Distance of 0.45—followed by the logistic regression at ED = 0.50, Neural Net at ED = 0.51, and SVM ED = 0.53.

The AUCs (Fig. 6.27) of the four classifiers were plotted against different number of groups in the training sets that had been used to train the classifier. Groups were replications of model-fitting procedures, where 10% of the training points of risk and no risk groups were used, followed by averaging the predicted probabilities from all models to obtain the final predicted classification.

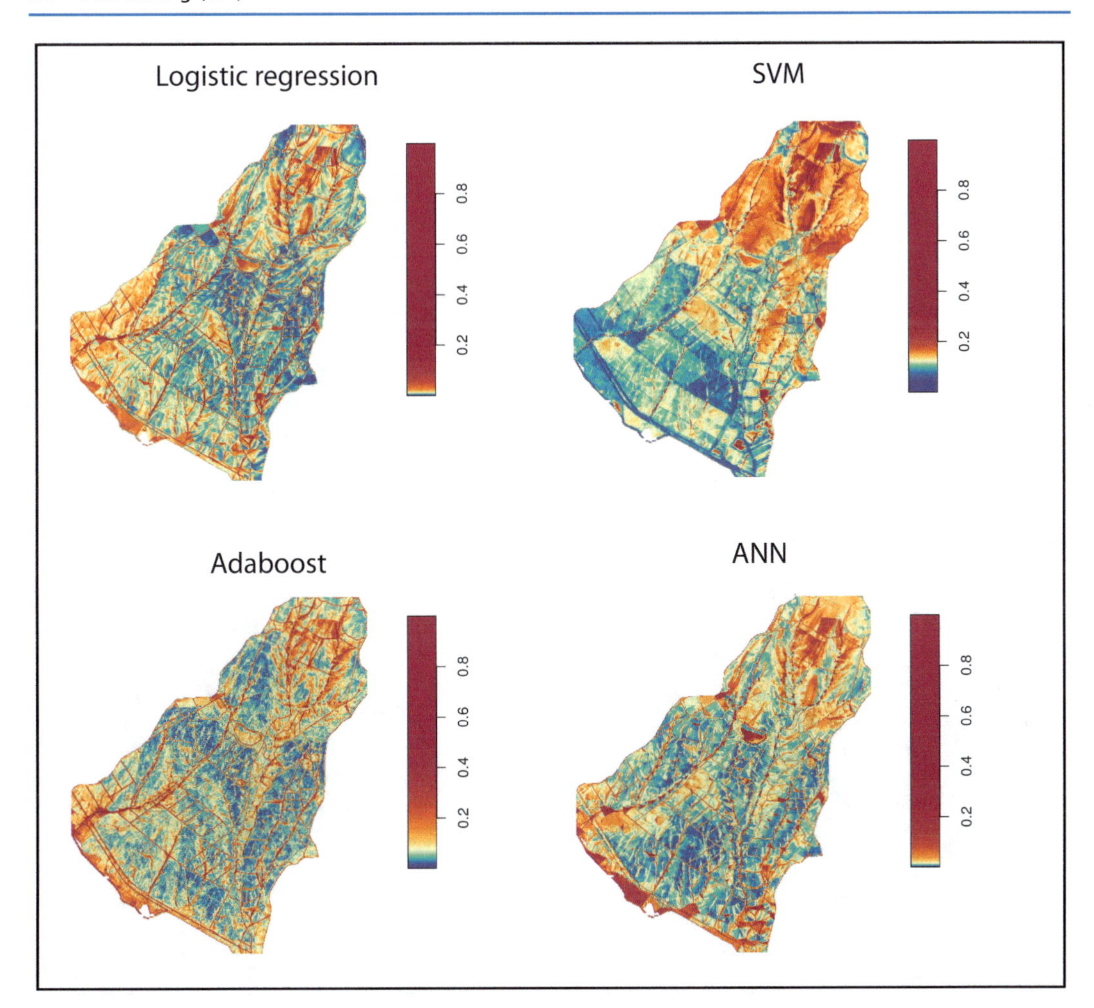

Fig. 6.24 Predicted erosion probability in the Yehezkel catchment based on the ten clorpt variables, using four data-mining methods. The maps show the probability for risk between 0 and 1. The color scale breaks were calculated according to quantiles, so that the number of grid-cells represented by each color is equal

Table 6.7 Statistical moments of the four data-mining classifiers applied to the Yehezkel catchment, based on grid-cell-based analysis

Model	Mean	Variance	Kurtosis	Skewness
Logistic reg	0.022	0.0015	86.76	7.50
Adaboost	0.106	0.0090	9.22	2.50
ANN	0.035	0.0041	43.86	5.72
SVM	0.129	0.0024	24.67	3.13

Overall, the AUC curve exhibits high stability of the classifiers' performances, regardless of the number of model-fitting replications being used. In most instances, the Adaboost provides the best performances, and the logistic regression is second best. Third is the Neural net, and the last one —not very stable—is the SVM.

Due to the structure of the trees, the Adaboost may allegedly, provide us with a more interpretable way than the other classifiers to examine the data. However, note that the tree presented in Fig. 6.28 is just one of N = 100 trees being averaged in the Adaboost model. The complete Adaboost model actually cannot be presented in

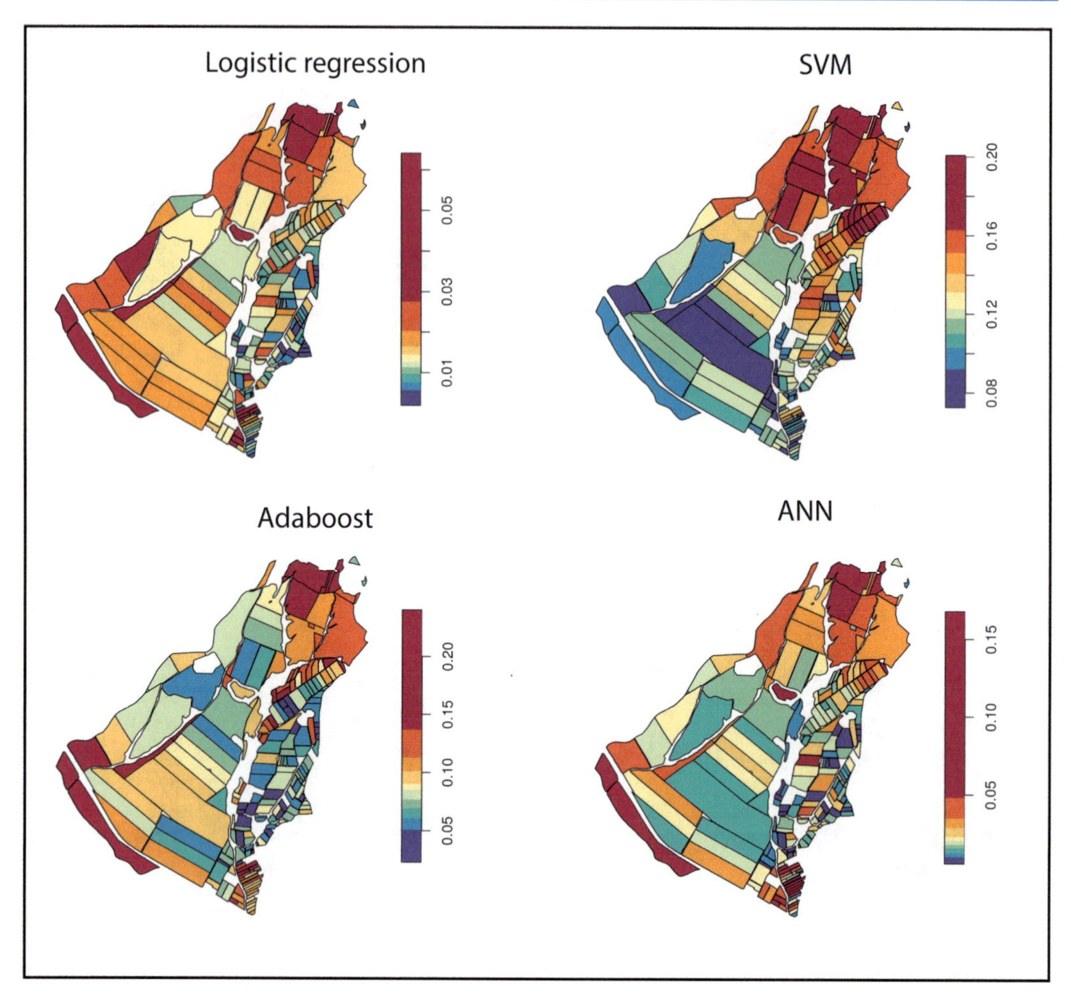

Fig. 6.25 Average predicted erosion probability in agricultural fields in the Yehezkel catchment based on the ten clorpt variables, using four data-mining methods. The color scale breaks were calculated according to quantiles, so that the number of agricultural fields represented by each class is equal

an interpretable way. The logistic regression—rather than Adaboost—is the one model of the four which can be easily interpreted in case it doesn't include interactions, such as here (ten layers of independent variable versus predicted probability). Adaboost, SVM and ANN, on the other hand, cannot be easily interpreted, only the importance of the ten variables can be calculated and displayed.

Figure 6.28 shows an individual decision tree for classifying a grid-cell. This example illustrates the rules for three variables only—but the trees that were actually used were far more complex. The advantage of Adaboost is that it is self-explanatory and allows the expert to follow the logic behind the classification. Again, this isn't the case—a single tree doesn't reflect the complete model since all trees are different (to unknown degree).

6.3.4 Summary

Various data-mining methodologies can be used to predict water erosion risks, by integrating multiple independent variables using machine-learning techniques and training sets. The various tools—including artificial neural nets, support

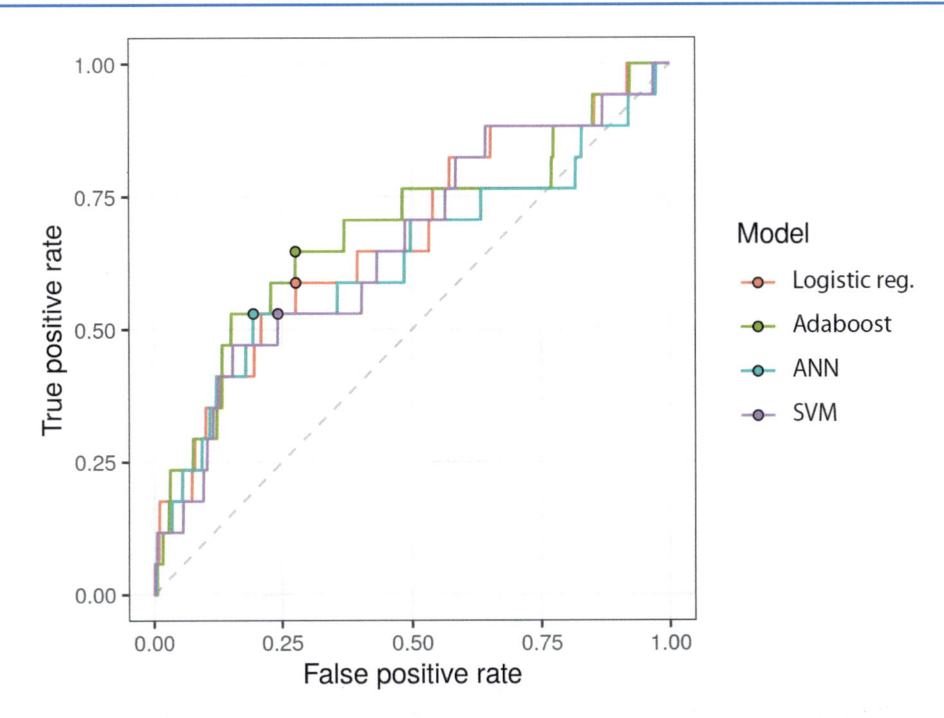

Fig. 6.26 The ROC curve of four data mining classifiers. All curves have larger AUC than the random classifier (i.e., AUC >0.5), while the Adaboost ROC curve was closest to the optimal point (0, 1)

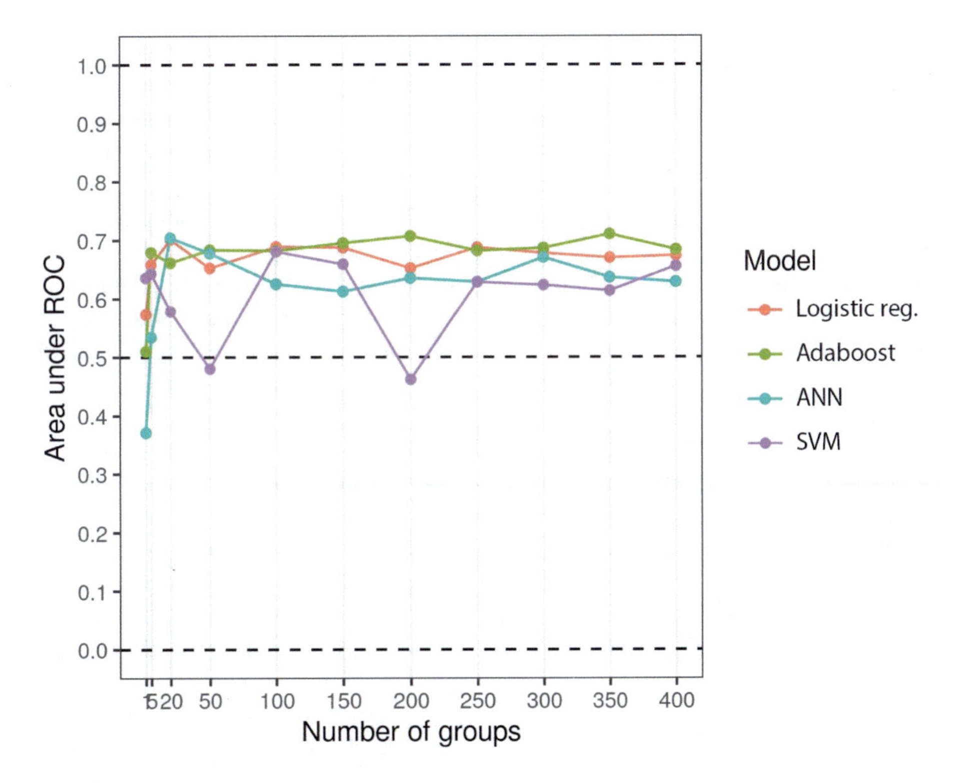

Fig. 6.27 The Area Under Curve (AUC) graph of the four classifiers plotted against the number of groups used for the classification procedure. Adaboost and Logistic regression were found as the best predictors, followed by ANN and finally SVM

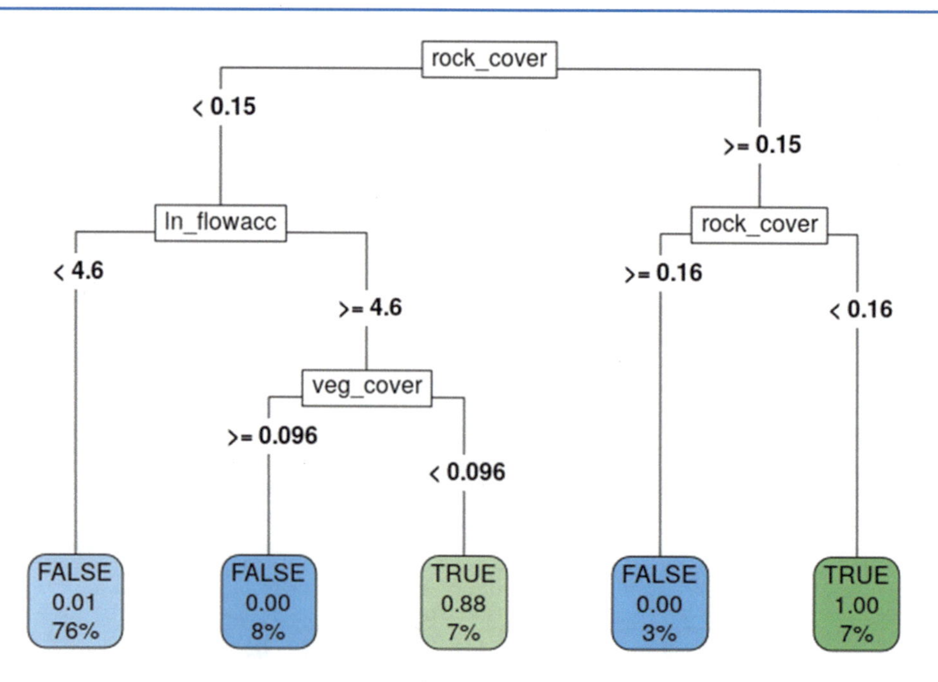

Fig. 6.28 An example of an individual decision tree which is part of the ensemble of (in this case, 100) decision trees that comprise the Adaboost model. The white rectangles represent the variables; the rules (conditions) refer to the threshold values; the top label in the leaves represents the prediction; the middle number in the leaves refers to the probability of the predicting the focal class (e.g., "risk"); and the lower number in the leaves is the percentage of observations in the branch

vector machines, decision trees and logistic regressions—have shown promising results in recent studies in predicting water erosion risks. The ROC and the AUC are standard measures for testing the performance of the classifiers, and can provide valuable analysis of the classifiers' performances. The limitation of the data-mining methods is that changing the catchment in question may impose a new configuration and normalization of the classifier upon the user (although the results achieved here were similar to those achieved in Svoray et al. (2012)—with slightly different classifiers, training sets, and a training procedure). So, adapting a data-mining classifier to a new catchment should be done with care. The creation of a large training set may help in this task.

Furthermore, in the study by Svoray et al. (2012), the data-mining approach was found to outperform the expert-based system that had been run for the same dataset. In other words, data-mining classifiers presented higher performance in predicting areas at risk than the expert-based system and much better performance than the TT that was associated with very large overestimation. An important component of the DM method is to choose the influential variables as input, and most importantly is to ensure that no important variable is left out. The physically based models and empirical models described in Chap. 3 can be used to ensure that all important factors are included in the data-mining procedure.

6.4 Fuzzy Logic

While dynamic physically based and empirical models (Chap. 3) have been used widely in predictive water erosion studies, advanced theories in soft computing were seldom used for this purpose. However, given that soil degradation processes are highly dynamic in time and space, new schools of thought suggest that erosion risks can be quantified using approximate equations that provide imprecise, but efficient, simulation of various surface processes. Such a soft-computing approach may release the user from the need to model the notoriously detailed and hard-to-compute sediment budget in physical units, and focus instead on estimations of Risk/No risk scores in the grid-cells.

Fuzzy logic is a form of soft-computing methodology in which the output values of the model equations are in the 0–1 range and are not confined by data type or physical units. In fact, equations used in fuzzy logic can be applied to a mixture of continuous and categorical data (Robinson 2003). Per se, fuzzy logic is used to predict the degree of truth rather than a means that provides a binary answer of Risk/No risk. The non-deterministic approach was found to be successful in previous studies, which showed that fuzzy methods perform better than crisp methods especi2003ally when the analyst attempts to simulate more multifaceted and composite environments (Robinson). In fuzzy logic methodology, the user selects the governing functions and their characteristics, hence she can bias, on purpose, the outcome. Therefore, some refer to fuzzy logic as a method that exploits tools of "subjective probability" (Fisher 1992).

Fuzzy logic was developed by Zadeh (1965) and later applied for numerous practices in the earth sciences. Recent applications focused on, for example, mapping vulnerability of multiple aquifers to contamination (Caniani et al. 2015; Nadiri et al. 2017), identification of groundwater quality (Vadiati et al. 2016), land use/land cover assessment (Bovkir and Aydinoglu 2018), and multi-hazard impact assessment (Araya-Muñoz et al. 2017). The unique capabilities of fuzzy logicto model "in exact, imprecise and ambiguous entities, and the relationships between them" (Burrough 1996) are especially apt for water erosion studies and to simulate the combined effect of the clorpt factors on water erosion processes because the soil is a complex and heterogeneous system that has no sharp boundaries (see review by McBratney and Odeh 1997).

The combination of functions for a multivariate analysis in fuzzy logic relies on weights that can be determined by experts using the pairwise comparison matrix (Sect. 6.2.2) or they can be extracted using data-driven methodology from a Sensitivity analysis that is based on data fitting (Vincenzi et al. 2006). In that sense, fuzzy logic can be applied either by using knowledge-driven or data-driven approach. Either way, the fuzzy logic approach can act as a heuristic model to study the importance of the different clorpt factors or climatic and soil processesesas will be shown in the next sections. Another important feature of the use of fuzzy logic paradigm in geoinformatics is the method's ability to provide formulation techniques to tackle system complexitywhich always exists in spatially and temporally explicit modelingusing relatively low computational cost.

This section describes fuzzy logic paradigms and main components: *Fuzzy sets*, *Membership Functions* (MFs) and *Joint Membership Functions* (*JMF*s), for estimating soil loss risk in the Harod catchment.

6.4.1 Theoretical Background

Fuzzy logic is a scientific discipline in the mathematical sciences that generalizes and expands binary logic (Yes/No, or 0/1) into a 0–1 continuumwith 0 as *Completely false* and 1 as *Completely true* (Zadeh 1965). This is done using MFs to normalize the output range of a single variable and *JMF*s that are used for multivariate analysis (Fisher 1996).

A fuzzy set is a class that does not have an explicit boundary between those members (e.g., grid-cells in a raster layer) that belong to it, and those that do not belong to it. Moreover, members can belong to more than one class (see next). Thus, fuzzy logic can be used to quantify erosion risks in a flexible manneras illustrated in the following toy example.

In a hypothetical field, covered by twelve raster grid-cells, with hillslope gradient in the range of 0°–10°, we ask whether each grid-cell is at water erosion risk, or not. According to binary logic, if the hillslope gradient in the grid-cell is, for example, $S \geq 5°$, the grid-cell is designated

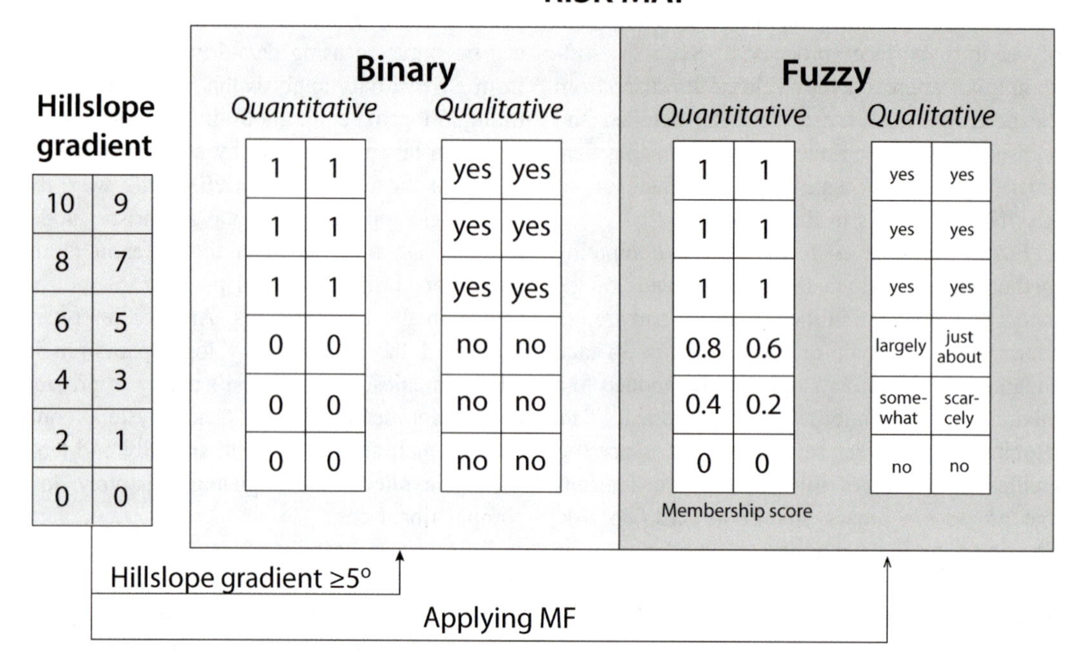

Fig. 6.29 Among grid-cells with a known hillslope gradient *S* (left-hand panel) ranging 0°–10°, which ones are at erosion risk? In the main panels we see binary (crisp) versus fuzzy analysis. In the binary approach, all grid-cells that are $S \geq 5°$ are designated at-risk. In the fuzzy logic approach, the answer is flexible, and the grid-cells are assigned a score which is the distance from full membership in the risk group (in this instance $S \geq 5°$). These computations can be translated into textual descriptions that may be more communicative to the decision-makers

at-risk. Using the fuzzy logic approach, however, the grid-cell is assigned a score, which is the membership score of the mathematical distance of the erosion risk from the value 5 (Fig. 6.29).

Fuzzy logic can therefore be used not only to determine whether or not there is an erosion risk in a given grid-cell, but also the level of that risk. Thus, factors such as rainfall intensity, soil erosivity, and available soil material to transportation, can be formulated mathematically not only by physically based models, but also based on how the expert perceives them, and translate this knowledge into functions of subjective probability.

A fuzzy set is defined (Eq. 6.13) as a subset **A** of a crisp set *X*, and the level of membership of **A** in *X* is described by the MF, which is denoted $\mu_A(x)$ (Montgomery et al. 2016). The type of MF is determined by an expert to represent any element *x* of *X* that partially belongs to **A**—or, in other words, the *grade of membership* of *x* of **A**. Thus, a candidate *x* is a full member of **A** if $\mu_A(x) = 1$ and is not a member of **A** if $\mu_A(x) = 0$. The higher a candidate's membership score, the more it belongs to the set *A* (Eq. 6.13).

$$\mathbf{A} = \{x, m_A(x)\} \text{ for each } x \in X, \qquad (6.13)$$

where $X = \{x\}$ is the set of observations, $\mu_A(\mathrm{x})$ is the MF of x in A.

The MF is therefore a crucial aspect of the fuzzy system, and the scientific literature provides numerous possible MFs for the analyst to choose. For a comprehensive survey of fuzzy logic MFs, used specifically in geoinformatics, see the review by Robinson (2003).

Figure 6.30 exemplifies a more advanced comparison between binary classification and membership to a level-of-risk group using fuzzy MF. In this example, the database is divided into

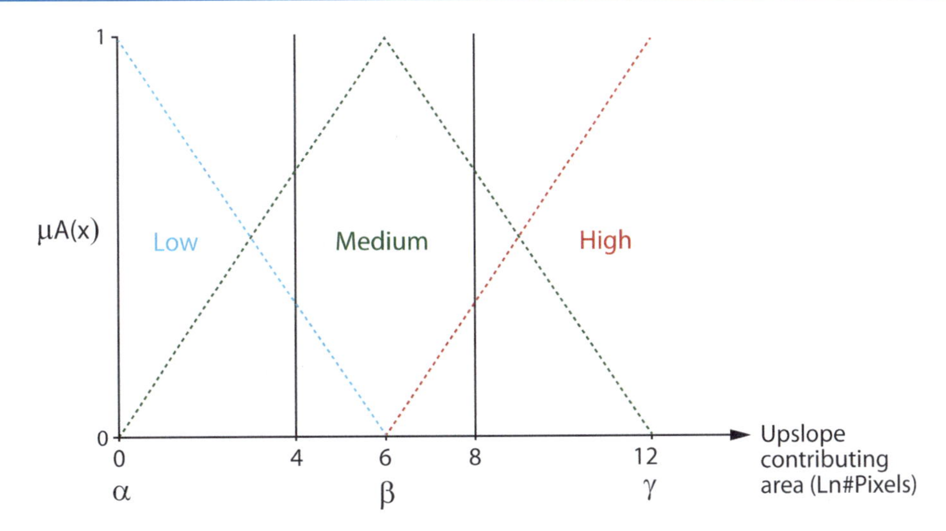

Fig. 6.30 Characteristic function of upslope contributing area as a determinant of water erosion risk, using binary (full line) versus fuzzy logic (dashed line)

three groups of erosion risk: Low, Medium, and High. The case of division by binary rules looks familiar (solid line): Grid-cells with contributing area values of between 0 and 4 are assigned exclusively to the Low erosion-risk group; grid-cells with values of between 4 and 8 to the Medium risk group, etc.... In the case of the fuzzy logic theory, the linear function increases from 0 to 1 as the contributing area value rises from 0 to 6 (where the membership score equals 1)—then decreases as the contributing area approaches high-erosion risk.

Using this computational strategy, one has a greater flexibility in expressing the effect of various variables on erosion risk. One can also quantify how much the grid-cell belongs to a given group (Medium risk), and how much to another (Low risk). That is the "fuzziness" of the measure. The grid-cell with the value 2 of upslope contributing area belongs ~0.8 to the Low risk group and ~0.3 to the Medium risk group—whereas in the crisp set, it belongs entirely (1) to the Low risk set, and not at all (0) to the Medium risk set. One can also choose between different MFs, as will be detailed further below, to enhance the ability to refine the impact of the various factors on the process.

An example of the MF described in Fig. 6.30 is presented in Eq. (6.14) with $\mu_A(x)$ as the subjective probability of belonging to the first category (i.e., A = low):

$$\mu_A(x) = max\left(\min\left(\frac{x-\alpha}{\beta-\alpha},\frac{\gamma-x}{\gamma-\beta}\right),0\right), \quad (6.14)$$

where $\mu_A(x)$ is the MF score; x is the actual value of the explaining factor in the given grid-cell (for example, hillslope gradient = 5); α and γ are the set boundaries (in the case of Fig. 6.30—$\alpha = 0$ and $\gamma = 12$); and β is the predicted value at the highest score ($\beta = 6$ in the case of Fig. 6.30).

Equation (6.14) can be exemplified with $x = 3$, then $\min\left(\frac{3-0}{6-0},\frac{12-3}{12-6}\right) = 0.5$, then max (0.5,0) = 0.5—i.e., $\mu_A(3) = 0.5$. This is the subjective probability to belong to the class (A). As we can see, MFs are functions used to convert a single variable values into membership scores for each variable. However, as in many other cases, several factors affect the soil erosion process, and these factors should be merged together to predict the final outcome (erosion risk in this case).

The computation of several variables together in a multivariate analysis in fuzzy logic is done using Joint Membership Functions (*JMFs*).

A typical *JMF* can be simply a compensatory rule—such as the weighted sum of the MFs of the participating variables (Eq. 6.15). The weights w_i are of major importance, because they represent the contribution of the variables to the final set—that is the overall score of erosion risk. In other words, the weights represent a scale (or order) of the contribution of the different variables to the *JMF* and, hence, to the overall predicted risk. They can be extracted either by using pairwise comparison matrices, or by applying a data-driven methodology, such as a sensitivity analysis against validation data.

$$JMF = \sum_{i=1}^{n} \mu_i(x_{ij}) \cdot w_i, \qquad (6.15)$$

where $i \ldots n$ is a running number representing the participating variables (input layers), $\mu_i(x_{ij})$ is the membership value of the *j*th grid-cell in data layer *i*, and w_i is the weight assigned to the variable (layer) *i* while $\sum_{i=1}^{n} x_i = 1$ and $w_i > 0$. *JMF*s can also be based on minimal or maximal scores of the MFs, when applying a bottleneck strategy—as in Eqs. (6.16)–(6.17). In a bottleneck strategy, the minimum or the maximum value among the membership scores of the different variables is the overall *JMF* score. Thus, for example, if we have a weighted value of 0.5 to hillslope gradient, 0.2 to contributing area and 0.1 to soil texture, the *JMF* of the three variables is 0.1 in the case of *JMF* min. Equivalently, in a maximal bottleneck, the *JMF* score will be 0.5.

$$JMF = \min\left(\mu_i(x_{ij}) \cdot w_i\right), \qquad (6.16)$$

or

$$JMF = \max(\mu_i(x_{ij}) \cdot w_i). \qquad (6.17)$$

And there is more. Urbanski (1999), for example, developed the *No Trade Off* (NTO) *convex combination JMF*, which is applied by means of Eq. (6.18):

$$JMF = \sum_{i=1}^{n} \mu_i(x_{ij}) \cdot w_i \wedge \sum_{i=1}^{m} \mu_i(x_{ij}) \cdot w_i, \qquad (6.18)$$

where ^ indicates the minimum between the two groups of MFs and *m* and *n* are the number of participating input layers in each group. In the NTO, the MFs are divided into groups of MFs joined together by the *JMF* in Eq. (6.15). Each MF may be assigned to several groups. The NTO *JMF* is then the score of the weakest link—i.e., the group with the minimal *JMF* score. That way, the rainfall *JMF*, for example, can force a zero score on the entire water erosion *JMF* if no rainfall occurs but if not, the final *JMF* score will be assigned as the value of the other group of MFs.

The functions and weights used in a fuzzy model to a given catchment will be the outcome of the interpretation by an expert of erosion processes in your area. Unlike the physically based models described in Chap. 3, the threshold values and the weights in fuzzy logic do not stand for accurate physical conversions between the input factors, but they are determined to create a general ranking of the input variables, as envisaged by the expert. This is useful, since fuzzy logic predicts the spatially and temporally explicit (subjective) probability of the erosion risk, rather than formalizing a rigorous physically based estimation of actual erosion yield. Furthermore, predictions at the catchment scale are very much needed for various land use applications and therefore generalizations in the determination of weights and functions are highly sought.

6.4.2 Procedure

6.4.2.1 Modeling Approach

Fuzzy Dynamic Soil Erosion Model (FuDSEM) by Cohen et al. (2008) is an example of a spatially and temporally explicit fuzzy rule-based model that simulates soil erosion risk at the hillslope scale, on a daily basis. The model is applied on a geoinformatics platform using a raster data model. All MFs and *JMF*s in FuDSEM are computed for the input variables, while the model translates into fuzzy logic some of the principal water erosion processes.

FuDSEM assumes the Hortonian runoff mechanism and is applied using four subroutines (Fig. 6.31): (1) The precondition of soil moisture content (*JMF1*) is quantified, based

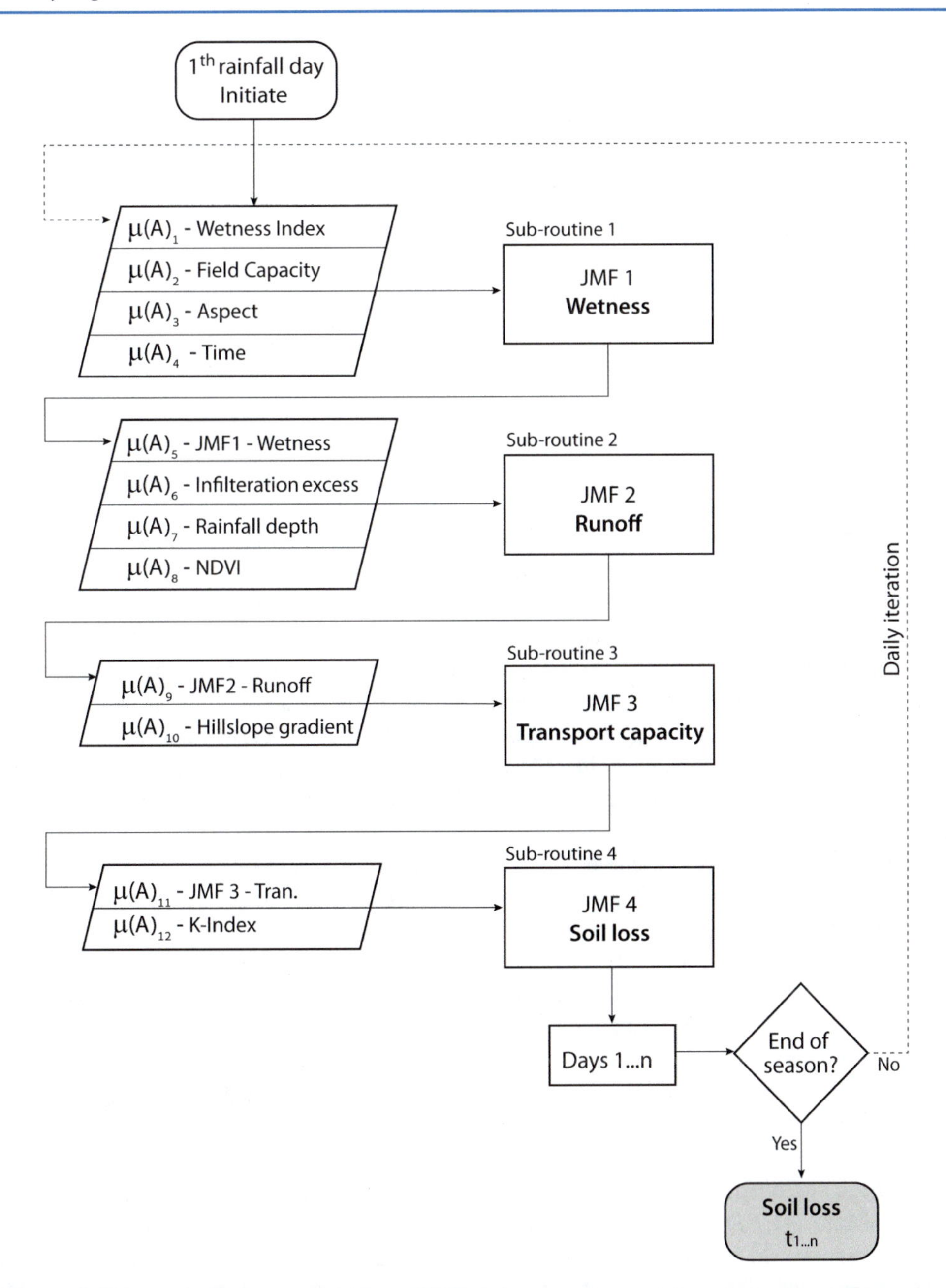

Fig. 6.31 A schematic flowchart of the application of FuDSEM to estimate soil loss risk. The μ(A)s are the MFs of the ten input variables. The *JMF*s represent the four sub-routines: *JMF1* is the potential antecedent surface moisture; *JMF2* computes the potential for runoff development; *JMF3* is the transport capacity of soil particles; and finally *JMF4* is the potential for soil loss. The model runs for every grid-cell using local operator on a daily basis. *JMF4* is the daily soil loss risk map (adapted from Cohen et al. 2008)

on topographic characteristics, soil properties, and the time elapsed from the last rainfall event; (2) The potential for overland flow generation (*JMF2*) is computed based on the antecedent soil moisture content (*JMF1*), daily rainfall depth, vegetation cover in the fields (*NDVI*) and infiltration excess which is the delta (Δ) between the saturated hydraulic conductivity of the soil and daily rainfall intensity. Both have dimensions of length per time; (3) Runoff from *JMF2* is accumulated using TauDEM (see Sect. 6.1.2), and soil transport capacity by overland flow (*JMF3*) is then computed based on flow accumulation and the hillslope gradient; and finally, (4) Subjective probability (*JMF4*) of soil loss is computed, based on transport capacity potential (*JMF3*) and soil erodibility factor—K-Index of the USLE (Sect. 1.1.3.2).

Thus, sub-routines 1…4 are mathematically represented by *JMFs* 1…4 with each of them assimilating the appropriate input variables—with each one quantified by the governing MFs $\mu_A(x)$ and the previous *JMF* that represents the preceding sub-routine. This is done on a daily basis, and after the final sub-routine is computed, the model moves on to the next day to read rainfall data. The model runs are repeated until the last day of the wet season.

The four *JMF*s comprise ten MFs, as detailed in Tables 6.8 and 6.9—where P_{max} and P_{min} are the maximum and minimum threshold values, whereas $\mu_A = 0$ for $x \leq P_{min}$ and $\mu_A = 1$ for x P_{max}. α is the crossover point ($\mu_A = 0.5$) and β controls the slope of the sigmoid curve.

The MFs provide the means for expressing one's understanding of the process, by selecting the most suitable function for each variable. Also, expert knowledge is needed to determine the coefficient of each function, and the maximal and minimal points.

Figure 6.32 shows the computed $\mu(A)_i s$ for the variables in Table 6.8.

6.4.2.2 Model Computation

Surface moisture (*JMF1*), a spatially and temporally dynamic factor, is crucial for the development of overland flow (Sela et al. 2015). FuDSEM estimates soil moisture conditions based on: Length of the dry spell, topographic Wetness Index (see Sect. 4.1.1.1), hillslope aspect, and field capacity. The MFs of the four variables are described in Tables 6.8 and 6.9. The exponential decrease in soil moisture with time, as observed by Hillel (1998), is the reason for using the left shoulder sigmoidal function for $\mu(A)_4$. The *S* function power parameters, α and β, were assigned by specialists based on their experience. Trigonometric computations of hillslope aspect were used to express the effect of topography and sun position on evaporation rate and soil moisture content (Oliphant et al. 2003). The 0–360 scale of the topographic aspect was rescaled to 0–180 and $\mu(A)_3$ was assigned with a decreasing function of radial distance from the south (180°) to the north (0°). The membership function of the topographic wetness index $\mu(A)_1$ was computed using a positive (an increasing version of) sigmoidal function (Table 6.9). Finally, the effect of soil properties on antecedent surface wetness was expressed by the field capacity $\mu(A)_2$—with high field capacity values increasing the grid-cell membership to the set A1. The function used is negative (decreasing) linear (Table 6.9) with P_{max} and P_{min} designated as the maximum and minimum values measured in the Harod's dataset. *JMF1*, which integrates the effects of all these four factors on surface wetness, in every grid-cell in the daily raster layer, is computed using the non-compensatory approach termed No Trade Off (NTO) convex combination (see Table 6.10). NTO was chosen because in semiarid areas, after long dry spell, the topsoil dries out. Namely, if the time elapsed $T_e = 0$, then $JMF1 = 0$. Moreover, the weight assigned to T_e (when $T_e \neq 0$) is considerably large while all the other factors were assigned an equal and smaller weight. This was done based on the assumption of their equal contribution to the soil moisture potential.

The function *JMF2* is used to compute the subjective probability of overland flow occurring in grid-cells where rainfall has exceeded soil infiltration capacity. This includes consideration of the antecedent soil moisture content (*JMF1*). Infiltration excess is computed as the Δ between the soil saturated hydraulic conductivity and the daily rainfall intensity. Sigmoidal function converted infiltration excess values into 0–1 scale, based on similar empirical relationship found in Valmis et al. (2005) while daily rainfall depth used a negative sigmoidal function based on USDA-SCS (1985). The *Normalized Difference*

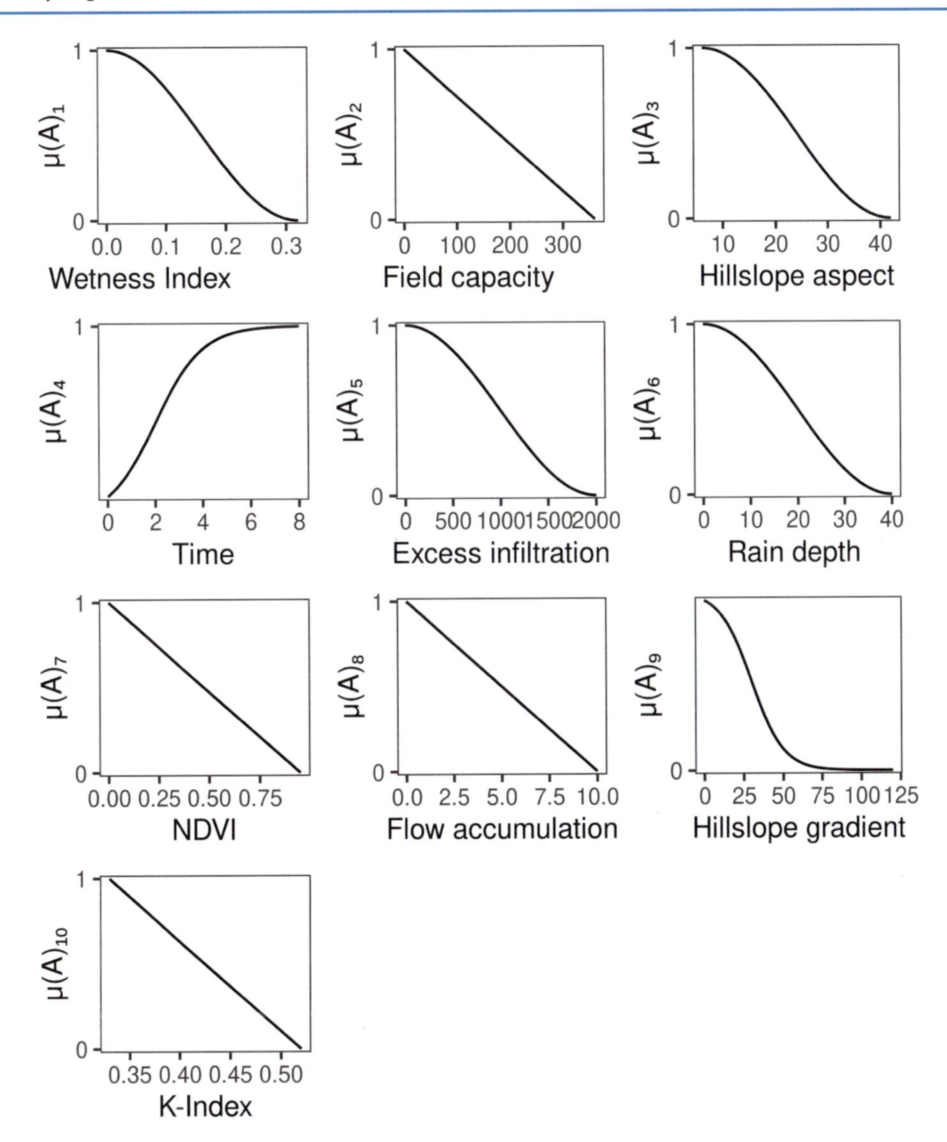

Fig. 6.32 The membership functions $\mu(A)_i s$ computed for the ten variables (Table 6.8) participating in the fuzzy rule-based modeling predictions of water erosion risks in the Harod catchment

Vegetation Index (*NDVI*—Tucker 1979) was used as a proxy for vegetation cover. As described in Chap. 2, the growth of crop plants substantially affects runoff generation and flow by reducing the raindrop impact, and halting water flow by the seedling roughness (Calvo-Cases et al. 2003). The MF of *NDVI* was assigned a linear function based on the FAO 1967 report. The synergy of the four components in the overall runoff membership score is applied using the No Trade Off combination form of *JMF* (Table 6.10). The infiltration excess is assigned as a threshold variable—i.e., only positive infiltration excess values were eligible for computing *JMF2*. Because of the importance of vegetation cover to the erosion process, the *NDVI* weight assigned by the experts was double the weights assigned to the other variables. Other variables were assigned an equal weight assuming equal contribution to overland flow generation.

Three variables were used in FuDSEM to compute *JMF3* (Transport capacity potential):

Table 6.8 The input variables involved in FuDSEM and the type of each MF assigned to each variable. Reprinted from: Cohen et al., Fuzzy-based dynamic soil erosion model (FuDSEM): Modelling approach and preliminary evaluation. Journal of Hydrology 356(1–2):185–198 (2008). doi: https://doi.org/10.1016/j.jhydrol.2008.04.010, with permission from Elsevier

Factor	Acronym	Type of MF	α	β	P_{min}	P_{max}
$m(A)_1$—Wetness Index	WI	Sigmoidal			0	0.32
$m(A)_2$—Field capacity	FC	Linear			0	360
$m(A)_3$—Hillslope aspect	SA	Sigmoidal			6.1	42
$m(A)_4$—Time	Te	Left shoulder	2	1		
$m(A)_5$—Excess infiltration	IE	Sigmoidal			0	2000
$m(A)_6$—Rain depth	RD	Sigmoidal			0	40
$m(A)_7$—*NDVI*	*NDVI*	Linear			0	0.95
$m(A)_8$—Flow accumulation	Acc	Linear			0	10
$m(A)_9$—Hillslope gradient	S	Right shoulder	30	0.1		
$m(A)_{10}$—K-Index	K	Linear			0.33	0.52

Table 6.9 The MFs being used to convert the values of the ten variables in Table 6.8 into membership scores between 0 and 1. Reprinted from: Cohen et al., Fuzzy-based dynamic soil erosion model (FuDSEM): Modelling approach and preliminary evaluation. Journal of Hydrology 356(1–2):185–198 (2008). doi: https://doi.org/10.1016/j.jhydrol.2008.04.010, with permission from Elsevier

Types	Functions	Graphs	Variables
Sigmoidal	$\mu A_i = \cos^2\left(\frac{x-P_{min}}{P_{max}-P_{min}}\right)\cdot\left(\frac{\pi}{2}\right)$	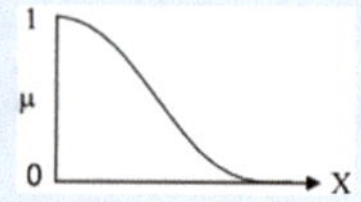	x—the variable original value. P_{min} and P_{max} are model parameters defined in Table 6.8 and α is the cross over point and β controls the way at which $\mu_{Ai}\rightarrow 1$
Linear	$\mu A_i = \frac{P_{max}-x}{P_{max}-P_{min}}$	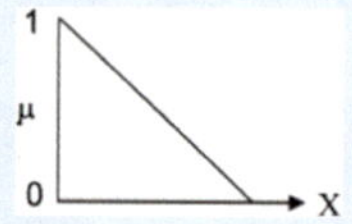	
Left shoulder	$\mu A_i = \frac{1}{1+e^{-\beta\cdot(x-\alpha)}}$	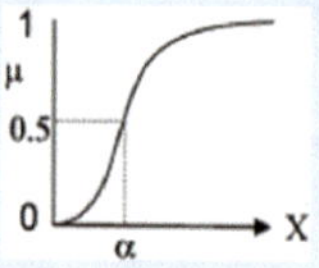	
Right shoulder	$\mu A_i = \frac{1}{1+e^{\beta\cdot(x-\alpha)}}$	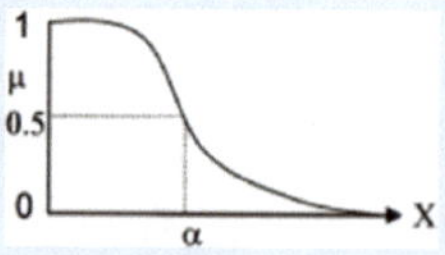	

Runoff potential as computed in *JMF2*; upslope contributing area; and hillslope gradient. Note that runoff accumulation in a given grid-cell is also governed by the land cover characteristics of upslope cells (Chen et al. 2013). The membership score of contributing area is determined by the experts as a linear function (Table 6.9). Hillslope gradient represents the effect of the gravitational force on overland flow: A steep hillslope gradient increases runoff discharge, resulting in a higher transport capacity (see Chap. 3). Right shoulder sigmoidal MF

Table 6.10 The four *JMF*s used to simulate erosion risk, using FuDSEM. See Table 6.8 for the definitions of membership functions. Reprinted from: Cohen et al., Fuzzy-based dynamic soil erosion model (FuDSEM): Modelling approach and preliminary evaluation. Journal of Hydrology 356(1–2):185–198 (2008). doi: https://doi.org/10.1016/j.jhydrol.2008.04.010, with permission from Elsevier

JMF	Function	Source	
1	$\mathrm{JMF} = \left(\sum_{j=1}^{m} \lambda_j \cdot \mu_{A_j}\right) \wedge \left(\sum_{j=1}^{n} \lambda_j \cdot \mu_{A_j}\right)$ $JMF1 = \begin{cases} 0.0 & Te = 0 \\ 0.4Te + 0.2SA + 0.2FC + 0.2WI & Te > 0 \end{cases}$	Urbanski (1999)	$/_{1,\ldots,n}$ are the MFs weights and ^ is the min between the two groups of *JMF*s
2	$JMF1 = \begin{cases} 0.0 & IE <= 0 \\ 0.2 \cdot IE + 0.2 \cdot RD + 0.4 \cdot NDVI + 0.2 \cdot JMF1 & IE > 0 \end{cases}$		
3	$JMF3 = 0.33 \cdot S + 0.33 \cdot Acc + 0.33 \cdot JMF2$		
4	$JMF4 = 0.10 \cdot K + 0.90 \cdot JMF3$		

(Table 6.9) was used to convert the hillslope gradient values to the 0–1 range, while the parameters α and β (Table 6.8) were evaluated based on Kirkby (1980). For their integration in *JMF3*, these three variables are assigned equal weights using a *convex combination operation function* (Table 6.10).

The final sub-routine—*soil erosion potential* (*JMF4*)—represents with convex combination operation function the soil erodibility and occurrence of erodible soil material for erosion. Thus, to express the soil erodibility, the transport capacity of *JMF3* is combined with the *soil erodibility index* (K-Index) (see Chap. 1). A high erodibility value means a higher erosion potential for a given set of runoff conditions. The membership score of K is described by the mirror version of linear MF (Table 6.9).

To summarize, fuzzy logic as implemented through FuDSEM combines various mathematical functions for simulating spatially and temporally dynamic processes, given an input of ten input layers representing ten variables. The MFs are independent from physical units and are used to compute the subjective probability score for water erosion risk as the expert perceives the erosion process. The superposition of the input variables (*JMF*s), the chosen MFs, and the function's parameters, allow both the field and catchment conditions, and the dynamics of the erosion processes to be simulated. The similarity between the dynamic fuzzy logic framework and the empirical and physically based models of soil erosion (see Chap. 3) is clear. The uniqueness of the fuzzy logic framework is its ability to express expert knowledge through (more specific/expressive) functions, coefficients, and weights, rather than relying on the crisp prediction of the classical models.

6.4.3 Results and Discussion

The fuzzy logic approach developed in FuDSEM for the Shiqmah catchment in Israel's central region, was applied, within the framework of this book, to the Harod catchment for the rainy season of 2017–2018. Figure 6.33 illustrates the eight GIS raster layers that were used to compute the MFs described in Table 6.8. The two other variables that were assigned MFs in this run—time elapsed from the last rainfall event and rainfall Intensity—were not spatially explicit.

The following layers were computed using a DEM extracted from a 30 m resolution SRTM data (that was also used in previous chapters of this book): Contributing area; hillslope gradient; wetness index; and hillslope aspect. The *NDVI* layer was computed from Landsat-8 OLI data, and averaged over time for the entire period (2017–2018). The source of information for the variables: Field capacity, hydraulic conductivity, and K-factor, was the 1:20,000 Soil Map of Israel, which was compiled by the Israeli Soil Conservation and Drainage Division.

The physical soil characteristics (in Fig. 6.33 e, g, and h) are relatively homogenous in space, due to the original scale of the soil map, which

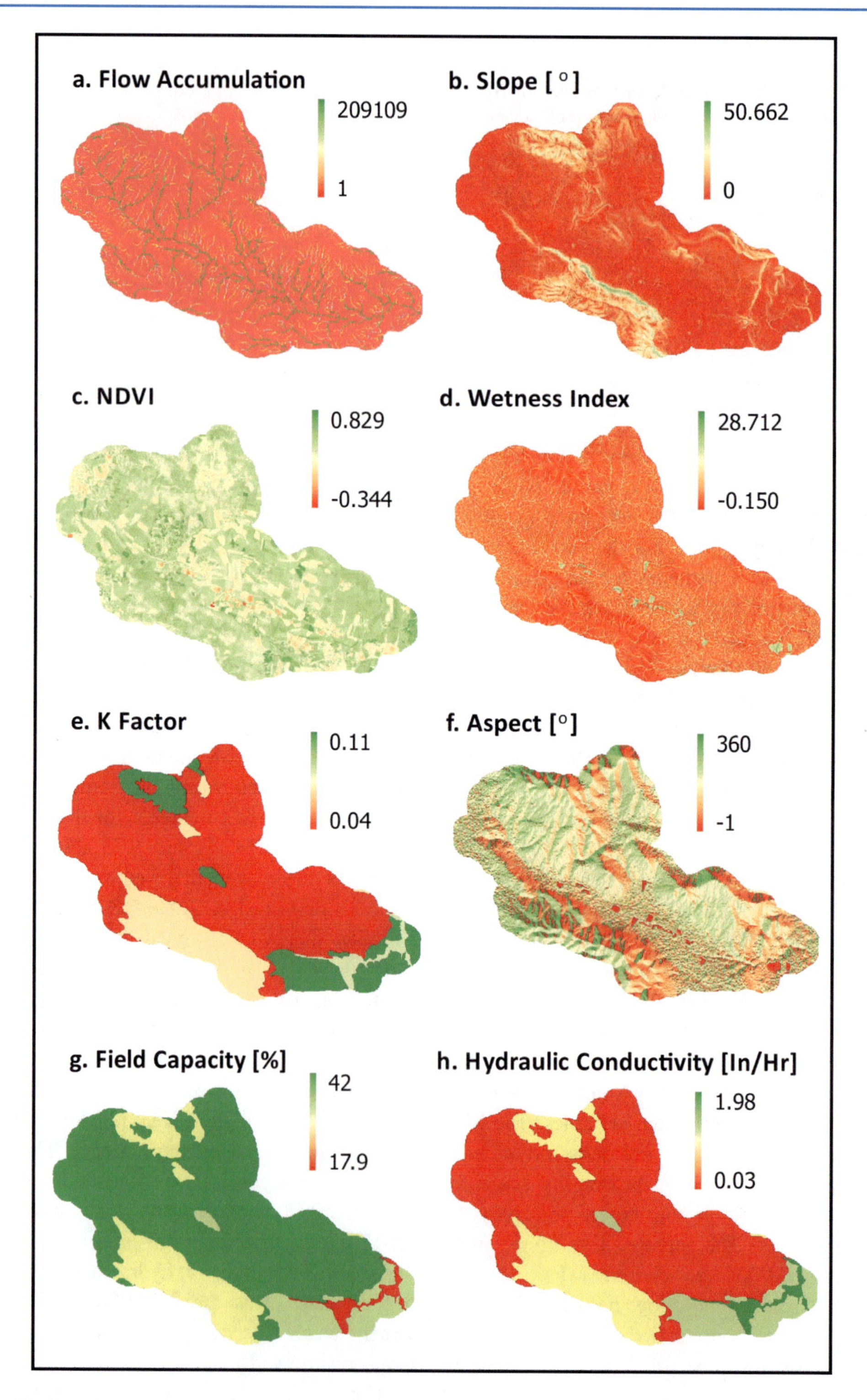

Fig. 6.33 Input layers to the fuzzy logic procedure of FuDSEM applied to the Harod catchment. Sources for the input layer are the SRTM-based DEM of the area, an averaged *NDVI* image based on Landsat data, and a soil map of Israel. Two other variables that were assigned MFs include (1) Time elapsed from the last rainfall event; and (2) Rainfall Intensity were not spatially explicit

was compiled using traditional methods. More detailed physical soil characteristics may be added to the model if the maps are created from high-density point measurement data. Such a research approach, however, requires a high sampling effort to express variation in soil properties at the local scale. Under the conditions of applying FuDSEM to the Harod catchment, most of the variation appears to have been the result of the topographic data and the *NDVI*. As such, the erosion potential (final *JMF4*) generally corresponds to the variables with the higher weights, while the MF themselves regulate this effect—especially when linear and exponential functions are compared.

Figure 6.34 shows the FuDSEM output. The model iterates on a daily basis, but the illustration shows the seasonal average. The grid-cell data, used as the basic unit for the computations of the MFs and the *JMF*s, were aggregated to the sub-catchment scale by means of the zonal average. Subcatchments are basic units that reflect the hydrological processes and topographic conditions at the landscape scale, which is well suited to the erosion processes described in FuDSEM (Sect. 6.4.2). As a result, the output estimate of risk maps is less noisy and can be interpreted for conservation planning purposes at the national or regional level, for example. But this is only an example, and the reader can similarly average the predicted data, based on hillslope catena units (see Sect. 4.1.1.1) or agricultural fields (see Sect. 4.1.2).

The results in Fig. 6.34 clearly show the spatial variation in the erosion risk in the catchment, while the effect of the contributing area was identified in the central part of the catchment, and the effect of hillslope gradient is evident in the footslopes of Mt. Gilboa in the southern part of the catchment, and of Hamoreh Hill in the northern section.

A good agreement was found by Cohen et al. (2008) between the runoff and erosion predictions by FuDSEM and overland flow volume and erosion/sediment yields measured in the field of the Bikhra catchment. In a further exploration of FuDSEM's accuracy in soil loss predictions, Cohen et al. (2008) have found that at a bigger agricultural catchment—the Shiqmah catchment in the center of Israel—an overestimation may occur in the infiltration excess sub-routine, and an underestimation may occur in the rainfall

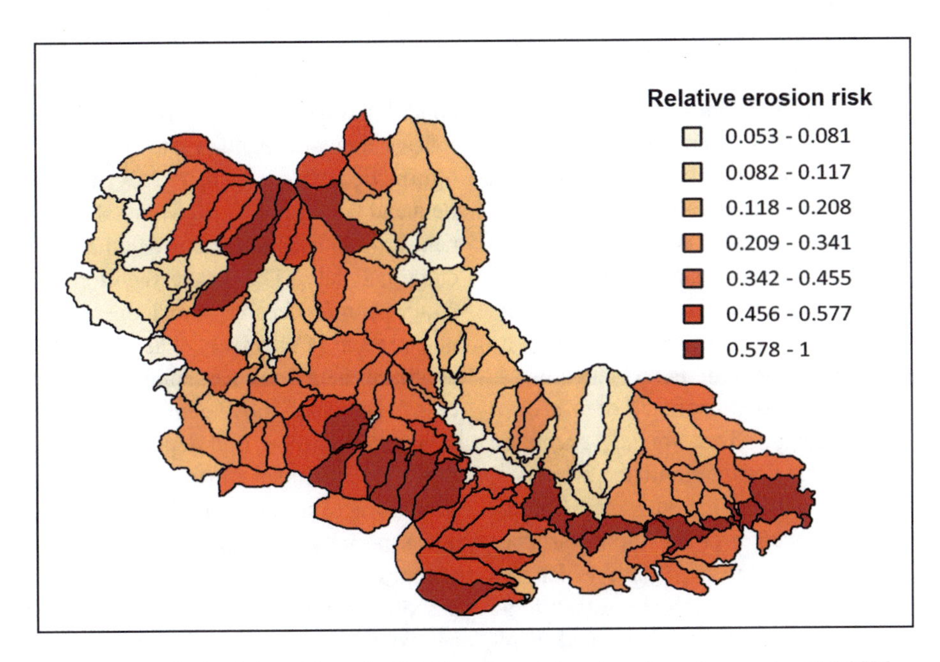

Fig. 6.34 Mapping gradual soil loss risk in the Harod catchment, based on fuzzy logic, using FuDSEM

intensity factor. However, despite these possible sources of bias, the association between the predicted results in the Shiqmah catchment and the validation data was significant ($R^2 = 0.72$; $p = 0.015$). Furthermore, FuDSEM has also outperformed the soil erosion model for Mediterranean regions (SEMMED—De Jong et al. 1999) in predicting erosion risk in the Shiqmah catchment.

The fact that FuDSEM has been validated against field data supports the use of soil erosion risk maps, such as Fig. 6.34, as an efficient tool for modeling water erosion processes in agricultural (and other) catchments. Using geoinformatics data, this can be done at different scales. Although fuzzy logic can be adapted to represent categorical data, it is efficient in representing the membership of continuous variables in various sets of erosion risk, for example, and soil map accuracy has been found to be improved through fuzzy logic compared with conventional methods (Moonjun et al. 2020). Fuzzy logic was developed to strike a balance between process understanding and representation of risk level, and the complexity of the process in time and space.

6.4.4 Summary

The theory of fuzzy logic offers principles that can be used for spatially and temporally explicit modeling of water erosion processes. The application of fuzzy modeling to the Harod catchment using daily time series involved developing a hierarchical structure of MFs and *JMF*s, based on ten geoinformatics layers, using R. These functions express the effect of the governing factors on the water erosion process, as determined by the experts. In that regard, fuzzy modeling is more a knowledge-based approach than a data-driven methodology. The model execution allows a backtracking of the simulated processes, as well as the reasoning of the outcomes. For any given output, the weight of the input variables and the MFs reflects the degree to which the variable affects the erosion process. Fuzzy logic is thus not a black-box methodology, and therefore may be preferred in cases where the user wishes to retrace the model simulations. Key to geoinformatics is the fact that the input variables are stored as geoinformatics layers. Thus, spatial and non-spatial data can be merged to compute the *JMF*s, using comparable units instead of physical ones. This requires the predictions of the fuzzy logic system to be compared with measurements that are proxy to erosion risks, and not the actual values of soil loss [ton · ha^{-1} · yr^{-1}].

6.5 Summary

This chapter offered a detailed description of four methods for classifying water erosion risks in agricultural catchments. In addition, it presents methods for analyzing the impact of input variables on the estimated risk (through the use of weights), and for estimating the catchment response to environmental changes, through simulations. The ability to apply these methods to basic units such as the hillslope catena and the subcatchments allow to extract knowledge on the principal water erosion processes, using experts' assessments or large datasets, and translating the processes—including space–time variation effects—into computer simulations. Spatial modeling concepts, such as MLC, DM, and fuzzy logic, can be used to model and simulate the effects of soil, atmospheric, and human-based conditions on erosion risk. Spatial MLC naturally reflects human capabilities to process long-term information that is gathered by generations of field and laboratory work, while DM allows us to gain an understanding of erosion risk based on empirical data. The latter, therefore, does not require comprehensive knowledge of the studied phenomena. Conversely, the fuzzy-rule-based approach is a tool for simulating actual processes of water movement, including soil detachment, transport, and deposition. This type of flexibility is crucial for investigating the consequences of various scenarios of environmental and especially atmospheric and human-induced changes in field conditions. MFs may be selected based on previous studies, and the weights may be extracted using MLC. The fuzzy model's ability to represent soil indeterminate boundaries makes it possible to investigate water erosion risks and emergence at various spatial and temporal resolutions. Perhaps the most important issue related to the methodologies presented here, is that despite the complex processes that they represent, they can be applied comparatively easily by every soil conservationist to their catchment, and to interpret the results in a bid to analyze risk in the catchment, and to predict the implications of future scenario risks.

Review Questions

1. Create a graph of the Topographic Threshold (TT) (e.g., Fig. 6.1) for your catchment. Compute the hillslope gradient and contributing area of at least 15 gully heads to prepare the graph. Use the lowermost line to extract the a and b coefficients for your catchment. Compare the coefficients you got with coefficients from other studies (Table 6.2), and try to explain the difference if there is any.
2. Use the coefficients you extracted in Question 1 to create a risk map for gully head incision in your catchment. If you do not have access to measure gully head characteristics in your catchment, use the Harod catchments coefficients as described in Fig. 6.1. What is the extent of areas at risk in your catchment? Is it reasonable compared with field observations? Based on your knowledge of the catchment, is the location of areas at risk reasonable?
3. Consider five variables that are the most influential on water ponding risks in your catchment. Compute the weights to these variables using the *pairwise comparison method* detailed in Sect. 6.2.2. Then, apply the coefficients to data layers of your catchment using WLC (don't forget to normalize the layers data to five levels of influence, as in Table 6.4). What is the extent of areas at risk, and their locations?
4. Use the weights from question 3 to create a risk map based on the five variables data normalized with CPT. Compare the two maps (WLC and CPT) against field data. Which one of the methods is more certain? Use *confusion matrices* and the *Kappa coefficient* to decide.
5. Compare the weights assigned by a single expert, and by several experts. What is their effect on the hierarchy of importance of the variables on soil erosion risks? Does it affect the predictability of the system? Who makes better predictions: A single expert, or several experts? Also compare the results to *Ranking* and *Rating* methods, as described in Eqs. (6.2)–(6.4).
6. The soil loss rates of agricultural catchments around the world have increased, due to climate change. Assuming that rainfall intensity grows, due to increased water vapor and the fact that warmer air can hold more water, try to estimate the increased risk area in your catchment, following 5, 10, 30, and 50% increase in rainfall intensity.
7. Use the same five variables to map soil loss risk in your catchment with data-mining tools. Implement the four classification techniques described in Sect. 6.3 to identify the areas at risk. Implement the classification technique to assess soil loss risk for each grid-cell, and plot the results as a color map classified into ten groups, using Quantile color breaks.
8. Develop a *fuzzy logic* based routine to map the water erosion processes in your catchment, by including the relevant controlling factors. Make decisions about the *Membership Functions* and the *Joint Membership Functions*, to map the erosion risks.

Strategies to Address the Review Questions

1. The TT graph is based on hillslope gradient S and contributing area A data extracted from a DEM (Eq. 6.1 from Sect. 6.1.1). You have to go to the field or to use high-resolution aerial photo, to locate the gully heads. Then you extract S and A data from $3 \cdot 3$ kernel window around the points and plot the TT line (note the physical units). A difference between the coefficients, if exists, can be explained by differences in environmental conditions (Fig. 6.1). One of the limitations of the TT approach is generalization of the coefficients. Namely, the TT graph represents

the empirically estimated conditions in the agricultural catchment in question.

2. The a and b coefficients extracted from the TT line of the lowest points can be used to compute the TT threshold for the site. See for details the example in Fig. 6.3. Figure 6.4 and Fig. 6.5 show how coefficients can change the output risk map tremendously.
3. Based on Sect. 6.2.2 on weighting, we need to compare between every two variables (using Satty scale, Fig. 6.9) and create the *pairwise comparison matrix* (Fig. 6.8). By applying the procedure steps, the weights are extracted (Fig. 6.10). Next step is to estimate the expert's consistency (Fig. 6.11) and compute lambada (Eq. 6.5) and the average consistency vector. Lambada can be used to compute the consistency index, CI (Eq. 6.6). Finally, CR is computed to indicate if the matrix is consistent. If it is, you can apply the coefficients to the catchment using WLC (Sect. 6.2.3, Eq. 6.8) to calculate the level of risk based on weights and the normalized variable's value.
4. CPT is computed using the same weights by the pairwise comparison matrix, but, in CPT, the multivariate data is normalized based on points of predetermined extreme reference in the input variables, and the "distance" of each grid-cell value (in the given variable) from those reference points (Eq. 6.9). Sections 6.2.4–6.2.6.2 describe an example for a WLC/CPT comparison.
5. Ask from five expert to fill in the pairwise comparison matrix and compute the weights based on each of their matrices (Sect. 6.2.6.1). Average the pairwise comparison matrices by the five experts for each intersection and compute the weights based on the average matrix. Next, you create a risk map based on the average matrix and based on one of the single experts weights (or all of them—separately). Use a validation data of gully heads to test the risk maps accuracy of the single expert and the average matrix and answer the final questions.
6. A simulation of the effect of an increase in rainfall intensity on risk areas is demonstrated in Sect. 6.2.6.5. To run the required simulation to your catchment, you can follow the procedure detailed in Sect. 6.2.4 and illustrated in Fig. 6.13.
7. The four classification techniques to assess soil loss risk are: *Decision trees* (Sect. 6.3.1.1, Fig. 6.20), *Artificial Neural Networks* (Sect. 6.3.1.2 and the architecture in Fig. 6.21), *Logistic regression* (Sect. 6.3.1.3, Fig. 6.22) and *Support Vector Machine* (Sect. 6.3.1.4, Fig. 6.23). Those four classifiers can be applied to map soil degradation risks (Figs. 6.24 and 6.25).
8. A soil loss map extracted using *fuzzy logic* techniques is illustrated in Fig. 6.31. First, decide about the water erosion processes to simulate. Then select the variables to model those processes. Choose the *Membership Functions* to be used (Tables 6.8, 6.9, Fig. 6.32) and the *Joint Membership Functions* (Table 6.10). Finally, you can use a GIS modeling tool such as Raster Calculator of Model Builder to execute the model. The expected outcome is the soil loss risk map.

References

Akgün A, Türk N (2011) Mapping erosion susceptibility by a multivariate statistical method: a case study from the Ayvalık region, NW Turkey. Comput Geosci 37 (9):1515–1524. https://doi.org/10.1016/j.cageo.2010.09.006

Alfaro E, Gamez M, Garcia N (2013) Adabag: an R package for classification with boosting and bagging. J Stat Softw 73(2):1–13. https://doi.org/10.18637/jss.v073.i02

Arabameri A, Pradhan B, Pourghasemi HR et al (2018) Identification of erosion-prone areas using different multi-criteria decision-making techniques and GIS. Geomat Nat Haz Risk 9(1):1129–1155. https://doi.org/10.1080/19475705.2018.1513084

Araya-Muñoz D, Metzger MJ, Stuart N et al (2017) A spatial fuzzy logic approach to urban multi-hazard impact assessment in Concepción, Chile. Sci Total Environ 576:508–519. https://doi.org/10.1016/j.scitotenv.2016.10.077

Baja S, Chapman D, Dragovich D (2007) Spatial based compromise programming for multiple criteria decision making in land use planning. Environ Model Assess 12 (3):171–184. https://doi.org/10.1007/s10666-006-9059-1

Basnet BB, Apan AA, Raine SR (2001) Selecting suitable sites for animal waste application using a raster GIS. Environ Manage 28:519–531

Baston D (2020) exactextractr: Fast extraction from raster datasets using polygons R package version 0.5.1

Bauder ET (2005) The effects of an unpredictable precipitation regime on vernal pool hydrology. Freshw Biol 50 (12):2129–2135. https://doi.org/10.1111/j.1365-2427.2005.01471.x

Begin ZB, Schumm SA (1979) Instability of alluvial valley floors: a method for its assessment. Trans Am Soc Agric Eng 22:347–350

Bennett K, Campbell C (2000) Support vector machines. SIGKDD Explorat 2(2):1–13. https://doi.org/10.1145/380995.380999

Bovkir R, Aydinoglu AC (2018) Providing land value information from geographic data infrastructure by using fuzzy logic analysis approach. Land Use Policy 78:46–60. https://doi.org/10.1016/j.landusepol.2018.06.040

Burrough PA (1996) Natural objects with indeterminate boundaries. In: Burrough PA, Frank AU (eds) Geographic objects with indeterminate boundaries. Taylor and Francis, London, pp 3–28

Calvo-Cases A, Boix-Fayos C, Imeson AC (2003) Runoff generation, sediment movement and soil water behavior on calcareous (limestone) slopes of some Mediterranean environments in southeast Spain. Geomorphology 50:269–291

Caniani D, Lioi D, Mancini I et al (2015) Hierarchical classification of groundwater pollution risk of contaminated sites using fuzzy logic: a case study in the Basilicata Region (Italy). Water (Basel) 7(12):2013–2036. https://doi.org/10.3390/w7052013

Chen L, Sela S, Svoray T et al (2013) The role of soil-surface sealing, microtopography, and vegetation patches in rainfall-runoff processes in semiarid areas. Water Resour Res 49(9):5585–5599. https://doi.org/10.1002/wrcr.20360

Choubin B, Moradi E, Golshan M et al (2019) An ensemble prediction of flood susceptibility using multivariate discriminant analysis, classification and regression trees, and support vector machines. Sci Total Environ 651(2):2087–2096. https://doi.org/10.1016/j.scitotenv.2018.10.064

Cohen S, Svoray T, Laronne JB et al (2008) Fuzzy-based dynamic soil erosion model (FuDSEM): Modelling approach and preliminary evaluation. J Hydrol (Amsterdam) 356(1–2):185–198. https://doi.org/10.1016/j.jhydrol.2008.04.010

Conoscenti C, Angileri S, Cappadonia C et al (2014) Gully erosion susceptibility assessment by means of GIS-based logistic regression: a case of Sicily (Italy). Geomorphology (Amsterdam, Netherl) 204:399–411. https://doi.org/10.1016/j.geomorph.2013.08.021

Dai FC, Lee CF (2003) A spatiotemporal probabilistic modelling of storm induced shallow landsliding using aerial photographs and logistic regression. Earth Surf Proc Land 28:527–545

Dalkey NC (1969) An experimental study of group opinion. Futures 1(5):408–426

De Jong SM, Paracchini ML, Bertolo F et al (1999) Regional assessment of soil erosion using the distributed model SEMMED and remotely sensed data. Catena 37(3):291–308. https://doi.org/10.1016/S0341-8162(99)00038-7

Dharumarajan S, Hegde R, Singh SK (2017) Spatial prediction of major soil properties using Random Forest techniques—a case study in semi-arid tropics of South India. Geoderma Reg 10:154–162. https://doi.org/10.1016/j.geodrs.2017.07.005

Fernandes MMH, Coelho AP, Fernandes C et al (2019) Estimation of soil organic matter content by modeling with artificial neural networks. Geoderma 350:46–51. https://doi.org/10.1016/j.geoderma.2019.04.044

Fisher PF (1992) First experiments in viewshed uncertainty: simulating fuzzy viewsheds. Photogramm Eng Remote Sens 58(3):345–352

Fisher PF (1996) Boolean and fuzzy regions. In: Burrough PA, Frank AU (eds) Geographic objects with indeterminate boundaries. Taylor and Francis, London, p 87Ð94

Freund Y, Schapire RE (1996) Experiments with a new boosting algorithm. In: Anonymous machine learning: proceedings of the thirteenth international conference, 3–6 July 1996. Morgan Kaufmann, Waltham, Massachusetts, pp 325–332

Fritsch S, Guenther F, Wright MN et al (2019) Neuralnet: training of neural networks R package version 1.33

Gholami V, Booij MJ, Nikzad Tehrani E et al (2018) Spatial soil erosion estimation using an artificial neural network (ANN) and field plot data. Catena (giessen) 163:210–218. https://doi.org/10.1016/j.catena.2017.12.027

Gilliams S, Raymaekers D, Muys B et al (2005) Comparing multiple criteria decision methods to extend a geographical information system on afforestation. Comput Electron Agric 49:142–158

Govers G (1991) Rill erosion on arable land in Central Belgium: rates, controls and predictability. Catena (giessen) 18(2):133–155. https://doi.org/10.1016/0341-8162(91)90013-n

Grekousis G (2019) Artificial neural networks and deep learning in urban geography: a systematic review and meta-analysis. Comput Environ Urban Syst 74:244–256

Gudino-Elizondo N, Biggs TW, Castillo C et al (2018) Measuring ephemeral gully erosion rates and topographical thresholds in an urban watershed using unmanned aerial systems and structure from motion photogrammetric techniques. Land Degrad Dev 29 (6):1896–1905. https://doi.org/10.1002/ldr.2976

Haklay M (2010) How good is volunteered geographical information? A comparative study of OpenStreetMap and ordnance survey datasets. Environ Plann B Plann Des 37(4):682–703. https://doi.org/10.1068/b35097

Hillel D (1998) Environmental soil physics. Academic Press, USA

Hoober D, Svoray T, Cohen S (2017) Using a landform evolution model to study ephemeral gullying in agricultural fields: the effects of rainfall patterns on ephemeral gully dynamics. Earth Surf Proc Land 42 (8):1213–1226. https://doi.org/10.1002/esp.4090

Hosseinalizadeh M, Kariminejad N, Rahmati O et al (2019) How can statistical and artificial intelligence approaches predict piping erosion susceptibility? Sci Total Environ 646:1554–1566. https://doi.org/10.1016/j.scitotenv.2018.07.396

Hounkpatin KOL, Schmidt K, Stumpf F et al (2018) Predicting reference soil groups using legacy data: a data pruning and Random Forest approach for tropical environment (Dano catchment, Burkina Faso). Sci Rep 8(1):1–16. https://doi.org/10.1038/s41598-018-28244-w

Jackson P (1998) Introduction to expert systems. Addison-Wesley, Reading, MA

Jankowski P (1995) Integrating geographical information systems and multiple criteria decision-making methods. Int J Geogr Inf Syst 9(3):251–273. https://doi.org/10.1080/02693799508902036

Joshi MV, Kumar V, Agarwal RC (2002) Evaluating boosting algorithms to classify rare cases: comparison and improvements. In: Anonymous 1st IEEE international conference on data mining, ICDM'01, San Jose, CA

Kadavi P, Lee CW, Lee S (2018) Application of ensemble-based machine learning models to landslide susceptibility mapping. Remote Sens (Basel, Switzerland) 10(8):1252. https://doi.org/10.3390/rs10081252

Karlik B, Olgac AV (2011) Performance analysis of various activation functions in artificial neural networks. J Phys: Conf Ser 1237:22030. https://doi.org/10.1088/1742-6596/1237/2/022030

Kirkby MJ (1980) Modelling water erosion processes. In: Kirkby MJ, Morgan RPC (eds) Soil erosion. Wiley, New York, pp 183–216

Knox JC (2001) Agricultural influence on landscape sensitivity in the upper Mississippi river valley. Catena 42(2):193–224

Lagos-Avid MP, Bonilla CA (2017) Predicting the particle size distribution of eroded sediment using artificial neural networks. Sci Total Environ 581–582:833–839. https://doi.org/10.1016/j.scitotenv.2017.01.020

Maimon O, Rokach L (2010) Data mining and knowledge discovery handbook. Springer

Malczewski J (1999) GIS and multicriteria decision analysis. Wiley, Canada

Malczewski J (2004) GIS-based land-use suitability analysis: a critical overview. Prog Plan 62:3–65

McBratney AB, Odeh IOA (1997) Application of fuzzy sets in soil science: fuzzy logic, fuzzy measurements and fuzzy decisions. Geoderma 77(2):85–113. https://doi.org/10.1016/S0016-7061(97)00017-7

Meyer D, Dimitriadou E, Hornik K et al (2019) Package "e1071" Misc functions of the Department of Statistics (e1071)

Montgomery B, Dragićević S, Dujmović J et al (2016) A GIS-based logic scoring of preference method for evaluation of land capability and suitability for agriculture. Comput Electron Agric 124:340–353. https://doi.org/10.1016/j.compag.2016.04.013

Moonjun R, Shrestha DP, Jetten VG (2020) Fuzzy logic for fine-scale soil mapping: a case study in Thailand. Catena (Giessen) 190:104456. https://doi.org/10.1016/j.catena.2020.104456

Mosavi A, Sajedi-Hosseini F, Choubin B et al (2020) Susceptibility mapping of soil water erosion using machine learning models. Water (Basel) 12(7):1995. https://doi.org/10.3390/w12071995

Nadiri AA, Sedghi Z, Khatibi R et al (2017) Mapping vulnerability of multiple aquifers using multiple models and fuzzy logic to objectively derive model structures. Sci Total Environ 593–594:75–90. https://doi.org/10.1016/j.scitotenv.2017.03.109

Ni JR, Li YK (2003) Approach to soil erosion assessment in terms of land-use structure changes. Soil Water Conserv 58:158–169

Oliphant AJ, Spronken-Smith RA, Sturman AP et al (2003) Spatial variability of surface radiation fluxes in mountainous terrain. J Appl Meteorol Climatol 42:113–128

Park SJ, McSweeney K, Lowery B (2001) Identification of the spatial distribution of soils using a process-based terrain characterization. Geoderma 103(3):249–272. https://doi.org/10.1016/S0016-7061(01)00042-8

Patton PC, Schumm SA (1975) Gully erosion, northwestern Colorado: a threshold phenomenon. Geology 3 (2):88–90

Robinson VB (2003) A perspective on the fundamentals of fuzzy sets and their use in geographic information systems. Trans GIS 7(1):3–30. https://doi.org/10.1111/1467-9671.00127

Rokach L, Maimon O (2008) Data mining with decision trees. Theory and applications. World Scientific Publishing Co Pte Ltd.

Rosas MA, Gutierrez RR (2020) Assessing soil erosion risk at national scale in developing countries: the technical challenges, a proposed methodology, and a case history. Sci Total Environ 703:135474. https://doi.org/10.1016/j.scitotenv.2019.135474

Saaty TL (1977) A scaling method for priorities in a hierarchichal structure. J Math Psychol 15(3):234–281

Saaty TL (1980) The analytic hierarchy process. McGraw-Hill, New York

Schields A (1936) Application of similarity mechanics and turbulence research for bed-load transport. Mitt. Preussichen Versuchsanstalt Wasserbau Schiffbau, Berlin

Sela S, Svoray T, Assouline S (2015) The effect of soil surface sealing on vegetation water uptake along a dry climatic gradient. Water Resour Res 51(9):7452–7466. https://doi.org/10.1002/2015WR017109

Shu C, Burn DH (2004) Artificial neural network ensembles and their application in pooled flood frequency analysis. Water Resour Res 40:1–10

Svoray T, Ben-Said S (2010) Soil loss, water ponding and sediment deposition variations as a consequence of rainfall intensity and land use: a multi-criteria analysis. Earth Surf Proc Land 35(2):202–216. https://doi.org/10.1002/esp.1901

Svoray T, Markovitch H (2009) Catchment scale analysis of the effect of topography, tillage direction and unpaved roads on ephemeral gully incision. Earth Surf Proc Land 34(14):1970–1984. https://doi.org/10.1002/esp.1873

Svoray T, Levi R, Zaidenberg R et al (2015) The effect of cultivation method on erosion in agricultural catchments: integrating AHP in GIS environments. Earth Surf Proc Land 40(6):711–725. https://doi.org/10.1002/esp.3661

Svoray T, Michailov E, Cohen A et al (2012) Predicting gully initiation: comparing data mining techniques, analytical hierarchy processes and the topographic threshold. Earth Surf Proc Land 37(6):607–619. https://doi.org/10.1002/esp.2273

Swets JA (1988) Measuring the accuracy of diagnostic systems. Science 240:1285–1293

Tarboton DG (1997) A new method for the determination of flow directions and upslope areas in grid digital elevation models. Water Resour Res 33(2):309–319. https://doi.org/10.1029/96wr03137

Tarboton DG, Bras RL, Rodriguez-Iturbe I (1989) Scaling and elevation in river networks. Water Resour Res 25 (9):2037–2051. https://doi.org/10.1029/WR025i009p02037

Tkach RJ, Simononic SP (1997) A new approach to multi-criteria decision making in water resources. J Geogr Inf Decis Anal 1:25–44

Torri D, Poesen J (2014) A review of topographic threshold conditions for gully head development in different environments. Earth-Sci Rev 130:73–85. https://doi.org/10.1016/j.earscirev.2013.12.006

Tucker CJ (1979) Red and photographic infrared linear combinations for monitoring vegetation. Remote Sens Environ 8:127–150

Urbanski JA (1999) The use of fuzzy sets in the evaluation of the environment of coastal waters. Int J Geogr Inf Sci 13(7):723–730. https://doi.org/10.1080/136588199241085

USDA-SCS (1985) National Engineering Handbook, Section 4—Hydrology. US Soil Conservation Service, USDA, Washington, DC

Vadiati M, Asghari-Moghaddam A, Nakhaei M et al (2016) A fuzzy-logic based decision-making approach for identification of groundwater quality based on groundwater quality indices. J Environ Manage 184(Pt 2):255–270. https://doi.org/10.1016/j.jenvman.2016.09.082

Valmis S, Dimoyiannis D, Danalatos NG (2005) Assessing interrill erosion rate from soil aggregate instability index, rainfall intensity and slope angle on cultivated soils in central Greece. Soil Tillage Res 80:139–147

Vandekerckhove L, Poesen J, Oostwoud Wijdenes D et al (2000) Thresholds for gully initiation and sedimentation in Mediterranean Europe. Earth Surf Proc Land 25(11):1201–1220. https://doi.org/10.1002/1096-9837(200010)25:113.3.CO;2-C

Vincenzi S, Caramori G, Rossi R et al (2006) A GIS-based habitat suitability model for commercial yield estimation of Tapes philippinarum in a Mediterranean coastal lagoon (Sacca di Goro, Italy). Ecol Model 193 (1):90–104. https://doi.org/10.1016/j.ecolmodel.2005.07.039

Wittenberg L, Kutiel P, Greenbaum N et al (2007) Short-term changes in the magnitude, frequency and temporal distribution of floods in the Eastern Mediterranean region during the last 45 years—Nahal Oren, Mt. Carmel, Israel. Geomorphology 84:181–191

Woodward DE (1999) Method to predict cropland ephemeral gully erosion. Catena 37:393–399

Zadeh LA (1965) Fuzzy sets. Inf Control 8:338–353

Zhou P, Luukkanen O, Tokola T et al (2008) Effect of vegetation cover on soil erosion in a mountainous watershed. Catena (Giessen) 75(3):319–325. https://doi.org/10.1016/j.catena.2008.07.010

7 The Health of the Remaining Soil

Abstract

This chapter discusses the formalization of spatial information on soil health, and its implications for the quality of the remaining soil in the wake of erosion processes. This chapter extends the notion of water erosion damage from soil budgets, to comprehensive soil health and the provision of ecosystem services. It describes spatial autocorrelation of soil properties in the Harod catchment, by various methods—Moran's I, Nugget: Sill ratio, and variogram envelope analysis. These are demonstrated using geoinformatics procedures, to show how GIS layers for the stratified random approach are produced. Spatial interpolation techniques—such as Ordinary Kriging, Universal Kriging and Cokriging—are discussed as tools for predicting spatial variation in soil health. The limited ability to scale up soil health mapping from point measurements to large agricultural areas is a major gap in soil research and is also discussed in this chapter. In this regard, this chapter is of major importance scientifically speaking, as it offers a methodology for studying the effect of water erosion on remaining soil in a spatially explicit fashion. From an applied standpoint, it provides farmers and professionals with a tool for estimating the state and dynamics of their field.

Keywords

Indexing · Moran's I · Kriging · Soil-health mapping · Soil properties autocorrelation

Chapter 6 described several sophisticated methods for identifying parts of the catchment that are at risk of water erosion. These methods can advance the study of various water erosion processes, based on data-mining or expert knowledge, but they do not provide information for estimating the state and change of soil properties and the functioning of the remaining soil on-site, after an erosion event occurs.

To address this research gap, the soil sciences community has developed tools of soil assessment. Specifically, for three decades, attention has been given to quantifying soil erosion effects on: Organic matter loss from agricultural fields, the reduction in water quality and quantity, and the decrease of *soil fertility*, productivity, and overall functionality (Doran and Parkin 1994). These efforts have enabled the development of soil indices that are sensitive to changes in soil properties over time, due to natural or human disturbances (Doran 2002; van Es and Karlen 2019).

Soil assessment is a broad topic that covers various aspects of the soil system, and as such, it attracts the attention of several professional groups—including, for example, agronomists,

T. Svoray, *A Geoinformatics Approach to Water Erosion*,
https://doi.org/10.1007/978-3-030-91536-0_7

pedologists, economists, farmers, and decision-makers. The involvement of different communities has resulted in a bewildering assortment of jargons and perspectives of the issue, from which, soil is assessed through four primary measures (Lehmann et al. 2020): (1) *Soil fertility*—which farmers usually use in reference to the soil's support for crop productivity and yield, i.e., the soil's contribution to plant growth which, in many cases, may be assessed by the amount of aboveground biomass; (2) *Soil security*—a measure that highlights the human aspect, and is related to the soil's well-being and soil governance. Inspired by similar measures of human rights, soil security encompasses human culture, capital, and legal aspects of soil management, and refers to the level of access of individuals and groups to soil services, and to soil-related policy; (3) *Soil quality*—a more wide-ranging measure that represents the soil's capacity to function toward an identified management goal. Specifically, soil quality effects have been studied on, for example, water quality, gas exchange, and fauna/flora health within the entire ecosystem; and finally (4) *Soil health*—is a more comprehensive approach, referring to the ecosystem level, that epitomizes the soil role as an infrastructure that supports life. Soil health links agricultural and soil sciences to policy, stakeholder needs, and sustainable supply chain management. Formal definitions of the two last measures and the difference between soil quality and soil health are discussed in Sect. 1.1.3.1.

Soil security and soil fertility are specific measures, and—as in many other studies of soil assessment in agricultural catchments—in this book we shall focus on the other two measures. Since the difference between soil quality (which relates to the soil's capabilities) and soil health (which relates to the soil's condition) is not very large, we shall use the concept and procedures of soil health. Soil health is the preferred among the two because it is the most widely used measure of the soil's function as a provider of *Ecosystem Services* (ES).

But how to quantify soil health? Bünemann et al. (2018) have reviewed numerous applications of soil health and have identified the following as key indicators of soil health: Crop roots and soil organisms, decomposition and element cycling, soil structure, water infiltration, retention and percolation, humus formation and carbon sequestration. These indicators were found reliable by Bünemann et al. to assess soil function in providing the following ES: Biomass production, maintaining agro-biodiversity, reducing soil erosion, pest, disease, and climate regulation.

Soil health indicators can therefore be used to quantify the effect of water erosion not only on the soil layer, soil minerals, and soil loss, but also on the wide range of contributions that an agricultural ecosystem might supply for the benefit and well-being of humans.

The ES concept is commonly used to designate the direct and indirect benefits that (agro) ecosystems provide to the human well-being (Adhikari and Hartemink 2016; Birgé et al. 2016). As such, they attract the attention of researchers in many fields while the variety of ES are grouped into four major categories (families of services), as set out in the Millennium Ecosystem Assessment (2005):

1. *Provisioning Services*—which, in practical terms, are the basis for the other three categories, as they encompass all those services that the ecosystem provides for human consumption—such as fruits and vegetables, drinking water, wood fuel, natural gas, fibers, etc.
2. The *Regulation Services* category refers to ES that moderate natural phenomena—such as flooding, decomposition, toxins control, and carbon storage.
3. *Supporting Services* are the ES that support other ES—such as soil formation, biodiversity, and nutrient cycling.
4. The final category, *Cultural Services*, includes non-materialistic benefits from the ecosystem—such as attention restoration, recreation, and aesthetics.

The way from a soil health score relying on a given soil indicator, to the soil capacity to provide an ES, passes through the level of soil functioning. Indeed, the wide range of soil traits that are used as soil health indicators, have already been linked with soil functioning and then, the soil functioning was linked with ES provision (e.g., Rinot et al. (2019), Kihara et al. (2020)).

Put differently, soil health indicators may say something about soil function, which in turn may indicate the ES that a given area can provide (Fig. 7.1).

The link between *Soil Health Index* (SHI), soil function, and ES, therefore, extends the assessment of the remaining soil beyond the mere quantification of soil mineral movement (soil loss and deposition)—as in the case of the physically based soil loss and soil evolution models described in Chap. 3. Such a paradigm shift is particularly important in agricultural catchments, as it results in a reference to the soil as a system, and as an intrinsic part of the entire agro-ecosystem.

We shall return to this link soil health-soil function-ES in a later stage; in the meantime, let us focus on the practical framework of the development of soil health indexing.

In a seminal work by three USDA researchers, Andrews et al. (2004) suggested a three-step framework for estimating the impact of soil erosion and management practices on soil function:

1. A selection of the most representative indicators;
2. An interpretation of the selected indicators into meaningful scales;
3. An integration of the various indicators into a composite index.

Since its development, the framework put forward by Andrews et al. has been adopted by many soil scientists, and has become a standard for soil health assessment (Moebius-Clune et al. 2017). However, the application of this framework to the hillslope and catchment scales requires quantifying soil variation in space, and for efficient soil conservation planning and implementation, it is necessary to create actual soil health maps.

Mapping soil health can be a challenge, due to logistical problems in measuring the large number of points necessary to represent spatial variation of soils over wide regions, and because the autocorrelation characteristics of soil properties are complex and not fully understood (see Sect. 4.2.2). Nonetheless, previous studies have shown that soil health estimates in points that were not measured (unknown points) can be successfully applied using spatial interpolation techniques at the field level (Wu et al. 2013), hillslope catena (Rosemary et al. 2017) and even at the catchment scale (Schloeder et al. 2001; Svoray et al. 2015a).

The theoretical background of interpolation methods needed for soil health mapping is widely presented in Chap. 4 and therefore will not be repeated here. This chapter shall therefore focus on soil health as a measure of soil function and will outline the science of soil health indices

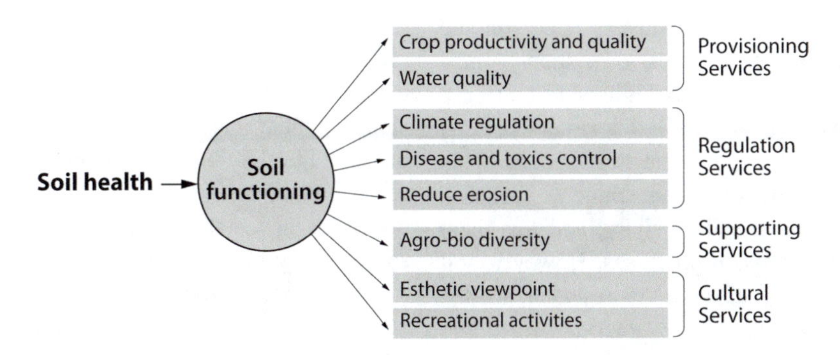

Fig. 7.1 A general schematic framework of the link between soil health, soil functioning, and the four categories of ES. For example, along a single trajectory: Soil health indicates on soil function in maintaining, or even increasing biodiversity, which is directly translated into a supporting service

and their application over agricultural catchments. The chapter begins with a description of the soil properties involved in soil health indexing and the development of soil health indices with minimized redundant data (Sect. 7.1). Then, Sect. 7.2 explores characteristics of *spatial autocorrelation* of soil properties related to soil health indexing. Finally, Sect. 7.3 describes interpolations of soil properties to soil health maps.

Such a multiphase approach to study the damage to the remaining soil characteristics allows supporting the practitioners with exploring underlying processes to stop the degradation process.

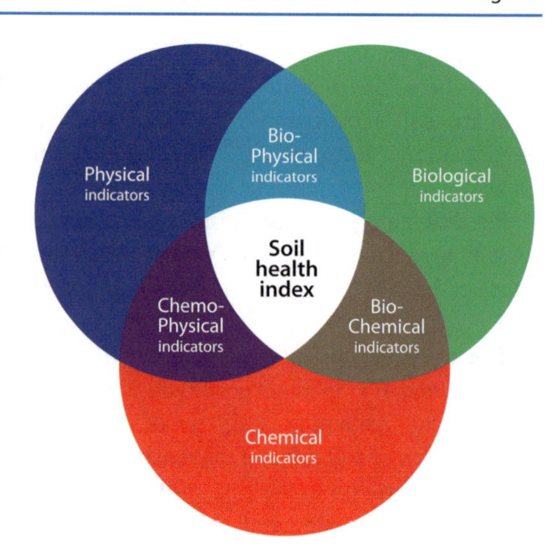

Fig. 7.2 The soil health framework: The three soil traits composited to assess soil health and their combinations (Adapted with permission from: Springer, *Plant & Soil*, Farmer-oriented assessment of soil quality using field, laboratory, and VNIR spectroscopy methods. Idowu, O.J., van Es, H.M., Abawi, G.S. et al. 307, 243–253, 2008, https://doi.org/10.1007/s11104-007-9521-0)

7.1 Soil Health Indicators

In the context of living organisms, the term *health* implies that those with good health function better. Using this analogy, it can be claimed that a soil with good health is expected to function better and to be able to provide the agro-ecosystem with various ES. To test if the soil functions well, a suitable soil health test must consider the effect of chemical, biological, and physical attributes of the soil (Fig. 7.2).

These three intertwined soil traits can be expressed in a quantitative soil health test involving measurements of soil properties that are converted into soil health indicators, with a high-low interval scale based on a process-based rationale. For example, high water holding capacity values increase microbial activity and support diverse soil organisms, store carbon and reduce greenhouse gas emissions, supply nutrients to plants and support high yield and crop quality (Erkossa et al. 2007). In another example, high values of organic matter content may increase the cohesion of the mineral particles, thereby decreasing soil erosivity, and may help regulate future erosion processes. However, soil health scale can also be represented in the opposite—low–high interval—direction. Namely, lower soil property values may indicate greater soil health. For example, high surface-hardness values may indicate a less healthy soil due to decreased aeration, lower infiltration and permeability, and increased surface runoff (Assouline and Mualem 2001).

The conversion procedure from *soil properties* values to *soil indicators* values must be therefore applied very carefully, since not all soil properties are indicative of soil health in the same direction. The synergy between the indicators is, however, crucial because they unravel a number of soil processes to formulate, in conjunction, a composite soil health index (de Paul Obade and Lal 2016). To do so, according to the framework proposed by Andrews et al. (2004), the user first computes soil indicators that are simply conversions of the values of soil properties that were measured in the field (or acquired from remotely sensed data), using scale functions, so that the converted new values indicate the effect of the soil property on soil health.

This general procedure of extraction of soil health indicators from soil properties raises, however, several questions: Which soil properties should be selected to characterize its physical, chemical, and biological traits and may act as the input for soil health indicators? Do any of these soil properties correlate with another, thereby providing redundant information? How can we keep the list of soil properties to be measured to a minimum? Answering these questions, and extracting a composite soil health score, requires the following three steps:

1. A set of soil properties is selected based on their response to changes in soil function, but also according to one's ability to measure, and to map them at the catchment scale. The selection of soil properties may depend on local clorpt factors in the catchment, but also may be context-dependent, based on the specific agricultural use of the catchment and, mainly, the requested ES.
2. *Scoring functions* are computed to convert the original soil property values to normalized values that are transformed based on whether the effect of the soil property on soil health is positive or negative—i.e., whether higher values are indicative of better soil health (as in the case of organic matter, for example), or of worse soil health (e.g., electrical conductivity).
3. A composite scoring function is designed to provide a single score of soil health for each point measurement. This composite score should express the soil health, based on physical, biological, and chemical traits as described in the soil health framework by many authors, e.g., Moebius-Clune et al. (2017).

Finally, note that the development of soil health indices should consider not only pure theoretical and scientific concerns, but also logistical issues—such as limited databases, which often characterize field campaigns of many research studies in the soil sciences (Pulido Moncada et al. 2014), and should also cover economic standpoints, because resources are always restricted (Bastida et al. 2008; Idowu et al. 2009).

7.1.1 Selection of Soil Properties

Many soil properties have been used as a basis for the development of soil health indicators (see examples in Table 7.1). The suitability of a soil property to serve as a soil health indicator depends on the ES that the soil health assessment is referring to. For example, many farmers would be happy to use tests that evaluate soil health for productivity and (even more likely) yield quantity, and quality. However, other users, such as governmental bodies, might wish to establish the soil function with regard to a broader range of aspects of ES—such as regulation services for gas exchange or soil conservation. This may involve other soil properties as soil health indicators than those used to indicate yield. Thus, the user must first decide on the ES they wish to represent, and then select the most suitable soil properties.

Beyond the target ES, the selection of soil properties from Table 7.1, or any other collection of soil properties, must take into consideration two arguments. *First,* the user must explore any inter-correlation that might exist between the selected soil properties, because such correlation may reduce interpretability: When the input properties correlate, coefficients fitted by statistical models become unstable and sensitive to small changes in the data. *Second,* the user usually strives for a shortlist of input soil health indicators, because logistically and economically it is impractical to rely on a long list, such as that described in Table 7.1. Thus, a property selection procedure is required to select a subset of soil properties, choosing only the most easily measurable variable in each group of collinear variables that contain the same information on the soil condition.

Identifying correlation between the soil properties and shortening the list is achieved through the following five statistical procedures:

First, a *correlation matrix* (Fig. 7.3) is computed using Eq. 7.1, based on Pearson correlation coefficient $\rho_{X,Y}$, between all possible pairs of soil properties (X, Y):

Table 7.1 A list of possible soil properties that can be used as soil health indicators. Assembled from previous studies of soil health indicators in different parts of the world: Idowu et al. (2008), Svoray et al. (2015a), Bünemann et al. (2018), Rinot et al. (2019), Lehmann et al. (2020)

Physical properties	Chemical properties	Biological properties
Aggregate stability	Aluminum	Active carbon test
Available Water Capacity (AWC)	Calcium	Beneficial nematode population
Bulk density	CEC	Decomposition rate
Dry aggregate size (<0.25 mm)	Copper	Glomalin content
Dry aggregate size (0.25–2 mm)	Electrical Conductivity (EC)	Microbial respiration rate
Dry aggregate size (2–8 mm)	Exchangeable acidity	Organic Matter content (OM)
Field infiltrability	Iron	Parasitic nematode population
Macro-porosity	Magnesium	Particulate organic matter
Meso-porosity	Manganese	PAR (Pest Assessment Report)
Micro-porosity	Nitrate Nitrogen	Potential mineralizable nitrogen
Penetration resistance at 15 kPa	pH	Root health assessment
Penetration resistance at 45 kPa	Available Phosphorus	Weed seed bank
Residual porosity	Available Phosphorus	Earthworms
Saturated hydraulic conductivity	Sulfur	Microbial biomass
Specific Surface Area (SSA)	Sodium Adsorption Ratio (SAR)	
Surface hardness (penetrometer)	Total Nitrogen	
Soil texture (% clay, silt, sand)	Heavy metals	
Wet aggregate stability (0.25–mm)		
Wet aggregate stability (2–8 mm)		

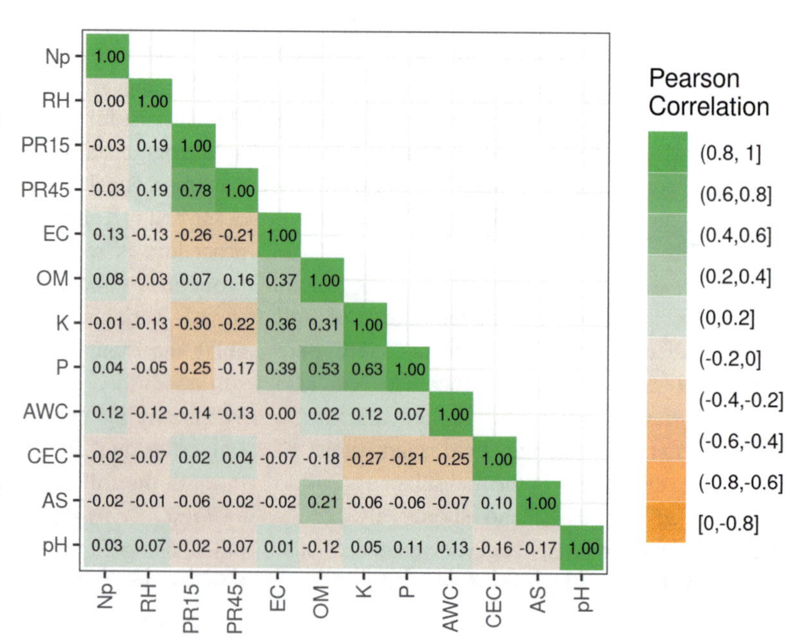

Fig. 7.3 A *correlation matrix* between twelve soil properties in the Harod catchment. The values in the matrix cells are the outcome of Eq. 7.1

$$\rho X,Y = \frac{\mathrm{cov}(X,Y)}{\sigma x \cdot \sigma y}, \tag{7.1}$$

where $\mathrm{cov}(X,Y) = E[(X - m_x) \cdot (Y - m_y)]$ in the numerator is the covariance between *X* and *Y* soil properties, and σ_X, σ_Y in the denominator are the standard deviation of *X* and *Y*, respectively.

The relationships tested are assumed linear and high $\rho_{X,Y}$ value between two soil properties means that the two are fitted. Generally speaking, under such conditions, one of the two soil properties may be eliminated from the soil indexing analysis. This point shall be further clarified in the following steps.

In the *second* procedure, the correlation matrix is transformed into a *dissimilarity matrix* (Fig. 7.4), by subtracting each absolute value achieved in the Pearson matrix from unity (i.e., dissimilarity = $1-|\rho_{X,Y}|$). The dissimilarity matrix expresses the "distance" between pairs of soil properties in terms of association; properties that are highly correlated have a small dissimilarity, or a small "distance".

In the *third* procedure, a *hierarchical cluster analysis* is applied, using the dissimilarity matrix formed in the second procedure. Hierarchical clustering allows similar observations to be grouped into clusters. It can be done using the function *hclust* in R (R Core Team, 2017), for example. The result of the hierarchical cluster analysis is a *dendrogram* (Fig. 7.5)—a diagram representing a tree with branches representing the soil properties under analysis—that shows the hierarchical association between the soil properties. In our case, a diagrammatic representation is used to illustrate the output clusters.

In the *fourth* procedure, the dendrogram is "cut" at a height of 0.45 to produce clusters of soil properties with mutual dissimilarities of <0.45—i.e., correlation coefficients of >0.55. In the example in Fig. 7.5, the dashed line illustrates a 0.45 cutoff point and onward. All soil properties below the dashed line are therefore of a dissimilarity distance of 0.45 or less. In this case, PR15 and PR45 are members of a single cluster (C_3), and K and P are members of a single cluster (C_6). All other properties are independent clusters ($C_1 \ldots C_{10}$). Note that the cutoff point may be user-defined, based on process-based considerations.

In the fifth procedure, a single representative soil property is selected from each cluster of correlated soil properties. This is chosen based

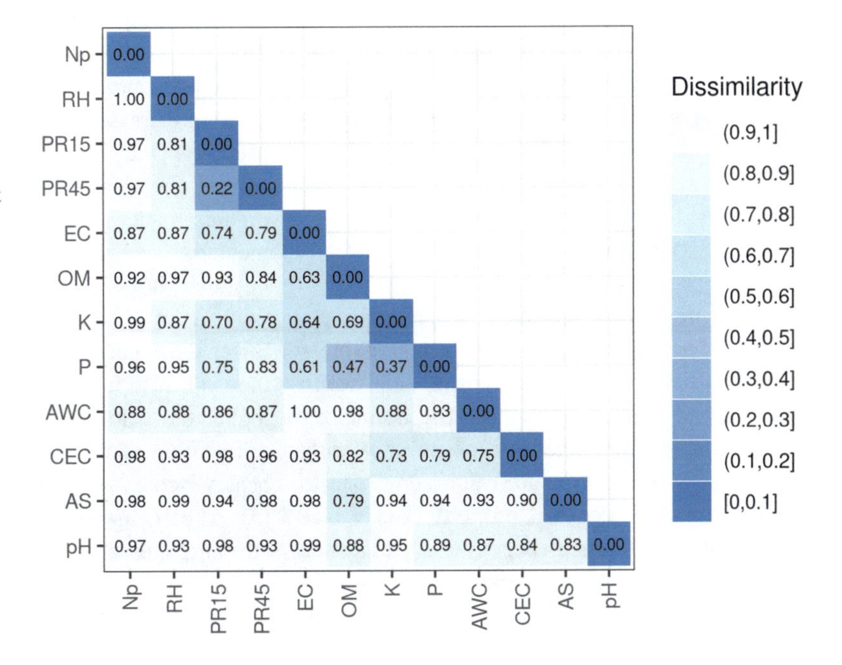

Fig. 7.4 A *dissimilarity matrix* to the correlation matrix in Fig. 7.3. White color represents small distances from similar observation and dark red colors represent the largest differences

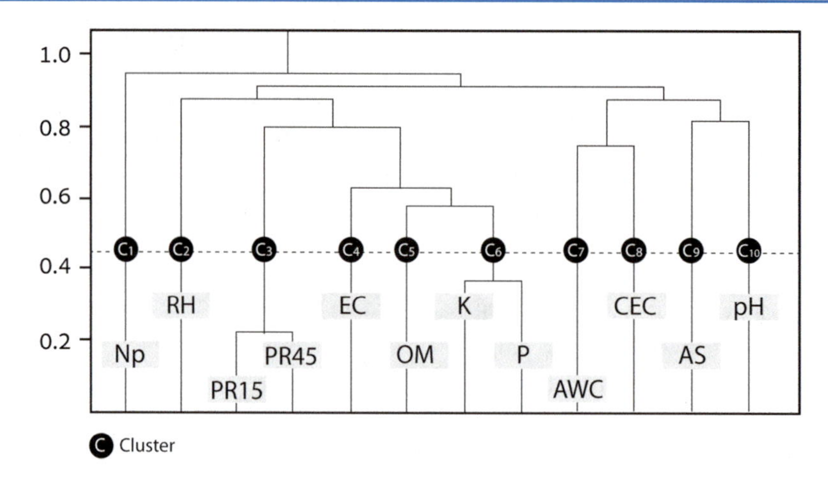

Fig. 7.5 A cluster *dendrogram* applied to various random soil properties of the Harod catchment. The cutoff point assigned in this example is 0.45, and the clusters are determined by the dashed line and the tree branches. The results show that PR15 and PR45 (C_3), as well as K and P (C_6) can be united and only one of these two variables must be used

on ease of reproducibility in the entire catchment. For example, properties that are cheaper to measure—i.e., do not require specialized lab equipment, and the like—may be preferred. As we shall see later, the autocorrelation characteristics of the soil property may also be a consideration—as if it is not autocorrelated in space, it may not be mapped.

In summary, the end result of this five-step process is the selection of the soil properties that best fit for an application as soil health indices, based on inter-correlation between the soil properties. Overlapping results can be removed from the process, and data redundancy is avoided.

7.1.2 Scoring Functions

Once the most suitable soil properties are selected, it is necessary to mathematically express their contribution to a composite soil health index, using normalized values. The normalization process is conducted by means of scoring functions (Idowu et al. 2008) that allow one to convert the measured soil property data values (for example, surface penetration resistance [kPa]) into a dimensionless range—usually between 0 and 100, or 0–10 (Lilburne et al. 2004). The conversion function is chosen based on an expert's interpretation of the effect of the soil property on soil health. Because some soil properties are more important in some environments than in others, the scoring functions of the soil health indices may vary between soil formations with respect to the local soil texture, for example (more details below).

Three types of scoring functions are customarily used to convert soil properties values into the soil health index range (Karlen and Scott 1994): (1) The *More-is-Better* type (upper asymptote curve) simply allows the soil health score to increase in line with the value of the soil property. One example of a suitable soil property for this function is OM [%]: The higher the soil organic matter content, the more the soil is "cemented," and less eroded, and therefore, the healthier it is. (2) The *Less-is-Better* function (with a lower asymptote curve) means that the lower the value of the soil property, the higher the soil health score. Thus, for example, the higher the value of EC [ds·m^{-1}] as a proxy for salinization of the study area, the lower the soil health scores of that area—and vice versa. (3) The *Optimum range function* (Gaussian curve) makes it possible to set a range of values in the original soil property that simulates the optimal conditions for soil health—beyond that point, upper and lower values are set by a continuous decrease of the soil health score. An example of the use of the optimum range strategy using a trapezoid function is the pH with optimum values around the value 7—Values decrease as the soil becomes more alkaline, and similarly they decrease as the soil becomes more acidic.

An illustration of the three scoring function types, based on data from the Harod catchment, is presented in Fig. 7.6.

Equations 7.2 and 7.3 are the two scoring functions based on Cherubin et al. (2016).

$$MIB = \frac{a}{1 + \left(\frac{B-UB}{x-UB}\right)^{s}}, \quad (7.2)$$

$$LIB = \frac{a}{1 + \left(\frac{B-LB}{x-LB}\right)^{s}}, \quad (7.3)$$

where *MIB* and *LIB* are, respectively, the More-is-better and Less-is-better scores of the soil indicator, with 0 as the minimum score and 100 as the maximum score (and the value of *a* in the numerator of Eqs. 7.2–7.3). *B* denotes the soil property value that leads to the score determined by the user as equaling 50; *LB* is the soil property at the lower cutoff point; *UB* is the value of the soil property at the upper cutoff point; *x* is the soil property value; and the s coefficient is set to –2.5 as a default that was set in previous studies (Cherubin et al. 2016). See the location on the graph of *B*, *LB*, and *UB* in Fig. 7.6. For the application of optimum range function, the two equations are merged. Table 7.2 illustrates a simulation of soil properties values converted to soil health indicators.

Previous studies, e.g., Moebius-Clune (2010), Lima et al. (2013), Fine et al. (2017) have shown that scoring functions can vary with soil textural class, and therefore the authors of those studies divided the soils in the sites by textural grouping (e.g., Coarse, Medium, or Fine). The assignment of conversion functions to specific soils can be crucial in cases of soil health mapping at the continental level, or even at the national or regional level—but, since in the case of the Harod catchment, the average clay content was 56%, with very low variance, the sorting of scoring functions by different texture types was not needed.

To fit the conversion of soil properties values to soil health scores based on the local texture conditions, the *Cumulative Normal Distribution* (CND) curve of the soil property within a polygon of a specific soil texture is computed. In this approach, if the soil property follows normal distribution, the analyst computes the CND for the soil property values measured in known points within the confines of an area that is characterized by a specific soil texture. First, a histogram value is established from the observed values in the field, to plot the percentage of observations of each group of values of the soil property. The output bins depict the frequency of the measured values that fall within each range of values. The normal distribution curve is then

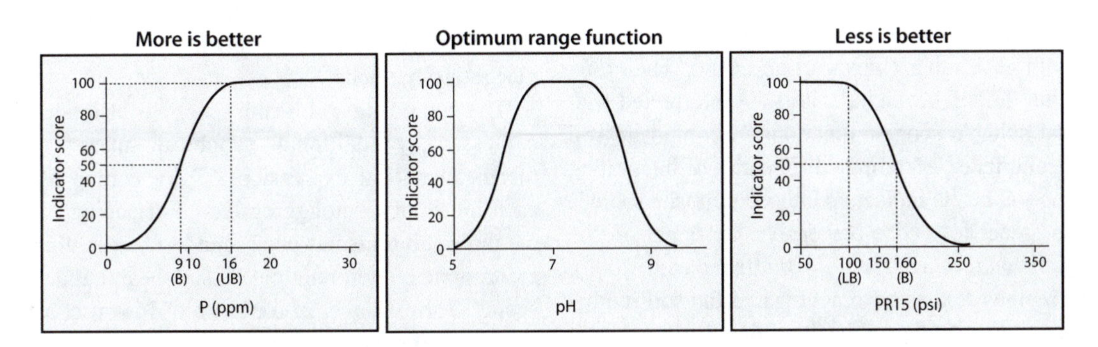

Fig. 7.6 An illustration of the three types of scoring functions for soil health indicators: *More-is-better* (left panel); *Optimum range function* (center panel); and *Less-is-better* (right panel) The values of phosphorus, pH, and surface penetration resistance are extracted from the Harod catchment database, adapted from Hassid (2012)

Table 7.2 Simulation of soil properties values converted to soil health indicator scores using the More-is-better and Less-is-better scoring functions of Eqs. 7.2–7.3

Given values					Computed values	
Soil property range of values	x	B	LB	UB	MIB	LIB
0–30	9	9		16	50	
0–30	5	9		16	24.41	
50–250	160	160	100			50
50–250	193	160	100			25.06

extracted, using the mean μ and standard deviation σ values of each soil property (Moebius-Clune 2010). The mean and std are also used to determine the CND described in Eq. 7.4 by assuming that the soil property values follow normal distribution:

$$p = f(x, \mu, \sigma) = \frac{1}{\sigma \cdot \sqrt{2 \cdot \pi}} \int_{-\infty}^{+\infty} e \cdot \frac{-(x-\mu)^2}{2 \cdot \sigma^2} dx, \quad (7.4)$$

p denotes the probability value in the CND that a soil property value x is placed within the $+\infty$, $-\infty$ range of values and is multiplied by 100 to reach the 0–100 range. Once a CND is established, the soil property values can be converted to the soil health score—either based on CND estimation in the More-is-better case, or 1-CND in the Less-is-better case (Moebius-Clune et al. 2017).

7.1.3 Composite Soil Health Index

The soil is a complex system, whose functioning depends on various processes that may interact with each other (Idowu et al. 2009). Therefore, after the soil health indicators are interpreted with the suitable scoring functions, a composite soil health index is computed. The aim of integrating the soil health indicators into a composite score is to generate a comprehensive measure of the *functional role of the soil* (Bindraban et al. 2000). Previous studies have aggregated the soil indicators into a single soil health index using a simple summation operator. This *Simple Composite Summation Index* (SCSI) is shown in Eq. 7.5:

$$SCSI = \frac{\sum_{i=1}^{n} IS_i}{n}, \quad (7.5)$$

where *IS* denotes the soil health index score of each of the soil properties as computed using the various methods described in Sect. 7.1.2, and i is the index of the specific soil property, ranging from 1 to n soil properties.

However, in many agricultural catchments, the simple summation method may be insufficient, because it (perhaps even unintentionally) attributes the same weight to each soil health indicator. Nevertheless, some indicators may be more important than others in estimating the composite soil health score and the weight of each soil health indicator—while, in this case, the final score may be determined by a *Weighted Composite Summation Index* (WCSI), such as the one in Eq. 7.6.

$$WCSI = \sum_{i=1}^{n} IS_i \cdot W_i, \quad (7.6)$$

where W_i is the weight assigned for each soil health indicator, by using the pairwise comparison matrix exemplified in Sect. 6.2 or by simpler rating methods.

The composite soil health score has also been expressed by additional—graphical—means to the mathematical expressions. Color coding for each indicator or an aggregated color coding for the three soil traits has been found to be useful in incorporated interpretation of soil health index scores. For example, in the work of Idowu et al. (2008), a three-color code of the *Cornell Soil Health Index* (CSHI) was used to represent physical, biological, and chemical indicators.

This was done in addition to the indicators scores that were provided on a percentile rating, so that the distinction between three levels of scores (high, medium, and low) will be clear. Figure 7.7 is the Cornell Soil Health Report for a given field, using twelve soil health indicators.

The mean measured value and its associated score are reported for each soil health indicator. The score report is color coded, so that scores below 3 are colored red; scores between 3 and 8 are colored yellow; and scores above 8 are colored green. The associated soil constraints are added to the low-rate red indicators. The percentile rating of the accumulated indicators ranking is provided in the right-hand column. The final score (in this case 49.2) and its verbal description (in this case low) are also presented.

Another possible graphic representation of soil health is suggested in Halpern et al. (2012), for an *Ocean Health Composite Score* (OHCS). The Halpern et al. graphic representation features three circles—with the outer ring being the groups of properties; the center ring being the properties; and the inner ring showing the sectors, with the attribute score and weight represented by the sector's length and width, respectively.

Ultimately, the purpose of the soil health indicator is to help soil users to manage the soil in a sustainable manner. An aggregated expression of the plethora of soil indices—especially by

Fig. 7.7 An illustration of a summary report for the *Cornell Soil health Test*. The red-green-yellow colors represent the soil health score group for each indicator. The overall soil health score (49.2 in this case) is categorized into groups (Low soil health in this case). The full report could be found in Moebius-Clune et al. (2017), freely available online (Adapted with permission from: Springer, *Plant & Soil*, Farmer-oriented assessment of soil quality using field, laboratory, and VNIR spectroscopy methods. Idowu, O.J., van Es, H.M., Abawi, G.S. et al. 307, 243–253, 2008, https://doi.org/10.1007/s11104-007-9521-0)

CORNELL SOIL HEALTH REPORT

PLOW TILL CORN GRAIN/POTATO ROTATION SILT TEXTURED SOIL	Sample ID: Email: Agent: Drainage: Soil texture:	Date: 4/18/2004 Phone: Slope: 0.2% Soil series:

	INDICATORS	VALUE	RATING	CONSTRAINT	PERCENTILE RATING
PHYSICAL	Aggregate Stability (%)	6.0	4	aeration, infiltration, rooting	
PHYSICAL	Available Water Capacity (m/m)	0.19	4		
PHYSICAL	Surface Hardness (psi)	162	5		
PHYSICAL	Subsurface Hardness (psi)	313	1	Subsurface Pan Deep Compaction	
BIOLOGICAL	Organic Matter (%)	1.5	1	energy storage, carbon sequestration water retention	
BIOLOGICAL	Active Carbon (ppm)	284	1	soil biological activity	
BIOLOGICAL	Potentialiy Mineralizable Nitrogrn (μgN/gdmsoil /week)	1.7	1	N supply capacity, N leaching potential	
BIOLOGICAL	Root Health Rating (1-9)	2.0	9		
CHEMICAL	pH (see chemical test report)	6.0	6		
CHEMICAL	Extractable Phosphorus (see chemical test report)	21.8	10		
CHEMICAL	Extractable Potassium (see chemical test report)	115	10		
CHEMICAL	Minor Elements (see chemical test report)		10		50th percentile ➤ BETTER
	Overall quality score (out of 100)			low	49.2

Ratings on this report are based on generalized crop production standards for New York. For crop specific market interpretations and recommendations see attached chemical test report.

computational means—can serve the users in multi-temporal and spatially explicit monitoring of the catchment status and changes.

7.1.4 Summary

Soil health indicators can be used as tools to assess the remaining soil function of a given ES, after an erosion event. The indicators are based on physical, biological, and chemical soil traits. To transform measured soil properties data into soil health indicators, Andrews et al. (2004) have proposed the following steps:

1. A finite list of soil properties must be selected. The decision as to which soil properties should be chosen may depend on the local conditions of the specific catchment, based on its climatic, land use, and geomorphometric characteristics. The database to be used for the selected indicators can be minimized, by applying the statistical methods described in Sect. 7.1.1, to reduce data redundancy. The soil properties data is usually collected in the field, and analyzed in a soil laboratory by well-established protocols;
2. A scoring function for each soil property must be chosen—among the most common are More-is-better, Less-is-better, and Optimum range, as described in Sect. 7.1.2;
3. Once the functions are chosen and computed, they must be combined into a single score value. This can be done either as an algebraic term, or as a graphic chart, as presented in Sect. 7.1.3.

Once we have an efficient tool for quantifying the soil health of a soil sample from a given point location, we can use geostatistical tools to explore spatial variation of each of the soil health indicators, and finally of the composite score. As this will require an input data of a single value per point (either of a composite score, or a soil health score for a given indicator), the algebraic term is required in this case, rather than the graphic tool, which can be very useful for visual interpretation, but not for geostatistical analysis.

7.2 Autocorrelation in Space

Clearly, if we wish to shift from a composite soil health index measured at a known point into a soil health map, we must be able to quantify spatial variation in soil properties (Winowiecki et al. 2016). This may allow to estimate soil properties at unknown points. However, doing so, is not an easy task, because soil processes operate within complex and intertwined extrinsic (clorpt) and intrinsic soil factors, be they natural or man-made (Sect. 1.1.2). Two notable examples of surface processes that affect soil variation under extrinsic and intrinsic soil factors are runoff flow and soil cultivation.

Runoff transports soil minerals, and chemical/biological soil components downslope, along hillslopes, between fields, catenary units, or subcatchment units, depending on the water flow spatial continuity. Likewise, vertical water flow increases variation in soil properties in the z-dimension. Conventional tillage is a notable example of a man-made process which, as an intrusive action, changes soil structure, and consequently several soil health indicators (Karlen et al. 2008). For example, significant differences were observed in soil particle size between fields cultivated in the Harod catchment by conventional, reduced, and no-tillage practices (Svoray et al. 2015b).

The resulted spatial distribution pattern in soil properties due to these two processes may be continuous along the hillslope (due to overland flow), or discrete and uniform at the field-unit level, where the farmers use the same cultivation method. *Spatial autocorrelation* analyses may help to unravel such complexities, thereby allow to predict soil properties values at unvisited locations. This may make it possible to produce soil health maps at the hillslope and catchment scales.

Spatial autocorrelation occurs whenever values of a (soil) property measured at a given location X, Y are related with this property values of data points in close proximity (Koenig 1999). In that regard, spatial autocorrelation analysis is a quantitative method that provides a mathematical framework to Waldo Tobler's First Law of Geography: "Everything is related to everything else, but near things are more related than distant things" (Tobler 1970). Accordingly, spatial autocorrelation analysis makes it possible to measure the degree to which a set of data points of similar soil properties values tends to cluster together (*positive autocorrelation*), or be evenly dispersed (*negative autocorrelation*) across a given area.

The following two sections provide evidence on spatial autocorrelation of soil properties that can act as soil health indicators. Such information is needed to explore how feasible soil health mapping is at the catchment scale. The first section reviews studies from different parts of the world (Sect. 7.2.1) while the second section illustrates a detailed application of autocorrelation analysis of soil health indicators in the Harod catchment (Sect. 7.2.2).

7.2.1 Spatial Variation in Soil Properties

One rudimentary statistical approach to compute spatial variation in soil properties is the *coefficient of variation* (C_V, Eq. 7.7). This is a standardized measure of dispersion and can be used to compare the variation in soil properties between discrete units, such as agricultural fields or hillslope catenary units.

$$C_v = \frac{\sigma}{\mu}, \quad (7.7)$$

where σ denotes the standard deviation, and μ is the mean of the soil property over the unit. As can be clearly seen from the ratio, the higher the value of C_V, the greater the soil variation. The fact that soil properties in landscape units are highly variable has long been known, with high C_v values being reported in soil properties in agricultural fields of diverse area sizes (Beckett and Webster 1971). Moreover, the authors found that much of the variation was evident even at the very small scale of a few square meters.

The discrete spatial unit approach (Sects. 4.1 and 4.2) has its roots in the biological and geological sciences, and it is based on the premise that the variation in soil properties within a given unit, even if substantial, is smaller than the variation between the units. The discrete model can be defined as in Eq. 7.8 (Heuvelink and Webster 2001):

$$Z(x) = \mu + \sum_{k=1}^{k} \delta_k(x) \cdot \alpha_k + \varepsilon(x), \quad (7.8)$$

where *Z* is the soil property as a random variable *x*; μ is the global mean; *k* is the class varying 1.. *k* within the region; δ_k is a binary function assigned 1 for the stratum of interest and 0 elsewhere; *a* is the mean variation within the class *k*; and the residual ε(x) represents the within-class variation.

The use of descriptive statistics measures to quantify spatial variation in soil properties was further advanced by Burgess and Webster (1980), and many others, after the formulation of kriging interpolation and geostatistics (see Sect. 4.2.2). The geostatistical approach uses mathematical equations to describe the soil stratum as a collection of soil properties whose variation is continuous in space. To this end, we estimate the variable *Z(**x**)* and the semivariance γ(**h**) as follows (Eqs. 7.9–7.10):

$$Z(\mathbf{x}) = \mu + \epsilon(\mathbf{x}), \quad (7.9)$$

$$\gamma(\mathbf{h}) = \frac{1}{2} \cdot E \cdot \left[\{Z(\mathbf{x}) - Z(\mathbf{x}+\mathbf{h})\}^2\right] = \frac{1}{2} \cdot E \cdot \left[\{\epsilon(\mathbf{x}) - \epsilon(\mathbf{x}+\mathbf{h})\}^2\right], \quad (7.10)$$

where we assume that *Z(**x**)* is randomly distributed but also spatially autocorrelated (Heuvelink and Webster 2001); μ denotes the global mean; and ε(**x**) is a random residual. *E* denotes expectation, and ***h*** denotes the point locations—***x*** and ***x* + *h***.

Geostatistics has a profound theoretical foundation (see Sect. 4.2) and has been found to be efficient in quantifying spatial variability in soil properties over large regions, and even at the continental scale. Paterson et al. (2018) estimated spatial variation in soil texture over the entire continent of Australia using empirical semivariograms as continuous functions. Continental scale variograms of texture showed high variability in soil texture, even over short distances. Approximately 50% of the continental scale variability of soil texture was found in as little as 10 km^2, and ~33% of the variability in soil texture was observed in the first 1 km^2.

Nevertheless, more research on the subject reveals that not all soil properties have the same autocorrelation characteristics and spatial distribution pattern. In a farmland in northern Iran—covering an area of 8300 ha with n = 283 points and lag size distances of 200 m—large differences in autocorrelation characteristics were found between Electrical Conductivity (EC), pH, $CaCO_3$, organic carbon, and available potassium (Mousavifard et al. 2013). In this work, the highest variation was observed for EC, and the lowest for pH. The minimum effective *Range* (see Sect. 4.2.2) was 1.5 km for organic carbon, and the maximum was 4 km for available potassium. The autocorrelation characteristics of these soil properties were strongly influenced by environmental factors. This, of course, has influenced sampling efforts, and in that respect, researchers have shown that soil properties should be measured by hundreds of points, at intervals of tens of meters, to cover areas in the size of dozens of hectares (Karlen et al. 2008). In another work, Bogunovic et al. (2017) found that spatial variability was very high for EC, and low for pH in the agricultural fields of the Istria Peninsula in Croatia. However, the best-fit theoretical variogram functions in this work differed between the soil properties. For organic matter, available potassium and available phosphorus best-fit functions were exponential; while for pH and electric conductivity they were spherical and Gaussian, respectively. The Spatial Dependence and Range were high for potassium, and lower for phosphorus. As in many other studies, the use of covariates improved the estimation of the studied soil properties, although the differences between the tested methods were not substantial.

Covariate is an important issue when quantifying variation in soil properties but not always consistent in its effect on the estimation of soil variance. Among the environmental factors that most noticeably affect spatial autocorrelation characteristics in soil properties is the relief: As the topography becomes sharper, the spatial variability in soil health indicators is expected to be greater than that of flatter areas. It is therefore expected that the characterization of spatial variation will require a greater number of samples in heterogeneous areas than in homogenous ones. However, a geostatistical analysis of acidic soils in two hilly regions in India (Chhipa et al. 2019) with respect to three soil properties—pH, EC, and Total Organic Carbon (n = 100)—found that geostatistics can be used successfully to estimate spatial variation in soil health, even based on small number of data points on pH as the key variable of soil health assessment. Thus, the recommendation with regard to the number of points is not straightforward.

The aforementioned studies on spatial autocorrelation in soil properties are summarized in Table 7.3, which provides evidence for the application of the theoretical geostatistical tools presented in Chap. 4 to soil properties, in particular, to those that can act as soil health indicators.

Table 7.3 shows that several soil properties—which may act as soil indicators—show sill value to be considerably larger than the nugget. Large *partial sill* (Δ sill-nugget) is required to interpolate soil health indices into soil health maps. The Range values in different environments have also been found to extend over a considerable distance. From these observations, we may deduce that the application of geostatistical tools to estimate soil health indicators values at unknown points is reasonable. Thus, collecting data points from sites and analyzing them in the lab can indeed serve very well for soil health mapping. Spatial autocorrelation of soil properties is among the most advanced tools for testing how feasible it is to map soil health over wide regions.

Table 7.3 Variogram characteristics of soil properties observed in agricultural catchments in different parts of the world. OC is total Organic Carbon; SWC is Soil Water Content; AK is Available potassium (K); FC is Field Capacity; WP is Wilting Point; AWC is Available Water Capacity; and AP is Available Phosphorus. Note the relatively small N:S ratio, and the large variability in the Range values

Source	Location	Soil property	Sill	Nugget	Range [m]
Mousavifard et al. (2013)	Iran	EC	0.15	0.10	2900
		pH	0.11	0.04	3800
		K	0.28	0.10	4000
Panday et al. (2018)	Nepal	OM	0.57	0.16	4951
		N	0.53	0.16	5209
		P2O5	0.29	0.14	5038
Schloeder et al. (2001)	Sudan	Na	~7.00	~0.17	~20,000
		Ca	~28.00	~12.5	~19,000
		Mg	~5.5	~4.0	~19,000
Gamage et al. (2019)	Canada	OC	0.18	0.14	17.16
		SWC	4.70	3.20	20
Bogunovic et al. (2017)	Croatia	AK	7365.8	0.00	1377
		AP	0.00002	0.000007	545
Iqbal et al. (2005)	N. England	AWC	12	31	93
		FC	60	27	741
		WP	70	18	425
		Ks	1.5	0.46	94

7.2.2 Autocorrelation of Soil Properties in the Harod Catchment—Procedure

Three methods to quantify autocorrelation in the Harod were applied to the soil properties of the catchment: (1) The Moran's I; (2) The variogram envelope; and (3) The N:S ratio. The theory behind these measures is described in Sect. 4.2.2.1 and will therefore not be repeated here. Computation of these measures to the Harod was executed using a four-step procedure.

1. Soil sampling—twelve soil properties that act as soil health indicators were measured at 130 points in the Harod catchment. Point locations were established using the *stratified random technique*. Three GIS layers including soil formations, hillslope gradient, and contributing area made up the strata (Svoray et al. 2015a). Stratified random can provide an alternative to random sampling, which requires many points to represent the spatial distribution pattern of soil properties (McBratney et al. 2003). Thus, when the number of sampling points available is restricted—even at the field scale—stratified random is more efficient than random sampling (McBratney and Webster 1983).
2. Compute semivariograms of the soil properties in the Harod catchment, with the gstat-I automap libraries in R studio (Pebesma 2004) to extract the N:S ratio.
3. Compute Moran's I, using Eq. 7.11 (Moran 1950):

$$I(d) = \left[\frac{1}{W(d)}\right] \frac{\sum_{\substack{i=1 \\ i \neq j}}^{n} \sum_{\substack{j=1 \\ j \neq i}}^{n} w_{ij}(d) \cdot (x_i - \bar{x}) \cdot (x_j - \bar{x})}{\frac{1}{n}\sqrt{\sum_{i=1}^{n} (x - \bar{x})^2}}, \quad (7.11)$$

where $W(d)$—within the given band d—denotes the sum of W_{ij}—the pairs of known points within an explicit distance; $W_{ij}(d)$ is the distance matrix between known points x_i and x_j with values of soil property x at coordinates i and j. $I(d)$ denotes the Moran index, varying between -1 to +1, with values at the extremes indicating high—positive or negative—spatial autocorrelation (Cliff and Ord 1981). Computation was applied with the *Spatial Dependence: Weighting Schemes, Statistics and Models* (SPDEP) library of R—a library that makes it possible to compute other spatial measures, such as Geary's C and Getis-Ord indices (Bivand et al. 2013). See User Manual for Package 'spdep' at: https://cran.r-project.org/web/packages/spdep/spdep.pdf.

4. Variogram envelopes were computed for each of the twelve soil properties from 99 random permutations of the point values, with the corresponding location held constant in independent permutations of the values. In the next step, the semivariogram of the actual data values was located in the graph between the minimum and maximum plots. If the points of short distances from the actual semivariogram depart from the minimal envelope, it is assumed to maintain autocorrelation. The computation was conducted using the geoR library in R studio (https://CRAN.R-project.org/package=geoR), as set out by Ribeiro and Diggle (2020). See more on the theory of variogram envelope and permutations in Chap. 4.

7.2.3 Autocorrelation of Soil Properties of the Harod—Results

C_V values of the twelve soil health indicators in the Harod were computed by Svoray et al. (2015a). The authors have found that the coefficient of variation of Available Water Capacity (AWC), pH, and Root Health (RH) was observed as small ($C_V < 15\%$). The Cation Exchange Capacity (CEC), Aggregate Stability (AS), K, Organic Matter (OM), and Penetration Resistance at 45 cm soil depth (PR45) yielded values of $15\% < C_V < 35\%$. The PR15, P, and Electrical Conductivity (EC) have shown $C_V > 35\%$.

The semivariograms of PR15, P, K, EC, PR45, Np, and AS show a relatively large sill, with wide-range values, and a relatively small nugget effect (Fig. 7.8). The spatial distribution of P shows high variability in the catchment, while its semivariogram is similar to the results of autocorrelation analysis of P values observed by Zhao et al. (2009) in the Yangtze River Delta Region, China. It is also similar to the results observed in the savannas of the Colombian "Llanos" (Jiménez et al. 2011). The semivariogram observed in the case of EC agrees with what was observed by Utset and Castellanos (1999), who studied drainage effects on the autocorrelation of soil EC at a Cauto River Valley vertisol in Cuba, and Schepers et al. (2004) in an irrigated cornfield in Gibbon, Nebraska. Penetration resistance depends markedly on soil water content (Utset and Cid 2001). This may suggest that PR15 is highly variable in the Harod catchment, and possibly biased by the catchment drainage structure that may lead to variation in soil water content between topographic sinks and sources. The K C_v values show moderate variability, and the semivariogram shows autocorrelation in the Harod catchment in a similar fashion as the results by Tesfahunegn et al. (2011) in their soil quality assessment in Northern Ethiopia. PR45—subsurface penetration resistance—depends on soil moisture at lower parts of the soil profile, and therefore the subsurface flow was less continuous with topography than the upper soil layer (PR15) that is much more influenced by the surface runoff flow. In the study of Panday et al. (2018), applied in the agricultural catchment of the Bara district in Nepal, Np was also found to autocorrelate in space but with a much larger sill than nugget and >5000 m Range.

The semivariograms of OM, CEC, AWC, pH, and RH did not show the standard semivariogram curve, while the nugget effect was comparatively high compared with the sill (Fig. 7.8). Spatial variation in OM can be caused by cultivation methods in agricultural catchments. Residue in no-tillage plots can increase OM variability in the long term, and these are probably the reasons for the lack of autocorrelation. As it was for Mueller et al. (2001) with sill = 3.4, nugget = 1.1 applied

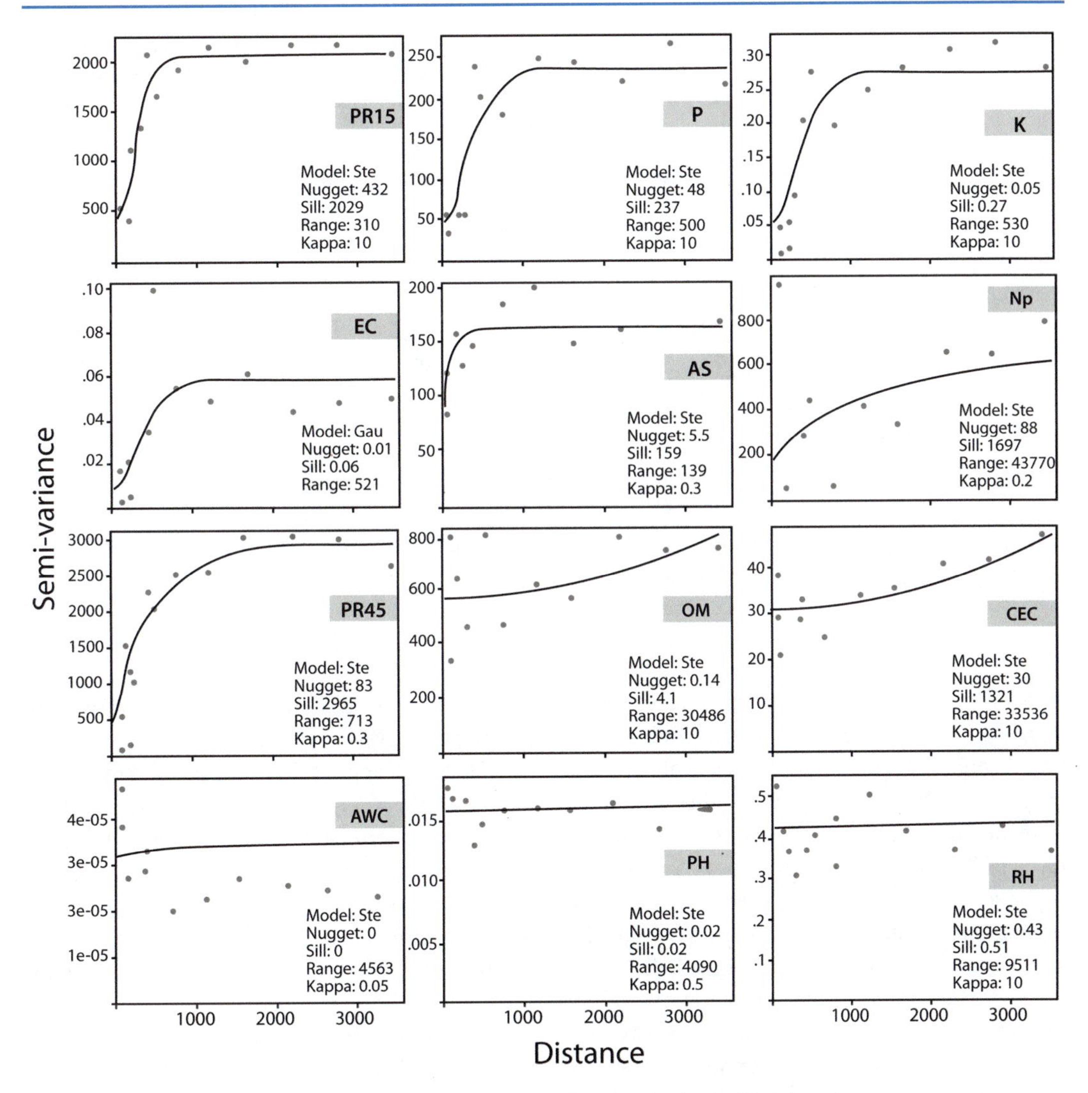

Fig. 7.8 Empirical and theoretical variograms for twelve soil properties in the Harod catchment

in Michigan, CEC was autocorrelated in the Harod. pH was affected by the bedrock with relatively low spatial variability (Dane and Topp 2002). The low variability observed in AWC ($C_V < 15\%$) can be the reason why autocorrelation was low in AWC, despite contradictory results in other studies that did observe autocorrelation in AWC. Another study of a large agricultural region (area of 800 km^2 and n = 113 points) in northern Tunisia (Annabi et al. 2017) showed spatially structured soil aggregate stability at the regional scale, with large sill variograms, and approximately 5 km Range distance. AWC showed low variability in the Harod catchment which was explained by the high (~34%) clay content of the Harod soils. The pH result of the Harod is at odds with previous research that showed spatial autocorrelation of this soil property in other areas (Cambardella et al. 1994; Mueller et al. 2001). Nematode activity was not measured in the research site, so no autocorrelation in RH was measured.

Continuing the semivariograms in Fig. 7.8, Table 7.4 describes a comparative dataset of the three aforementioned measures of spatial

Table 7.4 The results of autocorrelation tests for the Harod catchment. *N:S* is the nugget-sill ratio, *Moran's_I_est.* is the Moran's I estimate, and *Moran's_I_p* is the significance level. The *variogram_env.* column shows cases where the variogram crosses the lower boundary of the envelope, and the Range is in meters

Property		N:S	Moran's I est	Moran's I p	Variogram env	Range [m]
AS	Aggregate stability	0.0344	0.009	0.2555	TRUE	138.97
AWC	Avail. Water capacity	0.7649	0.0355	0.0442	FALSE	4562.67
PR15	Penetration resistance	0.213	0.1783	0	TRUE	309.94
PR45	Penetration resistance	0.028	0.1631	0	TRUE	713.22
RH	Root health	0.8307	0.0096	0.2438	FALSE	9510.88
OM	Organic matter	0.0346	0.0541	0.0063	FALSE	30,485.96
EC	Electrical conductivity	0.1719	0.1265	0	TRUE	520.68
CEC	Cation exchange cap	0.0226	0.2653	0	TRUE	33,535.62
K	Potassium	0.1847	0.2139	0	TRUE	529.72
pH	Acidity	0.9885	0.0355	0.0438	FALSE	4090.31
Np	Nitrogen	0.0518	0.0221	0.0191	FALSE	43,770.34
P	Phosphorus	0.2014	0.1358	0	TRUE	499.74

autocorrelation of the twelve soil health indicators. The *N:S* measure is the nugget to sill ratio of the theoretical variogram of each of the soil properties. The *Moran's_est.* is the Moran's I estimate, and *Moran's_I_P* is the significance level of Moran's I, with values <0.05 representing significant spatial autocorrelation. The *variogram env.* values represent cases where the empirical semivariogram has crossed the lower boundary of 100 permutations iterations of the semivariogram in the catchment. Finally, the Range represents the size, in the distance axis, of the active autocorrelation area.

In keeping with the theoretical semivariograms in Fig. 7.8, PR15, PR45, EC, P, As, CEC, and K crossed the lower boundary of the variogram envelope, and were found to bear significant autocorrelation. The sill value of Np, however, was not sufficiently long to cross the envelope boundary, although the soil property bore a very large difference between the sill and the nugget. All other soil properties that in Fig. 7.8 had unsuitable variograms (OM, pH, AWC, RH) were found to be FALSE in crossing the lower boundaries of the variogram envelope, and therefore bearing insignificant autocorrelation according to the variogram envelope analysis.

Moran's I appeared as a more flexible measure because a larger number of soil properties yielded autocorrelation than through the semivariogram analyses. Indeed, there was agreement about autocorrelation between Moran's I and semivariogram analysis of the PR15, EC, K, and P—however, Moran's I shows significant autocorrelation also in the case of PR45, OM, CEC, and a near-significant autocorrelation in the case of pH and AWC—i.e., despite a high N:S ratio in these two soil properties. Thus, Moran's I is slightly more tolerant than the variogram analyses.

According to Conacher and Dalrymple (1977) *soil-landscape theory*, the physical soil properties are expected to maintain higher autocorrelation, because they are supposed to be linked with the landscape units along the hillslope catena, due to long-term water flow processes. This estimation was found to be valid here only partially, probably due to the effect of soil cultivation that changes the autocorrelation characteristics of the physical (and other) soil properties.

7.2.4 Summary

Previous studies show that many environmental variables are continuous in space and that the semivariograms, that we believe that represent them, show high spatial autocorrelation as well (Webster and Oliver 2007, p. 57). That is, the semivariogram model, in its basic form, can quantify Tobler's law and the estimated relationship between soil properties and distance between measurement points, but, with varying size in the N:S ratio. The results of the Harod catchment, for example, show that semivariograms of some soil properties can capture the spatial variation better than others. But, in the end of the day, the 130 observation points processed here suggest that soil health mapping can be applied in large agricultural catchments. This sample size agrees partially with the results of Webster and Oliver (1992) who found that in the case of a soil property that follows a normal distribution, the semivariogram computed from a sample of 150 data points is *satisfactory*, while an addition of 75 points (up to a total of 225 points) yields *reliable* results.

The application of the three techniques—nugget:sill ratio, variogram envelope, and Moran's I—shows agreements and discrepancies between the measures of autocorrelation. We may divide the twelve soil properties into three groups: (1) Soil properties on which the three methods agree that they maintain spatial autocorrelation—P, PR15 and PR45, K, and EC; (2) Soil properties that the three methods agree that they do not maintain autocorrelation with a single soil property as a member: RH; and (3) Soil properties on which there is partial agreement—Np, AS, OM, CEC, and AWC.

The application of kriging interpolation to RH is therefore in question. In the case of the third group (3), the analyst should be more cautious. If these results are repeated in other regions, the soil properties in this group may benefit from better estimations using the IDW or Thiessen polygons (see Sect. 4.2.3.2). Using d^{-2} (that is $\frac{1}{distance^2}$) as the weight may be more certain than creating empirical weights that do not represent the spatial distribution pattern of this soil property as extracted at a particular spatial configuration of the point measurements. Another insight may be that these soil properties may need more empirical evidence to realistically represent their autocorrelation characteristics.

In future work, therefore, it may be useful to apply the three methods to classify the soil properties into the three groups and in the case of the third group to test if IDW provides with more certain estimations than kriging. It could be also beneficial to apply the autocorrelation tests in a multiscale approach because the results of this data (and others) show that the Range can vary between hundreds to thousands of meters.

The autocorrelation methods described in this section can be applied to various agricultural catchments, whether they are located in semiarid environments or affected by topographic features. The high level of spatial autocorrelation observed in soil properties allows them to be assessed quantitatively with relatively high certainty. The variogram envelope, in particular, suggests soil conservationalists with a useful mapping tool, especially where mapping of soil indices over wide regions is required.

7.3 Soil Health Maps

From the scale of the individual field to the scale of a large catchment, reliable soil health mapping provides the infrastructure of information for implementing spatially explicit soil conservation practices—such as terracing, crop rotation, reduced tillage, and many others—using the *precision agriculture management* concept (Amirinejad et al. 2011). We will further elaborate on this in Chap. 8 of this book.

Kriging interpolation techniques may be used to exploit semivariogram models for soil health mapping. However, even though kriging interpolation techniques were formulated more than fifty years ago (Matheron 1965), there is a notable lack of empirical studies that have applied these methods to soil health mapping. The reason for this gap may be the need for at list ~150 observation points to compute a reliable semivariogram (Webster and Oliver 1992) that is required for extracting the kriging weights. Despite this lacuna in the specific use of kriging for mapping soil health—which has resulted in a notable scarcity in soil health maps—this method has been widely used for mapping various soil properties.

The simplest and most widespread kriging interpolation method is *Ordinary Kriging* (OK). OK has been successfully applied in many soil studies, but the use of covariates, as was previously mentioned, can considerably improve kriging estimates of soil properties. For example, improved estimates by OK were achieved by adding categorical ancillary information on environmental properties (Liu et al. 2006). In Liu's et al. work, soil map information reduced estimation error for sand, silt, and clay contents, pH, and Mehlich-3 Ca. Similarly, Pei et al. (2010) have shown that *Cokriging* that incorporates the *Topographic Wetness Index* (*TWI*) layer achieved better performance than OK—which, in turn, produced better estimates than those of Simple Kriging (SK). Moreover, in a review study by Keskin and Grunwald (2018) of three top journals (*Catena*, *Geoderma*, and *Soil Science Society of America*) in the years 2004–14, the authors found Regression Kriging (RK) as one of the most efficient interpolation techniques for mapping soil properties at multiple spatial scales, in different geographic regions and areas of extent, spatial resolution, target soil properties and sampling design, size, and density and more characteristics.

Section 7.3 compares five interpolation techniques to map soil health in the Harod catchment: OK, Cokriging (CK), *Universal Kriging* (UK) with East–West coordinates surface and the contributing area as drifts and finally the Inverse Distance Weighted (IDW). The East–West direction was chosen due to the effect of rainfall gradient in this direction in the Harod. The case study provides an opportunity to explore the suitability of the different interpolation techniques to soil health mapping in a large catchment.

7.3.1 Procedure

Kriging interpolation techniques were applied for the purposes of this book using the R extension package automap based on the procedure developed in Hiemstra et al. (2009). The automap package makes use of the R package *gstat* developed in Pebesma (2004). IDW was applied using the gstat extension package. The input data are the same 130 point measurements of the 12 soil health properties, located using the stratified random technique described in Sect. 7.2.2.

7.3.2 Results and Discussion

7.3.2.1 RMSE Error

We begin the comparison between the interpolation methods with a very brief reminder of their principles: OK theory assumes that the user does not know the expected value (E or μ) (Webster and Oliver 2007, p. 155); UK is a kriging with an internal drift that is usually applied with the X- or Y-coordinates of the layer. A drift is set when a single or several spatial processes are subject to a trend—i.e., their expected mean is non-stationary (Webster and Oliver 2007, p. 229); Cokriging is an extension of OK that adds to the computation correlated information from ancillary data; and IDW is a deterministic way to use weighted average with the distance as a weight. For more on these interpolation techniques, see Chap. 4.

The validation of these interpolation techniques was applied using the *Leave-One-Out Cross-Validation* (LOOCV) procedure. In this procedure, the user computes RMSE (Root Mean Square Error) values from samples excluded from the databases. These samples act as the validation sets, while all other measurements are used as the training sets for the variogram computation. Table 7.5 presents the RMSE values for the interpolation methods applied to the twelve soil properties. Because RMSE values are a function of the units of the soil property being mapped, the RMSE/Mean was computed to allow a comparison between the methods (see Table 7.6).

Table 7.5 RMSE values of five kriging interpolation methods applied to the twelve soil properties measured in the Harod catchment

	AS	AWC	PR15	PR45	RH	OM	EC	CEC	K	pH	Np	P
OK	13.09	0.00	41.07	50.83	0.63	0.38	0.19	5.56	0.49	0.12	25.05	13.30
UK (X-Axis)	12.94	0.00	41.43	50.96	0.63	0.38	0.19	5.56	0.49	0.13	23.75	13.34
UK (Fa)	13.47	0.00	42.88	51.23	0.63	0.41	0.21	5.74	0.49	0.13	25.05	13.28
Cokriging (Fa)	12.91	0.00	43.14	50.99	0.61	0.38	0.18	4.93	0.47	0.13	27.40	13.27
IDW	13.77	0.01	41.97	49.42	0.64	0.36	0.20	4.62	0.45	0.13	27.53	12.46
Mean	77.63	0.14	104.6	172.2	2.08	1.18	0.55	40.76	1.84	8.23	21.16	19.04

Table 7.6 Normalized RMSE values (RMSE/Mean) for the twelve soil properties. The text inside the parentheses refers to X-Axis = the coordinate surface as a drift; and Fa = flow accumulation matrix as a drift. The gray boxes indicate the *minimal* RMSE among the interpolation methods for each soil property. Note that in the case of pH and AWC the lowest RMSE values are identified by lower decimals than those shown in the table

	AS	AWC	PR15	PR45	RH	OM	EC	CEC	K	pH	Np	P
OK	0.169	0.034	**0.393**	0.295	0.300	0.325	0.343	0.136	0.266	0.015	1.184	0.699
UK (X-Axis)	0.167	0.034	0.396	0.296	0.301	0.325	0.341	0.136	0.268	0.015	**1.122**	0.701
UK (Fa)	0.174	0.035	0.410	0.297	0.301	0.351	0.378	0.141	0.263	0.016	1.184	0.698
Cokriging (Fa)	**0.166**	**0.034**	0.412	0.296	**0.294**	0.321	**0.333**	0.121	0.255	**0.015**	1.294	0.697
IDW	0.177	0.036	0.401	**0.287**	0.307	**0.308**	0.364	**0.113**	**0.244**	0.016	1.301	**0.655**
Mean	0.171	0.035	0.402	0.294	0.301	0.326	0.352	0.129	0.259	0.015	1.217	0.690
"Best" Estimator	Co	Co	OK	IDW	Co	IDW	Co	IDW	IDW	Co	UKx	IDW

Table 7.5 shows that, in comparing the interpolation methods for each soil property, the difference between the RMSE values is relatively small across all soil properties. However, a comparison of the RMSE values with the mean value of each soil property reveals more clear differences between the soil properties. While in some soil properties (e.g., AS, pH) the RMSE is much smaller than the Mean, in others, the RMSE observed is much closer to the Mean (e.g., P) and in the case of Np the RMSE gained by all methods is even larger than the Mean. The RMSE results were obtained using the best-fitting exponential model, lag size, and number configuration. In other words, other variogram configurations produced higher RMSE values.

Table 7.6 shows normalized RMSE values of the twelve soil properties. The gray boxes indicate the *minimal* RMSE among the interpolation methods for each soil property. Note that the differences in normalized RMSE values between the methods can be relatively small in some cases—but, such results, if replicated in more sites, can be used to select the most appropriate method for each soil property. The last record in Table 7.6 ("Best Estimator") indicates the interpolation method that gained the lowest RMSE value for each soil property.

As a follow-up to Tables 7.6 and 7.7 shows the deviations from the Mean value of the normalized RMSE of the five interpolation methods (record 6 in Table 7.6). Such data can help in comparing between the interpolation methods. Deviation values greater than 0 imply that the given interpolation technique (OK, UK (X-Axis), UK (Fa), etc.) provides less accurate estimations than the mean of all interpolation methods—and vice versa.

Table 7.7 shows that, in almost all cases (excluding Np that showed unique autocorrelation characteristics), the Universal Kriging (UK), with the *flowaccumulation* matrix (Fa) as the

Table 7.7 Deviation of Normalized RMSE (RMSE/Mean) of each interpolation technique from the Mean Normalized RMSE values of all five techniques. Positive values mean that the RMSE value cultivated by the interpolation technique is higher than the mean, and negative values mean that it is below the mean

	AS	AWC	PR15	PR45	RH	OM	EC	CEC	K	pH	Np	P
OK	−0.0016	−0.0006	−0.0094	0.0008	−0.0006	−0.001	−0.0088	0.0066	0.0068	−0.0004	−0.033	0.009
UK (X-Axis)	−0.0036	−0.0006	−0.0064	0.0018	0.0004	−0.001	−0.0108	0.0066	0.0088	−0.0004	−0.095	0.011
UK (Fa)	0.0034	0.0004	0.0076	0.0028	0.0004	0.025	0.0262	0.0116	0.0038	0.0006	−0.033	0.008
Cokriging (Fa)	−0.0046	−0.0006	0.0096	0.0018	−0.0066	−0.005	−0.0188	−0.0084	−0.0042	−0.0004	0.077	0.007
IDW	0.0064	0.0014	−0.0014	−0.0072	0.0064	−0.018	0.0122	−0.0164	−0.0152	0.0006	0.084	−0.035

drift, yielded a higher value than mean normalized RMSE values. This means that according to the Harod data, the method overall is consistently less successful than the other methods in estimating the validation set at the known points. The cokriging method was more successful with eight cases of below mean, and four above mean. OK provided similar results to UK, with eight cases below mean, and four above. The two other interpolation techniques produced more mixed results—with UK (X-Axis) seven cases below mean, and five above; and IDW with equal results of six above mean, and six below.

7.3.2.2 Output Layers

The comparison between RMSE values of sample points is an important step in validating interpolation methods. However, exploration of the interpolation maps—though more qualitative in nature—can broaden the perspective to the areal extent, and provide more insights into the performance of the interpolation methods. OK output layers of the twelve soil properties in the entire Harod catchment are presented in Fig. 7.9a, b, c.

As expected, the soil properties create different spatial distribution patterns that can be roughly divided into two main groups.

The first group includes maps with a spatial pattern ranging between two extreme values observed in the map margins, with a gradual change in the soil property values between those two poles. This polarized pattern can be observed through sequences of elongated straight (or sometimes curved) lines, and the gradual change of colors (from deep purple for high estimated value of each soil property, to cardinal red for low value of this soil property). Such a pattern is observed in the case of the CEC map in Fig. 7.9a between the northern margin of the map to the southern margin of it. A polarized pattern means that there is a strong directional influence of the catchment on the soil indicator. In this case, the direction is from north to south, but it can be

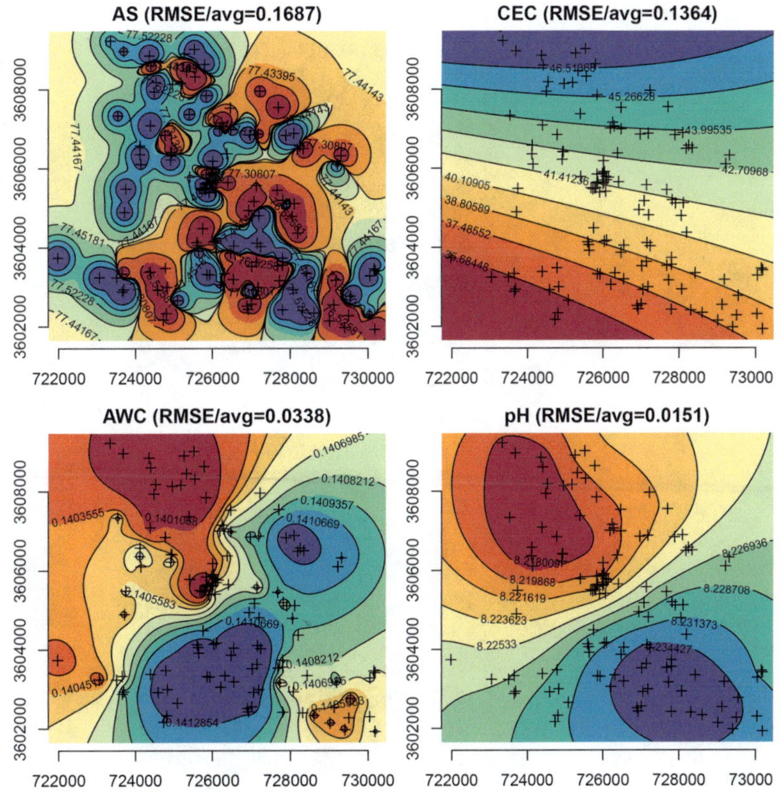

Fig. 7.9 **a** OK maps of the soil properties with lowest RMSE values. **b** OK maps of the soil properties with intermediate RMSE values. **c** OK maps of the soil properties with highest RMSE values

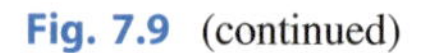
Fig. 7.9 (continued)

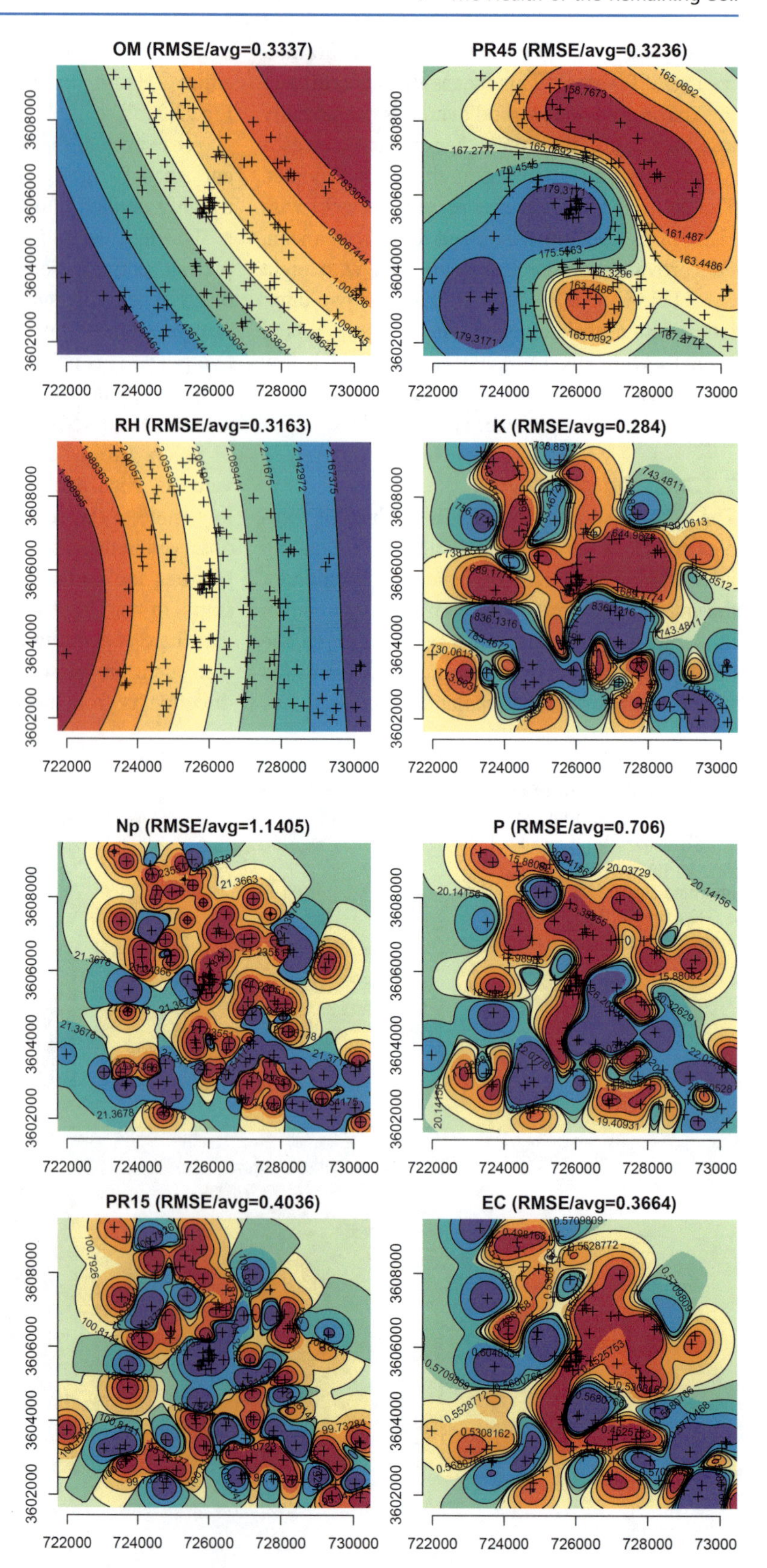

from east to west as observed in the case of the RH map. Members of this group of polarized patterns include five soil properties with CEC, pH and AWC showing a gradual north–south pattern, and RH and OM displaying an east–west pattern.

In the second group, the soil properties (AS, PR45, PR15, K, P, EC, Np) show a mixed/heterogeneous spatial distribution pattern. Four of them (AS, PR45, PR15, EC) display a local *bullseye effect* which is a pattern of circular polygons each wrapped around a known point, or several close points clumped together. Bullseyes are artefacts resulted from too sparsely distributed points to express spatial autocorrelation by the neighboring known points. A map of soil property with few large circles (bullseyes) means that a group of known points of similar values are clustered together (e.g., reddish clusters in AS, EC). Many small circles with different values (e.g., PR15, PR45) mean that the points are notably different from their surroundings and the effect of neighboring points is minimized. Different bullseye patterns are shown in AS vs. Np in Fig. 7.9a, c. The patterns do not differ due to differences in point density, as all soil properties were measured at the same points, but due to differences in autocorrelation characteristics between the soil properties.

A question that arose earlier is whether the use of a covariate may improve the representation of the spatial distribution pattern of the output layers. Figure 7.10a, b, c show the results of a cokriging interpolation procedure, with the topography (contributing area, or in other words, *flowaccumulation* matrix) as a covariate. Previous studies have already shown the effect of clorpt factors, and topography in particular, on the spatial distribution pattern of soil properties. Geomorphometric variables have already been used successfully as covariates in estimating several soil properties in both the hillslope and catchment scale (Bishop and McBratney 2001; Schepers et al. 2004). It has therefore been hypothesized here that an interpolation technique that uses topographic characteristics as covariates may improve soil health mapping in the Harod catchment. Svoray et al. (2015a) tested the effect of topography on the twelve soil properties values using traditional statistics. pH, CEC, and P were related to the DEM original data ($p < 0.05$), and phosphorus, pH, AS, PR15, PR45, EC, K, and CEC were related to the hillslope gradient ($p < 0.05$). Here, cokriging, with the *flowaccumulation* layer as the covariate, increased the accuracy of estimation for OM, CEC, K, P, As, AWC, pH, EC and RH over OK applied without considering the topographic effect. The cokriging maps in Fig. 7.10a, b, c also show a considerable bullseye effect, and the effect of topography is well reflected, especially in the case of the soil properties that have shown gradual change (CEC, pH, AWC, RH, OM).

The purpose of using contributing area data (represented by the *flowaccumulation* matrix) as a covariate is to express the water movement in the catchment in the interpolation process. The assumption behind this strategy is that hydrological sinks and sources may have similar soil properties due to the transportation of soil materials with water. The contributing area layer of the Harod catchment (Fig. 7.11) shows the main hydrological sinks (purple-like tentacles) and hydrological sources (reddish areas). We can see how, to some extent, the main flow lines affect the estimates of the cokriging procedure in Figs. 7.10 and 7.12a UK maps of the soil propertiesSoil properties with lowest RMSE. b UK maps of the soil propertiesSoil properties with indeterminate RMSE values. c UK maps of the soil propertiesSoil properties with highest RMSE values. Note in particular the flow lines of the Harod River in the lower left quarter of the layer.

UK is applied frequently with the X-, Y-coordinates as an internal drift. In the case of Fig. 7.12a, b, c, the X-direction (eastward) was used as the drift, because the rainfall gradient in the Harod catchment extends most frequently eastward from the town of Afula to the Jordan River. The results indeed show a strong bias in most of the maps toward the west, however, the RMSE values predicted with UK and the coordinates drift were lowest among the interpolation methods. The CEC output layer, however, shows a very strong north–south orientation—a rare occurrence, due to UK with X-direction interpolation.

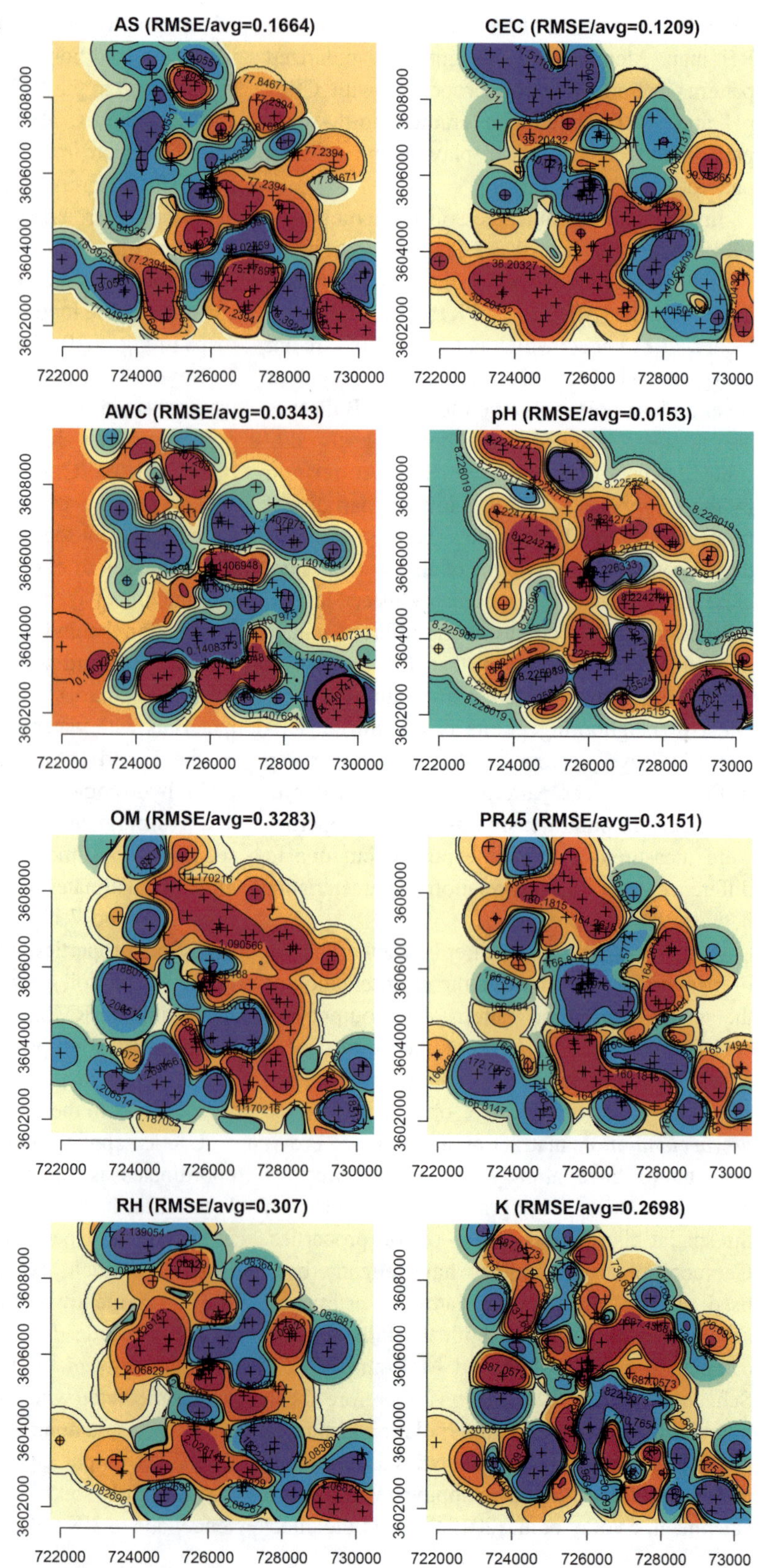

Fig. 7.10 **a** Cokriging maps of the soil properties with lowest RMSE. **b** Cokriging maps of the soil properties with indeterminate RMSE values. **c** Cokriging maps of the soil properties with highest RMSE

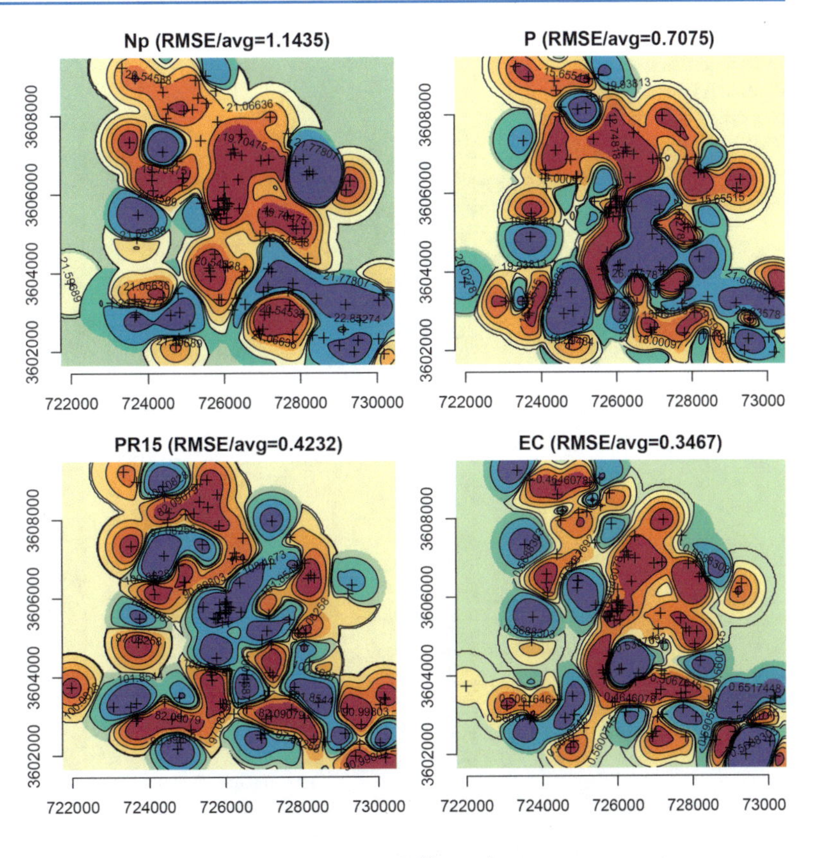

Fig. 7.10 (continued)

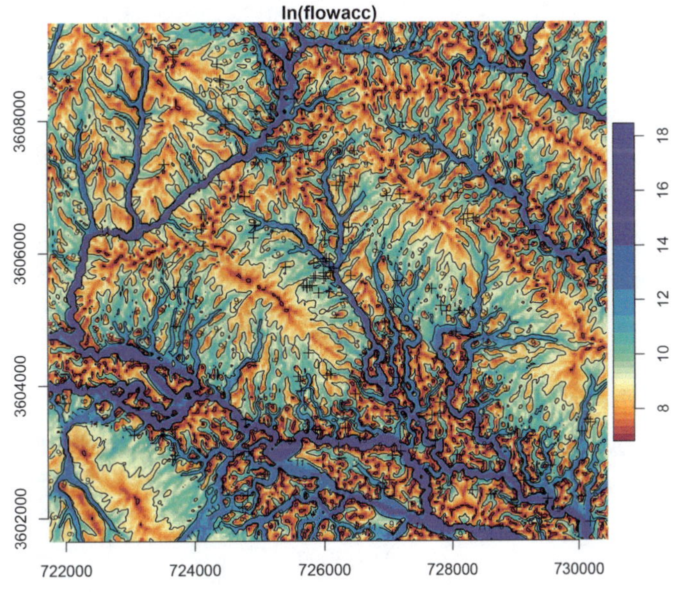

Fig. 7.11 Contributing area layer of the Harod catchment. The matrix was used as the covariate for cokriging. *ln* function was used for visualization purposes. The color scheme is quantile, and the black lines represent *ln flowaccumulation* values of 7, 9, and 11

A clear difference is observed between cokriging and UK maps. In most of the UK maps, the change in values from west to east is apparent, and in many cases with the aforementioned elongated artificial lines. This is most likely not the case in the field, and this

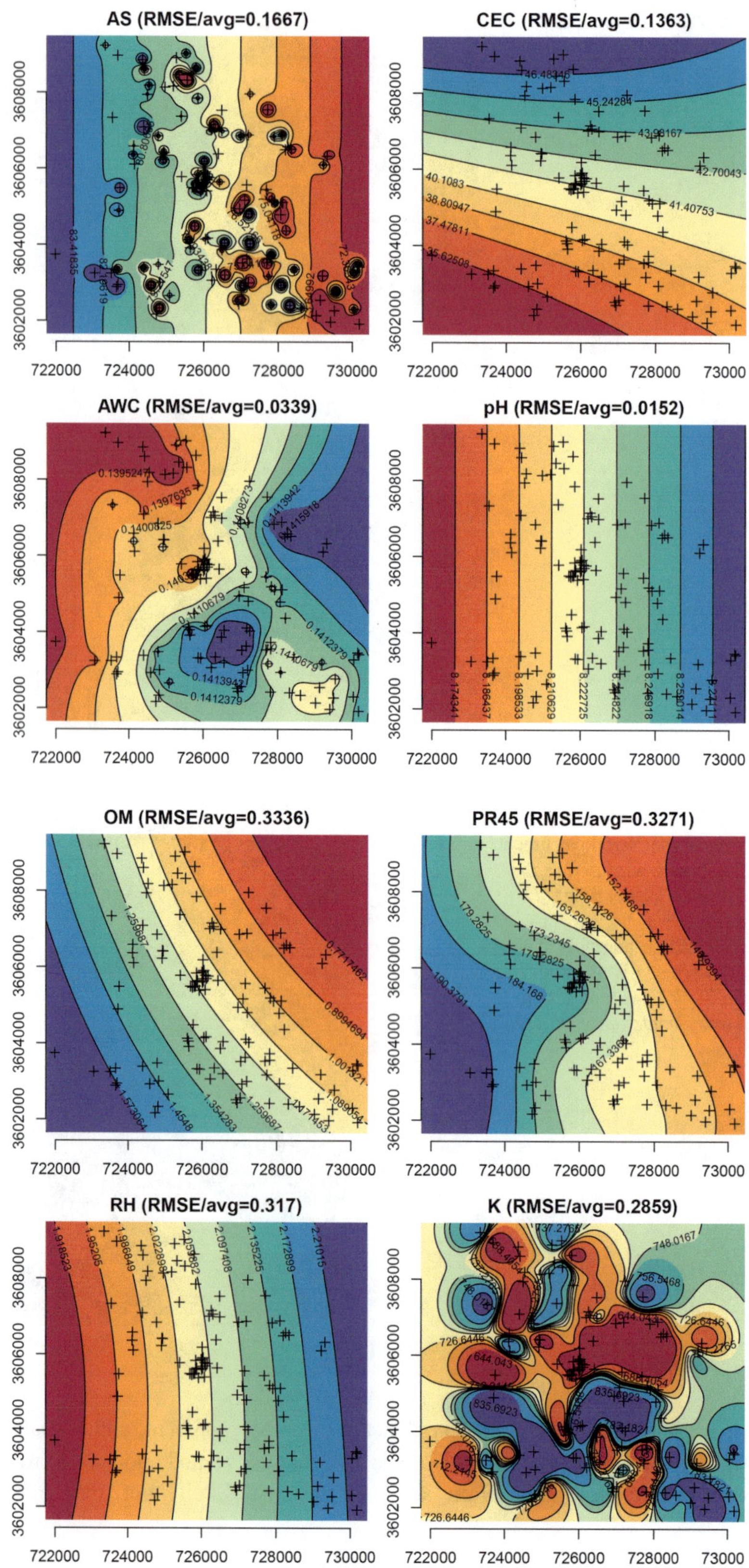

Fig. 7.12 **a** UK maps of the soil properties with lowest RMSE. **b** UK maps of the soil properties with indeterminate RMSE values. **c** UK maps of the soil properties with highest RMSE values

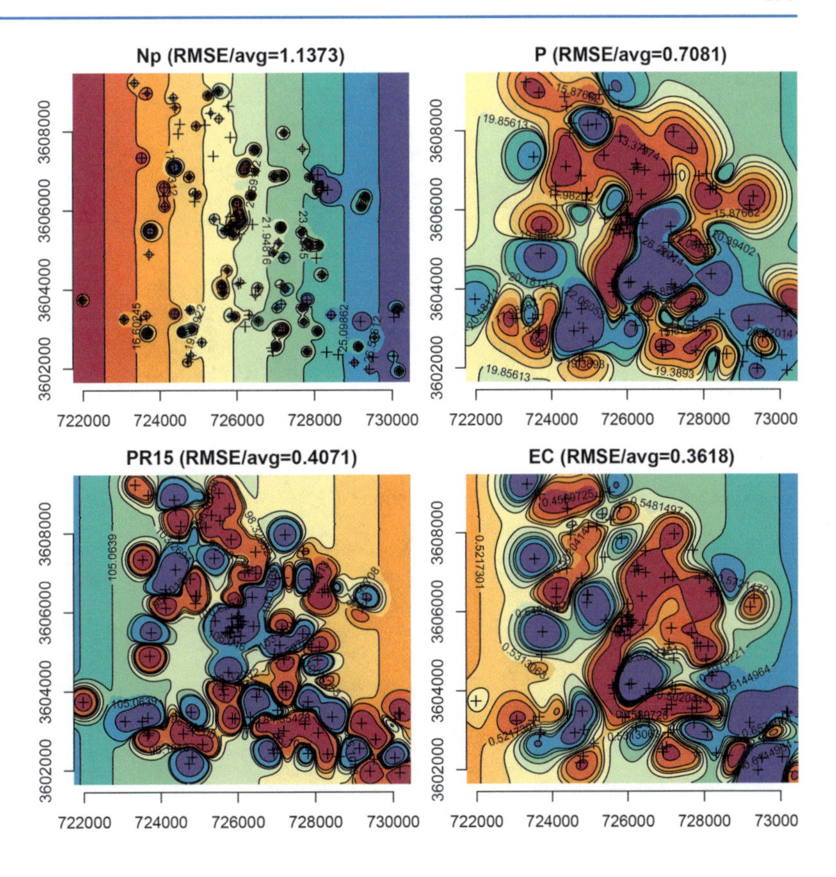

Fig. 7.12 (continued)

discrepancy is reflected in the RMSE values of the UK. However, in many cases we see the bullseye effect that, to some extent, masks the east–west trend. This tendency of the dots on the AS indicator maps to be more clustered suggests that the AS values are slightly more continuous over short distances. Further on, we shall see that the covariance functions quantitatively verify this observation.

IDW computations can provide spatial estimations of the soil properties without the need to compute a semivariogram model. That is because IDW uses distance as a weight to estimate the soil property. Previous studies have shown that in some cases, despite its simplicity, IDW can, on occasion, perform just as well as kriging (Erdogan 2009). Figure 7.13 presents estimates of the twelve soil properties using IDW with d^{-2} as the weights. Note that IDW can also be applied with a weight of d^{-1} or d^{-3} etc. with a try-and-error test to find the weight with the best fit. The results of IDW with d^{-2} were found to be similar to those of the OK maps, but with a more intensive bullseye impact. Note mainly the map of CEC with single points bullseye effect.

The main goal of all these geostatistical analyses is the soil health map—i.e., the interpolation of the composite score to the catchment (Fig. 7.14). To test the effect of color coding on visual interpretation by the analyst, the soil health map is represented here by four color schemes. The results show that interpretation of the final map is not color-scheme dependent and that the four maps show similar spatial patterns.

In all four color schemes, the central section of the Harod shows a large patch of low soil health values that needs more attention, and just below it, a patch of high soil health values. Just below that and to the western section of the catchment, we see a cluster of relatively high values that need no further treatment—and so on, with all four maps displaying similar patterns.

Figure 7.15 shows an IDW map based on the same data. For a detailed planning of

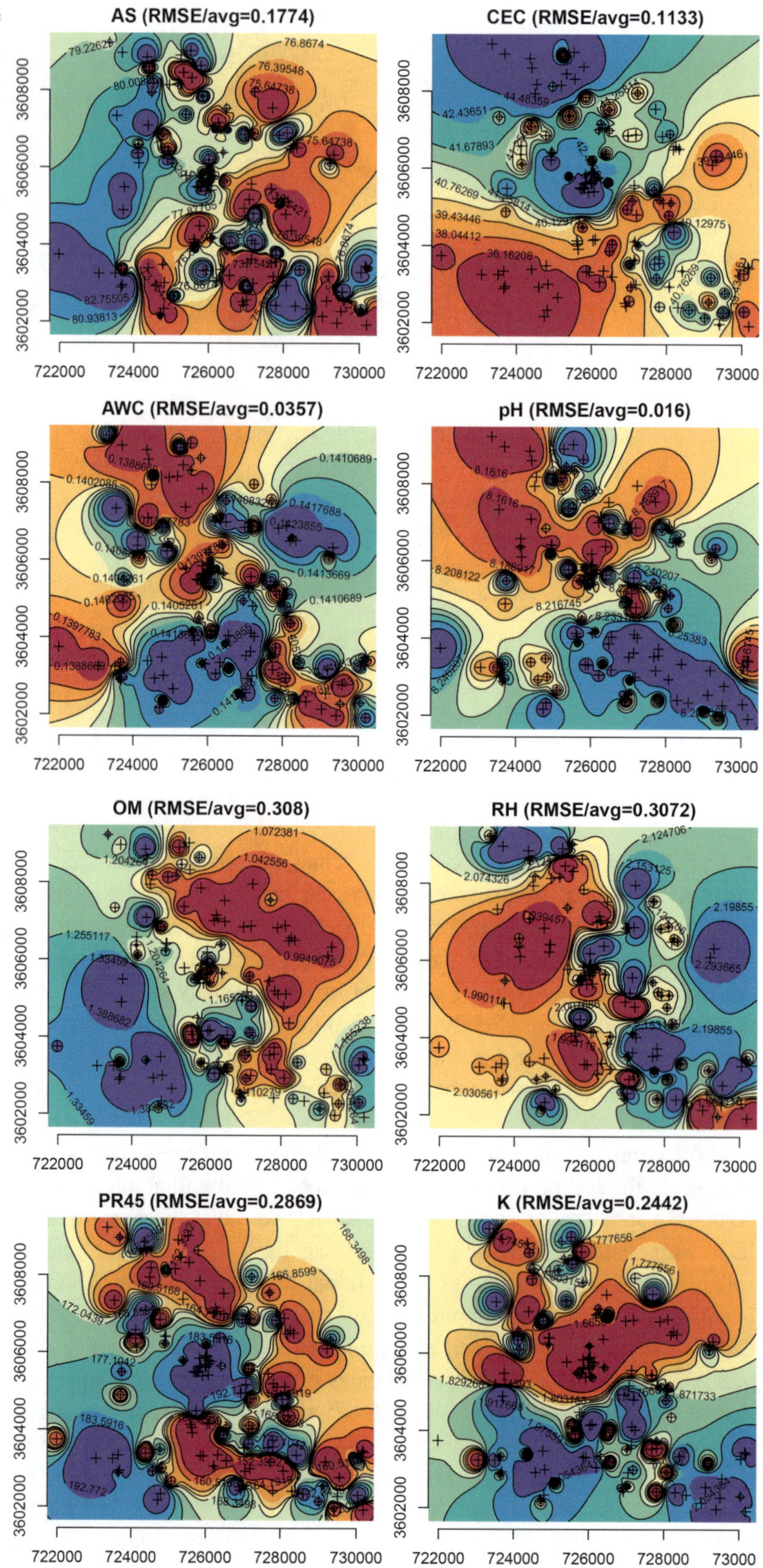

Fig. 7.13 **a** IDW maps of the soil properties with lowest RMSE, **b** IDW maps of the soil properties with indeterminate RMSE, **c** IDW maps of the soil properties with highest RMSE values

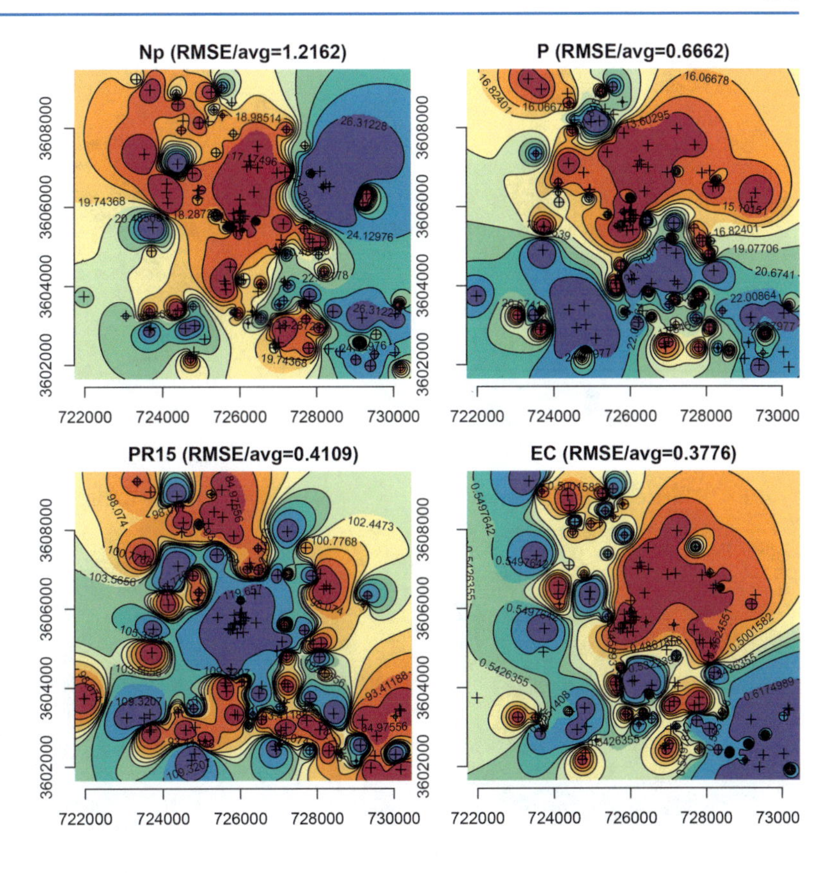

Fig. 7.13 (continued)

conservation actions, we need a more detailed database, but at this scale the results seem similar to the cokriging in Fig. 7.14. Therefore, IDW should not be rejected as an option for mapping soil health if kriging interpolation tools are not available. Thus, although other studies, e.g., Erdogan (2009) have found that the IDW algorithm produced more errors overall than kriging (probably due to its inability to model the abrupt changes in the input variables, that may be common in agricultural areas), in our case the IDW maps of the Harod were found very similar to those produced by OK, and other interpolation methods.

Moreover, the fact that the IDW map depicts the main features of the spatial distribution pattern of soil health in the Harod opens the option to mapping soil health without the need for variogram modeling, or by using GIS software that does not provide advanced geostatistical tools.

7.3.3 Summary

Application of geostatistics to map soil health in the Harod catchment reveals several insights, which may help the reader when testing for the most suitable spatial interpolation procedure for their own catchment. Although catchments may differ in local conditions, which may lead to different results, the following statements hold true:

1. Kriging interpolation is a useful tool for mapping soil health at the field, hillslope, and catchment scales of observation. The various families of kriging testing should be tailored specifically to each site, but overall, the method works, and should be further used and tested. Furthermore, the results by Webster and Oliver (2007) and our results show that most soil properties are continuous in space, so a sufficient number of known points, correctly distributed in space, is needed to express this continuity;

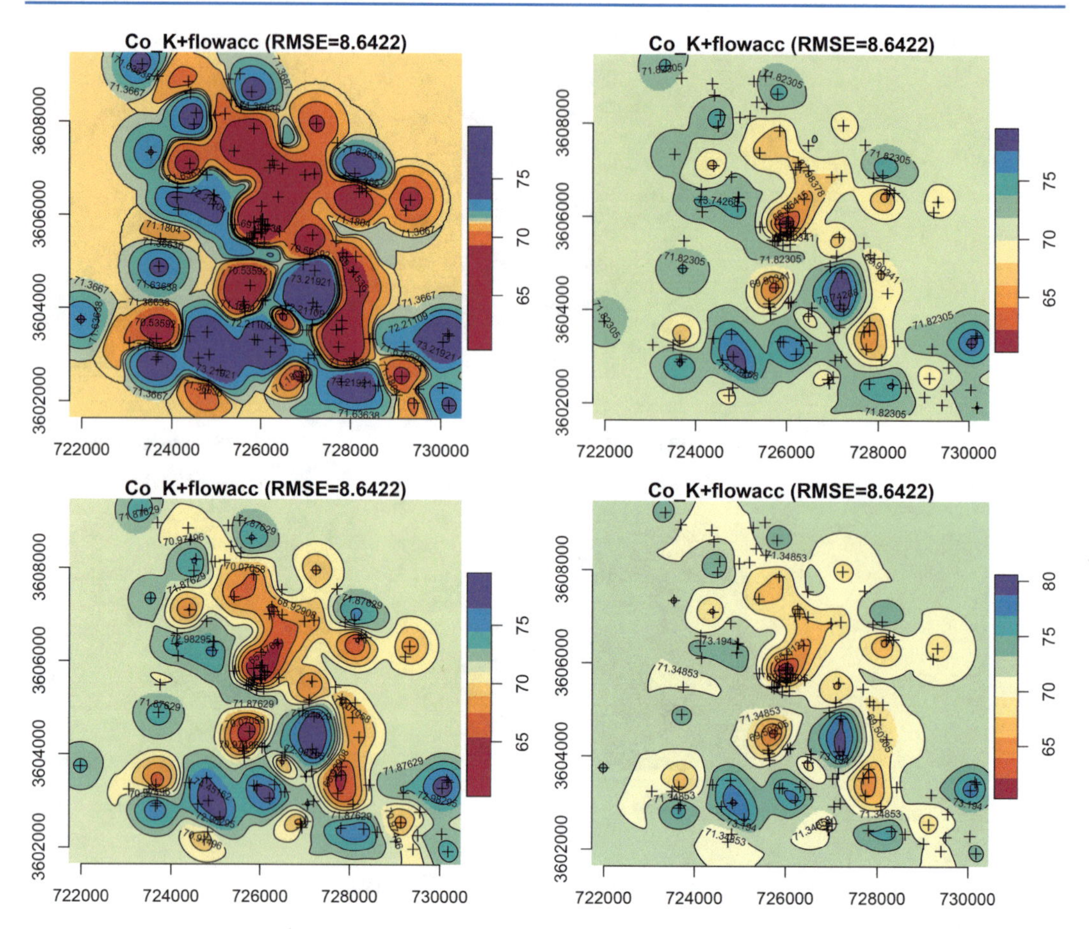

Fig. 7.14 Final soil health score map, rendered by different classification methods. In all cases, the interpolation method is cokriging, with *flowaccumulation* matrix as a covariate

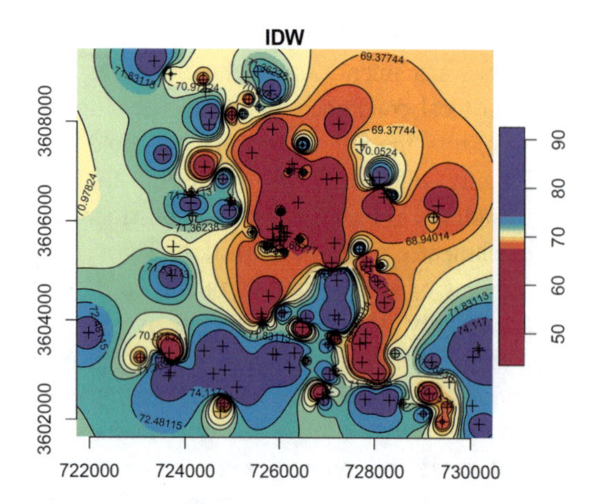

Fig. 7.15 Final soil health score map represented by a quantile color scheme. The interpolation method is IDW, with d^{-2} as the weight function

2. In many cases, the addition of topography as a covariate improves the estimation—but not for all soil properties, because not all soil properties distributions depend on topography. Thus, the use of topography as a covariate must be tested specifically for each soil property;
3. The number of points for the kriging interpolation is crucial. According to Webster and Oliver (1992) the user needs at least 150 data points for satisfactory results, and 225 for reliable data. In this case, we applied 130 points—a number that was found to be sufficient for the Harod;
4. IDW can achieve reasonable results—on occasion, even very similar to those of OK

and other kriging interpolation techniques. Thus, when modeling variograms is not available due to insufficient number of points or software tools, the user must not exclude IDW from the toolbox.

The main problem with the use of kriging interpolation for soil health mapping may be the fact that it is a demanding task (due to the cost of measuring many points), and the difficulties encountered with a formal investigation of the autocorrelation analysis. As a result, there is a notable lack of evidence regarding the spatial interpolation of soil health indicators, and soil health mapping is scarce, so the understanding of soil health in the landscape scale is overlooked. However, with the emergence of geoinformatics tools, more studies have been conducted, and it is reasonable to expect that mapping soil health across wide regions will attract much more attention in the next decade.

7.4 Summary

Water erosion processes increase the pressure on agricultural soils—resulting in accelerated soil degradation, and reduced provision of Ecosystem Services (ES). One commonly explored ES is the provision of crop yield because food security is a critical problem in a world with a growing population subject to global events and climatic changes.

Soil health indicators offer a platform for quantifying the relationship between soils as multifunctional entities and ES as supporters of well-being through the assessment of soil function. Soil health indicators may indicate the soil status and changes after water erosion events, and can be also used to monitor the efficiency of soil conservation actions in mitigating soil degradation (Andrews et al. 2002). The added value of soil health indicators in doing so is the fact that they are based on a comprehensive approach that expresses the physical (Castellini et al. 2019), chemical (Bünemann et al. 2018), and biological (Doran and Zeiss 2000; Bastida et al. 2008) traits of the soil when attesting to the soil function.

Mapping soil health over large regions is required by practitioners for the management of agricultural catchments (de Paul Obade and Lal 2016), because sustainable agricultural catchments entails consistent soil conservation treatment. For the interpolation of such maps from point measurements of soil properties, there is a crucial need to (1) Locate the point measurements in a spatially representative manner (Nortcliff 2002); and (2) To select a suitable interpolation method to estimate the soil properties in unknown points. Previous studies have clearly shown that soil properties can be mapped over large regions using both geostatistical paradigms and remotely sensed data (Forkuor et al. 2017; Hartemink and Minasny 2014; Liu et al. 2006; Marchetti et al. 2012).

One burning question is how certain is the information of the indices when they are interpolated in space to create maps—and consequently, how reliable are those maps for real-world applications? Spatial interpolation depends on the autocorrelation characteristics of the soil properties that make up the soil health index. If a given property is discontinuous in space, its interpolation may be uncertain. This is why empirical evidence on the autocorrelation of soil properties is so critical to soil health mapping. Our study in the Harod (Svoray et al. 2015a), and previous studies (e.g., Webster and Oliver (2007)) show that most soil properties are indeed continuous in space—but for applied drainage engineering and soil conservation purposes, more evidence may be very effective in determining local conditions of sill, Range and nugget characteristics of soil properties.

Mapping time series of soil health scores may make it possible to test if specific soil-management tools change the soil's function. The use of non-intrusive cultivation methods, for example, may be found to outperform the use of conventional cultivation methods. Thus, it is expected that in the fields that have made the transition to less intrusive methodology, the soil health score will improve. Therefore, the merit of mapping soil health, if done routinely, provides an ongoing monitoring of the remaining soil in the catchment.

The wealth of interpolation methods available today allow for spatial interpolation of soil properties with relatively high certainty. In many cases (though not in all), the use of a covariate may increase efficiency in estimating soil properties in unknown points. This assertion can be tested for local conditions: While in hilly catchments, topographic characteristics as a covariate can be expected to improve estimations; in catchments with small, heterogeneous fields, a layer of cultivation method will do so.

Finally and most importantly, recent studies show that indeed soil health may act as a principal tool to achieve sustainability goals and it is not merely a tool for local decisions by the farmer. Moreover—albeit, with some difficulties—soil health may be defined as a tool for quantitative assessment of the ecosystem functioning and the provision of various ecosystem services for human consumption (Lehmann et al. 2020). The strong relationship between soil properties and processes, and ecosystem services and human security and social needs (Baveye et al. 2016), highlights the importance of studying the effect of water erosion on the remaining soil.

Review Questions

1. Compute a *dissimilarity matrix* for the database below, and determine which of the variables correlates with each other to the degree that one of them can be excluded from the *soil health index*. Which one of them would you choose to exclude, and why?

AS	OM	pH	P
51.98	0.00	7.9	0.77
74.33	0.34	7.8	1.53
101.04	0.87	8.4	3.37
98.04	0.68	8.1	2.29
89.07	0.03	7.9	0.99
67.04	0.98	8.3	4.15
79.05	0.77	8.1	2.77

2. Convert the values of the four soil properties in question 1, as though they need to serve in calculating a soil health index. Do this by using *MIB* in the case of AS, OM, and P and *LIB* in the case of pH. You can decide on the values of UB (in the case of MIB) and LB (in the case of LIB) in accordance with the description in Fig. 7.6.
3. Compute a composite soil health score to reflect crop production from three soil properties that you choose from the database in question 1, by using the soil health scores extracted from question 2. The weights are divided equally between the three soil properties (0.3 for each soil health indicator).
4. In this simple example, would you decide on different weights if you had to determine a soil health index based on other ecosystem services? Give an example and explain your answer.
5. Rank into three groups the autocorrelation level of the soil properties in Table 7.4 based on the *N:S ratio*, and identify which of the soil properties belongs to each group.
6. Use *Moran's I* to compute autocorrelation in the table below. Based on the result, which of the two properties autocorrelate in space, and which do not? Explain why.

Point	X	Y	OM	AS
A	0.7	1.0	0.00	51.98
B	1.25	3.0	0.34	74.33
C	2.5	3.7	0.87	101.04
D	3.3	2.75	0.68	98.04
E	4.0	4.0	0.03	89.07
F	3.8	1.0	0.98	67.04
G	1.1	4.4	0.77	79.05

7. In Table 7.6, we see that cokriging outperforms other methods in estimating the soil properties at unknown points. Why do you think this happened? What are its implications for future use?
8. Which of the clorpt variables can create a geographic trend (e.g., from east to west, or north to south), and which kriging

interpolation technique can express this trend in the interpolation process?

Strategies to address the Review Questions

1. Computing a *dissimilarity matrix* can be applied as follows (Sect. 7.1.1): First, a *correlation matrix* is computed (Fig. 7.3, using Eq. 7.1). Then, using a simple subtraction (dissimilarity = $1-|\rho_{X,Y}|$) a dissimilarity matrix is computed (Fig. 7.4).

Correlation Matrix:

	AS	OM	pH	p
AS	1			
OM	0.85997	1		
pH	0.982438	0.854435	1	
p	0.899461	0.829241	0.90888	1

Dissimilarity Matrix:

	AS	OM	pH	p
AS	0			
OM	0.14003	0		
pH	0.017562	0.145565	0	
p	0.100539	0.170759	0.09112	0

To determine which of the variables correlates with other variables to the degree that one of them can be excluded from the *soil health index*, we apply a hierarchical cluster analysis using the dissimilarity matrix. Creating a *dendrogram* (Fig. 7.5) allows to set a cutoff point which produce clusters of soils properties that only one representative of them should be included in the index.

2. Converting the values of the four soil properties in question 1 into soil health indicators is applied using Eqs. 7.2–7.3 (Sect. 7.1):

$$MIB(\mathrm{OM})\frac{a}{1+\left(\frac{B-UB}{x-UB}\right)^{s}}=\frac{100}{1+\left(\frac{0.49-0.98}{0.34-0.98}\right)^{-2.5}}=72.91$$

$$MIB(\mathrm{AS})=\frac{a}{1+\left(\frac{B-UB}{x-UB}\right)^{s}}=\frac{100}{1+\left(\frac{76.51-101.04}{67.04-101.04}\right)^{-2.5}}=30.66$$

$$MIB(\mathrm{P})=\frac{a}{1+\left(\frac{B-UB}{x-UB}\right)^{s}}=\frac{100}{1+\left(\frac{2.46-4.15}{1.53-4.15}\right)^{-2.5}}=25.05$$

$$LIB(\mathrm{pH})=\frac{a}{1+\left(\frac{B-LB}{x-LB}\right)^{s}}=\frac{100}{1+\left(\frac{8.1-7.8}{7.9-7.8}\right)^{-2.5}}=93.97$$

3. Computing a composite soil health index to reflect crop production from three soil properties can be done using Eq. 7.6, Sect. 7.1.3. The given weights are used to compute the *Weighted Composite Summation Index* (WCSI) for three soil properties (AS, OM, P), where IS_i, are the soil indicators computed in question 2 and W_i are the given equal weights:

$$WCSI=\sum_{i=1}^{n} IS_i \times W_i = 72.91\cdot 0.3+33.90\cdot 0.3+35.82\cdot 0.3 = 42.789$$

4. The key to answer question 4 is the assumed importance of each of the indicators in computing the composite soil health index. For example, if you consider AS to be more important than OM or vice versa. If you think that the importance of the indicators is

unequal—rank them, rate them, or use the *pairwise comparison matrix method* (Sect. 6.2.2) to extract the weights that quantitatively represent the importance of the indicator to the composite index.

5. *N:S ratios* for the 12 soil properties is presented in Table 7.4. The N:S ratio values can be divided into three groups based on the scale of the ratios. According to such approach, we can subject the soil properties into the following groups:

   AS, PR45, OM, CEC, Np; N:S < 0.1
   PR15, EC, K, P; 0.1 < N:S < 0.5
   AWC, RH, pH; 0.5 < N:S
6. *Moran's I* can be computed using Eq. 7.11. Note that you compute separately the index for each of the two soil properties. The weights—wi = d^{-1} based on the distance between the points that can be computed using the *Pythagorean theorem* between all points located at the given X-, Y-coordinates. Create a d^{-1} matrix between all points, where the x_i and x_j are the values of the soil property at location i and j, respectively.
7. Cokriging improves estimates for a poorly sampled variable by using a more widely sampled variable (the covariate). The estimated variable and the covariate should be highly correlated (Sect. 7.3.2.1). Yet, a covariate can also add an estimation error. To assess the contribution of the covariate, empirical testing is needed. The development of new RS platforms may substantially advance the extraction of useful covariate data.
8. The following clorpt model components (Sect. 7.3.2.2) have previously shown geographical trends and have already been used for cokriging interpolation as layers of trend: Rainfall amount, topography (elevation, hillslope gradient, etc.), and vegetation cover.

References

Adhikari K, Hartemink AE (2016) Linking soils to ecosystem services—a global review. Geoderma 262:101–111. https://doi.org/10.1016/j.geoderma.2015.08.009

Amirinejad AA, Kamble K, Aggarwal P et al (2011) Assessment and mapping of spatial variation of soil physical health in a farm. Geoderma 160(3):292–303. https://doi.org/10.1016/j.geoderma.2010.09.021

Andrews SS, Karlen DL, Mitchell JP (2002) A comparison of soil quality indexing methods for vegetable production systems in northern California. Agr Ecosyst Environ 90(1):25–45. https://doi.org/10.1016/S0167-8809(01)00174-8

Andrews SS, Douglas KL, Cambardella CA (2004) The soil management assessment framework: a quantitative soil quality evaluation method. Soil Sci Soc Amer J 68(6):1945–1962

Annabi M, Raclot D, Bahri H et al (2017) Spatial variability of soil aggregate stability at the scale of an agricultural region in Tunisia. Catena 153:157–167. https://doi.org/10.1016/j.catena2017.02.010

Assessment ME (2005) Ecosystems and human well-being 5:563. United States of America: Island press

Assouline S, Mualem Y (2006) Runoff from heterogeneous small bare catchments during soil surface sealing. Water Resour Res 42(12):W12405. https://doi.org/10.1029/2005WR004592

Bastida F, Zsolnay A, Hernández T et al (2008) Past, present and future of soil quality indices: a biological perspective. Geoderma 147(3):159–171. https://doi.org/10.1016/j.geoderma.2008.08.007

Baveye PC, Baveye J, Gowdy J (2016) Soil "ecosystem" services and natural capital: critical appraisal of research on uncertain ground. Front Environ Sci 4(41)

Beckett PHT, Webster R (1971) Soil variability: a review. soils fert 34(1):1–15

Bindraban PS, Stoorvogel JJ, Jansen DM et al (2000) Land quality indicators for sustainable land management: proposed method for yield gap and soil nutrient balance. Agr Ecosyst Environ 81(2):103–112. https://doi.org/10.1016/S0167-8809(00)00184-5

Birgé HE, Bevans RA, Allen CR et al (2016) Adaptive management for soil ecosystem services. J Environ Manage 183(Pt 2):371–378. https://doi.org/10.1016/j.jenvman.2016.06.024

Bishop TFA, McBratney AB (2001) A comparison of prediction methods for the creation of field-extent soil property maps. Geoderma 103:149–160

Bivand RS, Pebesma EJ, Gömez-Rubio V (2013) Applied spatial data analysis with R, 2nd edn. Springer, New York

Bogunovic I, Pereira P, Brevik EC (2017) Spatial distribution of soil chemical properties in an organic farm in Croatia. Sci Total Environ, 584–585:535–545
Bünemann EK, Bongiorno G, Bai Z et al (2018) Soil quality—a critical review. Soil Biol Biochem 120:105–125. https://doi.org/10.1016/j.soilbio.2018.01.030
Burgess TM, Webster R (1980) Optimal interpolation and isarithmic mapping of soil properties: I the semi-variogram and punctual kriging. J Soil Sci 31(2):315–331
Cambardella CA, Moorman TB, Novak JM et al (1994) Field-scale variability of soil properties in central Iowa soils. Soil Sci Soc Am J 58(5):1501–1511
Castellini M, Stellacci AM, Barca E et al (2019) Application of multivariate analysis techniques for selecting soil physical quality indicators: a case study in long-term field experiments in Apulia (southern Italy). Soil Sci Soc Am J. https://doi.org/10.2136/sssaj2018.06.0223
Cherubin MR, Karlen DL, Cerri CEP et al (2016) Soil quality indexing strategies for evaluating sugarcane expansion in Brazil. PLoS One 11(3):e0150860. https://doi.org/10.1371/journal.pone.0150860
Chhipa V, Stein A, Shankar H et al (2019) Assessing and transferring soil health information in a hilly terrain. Geoderma 343:130–138. https://doi.org/10.1016/j.geoderma.2019.02.018
Cliff AD, Ord K (1981) Spatial processes: models & applications. Taylor & Francis, London
Conacher AJ, Dalrymple JB (1977) The nine unit landsurface model and pedogeomorphic research. Geoderma 18(1):127–144. https://doi.org/10.1016/0016-7061(77)90087-8
Dane JH, Topp GC (2002) Methods of soil analysis: part 4 physical methods, 5.4. SSSA Book Series, Madison, WI
de Paul OV, Lal R (2016) Towards a standard technique for soil quality assessment. Geoderma 265:96–102. https://doi.org/10.1016/j.geoderma.2015.11.023
Doran JW (2002) Soil health and global sustainability: Translating science into practice. Agr Ecosyst Environ 88(2):119–127. https://doi.org/10.1016/S0167-8809(01)00246-8
Doran JW, Parkin TB (1994) Defining and assessing soil quality. Def Soil Qual Sustain Environ (35):3–21
Doran JW, Zeiss MR (2000) Soil health and sustainability: managing the biotic component of soil quality. Appl Soil Ecol 15(1):3–11. https://doi.org/10.1016/S0929-1393(00)00067-6
Erdogan S (2009) A comparision of interpolation methods for producing digital elevation models at the field scale. Earth Surf Proc Land 34(3):366–376. https://doi.org/10.1002/esp.1731
Erkossa T, Itanna F, Stahr K (2007) Indexing soil quality: a new paradigm in soil science research. Aust J Soil Res 45(2):129–137. https://doi.org/10.1071/SR06064
Fine AK, van Es HM, Schindelbeck RR (2017) Statistics, scoring functions, and regional analysis of a comprehensive soil health database. Soil Sci Soc Am J 81(3):589–601. https://doi.org/10.2136/sssaj2016.09.0286
Forkuor G, Hounkpatin OKL, Welp G et al (2017) High resolution mapping of soil properties using remote sensing variables in south-western Burkina Faso: a comparison of machine learning and multiple linear regression models. PloS one 12(1):e0170478. https://doi.org/10.1371/journal.pone.0170478
Gamage DNV, Biswas A, Strachan IB (2019) Spatial variability of soil thermal properties and their relationships with physical properties at field scale. Soil Tillage Res 193:50–58
Halpern BS, Longo C, Hardy D et al (2012) An index to assess the health and benefits of the global ocean. Nature 488(7413):615–620. https://doi.org/10.1038/nature11397
Hartemink AE, Minasny B (2014) Towards digital soil morphometrics. Geoderma 230–231:305–317
Hassid I (2012) Soil quality assessment over wide regions, Ben-Gurion University of the Negev
Heuvelink GBM, Webster R (2001) Modelling soil variation: past, present, and future. Geoderma 100(3):269–301. https://doi.org/10.1016/S0016-7061(01)00025-8
Hiemstra PH, Pebesma EJ, Twenhofel. CJW et al (2009) Real-time automatic interpolation of ambient gamma dose rates from the Dutch radioactivity monitoring network. Comput Geosci 35:1711–1721
Idowu OJ, Van Es HM, Abawi GS et al (2008) Farmer-oriented assessment of soil quality using field, laboratory, and VNIR spectroscopy methods. Plant Soil 307(1/2):243–253. https://doi.org/10.1007/s11104-007-9521-0
Idowu OJ, van Es HM, Abawi GS et al (2009) Use of an integrative soil health test for evaluation of soil management impacts. Renew Agric Food Syst 24(3):214–224. https://doi.org/10.1017/S1742170509990068
Iqbal J, Thomasson JA, Jenkins JN et al (2005) Spatial variability analysis of soil physical properties of alluvial soils. Soil Sci Soc Am J 69(4):1338–1350
Jiménez JJ, Decaëns T, Amézquita E et al (2011) Short-range spatial variability of soil physico-chemical variables related to earthworm clustering in a neotropical gallery forest. Soil Biology & Biochemistry 43:1071–1080
Karlen DL, Scott DE (1994) A framework for evaluating physical and chemical indicators of soil quality. In: Anonymous defining soil quality for a sustainable environment SSSA Special Publishing, Madison, WI, pp 53–72
Karlen DL, Tomer MD, Neppel J et al (2008) A preliminary watershed scale soil quality assessment in north central Lowa, USA. Soil Tillage Res 99(2):291–299. https://doi.org/10.1016/j.still.2008.03.002
Keskin H, Grunwald S (2018) Regression kriging as a workhorse in the digital soil mapper's toolbox. Geoderma 326:22–41

Kihara J, Bolo P, Kinyua M et al (2020) Soil health and ecosystem services: lessons from sub-Sahara Africa (SSA). Geoderma 370(114342)

Koenig WD (1999) Spatial autocorrelation of ecological phenomena. Trends Ecol Evol 14(1):22–26

Lehmann J, Bossio DA, Kögel-Knabner I et al (2020) The concept and future prospects of soil health. Nature Reviews Earth & Environment 1(10):544–553

Lilburne L, Sparling G, Schipper L (2004) Soil quality monitoring in New Zealand: interpretative framework. Agric Ecosyst Environ 104(3):535-544

Lima ACR, Brussaard L, Totola MR et al (2013) A functional evaluation of three indicator sets for assessing soil quality. Appl Soil Ecol 64:194–200. https://doi.org/10.1016/j.apsoil.2012.12.009

Liu TL, Juang KW, Lee DY (2006) Interpolating soil properties using kriging combined with categorical information of soil maps. Soil Sci Soc Am J 70 (4):1200–1209

Marchetti A, Piccini C, Francaviglia R et al (2012) Spatial distribution of soil organic matter using geostatistics: A key indicator to assess soil degradation status in central Italy. Pedosphere 22(2):230–242

Matheron G (1965) Les variables régionalisées et leur estimation: une application de la théorie des fonctions aléatoires aux sciences de la nature

McBratney AB, Mendonça Santos ML, Minasny B (2003) On digital soil mapping. Geoderma 117(1):3–52. https://doi.org/10.1016/S0016-7061(03)00223-4

McBratney AB, Webster R (1983) How many observations are needed for regional estimation of soil properties? Soil Sci 135(3):177–183. https://doi.org/10.1097/00010694-198303000-00007

Moebius-Clune BN (2010) Development and evaluation of scoring functions for integrative soil quality assessment and monitoring in western Kenya. In: Applications of integrative soil quality assessment in research, extension, and education, Cornell University

Moebius-Clune BN, Moebius-Clune DJ, Gugino BK et al (2017) Comprehensive Assessment of Soil Health, Third Edition edn. Cornell University, New York

Moran PA (1950) Notes on continuous stochastic phenomena. Biometrika 37(1):17–23. https://doi.org/10.2307/21332142.JSTOR2332142

Mousavifard SM, Momtaz H, Sepehr E et al (2013) Determining and mapping some soil physico-chemical properties using geostatistical and GIS techniques in the Naqade region. Iran. Archiv Agron Soil Sci 59 (11):1573–1589. https://doi.org/10.1080/03650340.2012.740556

Mueller TG, Pierce FJ, Schabenberger O et al (2001) Map quality for site-specific fertility management. Soil Sci Soc Am J 65(5):1547–1558

Nortcliff S (2002) Standardisation of soil quality attributes. Agr Ecosyst Environ 88(2):161–168. https://doi.org/10.1016/S0167-8809(01)00253-5

Panday D, Maharjan B, Chalise D et al (2018) Digital soil mapping in the Bara district of Nepal using kriging tool in ArcGIS. PloS One 13(10):e0206350

Paterson S, Minasny B, McBratney A (2018) Spatial variability of Australian soil texture: a multiscale analysis. Geoderma 309:60–74. https://doi.org/10.1016/j.geoderma.2017.09.005

Pebesma EJ (2004) Multivariable geostatistics in S: The gstat package. Comput Geosci 30(7):683–691

Pei T, Qin CZ, Zhu AX et al (2010) Mapping soil organic matter using the topographic wetness index: a comparative study based on different flow-direction algorithms and kriging methods. Ecological Indicators 10:610–619

Pulido Moncada M, Gabriels D, Cornelis WM (2014) Data-driven analysis of soil quality indicators using limited data. Geoderma 235–236:271–278. https://doi.org/10.1016/j.geoderma.2014.07.014

Ribeiro PJ, Diggle JPJ (2020) geoR: analysis of geostatistical data. Vienna, r Package Version 1(7–5):1

Rinot O, Levy GJ, Steinberger Y et al (2019) Soil health assessment: a critical review of current methodologies and a proposed new approach. Sci Total Environ 648: 1484–1491. https://doi.org/10.1016/j.scitotenv.2018.08.259

Rosemary F, Vitharana UWA, Indraratne SP et al (2017) Exploring the spatial variability of soil properties in an Alfisol soil catena. Catena 150:53–61

Schepers AR, Shanahan JF, Liebig MA et al (2004) Appropriateness of management zones for characterizing spatial variability of soil properties and irrigated corn yields across years. Agron J 96(1):195–203

Schloeder CA, Zimmerman NE, Jacobs MJ (2001) Comparison of methods for interpolating soil properties using limited data. Soil Sci Soc Am J 65(2):470–479

Svoray T, Hassid I, Atkinson PM et al (2015) Mapping soil health over large agriculturally important areas. Soil Sci Soc Am J 79(5):1420–1434. https://doi.org/10.2136/sssaj2014.09.0371

Svoray T, Levi R, Zaidenberg R et al (2015) The effect of cultivation method on erosion in agricultural catchments: Integrating AHP in GIS environments. Earth Surf Proc Land 40(6):711–725. https://doi.org/10.1002/esp.3661

Tesfahunegn GB, Tamene L, Vlek PL (2011) Catchment-scale spatial variability of soil properties and implications on site-specific soil management in northern Ethiopia. Soil Tillage Res 117:124–139

Tobler W (1970) A computer movie simulating urban growth in the Detroit region. Econ Geogr 46:234–240

Utset A, Castellanos A (1999) Drainage effects on spatial variability of soil electrical conductivity in a vertisol. Agric Water Manag 38:213–222

Utset A, Cid G (2001) Soil penetrometer resistance spatial variability in a Ferralsol at several soil moisture conditions. Soil and Tillage Research 61(3–4):193–202

van Es HM, Karlen DL (2019) Reanalysis validates soil health indicator sensitivity and correlation with long-term crop yields. Soil Sci Soc Am J 83(3):721–732

Webster R, Oliver MA (1992) Sample adequately to estimate variograms of soil properties. J Soil Sci 43 (1):177–192

Webster R, Oliver MA (2007) Geostatistics for environmental scientists, 2, ed. Wiley, England

Winowiecki L, Vågen TG, Massawe B et al (2016) Landscape-scale variability of soil health indicators: effects of cultivation on soil organic carbon in the Usambara Mountains of Tanzania. Nutr Cycl Agroecosyst 105(3):263–274. https://doi.org/10.1007/s10705-015-9750-1

Wu YH, Hung MC, Patton J (2013) Assessment and visualization of spatial interpolation of soil pH values in farmland. Precision Agric 14(6):565–585

Zhao Y, Xu X, Darilek JL et al (2009) Spatial variability assessment of soil nutrients in an intense agricultural area, a case study of Rugao County in Yangtze River Delta Region. China Environ Geol 57(5):1089–1102

Spatial Decision Support Systems 8

Abstract

This chapter is the most application-oriented section of the book, as it provides the reader with up-to-date study materials on Spatial Decision Support Systems (SDSS) methodologies. These may be used to prioritize soil-management practices, and to allocate agricultural machinery for combating soil degradation. The chapter begins with a review of SDSS that have been developed for water erosion studies, and a description of the factors to be taken into account when developing SDSS for agricultural catchment analysis. To this end, existing SDSS may provide recommendations for soil conservation treatments that can reduce erosion risks under various scenarios, while enhancing the ecosystem services provided in the catchment. The theory behind the decision-making process, and the well-known MCDM (Multi-Criteria Decision-making) procedure and equations, are described. This description is followed by an example that demonstrates the development of just such an SDSS for an at-risk agricultural catchment, by means of the German decision-making system GISCAME (Geographic Information System, Cellular Automaton, Multicriteria Evaluation)—which is available online, and can easily be applied by the reader in at-risk catchments. Finally, the pros and cons of SDSS are discussed, to show the reader the risks and benefits that the automatic application of MCDM and GISCAME can offer.

Keywords

Ecosystem services · GISCAME · Recommendations · SDSS · Soil conservation

The *soil health* concept described in Sect. 7.1 is based on the interpretation of physical, chemical, and biological soil traits as indicators of soil function and the provision of *Ecosystem Services* (ES). A composite soil health index can provide the analyst with comprehensive and quantitative information about the state of a given field after an erosion event. Estimated changes in soil health can be used as criteria by the farmers when prioritizing soil conservation practices for protecting their fields.

Spatial Decision Support Systems (SDSS) are among the useful and modern software tools available for structured decision-making when solving problems with a spatial dimension (Tayyebi et al. 2016; Crossland et al. 1995). Broadly speaking, SDSS use geoinformatics representations of real-world environments and ranking methods to recommend courses of action that allow decision-makers to meet their goals (Keenan 2003). By exploiting such computational capabilities, SDSS can be used to extend the concept of soil conservation beyond attempts at soil loss reduction, into a more systematic approach of quantifying soil function, based on the soil health indexing concept.

SDSS are a well-established methodology in the information sciences, and are applied in

T. Svoray, *A Geoinformatics Approach to Water Erosion*,
https://doi.org/10.1007/978-3-030-91536-0_8

various fields, including, for example, The prevention of environmental hazards, agricultural, water resources, and forest management, soil studies, public participation and business geodemographics. For many more applications of SDSS, see the review study by Keenan and Jankowski (2019).

This chapter describes the use of SDSS in water erosion studies. It begins with the description of the theory of decision-making processes by humans, and the basic terms of SDSS discipline. Next, it reviews several advanced examples of SDSS being applied to soil conservation studies. Finally, it describes two well-known SDSS tools: MCDM (*Multicriteria Decision-Making*) and GISCAME (*Geographic Information System Cellular Automaton Multicriteria Evaluation*).

8.1 Introduction

8.1.1 The Decision-Making Process

Decision-making by human experts is a challenging task, because it relies on complex cognitive processes that are bounded by several factors. These may include, for example, Time limits, a host of physical and mental conditions, and limited availability of information on the problem at hand (Simon 1991; Kahneman et al. 1974). Within the decision-making process, the decision-maker—or, as often is the case, a group of decision-makers—chooses a particular course of action from among several possible *alternatives*, based on adequacy *criteria*. The recommended course of action, or the final decision, is the output of the process. The decision-maker can also choose a second-best course of action, a third one, etc. These feasible courses of action can be ranked according to the decision-maker's preferences.

Computerized SDSS are founded on the human decision-making process, as widely studied by psychologists and economists in the twentieth century. One luminary of the field, Herbert Simon, won both a Turing Award and a Nobel Prize for his multi-disciplinary work on bounded rationality in decision-making within organizations. Simon (1976) identified and coined three levels of the decision-making process by humans with regard to problem-solving:

1. *Intelligence* is the level at which the decision-maker defines her goals and collects information on the environment in a bid to document the conditions. For example, if she wishes to increase crop production, then at the Intelligence level, she collects information on Governing processes, clorpt and intrinsic factors, and extent of erosion consequences.
2. *In the design* level, the decision-maker establishes: (1) A list of several possible courses of action to achieve the goals, and (2) A list of suitable criteria to test the performance of these courses of action in achieving the goals. The criteria are ranked by importance using weights. For example, the performance of reduced tillage (a course of action) in increasing organic matter (a *criterion*) may be used to improve crop production (the goal).
3. At the *choice* level, the possible courses of action are compared, based on the weighted criteria, to rank the entire list of courses of action to identify the best performer, second best, third, etc. The courses of action are ranked according to their ability to meet the decision-maker's goal. This is done in four steps (Keeney 1982): (1) Formalizing the goal in a structured manner; (2) Identifying the outcomes of the application of every course of action; (3) Detecting possible limitations; and (4) Suggesting the list based on performance.

The decision-making process as prescribed by Herbert Simon can be exemplified on a water erosion problem in an agricultural catchment, as shown in Fig. 8.1.

The human decision-making process can be translated into a computer language and help humans with solving complex (spatial) problems. The conversion of the decision process into an

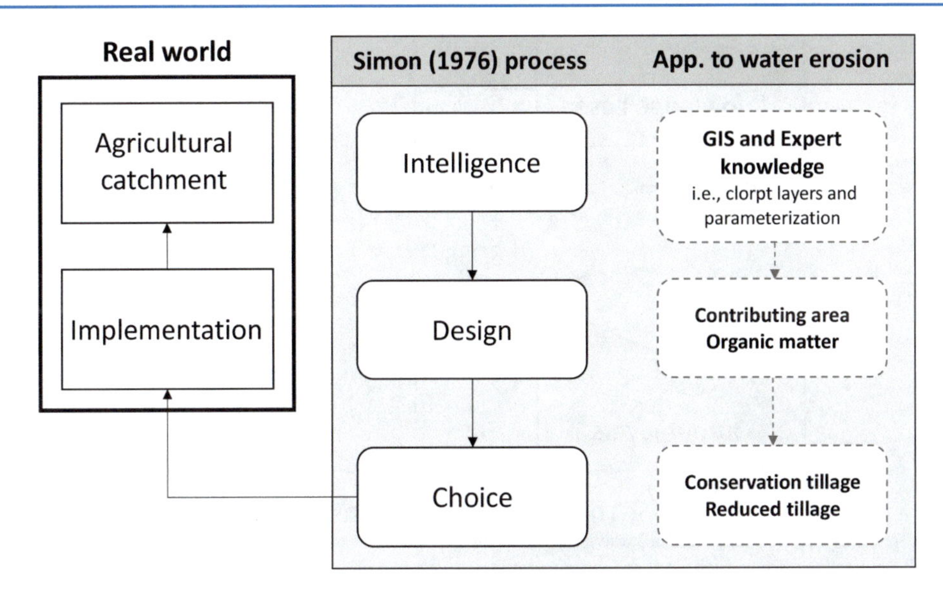

Fig. 8.1 The decision-making process based on Simon (1976) adapted to a water erosion problem in an agricultural catchment. The Intelligence level refers mainly to the data collection using GIS. The Design level is the selection of criteria and their ranking with weights (in this case contributing area and organic matter). The Choice level is the selection of the appropriate course of action in the catchment (in this case between conservation and reduced tillage)

SDSS is done by means of several basic terms, as described in the following section.

8.1.2 Basic Terms in SDSS

An SDSS is formally defined as "an interactive, computer-based system designed to support a user or group of users in achieving higher effectiveness in decision-making while solving a semi-structured spatial decision problem" (Malczewski 1999). Accordingly, SDSS are a subset of a family of computerized tools known as *Decision Support Systems* (DSS) that were developed as information systems in support of decision-making in major fields such as medicine, engineering, business administration, military studies, and many others (Newman et al. 2017). The uniqueness of SDSS within the DSS family lies in its spatial nature. A general structure of a typical SDSS is described in Fig. 8.2.

An SDSS is fed by input spatial layers from GIS datasets, and the knowledge is extracted from experts, and often formulated by *decision rules*. The input layers—and further non-spatial data if necessary—represent the conditions that help to suggest recommendations, or a ranked list of alternative solutions, in the form of reports or, more often, output GIS layers. Within such a conceptual framework, in practice, SDSS draw their jargon and computational approach from both fields of DSS and GIS. Four basic terms of that jargon are explained here as a reference for further reading in this book. These form the basis for a commonly used SDSS known as MCDM (Multicriteria Decision-Making) but can also serve in the study of other SDSS:

1. *Goal* is the state that the decision-maker wishes to achieve for the site (catchment) in question.
2. A *criterion* is the standard by which the selection of a course of action is being made. A criterion includes an *objective function* that indicates the desired state, and an *attribute* that is a quantitative value of the criterion in question.
3. An *alternative* is a possible course of action that has an actual outcome in the problem-solving process.

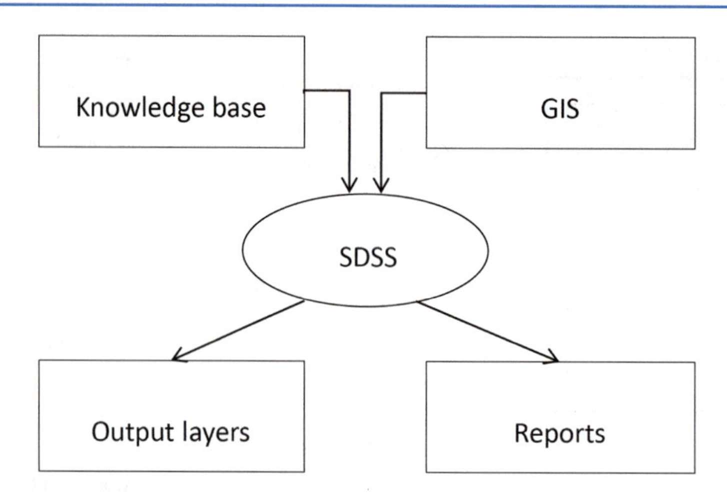

Fig. 8.2 A conceptual structure of a Spatial Decision Support System (SDSS)—with a spatial database (extracted from a GIS and remotely sensed data) and a knowledgebase (by experts, literature, or other) as input, and maps/reports as outputs. SDSS use the knowledgebase for recommendations on courses of action for a wide range of scenarios or threats

4. *An optimization* is a process of choosing the most preferred alternative, based on a well-defined criterion, from a set of *feasible alternatives*.

For a more comprehensive introduction to SDSS and specifically MCDM systems, the reader is referred to Malczewski and Rinner (2015). Here, we focus on the application of SDSS to water erosion studies—while several examples of applied goals, criteria, alternatives, and optimization tools, from SDSS in this field, are presented in Table 8.1.

SDSS therefore stem from Simon's rational decision-making process, and comprise a set of possible courses of actions (the alternatives) and evaluation criteria (Malczewski 2006). Most environmental problems have multiple criteria functions (the MCDM systems described in the following section), but some systems were developed based on a single criterion. For the sake of clarity, we will begin by exemplifying a simple system of this type.

8.1.3 The Single-Criterion System

A typical SDSS problem may be defined by the difference between a given state of the catchment, and its desired state (the goal). For example, the SDSS analyst might compare the current spatial pattern of organic matter content in the catchment with the organic matter spatial pattern needed to support the ES of Crop Production. To make rational decisions needed to achieve the desired state, alternatives for action (such as fertilization, cultivation, mulching land use changes, etc.) are tested and ranked based on their contribution to the optimization of a given objective function.

To implement such a strategy, each alternative from the set of alternatives (X) is assigned a criterion with a *decision variable* (x) that is used to measure the performance of the alternative in relation to the system goal (Malczewski 1999). The objective function indicates the alternative status or distance from the optimal point—i.e., the desired state. Maximizing soil health in agricultural fields may be determined by decision variables such as Cultivation method, hillslope gradient, rainfall intensity, or any other clorpt factor. The decision variable is extracted from a decision space which must be restricted by *constraints* that determine the set of feasible alternatives. These constraints may be cost limitations, extreme environmental conditions, lack of education/knowledge of the farmer, or something else. The feasible alternatives are then

Table 8.1 Goals, criteria, alternatives and optimization tools in environmental SDSS. The alternatives in this table suggest different recommended land uses, based on the existing criteria

Source	Goal	Criterion	Alternative	Optimization
Massei et al. (2014)	Management of wastewater in agricultural areas	1. Soil purification and protective capacity (with pH as system variable) 2. Avoiding runoff risk (with Hyd. Conductivity as system variable)	Ranked classes of suitability of the area for the waste-water application	Five methods were compared: ELECTRE I, Fuzzy set, REGIME analysis, Analytic Hierarchy Process (AHP) and Rough Set Approach
Meshram et al. (2020)	Prioritization of sensitive area at risk of water erosion	Morphometric attributes: linear, aerial and relief aspects of the catchment	Priority levels for treatment: Low, medium, high, very high	Three methods were compared: Compound Factor (CF), ELECTRE I, and VICOR
Souissi et al. (2020)	Flood susceptibility reduction	Elevation, Rainfall intensity, Drainage density, Groundwater level, Hillslope gradient, Lithology	Flood Hazard Sensitivity level (Very High, High, Moderate, Low, Very Low)	MCDM- Analytic Hierarchy Process
de Azevedo Reis et al. (2020)	Reduction of drought vulnerability	An assemblage of social, economic, sanitation, hydrology, drought, and adaptation indices	Drought vulnerability groups	MCDM weighting schemes

ranked, based on their ability to meet the objective function, using a set of decision rules. The decision rules are programmed to rank the possible courses of action (alternatives) based on their expected outcome, as measured against the objective function. The entire decision process is therefore targeted at the possible recommendations of alternatives. In other words, the recommendations are, in effect, a set of alternatives that are ranked by their respective contribution to meet the system goals.

Decision-making in water erosion studies in agricultural environments calls for optimizing the use of soil treatment, land use and conservation actions to minimize the damage to soil health on-site and off-site, under specific environmental conditions (Kangas et al. 2015). The objective function's direction of improvement can be computed as either a *More-is-better* (maximization) function, or a *Less-is-better* (minimization) one. Mathematically, such an optimization process is described as follows (Malczewski and Rinner 2015): Minimize/maximize the criterion objective function f(x), subject to $x \in X$, where x is a value of a decision variable, and the range X is a set of alternatives (Fig. 8.3).

The analyst begins the optimization process by selecting a criterion and the objective function, and then a decision variable. Next, she must identify the alternatives that provide the optimal ("best") answer to the system demands. Once they have an optimum (critical) point of the best "answer" to the "problem", the values near the optimum can be further explored, to study the effect of small variations and noise on the predictions of the SDSS. This is done by means of sensitivity analysis. Because small variations in the input data can occur and undermine the selection of alternatives, an exploratory process of *sensitivity analyses* must be applied to test the robustness of the alternatives ranking. For example, the analyst needs to test if small differences in computations of hillslope gradient can lead to changes in ranking (and as a result in recommendations)—e.g., as a result of data uncertainty. In this way, one can ensure that the suitable course of action will not be ruled out because of noisy data. Thus, in an additional and different example to that presented in Fig. 8.3, our aim might be to minimize the contributing area f(x), as the criterion function of the summation of the contributing area ($\sum a_i$)—where x

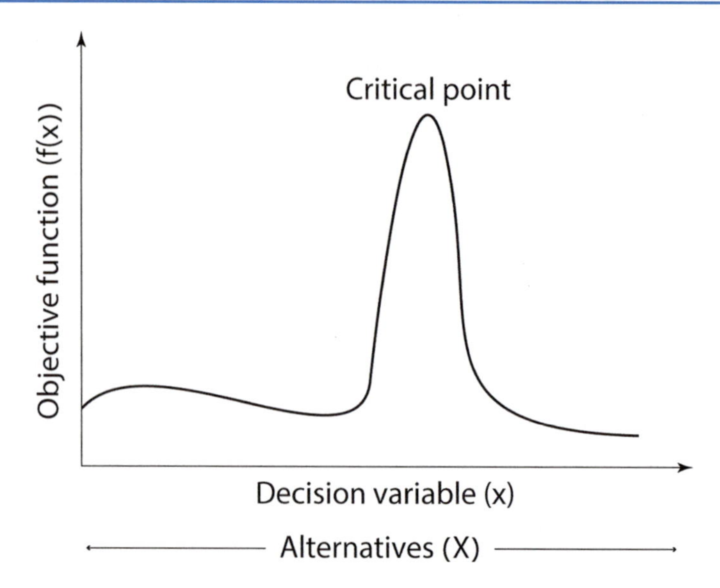

Fig. 8.3 Optimization of a single-criterion SDSS. The objective function might, for example, be the increase of soil resistance to erosion in the catchment, defined by the organic matter content [%]. The decision variable is the measured Organic Matter (OM_i). The alternatives are possible groups of actions of some sort—such as no-tillage cultivation method, reduced tillage, and terracing and they are ranked according to their respective performance of the objective function. Accordingly, we ask which values—from the set of available alternatives—are the critical points of the function f(x)

is the contributing area decision variable, and X is the space of several alternatives to minimize $\sum a_i$. If the optimum is a contributing area of 30 m^2 and the chosen method to achieve that goal is terracing, we can change the input data slightly and consistently, to see if this alternative continues to achieve the optimum under different conditions.

8.2 SDSS in Soil Conservation

8.2.1 Applications of SDSS in Water Erosion Studies

In the modern era, agricultural fields cover large areas that are usually heterogeneous in terms of their clorpt properties. Extensive agriculture therefore poses a major challenge for effective planning and practice of conservation management over wide regions. Crop rotation, reduced tillage, mulching, cover cropping, crops sequence, cross-slope farming, and other soil conservation procedures must be applied—usually under tight budget constraints—while striving for maximal provision of ES. This task makes SDSS a much sought-after tool by farmers and decision-makers at the national level. Not surprisingly, therefore, many examples can be found in the literature of beneficial uses of SDSS for planning soil conservation allocation (e.g., Bhattacharya et al. (2020); Bodini and Giavelli (1992)).

One very basic approach for SDSS in conservation management is the use of USLE (see Sect. 1.1.3.2) as a simulation engine to quantify the effect of soil cultivation and management on estimated soil loss. This can be done by controlled adjustments of the C-factor of the USLE that quantifies the crop-cover effect, and the P-factor that quantifies soil conservation practice. Such simulations can be used to make decisions about cultivation and conservation management tools, in light of the expected soil loss consequences according to the USLE estimation. Nevertheless, Cohen et al. (2005) showed that although USLE can be applied at the catchment scale using geoinformatics, it may be limited in its accuracy in many areas, and consequently lead to inaccurate identifications of part of the catchment in need of tailored treatment.

In a more structured SDSS for water erosion study, Dragan et al. (2003) proposed a multicriteria decision analysis process that was developed for crop-rotation strategies, to protect the fields from soil loss. In this paper, S_j was defined as the suitability of a given grid-cell to a crop type, using a weighted linear combination of various estimating variables as the criteria, such as—proximity from roads (0.0375), crop type (0.1521), water availability (0.1079), altitude (0.2580), potential erosion (0.3590), and croplands (0.0855). In parentheses are the criteria weights W_i extracted by means of pairwise comparison. This proposed system proved to be successful, resulting in a substantial decrease in soil loss from 4.5 to 1 $t{\cdot}ha^{-1}{\cdot}yr^{-1}$.

In the past twenty years, other SDSS methodologies have become available to complement the simple, but effective one, proposed by Dragan et al. due to the complexities of allocating soil conservation facilities, drainage devices, and other cropping systems and cultivation methods to reduce water erosion; optimization processes have proven to improve substantially the status of agroecosystems in many ways (see examples in Table 8.2).

Table 8.2 shows that SDSS have been applied to various cropping systems, with diverse aims, such as (in the main): Reducing soil loss, preventing overland flow, increasing crop productivity and other environmental outcomes, and selecting the most suitable plant species to avoid soil detachment. Several types of SDSS have been used, but many are based on a relatively simple optimization process, using algebraic expressions. In other words, Table 8.2 shows that MCDM systems are a commonly used SDSS mechanism in water erosion studies.

The value of SDSS in water erosion studies lies in its simulation of the impact of soil management, under varying and dynamic clorpt conditions, on off-site and on-site consequences. Until recently, many water erosion studies used SDSS to propose various cultivation practices—mainly in a bid to reduce soil loss, and runoff flow.

While soil loss is, of course, a very important factor in the degradation of soils (see Chap. 1), it is, however, not the only one affecting soil health, as made clearly evident in Sect. 7.1. The goal of protecting agricultural catchments should be therefore more comprehensive, with soil health as the criterion, and ES as the goal. In such an SDSS, the output layer generated could then be fed back into the system, and the status of the catchment then tested before the next erosion event (see Sect. 8.3).

8.2.2 Beyond Soil Loss

Healthy soils in agricultural catchments are life-supporting systems that can provide humans with various services—such as food and water supply, climate regulation, and nutrient cycling—that are crucial to our welfare and have large economic value (Costanza et al. 1997). Accordingly, consequences of water erosion processes go beyond the damage of the loss of the soil itself, which is severe in and of itself, and perhaps most critical. Therefore, in addition to reducing soil loss, an SDSS should consider the criteria of various soil health indices, and the goal of maximizing various ES, as well as Food Supply and increased Biomass and Grain Yield (Hein et al. 2006).

Figure 8.4 suggests a possible framework for an SDSS aimed at extending the spatial decision-making process beyond merely minimizing soil loss. To this end, it recommends soil treatment, in a manner that reduces possible erosion damage to soil health and to subsequent provision of various ES. In such an SDSS, the soil health indicators may act as the criteria for choosing the alternatives among possible soil-management practices.

The general idea of using SDSS to facilitate ES has been suggested before. Grêt-Regamey et al. (2017) reviewed the deployment of the ES concept in decision-making in various fields of environmental studies. They found mathematical tools for integrating ES into a spatial decision-making process, both in the local and the regional scale. Specifically, SDSS in agricultural

Table 8.2 A collection of research studies that applied SDSS tools to examining water erosion processes. While many of these were focused on recommendation systems to reduce soil loss, some extended their recommendations to improve the provision of ecosystem services

Source	Objective	Method	Cropping system
Dragan et al. (2003)	To reduce erosion rate by reallocating crops according to their capacity to protect the soil	MCDM	Tef, wheat, Sorghum, maize, wooded cropland
Kaur et al. (2004)	Optimized land use planning, to reduce soil loss	Linear Programming	Paddy and corn, wheat germ, and mustard
Shahbazi and Jafarzadeh (2010)	Agro-ecological land evaluations, based on clorpt data, for recommendations for sustainable soil management	MicroLEIS, DSS system	Wheat, alfalfa, sugar beet, potato, and maize
Lorenz et al. (2013)	Crop-rotation strategies for assessing and visualizing the impact of possible agricultural land use strategies on soil loss and provision of ecosystem services	GISCAME	Cereals, corn, legumes, oilseeds, root crops
Macary et al. (2014)	Classification of clorpt data into risk levels, to help farmers in decision-making for parcels	MCDM based on ELECTRE TRI-C and GIS	Mainly maize and wheat
Frank et al. (2014)	Recommended catchment-scale planning, to reduce water erosion	GISCAME	Short crop rotation: "Rape–Wheat–Rye" and "Rape–Wheat–Corn silage–Rye"
Jaiswal et al. (2015)	Prioritizing vulnerability levels of subcatchments, based on clorpt parameters	MCDM based on AHP	Large-scale model
Patel et al. (2015)	Ranking priority to build dams to prevent runoff and erosion in subcatchments	An SDSS	Large-scale model
Perez et al. (2017)	To aid the site manager choose the most appropriate species	A database of species sorted by their utility for retaining soil on slopes subject to shallow landslides, wind and water erosion	Agro-ecosystems
Lilburne et al. (2020)	A framework for providing land resource information to support policy development to issues relating to productivity and environmental outcomes	"The Land Resource Circle"	Ryegrass, white clover, wheat (hypothetical example)

catchments were implemented for the provision of food, water, wood, and for erosion prevention. However, these applications of SDSS in support of ES in agricultural catchments are not straightforward, and they face five challenges that must be addressed (De Groot et al. 2010):

1. The analyst needs to: (1) Identify the ES that the catchment can potentially provide; (2) Detect the processes that allow the provision of these ES; and (3) Formulate indicators that measure the catchment capacity to provide the ES;

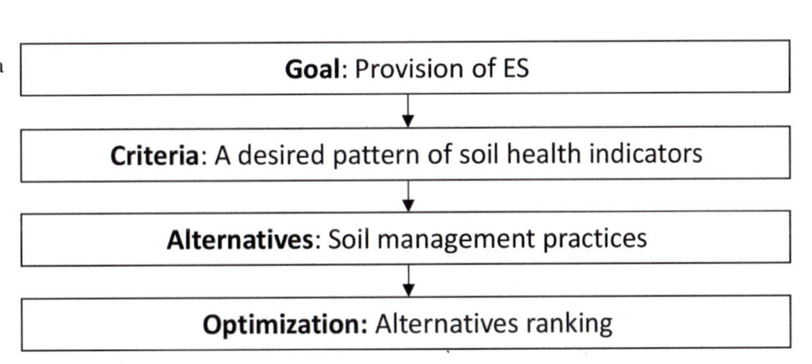

Fig. 8.4 A conceptual framework of an SDSS, exemplified for provision of ES, using GIS layers of soil health indicators as the criteria for ecosystem services provision, by selecting the optimized soil-management practices

2. The provision of the ES should be translated into socio-economic values;
3. The cost–benefit optimization procedure on the feasible alternatives is tested to see if the benefits serve the stakeholders;
4. The possible consequences of the treatments, and the adequacy of financing methods for investing in ES, and for communications about the ES and their social and economic importance to the stakeholders, are assessed;
5. Solving practical issues, such as data availability, space–time modeling, and evaluation of resilience of landscape functions.

For more details on these five general challenges, the reader is advised to refer to the paper of De Groot et al. (2010).

As we saw in Table 8.2, the literature suggests various options for implementing SDSS in water erosion studies, using the approach presented in Fig. 8.4. However, as has already been noted in the review studies by Keenan and Jankowski (2019) and by Ferretti and Montibeller (2016), the MCDM systems are a widely used and important conceptual tool.

In the following sections, we introduce an SDSS framework that considers clorpt factors as criteria, and various land use managements as alternatives, in an effort to increase the provision of ES. We begin with an explanation of the MCDM background, and extend the description to decision rules in MCDM systems. We end with a demonstration of the application of a more complex, cellular automata-based GISCAME system to the Harod catchment.

8.3 MCDM

8.3.1 Background

Decision-making problems—especially those of agroecosystems and water erosion—always involve more than one criterion. MCDM methods have been developed to address multivariate problems in a formal way, based on multiple criteria, using weights of influence for each criterion (Jelokhani-Niaraki et al. 2018). In this book, we address quantitative problems, but MCDM systems also offer the possibility of using qualitative information. Koschke et al. (2012), for example, suggested a spatial MCDM system to conduct qualitative estimations of potential to provide ES in support of regional planning in Germany, using expert experience.

MCDM methods were developed in the field of operational research to optimize the selection of alternative solutions, by considering criteria defined by objective functions and attributes (Malczewski 1999). For example, the soil health indices are used as attributes, to test how far a given grid-cell or a field is from the desired soil health status, after testing the implementation of an alternative—e.g., a particular soil cultivation method. This assumes that changes in the criteria attributes can lead to a desired pattern.

Due to their simplicity in implementation and execution, MCDM systems are widely applied as SDSS, using spatial databases with geoinformatics tools. For example, as we shall presently see, a spatial database of the Harod catchment, structured from several clorpt variables, can be used to make decisions about spatially explicit soil conservation actions. In such an application, reaching the optimal decision for combating water erosion must consider not only the variation in criteria, but also the consistency of the experts, the decision-makers' goals, and the complex outcome under various possible scenarios. The concept of MCDM—whereby alternative choices are analyzed by considering a set of multiple weighted (and thus ranked) criteria—can be useful when trying to solve this kind of problem.

But despite their advantages of simplicity and robustness, in MCDM systems the criteria weights depend on the experts' subjective—and always limited—knowledge. The adverse effect of expert's subjectivity can be mitigated using group knowledge of several experts, or by means of a sensitivity analysis of the weights (see de Azevedo Reis et al. (2020) and Sect. 6.2.6.1). In the former approach, several studies have shown that experts of different fields can yield more reliable results by compensating for each other's gaps in expertise (e.g., Svoray and Ben-Said (2010)). In addition, as previously noted, sensitivity analyses can be used to statistically test to what extent small changes in filling in the pairwise matrices change the final estimations.

MCDM are not merely theoretical systems but are recurrently used in water erosion geoinformatics for various purposes. We finalize this section with three current cases of practical use of MCDM systems for water erosion studies to demonstrate the diverse possibilities of this method for applications in water erosion geoinformatics.

In the first case study, a MCDM system was programmed to minimize the socio-economic and environmental consequences of flood damage by classifying flooded fields using RS data, and evaluating flood risk potential. An analytical hierarchical process was used to rank the criteria for flood risk—including a sensitivity analysis to compute the sensitivity rate of the criteria to changes of weight values (Souissi. et al. 2020). In the second case, Akbari et al. (2020) used MCDM to prioritize driving forces of desertification, and suggest management strategies. The Preference Ranking Organization METHod for Enrichment of Evaluations (PROMETHEE) was run, based on 113 experts and field studies, to identify and rank the top 15 (out of 29) factors of desertification—namely: "Overgrazing, land use change, improper land management, drought, reduced precipitation, soil salinity, overpopulation, erosion, waterlogging, overuse of pesticides and fertilizers, inappropriate tillage, improper irrigation, and decreased soil fertility." Categories of policies for sustainable management were then suggested, based on the prevailing environmental factors, field surveys, and expert opinions on how best to defend sensitive components at risk of desertification. In the third case, Meshram et al. (2020) grouped 20 subcatchments into four classes, by considering linear (i.e., drainage density, stream frequency) and shape (i.e., elongation ratio, form factor, circulatory ratio) parameters and relief aspects to be used in MCDM analysis of subcatchments in terms of their vulnerability to soil loss. Morphometric data has been found to be effective in prioritizing development and management of natural resources.

More generally, MCDM systems were used spatially for many other applications—such as (Malczewski 2006): Hydrology and water resources, transportation problems, environmental planning ecological conservation, forestry, and waste management. In the following two sections, we describe the workflow and computation of MCDM as an SDSS.

8.3.2 Procedure

From an operational point of view, MCDM methods are used to prioritize feasible alternatives based on attribute values of several criteria, using a combination of objective functions. Such an approach should, however, be used with caution—because some criteria may increase the value of one objective function, while reducing the values of another (Ghaleno et al. 2020). For example, sharp hillslope gradient may increase the risk of soil loss, but decrease risk of water ponding—which is also a soil degradation consequence that reduces aeration and degrades the soil. This point will be further discussed below (Sect. 8.3.3.2) in describing the algebraic expressions used by MCDM systems.

The general form of MCDM procedure consists of the following six operative steps (based on Kumar et al. 2017):

1. The analyst formulates a goal, or a set of goals, that the decision-makers wish to achieve. For example, one goal may be to keep the catchment in a state that provides crop yields in the quality and amount that allows the farm to sustain itself financially over the next decade;
2. The analyst determines all the criteria that affect the goal defined in Step 1. The objective function of each of these criteria must be minimized or maximized in pursuit of that goal. One such example of an objective of a criterion might be the upslope contributing area that must be minimized (e.g., by means of terracing) to maximize soil health while, conversely, organic matter content must be maximized to increase soil health as an objective function;
3. A set of available alternatives to fulfill the objective functions is tested using a well-established measure (such as the soil health score). Each alternative is a course of action—e.g., changing cropping system. In this way, the analyst can test if by applying a given cropping system, the soil health score improves;
4. The decision-maker's requirements should be expressed in terms of the criteria weights. These weights quantify the importance of each criterion in estimating the alternative's performance;
5. The MCDM is applied, using geoinformatics tools, to the entire catchment domain, to test the possible alternatives for each of the various spatial entities;
6. The ranked list of alternatives is suggested to the decision-makers, on a spatially explicit layer.

These six steps are illustrated in Fig. 8.5.

Within this MCDM framework, the decision rules are procedures that allow the analyst to decide which available alternative is the optimal course of action, which is second best, third best, etc. The MCDM literature (Malczewski 1999) offers a long list of decision rules—including the very well-known combination rules *weighted summation, ideal/reference points*, and *outranking methods*.

8.3.3 Decision Rules

8.3.3.1 Weighted Summation (WS)

This section reviews two common methods of decision rules for ranking the MCDM alternatives: (1) *Weighted Summation* (WS); and (2) The *Concordance–discordance Analysis* (CA), both of which are described by Feick and Hall (2001) and many others.

The WS is simply a weighted sum (Eq. 8.1):

$$U_i = \sum_{j=1}^{J} \left(w_j \cdot \hat{x}_{ji} \right), \tag{8.1}$$

where U_i is the alternative i score summarized across all criteria J; w_j is the weight assigned to criterion j; and $\widehat{x}_{ji}$ is the standardized score of criterion j for alternative i. WS was referred to above, in Sect. 6.2.3, as the *Weighted Linear Combination* in the AHP jargon used for the final score. The main shortcoming of WS is that it suffers from a compensatory effect: If one of the criteria is very small and another is very large, they may compensate each other, when in reality they cannot actually do so. For example, if the value of the criterion of rainfall intensity is extremely low, but the value of the criterion contributing area is high, the final U_i will be moderate—but when rainfall of extremely low intensity and depth occurs, no runoff whatsoever is expected, so U_i should actually be nil,

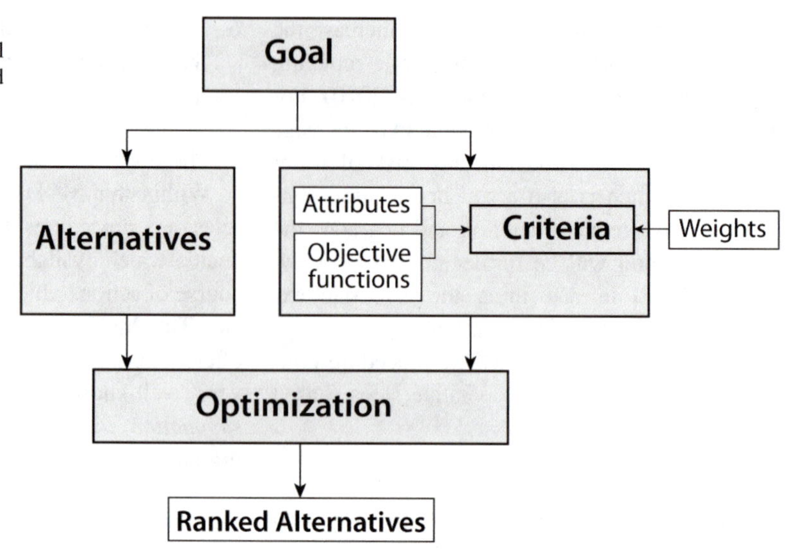

Fig. 8.5 Structural scheme of a MCDM system. The goal is set by the analyst; weighted criteria including attributes and objective functions are selected to rank the alternatives which must be optimized for a given goal

irrespective of the contributing area. Thus, in reality, rainfall is actually a prerequisite for runoff development, because when there is no rainfall there is no runoff no matter what the other criteria are, and therefore rainfall intensity is a bottleneck criterion when assigning alternatives and as such cannot be expressed solely by means of a compensatory method.

One example of the application of WS as an SDSS computed for a given grid-cell is presented in Tables 8.3 and 8.4. The SDSS suggests three alternatives (i) of soil treatments: A (Conventional Tillage), B (Reduced Tillage), and C (No Tillage), and three decision variables, to define three criteria of soil health (j): Nitrogen content (N), Available Water Capacity (AWC), and Organic Matter (OM). In the parentheses of these three variables we see the weights of the three criteria assigned $\sum w_i = 1$ by the analyst:

The $\widehat{x}_{ji}$ values are multiplied by the criterion weight and summarized for a full score value (Σ) for each alternative, as shown in Table 8.4.

In Table 8.4, we see the contribution of each alternative to increasing the criteria values. In this instance of soil health, we wish to maximize the value of each of the criteria. Also, the values (performances) of $\widehat{x}_{ji}$ must be normalized, in the case of Table 8.3, between 0 and 1.

Table 8.3 The score matrix—simply the $\widehat{x}_{ji}$ value at a grid-cell

Criteria (j)	N ($w_j = 0.3$)	AWC ($w_j = 0.5$)	OM ($w_j = 0.2$)
Alternative A	$\widehat{x}_{ji} = 0.8$	$\widehat{x}_{ji} = 0.9$	$\widehat{x}_{ji} = 0.4$
Alternative B	$\widehat{x}_{ji} = 0.5$	$\widehat{x}_{ji} = 0.4$	$\widehat{x}_{ji} = 0.7$
Alternative C	$\widehat{x}_{ji} = 0.2$	$\widehat{x}_{ji} = 0.3$	$\widehat{x}_{ji} = 0.1$

Table 8.4 The weighted normalized decision matrix is simply the $\widehat{x}_{ji}$ at each cell multiplied by the criterion weight: The rightmost two columns show the final (summation) score and consequent ranking

Criteria (j)	N ($w_j = 0.3$)	AWC ($w_j = 0.5$)	OM ($w_j = 0.2$)	Score (Σ)	Rank
Alternative A	0.24	0.45	0.08	0.77	1
Alternative B	0.15	0.20	0.14	0.49	2
Alternative C	0.06	0.15	0.02	0.23	3

The results are summarized for each alternative in the Score (Σ) field, then the outcomes are ranked from large to small in the Rank field. Based on this approach, Alternative A is the most recommended, Alternative B is second best, and Alternative C is in third place.

8.3.3.2 Concordance–Discordance Analysis (CA)

The Concordance–discordance Analysis (CA) offers a more detailed comparison of the alternatives in the MCDM, based on performance. In other words, in the CA method, the MCDM uses concordance and discordance indices to quantify the performance of a given alternative i in relation to an alternative i', using analytical hierarchal processes (see Sect. 6.2).

Before we describe the actual method and the computation of indices, a definition of concordant and discordant relationships is in order (Table 8.5). A *concordant relationship* occurs when in two pairs of observations (in our case, alternatives A and B), both criteria (in this example, contributing area and soil crusting) are higher in one alternative (B) than in the other (A): (5 > 3 and 2 > 1). *A discordant pair* is defined when the criterion scores are inverted—e.g., if alternative C has a higher value of one criterion (soil crusting) than alternative D (2 > 1), whereas in the other criterion (contributing area), the opposite is true: Alternative C's value is lower than that of alternative D (4 > 3).

The aim of the analysis is to test whether a given alternative outperforms any other that the system might suggest. Within these definitions, the Concordance Score $c_{ii'}$ is computed as the sum weights of the concordance matrix in a pairwise comparison—namely, only those criteria of the type alternative $i \geq$ alternative i'—(Eq. 8.2):

$$C_{ii'} = \sum_{j \in c_{ii'}} w_j, \tag{8.2}$$

—whereby $c_{ii'}$ is computed for every i compared with i'; $C_{ii'}$ is the concordance set of criteria for alternatives i and i'; and w_j is the criterion j weight. In other words, the summation operator is applied to the entire set of $C_{ii'}$.

The *discordance index* $d_{ii'}$ quantifies the difference between pairs of criteria—where alternative i' outperforms alternative i. In that regard, it provides further information to that of $c_{ii'}$. The index values are standardized between zero and one, and the discordance index is computed based on inter-criteria comparisons (Eq. 8.3):

$$d_{ii'} = \max_{j \in D_{ii'}} \left(\hat{x}_{ij'} - \hat{x}_{ji} \right), \tag{8.3}$$

whereby $d_{ii'}$ is computed for alternative i in relation to alternative i'; $d_{ii'}$ is the discordance set of criteria for alternatives i and i'; $\hat{x}_{ji}$ is the normalized score of i' on j; and $\hat{x}_{ji}$ is the normalized score of i on j.

The concordance and discordance indices are compared against $\bar{c}$ and $\bar{d}$ in Eq. 8.4, which are user-defined threshold values of concordance and discordance, respectively.

$$c_{ii'} \geq \bar{c} \text{ and } d_{ii'} \geq \bar{d}. \tag{8.4}$$

The thresholds can be systematically changed to eliminate alternatives, and to suggest rankings

Table 8.5 An illustration of concordant and discordant relationships. The upper two alternatives (A and B) represent concordant pairs, with consistent difference between the two criteria; and the lower two alternatives (C and D) represent discordant pairs, with inverse differences between the two criteria

	Contributing area	Soil crusting	
Alternative A	3	1	Concordant pair
Alternative B	5	2	Concordant pair
Alternative C	3	2	Discordant pair
Alternative D	4	1	Discordant pair

of less performing alternatives. The CA is then based on a systematic comparison between all pairs of alternatives and their counterparts, and is therefore a commonly used and a successful method in SDSS studies. The method of threshold values can be subjective and arbitrarily determined, which makes it less preferred by some users.

The term *net* c_i—the net concordance dominance value for i—is the delta between the degree to which alternative i dominates all other alternatives, and the extent to which they, in turn, dominate it (Eq. 8.5):

$$net\ c_i = \sum_{i'=1\ i'\neq i}^{I} c_{ii'} - \sum_{i'=1\ i'\neq i}^{I} c_{i'i}, \tag{8.5}$$

and, accordingly, in *net* d_i in Eq. 8.6:

$$net\ d_i = \sum_{i'=1\ i'\neq i}^{I} d_{ii'} - \sum_{i'=1\ i'\neq i}^{I} d_{i'i}. \tag{8.6}$$

The alternatives are ranked based on their net c_i and net d_i values—or the mean value of the two, with the highest value as the selected alternative.

MCDM are applied using various algebraic expressions, and the differences between the respective outcomes of the alternatives selection can be significant. Therefore, and since it is easy to apply, the reader is advised to test several methods against field validation. The differences in predictions may be due to the compensation effect that occurs with compensatory methods such as the WS, which can result in overestimation of the performance of some alternatives in extreme cases, as the performance score by some other criterion is increased. However, MCDM usually operate using local operators on a single grid-cell basis, without consideration of the neighborhood effect. Moreover, MCDM are not usually aimed at representing the dynamics of a system, but are more static tools. Other SDSS can be used to address these shortcomings.

The following example in Tables 8.6 through 8.10 illustrates how the ranking of alternatives is applied, using a CA analysis based on the choice problem ELECTRE I (ELimination Et Choice Translating REality) method (Roy 1968; Dortaj et al. 2020).

Table 8.6 is the dataset with three criteria (AWC, OM, N), and three possible alternative land uses (No Tillage, Reduced Tillage, and Traditional Tillage). The original data in the intersection cells shows the values of the indicators for these criteria. This data is normalized for a 0–1 range, using Eq. 8.7 in the center columns (Table 8.6b). Finally, the rightmost columns (Table 8.6b) present the normalized data, multiplied by the weight of the criterion.

$$x_{ij} = \frac{a_{ij}}{\sqrt{\sum_{i=1}^{m} a_{ij}^2}}. \tag{8.7}$$

With a as the indicator value from the original matrix (Table 8.6a), the index i as the running number of alternatives from 1 through m, and j as the running numbers of criteria.

The values in the *Concordance matrix* (Table 8.7a) are simply the sum of weights of the criteria by which a given alternative is higher

Table 8.6 Data matrices for concordance–discordance analysis based on ELECTRE. The left columns a are the original indicator data; b is the normalized data based on Eq. 8.7; and c is the normalized data x_{ij}, multiplied by the criterion weight

	Original data (a)			Normalized x_{ij} (b)			Normalized weight (c)		
Criteria →	AWC	OM	N	AWC	OM	N	AWC	OM	N
Weights →							0.3	0.25	0.45
Alternatives ↓									
No till	50	15	40	0.70	0.16	0.46	0.21	0.04	0.21
Reduced till	31	50	30	0.44	0.52	0.35	0.13	0.13	0.16
Traditional till	40	80	70	0.56	0.84	0.81	0.17	0.21	0.37

Table 8.7 The *concordance matrix* computed by the sum of weights by which this alternative normalized weight (Table 8.6c) is higher than the other two alternatives. For example to the upper right intersection (Traditional Till vs. No Till) in the Normalized weight table (Table 8.6c), the Traditional Till alternative is higher in two normalized weights (OM and N) of three criteria than the No Till one and therefore we assign in this intersection in the concordance matrix the value 0.25 + 0.45 = 0.7. The rightmost b is the cutoff table where a cutoff point is assigned by the analyst (in this case 0.5) to assign 1 for values larger than the cutoff point and 0 to those below

	Concordance matrix (a)			Concordance matrix—cutoff (b)		
	No till	Reduced till	Traditional till	No till	Reduced till	Traditional till
No till	0	0.25	0.7	0	0	1
Reduced till	0.75	0	1	1	0	1
Traditional till	0.3	0	0	0	0	0

than its pairwise alternative. For example, in the case of Traditional Till vs. No Till alternatives, the OM and N criteria are higher, so the value in the concordance matrix is the sum of their weights (0.25 + 0.45 = 0.7). The values in the *Concordance matrix—cutoff* (Table 8.7b) are assigned 1 when the sum of weights is larger than a user-defined cutoff (in this case, 0.5), and 0 when it is not.

The discordance index is computed by means of rescaling and normalization. First, the analyst creates the rescaling matrix as in Table 8.8. The rescaling procedure is aimed at changing the criteria values within a new range, usually 0–100. The 0–100 range is divided into three levels in accordance with the three alternatives and their respective assigned values. In practice, the AWC criterion is a *More-is-better* type of criterion for our purposes here. Therefore, from Table 8.6a, the maximum value of AWC is 50, and in the rescaling process is assigned the maximal value in the 0–100 range: 100. The next value of AWC is 40, assigned the value 66. The lowest value, 31, associated with the Reduced Till, is assigned the value 33 (see Table 8.8). The values for the criteria OM, and N in Table 8.8 are obtained by the same procedure.

In the next step, the rescaled matrix is normalized using Eq. 8.8, based on the importance level of the participating criteria, and the normalized discordance index d_{ij} is also computed by the same equation (Malczewski and Rinner 2015):

$$d_{ij} = \max\left(\frac{Interval}{Range}\right), \tag{8.8}$$

—whereby d_{ij} is the normalized discordance index, the *Interval* is the value of the rescaled matrix and *Range* is the range of the rescaled matrix (0–100). The discordance matrix (Table 8.9) is therefore computed for each pair of alternatives based on the following procedure:

We begin with the discordance index between No Till and Reduced Till: Since the values associated with No Till are greater than those of Reduced Till for the criteria AWC and N, Eq. 8.8 is applied for these two criteria, and the largest

Table 8.8 The rescaled matrix

	Rescaled interval matrix		
Criteria →	AWC	OM	N
Alternatives ↓			
No till	100	33	66
Reduced till	33	66	33
Traditional till	66	100	100

Table 8.9 The *discordance matrix*. The left columns a represent the difference between the best and worst alternatives, while the right columns b represent the largest differences based on a 0.5 cutoff value

	Discordance matrix			Discordance matrix—cutoff		
	No till	Reduced till	Traditional till	No till	Reduced till	Traditional till
No till	0	0.34	0.67	0	0	1
Reduced till	0.67	0	0.67	1	0	1
Traditional till	0.34	0	0	0	0	0

Table 8.10 The *Dominance matrix* that aggregates the Concordance and Discordance cutoff matrices. The cells with 1 value at the intersections of the dominance are those that satisfy both cutoff conditions in the concordance and discordance tables. According to this table, the Reduced Till and No Till both outrank Traditional Till—and Reduced Till outranks No Till

		Dominance matrix	
	No till	Reduced till	Traditional till
No till	0	0	1
Reduced till	1	0	1
Traditional till	0	0	0

difference is assigned to the discordance matrix. Thus, for AWC = (100–33) = 67/100 = 0.67, and N = (66–33)/100 = 0.33, so, max (0.67, 0.33) = 0.67. Similarly, in the case of No Till versus Traditional Till, AWC = (100–66) = 34/100 = 0.34, so, max (0.34) = 0.34.

The *Dominance matrix* (Table 8.10) basically summarizes the decision-making process by ranking the alternatives, based on the aggregation of the concordance and discordance matrices. We read the alternative in the row vs. the alterative in the column. In this case, the Reduced Till alternative clearly outperforms the other two, while the No Till alternative outperforms the Traditional Till one.

While the CA process—as demonstrated here using ELECTRE—is perhaps more complex than WS, it provides a more comprehensive approach that compares the alternatives on a pairwise basis. Another more complex and comprehensive SDSS is GISCAME, which is described in the following section.

8.4 GISCAME

8.4.1 Background

GISCAME is an object-oriented SDSS software tool developed to simulate land use changes and their effect on water erosion and the delivery of ES by agroecosystems. Two unique features of GISCAME compared with MCDM are: (1) The ability to temporally explore dynamic simulations of the catchment and the effect of neighboring grid-cells, using cellular automata modeling tools; and (2) Its landscape scale approach, that relies on spatial ecology indices as criteria for ES delivery (Koschke et al. 2012). These two capabilities allow the use of GISCAME to solve more complex problems than MCDM.

From an operational standpoint, GISCAME is available for download online at: https://www.giscame.com/giscame/english.html, as is the GISCAME 2.0 Manual (Fürst et al. 2015), which provides full instructions on its operation. Here, we will describe the GISCAME approach and its possible use for providing various ES beyond the reduction of soil loss. This is done in the following five steps (Fürst et al. 2010):

1. First, the analyst creates a spatial dataset, with land use/land cover layers, and selects the ES relevant to the catchment in question, based on socio-economic requirements and environmental characteristics;
2. Next, indicators are selected, to rank the impact of land uses (e.g., orchards, field crops, rangelands, etc.) and infrastructure elements (such as roads and fields borders) on the ES. The ranking is normalized in the range 0–100, based on the presumed effect of the chosen land uses on the selected ES (such as Carbon Mitigation, Yield Provision, and Reduced Erosion);
3. An *evaluation matrix* is then computed between the selected land uses of the catchment in question and all the selected ES that it is expected to provide. The scores at the intersection cells are normalized in the range 0–100;
4. Various environmental and planning decision rules are chosen to evaluate the extent to which the land use type, or infrastructural element, is distant from its potential to provide the selected ES;
5. The recommendations are then described in a spatially explicit manner, using GIS. The output layers allow the analyst to test if the system's state in providing ES is better than in its original state.

These steps are summarized in Fig. 8.6, which shows the overall flow of the GISCAME concept for supporting decision-making.

In an example from the field, in Step 1, a GIS database is designed including land use/land cover layers relevant to the decision-making and to the selection of ES to be analyzed in the GISCAME configuration. In Step 2 the authors compiled the selected land uses from the previous step in an *indicator table*. To complete it, the analyst must run a process of quantifying the effect of land uses on ES (Table 8.11). Three ES are selected for the analysis in each SDSS configuration, and the variables are selected to represent these ES under the conditions of five different land uses. For example, the Food & Fodder ES is represented by Grain Yield and Soil Erosion Protection by the C-factor of USLE.

The values in the intersection cells of Table 8.11 are normalized to the 0–100 range by means of a simple linear transformation equation (Eq. 8.9):

$$I_{i\,normalized} = \frac{(I_i - I_{min}) \cdot 100}{I_{max} - I_{min}}, \tag{8.9}$$

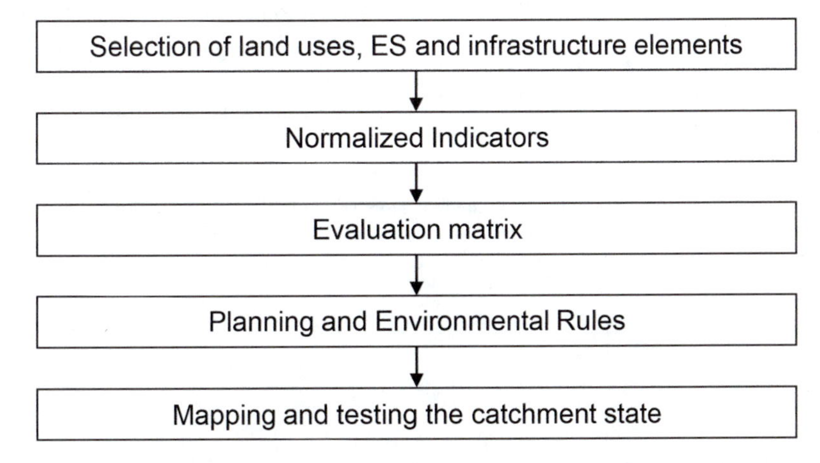

Fig. 8.6 Five steps within the iterative evaluation procedure in GISCAME, as demonstrated in Fürst et al. (2010): First, the land uses, ES, and input variables are chosen, then the indicators assigned to be used by the rules and the output of the rules are mapped as GIS layers

Table 8.11 Example of an *indicator table* prepared by the GISCAME analyst to relate land uses with the ES. The Table lists the indicator values in real-world units, based on the selected variables, to indicate the specific associated ES (GISCAME 2.0 Manual)

ES →	Food and fodder	Soil Erosion protection	Drought risk regulation
Variables → Land uses ↓	Yield [g·m^2]	C-factor - USLE	Water demand [mm]
Orchards	64	40	500
Field crop	60	20	300
Built area	0	0	0
Roads	0	0	0
Rangeland	40	0	434

—where $I_{i\ normalized}$ is the indicator score in the 0–100 range, I_i denotes the value of the indicator in the grid-cell, I_{min} denotes the lowest value of the indicator I in real-world physical units as in Table 8.11, and I_{max} is the maximum value.

In Step 3, the evaluation matrix (Table 8.12) is formed, from the data on the 0–100 normalized range. Again, the land uses are plotted against the ES—whereby 0 means no effect on the ES provision, and 100 means maximal effect on the ES provision. The evaluation matrix must be completed before the GISCAME SDSS is executed. When the measured data is not available for the catchment in question, expert knowledge can replace field measurements. Expert knowledge has limitations, given its absence of real-world evidence, but also advantages of generality and the fact that experts usually integrate knowledge from various sources, whereas measurements are always limited by extent.

In Step 4, the system is ready to apply the decision rules. The rules GISCAME provides the analyst are divided into two groups: *Planning rules* and *Environmental rules*. These two sets of rules are formulated, to dictate the possible land use changes in line with the experts' preferences, and the conclusion drawn from the observed relationships.

The Environmental rules affect the process of land use transformation, based on environmental conditions in the catchment—such as soil properties, hillslope gradient, rainfall intensity, and so on. An example of an environmental rule might be:

> Reduced Tillage can be transformed into Conservation Tillage, only if it is on a hillslope gradient < 10°

The rationale for such a rule is that cultivation with a No-Tillage plow in this catchment can only apply to a hillslope gradient < 10°, which in the study-area soil conditions is appropriate for the No-Tillage plow to operate. Finally, it is important to note that these rules restrict land use transformations in the catchment. To put it another way: If no rule is activated, all combinations of land use transformations are possible.

Table 8.12 An *evaluation matrix* computed from the indicators table (Table 8.11) to evaluate the effect of each land use on the provision of ES. This is a subset table, for illustration only. In a real-world case study, the analyst must prepare such a table before the software is run. The scores at the intersection are normalized from the indicator table, based on real-world measured data (GISCAME 2.0 Manual)

Land use	Food and fodder	Soil Erosion protection	Drought risk regulation
Orchards	60	50	60
Field crop	80	30	50
Built area	0	0	0
Roads	0	0	0
Rangeland	30	60	70

The second group of rules in GISCAME—*Planning rules* allow the analyst to estimate which land use transformation can actually occur in the catchment—i.e., to simulate the range of possible changes in land use. One important characteristic of planning rules is that they integrate the effect of neighboring grid-cells and/or linear elements on land use transformation. GISCAME also allows the analyst to consider future trends in the development of land uses, to test the system to identify conflicts between rules or outcomes of rules, and to add metadata for the user. For example, one planning rule that illustrates neighboring effect might be:

> Transformation of the No Tillage is only allowed if the neighboring land use is one of the selected crop-rotation types.

Or, in the case of a planning rule that considers futuristic trends:

> The proportion of No Tillage should generally diminish, until it reaches the level of 10%.

Thus, planning rules allow neighboring land uses and futuristic scenarios to be considered. An important advantage of GISCAME is that the two groups of decision rules are embedded in a *cellular automata modeling framework* (see Fig. 8.7). This framework is aimed at predicting for each grid-cell, for each time step, a new (or not) land use type based on the two groups of decision rules, and on the given grid-cell state and that of the grid-cells of the *Moore neighborhood* (D8) around it. The cellular automata framework is embedded in the fourth step of the iterative evaluation procedure in GISCAME described in Fig. 8.6.

The use of cellular automata in GISCAME allows for dynamic representation of the catchment and the effect of neighboring grid-cells, as described in the soil evolution models in Chap. 3. This provides GISCAME with a unique advantage over MCDMs, that are usually applied using local operators and static temporal representation. The *transition probabilities* can be assigned based either on statistical data, or on expert knowledge, in the following manner: (1) The current state is defined based on the catchment condition in a given year, for example; (2) Transition probabilities are assigned—based either on expert knowledge or available statistical data (e.g., Markov Chains); (3) Conditions are assigned for the neighborhood effect; (4) The model is executed, and a new recommendation formed.

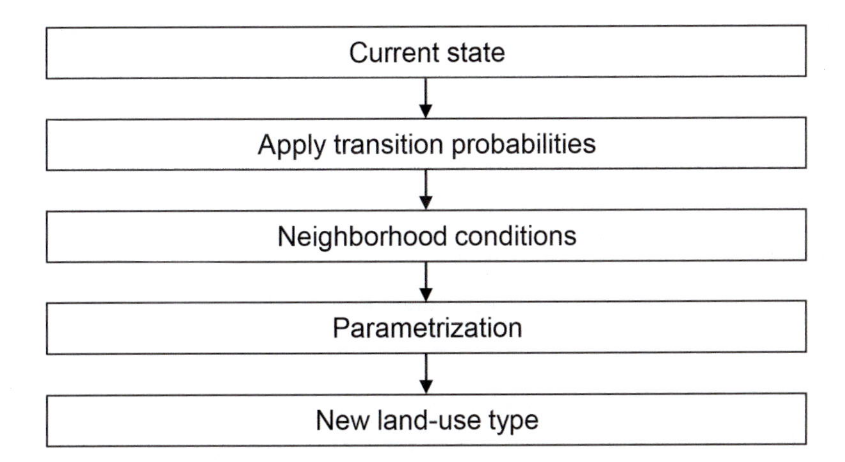

Fig. 8.7 The *cellular automata model* embedded in the fourth stage of GISCAME for suggesting new land use type. The model supports two-way interaction between the transition probability and the effect of neighboring grid-cells—making it possible to represent the field conditions more realistically. Reprinted from Inkoom, Frank, Greve, et al., A framework to assess landscape structural capacity to provide regulating ecosystem services in West Africa. Journal of environmental management, 209:393–408, (2018), with permission from Elsevier

The execution of the cellular automata model leads to Step 5, namely, that of the prediction output layer. The results of the recommendation layers are compared against the conditions prior to the application of the recommendation, using a *radar chart*, which illustrates data values from several variables by means of a graph of several axes emanating from a single origin. The connection of the dots on the axes (which may be variables or ES) creates the *radar diagram*, which illustrates the status of the phenomena in question, based on the variables (see Fig. 8.8). The radar diagram can be used to describe the catchment state before and after applying the system recommendations, to illustrate how the SDSS perform by measuring the system's response to the recommended changes in several aspects (Fig. 8.8).

Figure 8.8 shows the use of a radar chart to demonstrate the change in five soil erosion risks in the catchment following deployment of the recommendation of a hypothetical SDSS. We can clearly see that: (1) There is a large difference in the land degradation risks before and after application of the system recommendations, since the "Before" extents are much larger than those in the "After" diagram; and (2) The system recommendations were successful in reducing all land degradation risks except the drought. Continuing this line of thought, we can extend the use of the radar *diagram* to explore the system function in increasing the ES in the catchment. The illustration shows the state before and after applying the system recommendations: Here, the larger the polygon, the higher the ecosystem services score. In Fig. 8.8 we see how the system recommendations lead to an increase of all ES scores.

Radar *diagrams* can be used to compare the outcomes of different scenarios of system recommendations. A visual interpretation of this sort can often show the difference between the results before and after—but if the differences are minor, or the pentagons are configured such that visual interpretation may be insufficient, the reader can divide the polygons into triangles, and compute and compare the areas. Alternatively, the polygon areas may be computed by *the shoelace algorithm* (Braden 1986).

8.4.2 Application of GISCAME to the Harod Catchment

8.4.2.1 Method

GISCAME application to the Harod catchment was performed by Barshad (2019). The infrastructure element of the application of GISCAME to the Harod conditions was a soil loss risk layer, that was calculated using AHP and expert

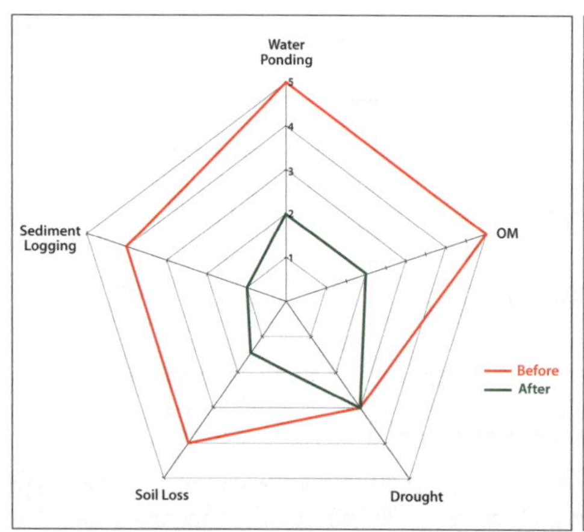

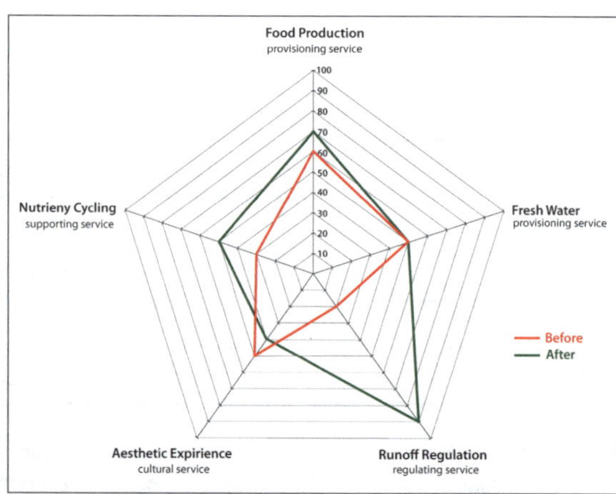

Fig. 8.8 A *radar diagram* comparing the catchment performance before and after the system recommendations. The left panel represents the status of soil degradation phenomena due to water erosion; the right panel represents the comparison before and after provision of ES

knowledge, with the system configuration described in Svoray et al. (2015) The environmental conditions used to predict this soil loss risk layer were Hillslope gradient and orientation; contributing area; vegetation cover; tillage direction; roads as runoff contributors and as barriers; land use; cultivation method; and rainfall intensity. In the output simulation, the cultivation method layer that was processed according to the GISCAME recommendations was tested to see if the recommendations did indeed reduce soil loss. In the next step, the effect of the GISCAME recommendations on three ecosystem services was tested. Three ES that were selected as the most relevant to the Harod were: (1) Food & Fodder; (2) Drought Risk Regulation; and (3) Erosion Regulation. The effect of the cultivation method recommendations on the provision of ES was examined under the conditions of three rainfall intensity layers: Low, moderate, and high. These layers were extracted from a meteorological radar, at a temporal resolution of five minutes, based on the bulk gauge-adjustment calibration method (Morin and Gabella 2007). For each of the three ES, quantitative indicator values were mapped to the individual land use types (before and after the recommendations) to represent their potential to provide the ES.

The data for Food & Fodder [$\text{ton}\cdot\text{ha}^{-1}\cdot\text{a}^{-1}$] and Drought Risk Regulation [$\text{m}^3\cdot\text{ha}^{-1}\cdot\text{a}^{-1}$] was taken from the Israeli Ministry of Agriculture and Rural Development website (https://www.moag.gov.il/shaham). The Erosion Regulation is based on the C-factor from the Universal Soil Loss Equation (USLE), ranging between 0 and 1 (Kuok et al. 2013). Normalized values of the ES (0–100) were then used to compute the differences in the ES contribution before and after recommendations.

GISCAME was applied using the cellular automata mechanism, with grid-cell states updated according to the environmental and planning rules applied to the grid-cell state, and the state of the grid-cells in the surrounding Moore Neighborhood. The evaluation procedure is as follows:

1. Land uses, influential spatial variables and ES are chosen by the analyst as the dataset;
2. Indicators to quantify the effect of land uses and other spatial variables on ES are determined, based on the expert's knowledge;
3. The indicators are tabulated against ES in an evaluation matrix, to gauge the effect of each land use on the provision of ES;
4. The decision rules are divided into two groups—Environmental and Planning—and are formulated to express the effect of the explaining variables on ES.

The environmental and planning rules basically determine if there will be or will not be a transition of the grid-cell state from one land use to another. The analyst configures rules that determine the grid-cell state in the next temporal iteration. GISCAME rules and restrictions are used to affect the evaluation results and the transition probability of a given land use type—which is a function of the previous land use type, the proximity effects of other land uses, and the environmental factors.

The decision-making system of GISCAME input was composed of two layers of the Harod at a 5.5 m^2 spatial resolution: Land use (including water bodies; non-irrigated and irrigated fields; orchards; grove; rangeland; forest; and built area), and hillslope gradient.

The first rule was to use only the grid-cells with agricultural land uses—such as non-irrigated field crops; irrigated field crops; and orchards. Cultivation methods were sub-categorized in field crops into three groups—No Tillage, Reduced Tillage, and Conservation Tillage—while orchards were divided into two groups: Conventional Tillage and Conservation Tillage.

The second rule was to allow only intra-land use changes—whereby orchards cannot be changed into irrigated/non-irrigated, and vice versa. GISCAME was designed to recommend the preferred cultivation method for each grid-cell, according to the hillslope gradient layer S: Whereby grid-cells with $S > 10°$ were assigned for Conservation Tillage; grid-cells with

$6^{\circ} < S < 10^{\circ}$ were assigned to Reduced Tillage; and grid-cells with low, or very low, hillslope gradient ($<6^{\circ}$) were assigned for Conventional Tillage.

The output of this GISCAME analysis was a layer with a recommendation of the most suitable cultivation method for each grid-cell. This output layer was used to replace the original cultivation method layer, representing the conditions in 2014, for the recalculation of the erosion risk maps. This was applied for each of the three rainfall intensities, with all other variables remaining constant. Zonal statistics operators were applied to average the risk on a per-field basis.

8.4.2.2 Results

As demonstrated in Chap. 3, spatial variation in rainfall intensity greatly affects water erosion processes, and thereby the spatial distribution of erosion risk. This is particularly true in a semi-arid catchment such as the Harod (Bauder 2005). The three scenarios of rainfall intensity below clearly show the effect of rainfall intensity on erosion risk (Fig. 8.9).

The left panel clearly shows how rainfall intensity affects erosion risk in the Harod. One can see how the case of moderate rainfall intensity shows far more yellowish hues, while the lower panel shows green fields with even lower

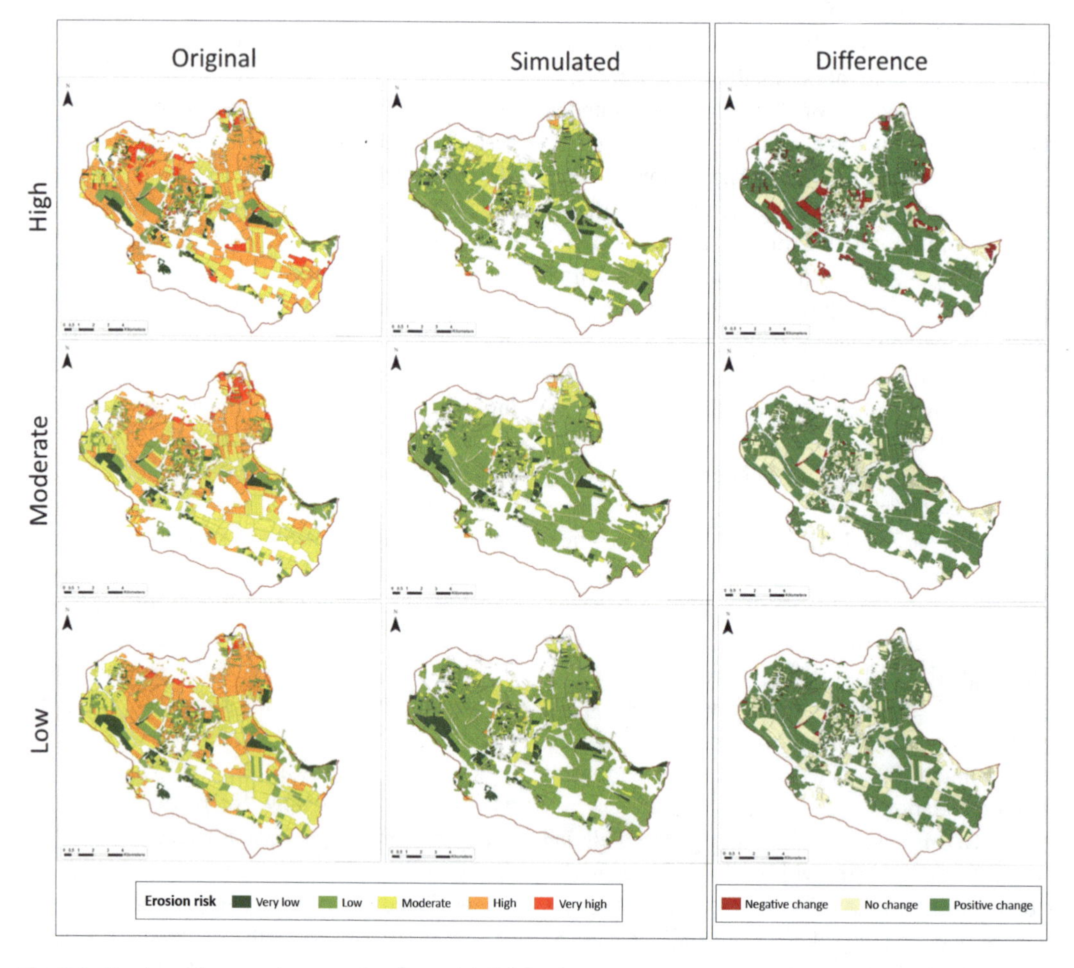

Fig. 8.9 Erosion-risk output layers under three rainfall intensity simulation scenarios. The left panels are original risk maps under the three rainfall intensity events; the middle column panels are the risk output layers after the recommendations by GISCAME; and the right-hand panels are the algebraic differences between the two. Each map is divided into five categories of erosion risk—from Very Low (dark green) to Very High (red) (from Barshad 2019)

water erosion risk. The central panel shows a dramatic decrease in water erosion risk following application of the recommendations in the central column. The colors show how recommendations for cultivation methods have much larger impacts on water erosion than the rainfall intensity effect (i.e., the middle column panels are greener than the those in the left column). The right-hand panels show the spatial distribution of change in water erosion risk in the wake of the GISCAME recommendations. This clearly show how the system improved the conservation performance under the conditions of the three rainfall events. The impact score for each of the three ES can be presented in a comparative manner, as in Fig. 8.10.

The impact score represents the level the agro-ecosystem can provide the service, and a comparison can be made between the pre- and post-recommendation on field cultivation by GISCAME. Erosion regulation is widely studied—mainly in reference to conservation techniques (Xiong et al. 2018). In Barshad (2019), Water Erosion Regulation was significantly higher after shifting to the cultivation method recommended by the system in the entire catchment (Wilcoxon's signed rank test $P < 10–200$). It was found that water erosion can be reduced through informed use of soil conservation actions (Montgomery 2007; Morgan 2009). Drought-Risk Regulations, however, did not change following the system recommendations (Wilcoxon's signed rank test $P = 1.0$). In other words, the impact score did not change in response to the conversion of the cultivation method, since no significant differences were found between the different rainfall intensities, and because the slope of the field does not affect the amount of irrigation water used in the field. As with climate regulation, Food & Fodder did show a significant increase in the impact values, as a positive by-product of the enhanced Erosion Regulation (or

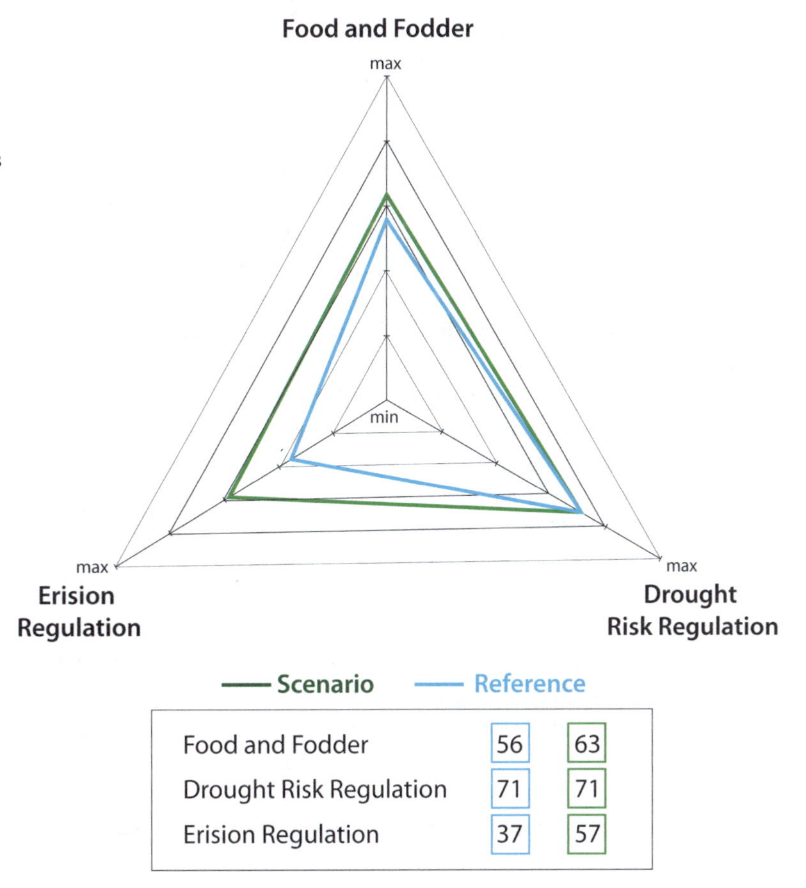

Fig. 8.10 The effect of the GISCAME recommendations on three ecosystem services. The radar *diagram* clearly demonstrates how the Drought-Risk Regulation was not affected, but the Erosion Regulation and Food & Fodder supply was improved by the GISCAME recommendations (from Barshad 2019)

the diminished erosion risk) (Wilcoxon's signed rank test $P < 10–170$). This parameter represents the yield [ton ha^{-1}] which the field can provide for each scenario. Thus, the transition to conservation tillage in light of the system recommendations led to an increase in yield.

From the farmer's point of view, the increased yield of the Food & Fodder ES improves profits from the field. Teklewold and Mekonnen (2017) and Jat et al. (2014) found that reduced tillage and residue management can increase profits from the cultivation of large areas of field crops. Higher net returns were also found in the conservation tillage scenario, and explained by a decrease in production costs (Jat et al. 2014). Farmers strive to increase yields while inflicting minimal environmental damage. In practical terms, that means efficient N, P, and water use, and reduced use of toxic pesticides (Holland 2004).

Since our stimulations predict a clear reduction in erosion risk associated with an increase in the Food & Fodder ecosystem service, farmers adopting the system will not only benefit from potential increases in crop yield, but also contribute to Israel's food security. Moreover, since many agricultural regions—both in Israel and around the globe—share similar characteristics with the Harod catchment, and experience extreme climate changes including desertification, the robustness of our system to such climate changes, as shown in the case of rainfall intensity, suggests that its recommendations can be used to reduce soil erosion risks both now and in the foreseeable future. Thus, applying decision-making systems and transitioning to conservation cultivation methods may arguably not only reduce soil erosion, but also benefit farmers in increasing yield productivity, and thereby profitability.

8.5 Summary

In practice, SDSS can be useful in simulating the catchment response to changing conditions based on possible land use alternatives (Srivastava et al. 2012). We can test the effect of soil treatment either by ranking the possible alternatives to the given environmental conditions, or by changing the criteria for assessing the alternatives. The output maps of the SDSS can be used as a guide to optimize ES delivery by the catchment at risk. Reliability of SDSS recommendations depends on the selection of criteria, indicators, and possible alternatives: If the criteria are not chosen wisely, the recommended alternatives will not be effective, and the recommendations will be less efficient in the provision of ecosystem services. With MCDM, the recommendations mechanism can be relatively simple and flexible in the decision-making process, and we might apply it over a wide array of conditions that require problem solving. The GISCAME comprises five components, that can provide a more complex SDSS based on cellular automation—including the effect of neighboring grid-cells. The estimates may therefore be more demanding than they appear. However, GISCAME opens an excellent opportunity to apply decision-making targeted directly to estimate the effect of changing land use alternatives or conservation treatments on the provision of ecosystem services with the use of soil health indicators as the system criteria. Soil conservation projects can apply this approach using soil sampling and laboratory tests. Other directions can include the use of remotely sensed data with the necessary adaptations to fit the landscape measures to the clorpt conditions in the catchment. It is expected that in the near future the estimation of soil health and ES will contribute substantially to the studies of water erosion consequence.

Review Questions

1. What are the components of an SDSS, and what is the role of each component? Specifically, what are objectives, alternative decisions, and criteria?
2. What is a *single-criterion system*—and how does it differ from a *multicriteria system*? Explain the functioning of a single-criterion system using Fig. 8.3.
3. The data in the table below represents measurements in a field in the Harod catchment. The weights W of the criteria j are as follows: Surface hardness $W_i = 0.2$; AWC $W_i = 0.3$; Aggregate stability $W_i = 0.5$. How would you

cultivate your field in the next season, using *Weighted Summation* (WS)? If this option is not possible, what is your second alternative?

Criteria (j)	Surface hardness	AWC	Aggregate stability	U_i
No Tillage	0.6	0.7	0.7	
Reduced Tillage	0.3	0.6	0.5	
Conservation Tillage	0.1	0.2	0.3	

4. Use the same table data and variable's weights of question 3, to compute *Concordance–discordance analysis* (CA) to decide on how you would cultivate your field in the next season. Rank the order of preference of the alternatives.
5. What are the five components of GISCAME? What is the role of each component? Explain the interaction between the components.
6. Display the data in the table below in *a radar diagram*, and explain which of the alternatives is more efficient, and why?

ES	Before recommendations	After recommendations
Pollination	0.5	0.7
Food	0.5	0.5
Pest control	0.2	0.7
Ecotherapy	0.4	0.1
Biogenic minerals	0.4	0.2

Note: The area of a pentagon can be computed by dividing it into triangles within the boundaries of the radar diagram coordinate system.

7. What is the role of the transition probabilities in the *cellular automata* component of GISCAME? Give an example to how to formulate such a transition probability.
8. Provide a single-line definition of each of the steps in the *cellular automata* component of GISCAME.

7. What is the role of the transition probabilities in the *cellular automata* component of GISCAME? Give an example to how to formulate such a transition probability.
8. Provide a single-line definition of each of the steps in the *cellular automata* component of GISCAME.

Strategies to address the Review Questions

1. SDSS is an interactive, computer-based system designed to support a user in achieving higher effectiveness in decision-making while solving a semi-structured spatial decision problem (Sect. 8.1.2). The four basic components of SDSS being used in *Multicriteria Decision-Making* include: *Goal*: the state that the decision-maker wishes to achieve. *Criterion/criteria*: the standard by which the selection of an alternative is being made. The criteria include an *objective function* which indicates the desired state and an *attribute* which is its values. *Alternative decision*: a course of action that has an actual outcome in the problem-solving process. *Optimization*: the process of choosing the most preferred alternative based on a well-defined criterion from a set of feasible alternatives.
2. *A single-criterion system* (Sect. 8.1.3) allows to identify the critical point of that criterion to choose the best alternative. To do so, a graph is designed where all different alternatives are considered while based on the criterion range of values, the optimal value and the course of action are selected (Fig. 8.3). In MCDM (Sect. 8.3.1) we use weights of influence for each criterion. Doing that, MCDM gives us a broader perspective than a single-criterion system which can only take one criterion in consideration at a time.
3. Using the *weighted summation* for ranking MCDM alternatives is done by following the procedure in Sect. 8.3.3.1. We can compute the weighted normalized decision matrix (Table 8.4, Eq. 8.1).

Criteria (j)	Surface hardness (wj = 0.2)	AWC (wj = 0.3)	Aggregate stability (wj = 0.5)	U_i	Ranking
No tillage	0.6	0.7	0.7	(0.6·0.2) + (0.7·0.3) + (0.7·0.5) = 0.68	1
Reduced tillage	0.3	0.6	0.5	0.49	2
Conservation tillage	0.1	0.2	0.3	0.23	3

Ranking is determined based on the size of U_i, from the highest score to the lowest.

4. Ranking the MCDM alternatives by *concordant-discordance analysis* is done according to the steps detailed in Sect. 8.3.3.2 and Table 8.6. To infer the normalized x_{ij} from the original data we use Eq. 8.7.

	Original data (a)			Normalized x_{ij} (b)			Normalized weight (c)		
Criteria →	AWC	OM	N	AWC	OM	N	AWC	OM	N
Weights →							0.2	0.3	0.5
Alternatives ↓									
No till	0.6	0.7	0.7	$\frac{0.6}{\sqrt{(0.6^2+0.3^2+0.1^2}}=0.88$	0.74	0.77	0.88·0.2 = 0.18	0.22	0.38
Reduced till	0.3	0.6	0.5	$\frac{0.3}{\sqrt{(0.6^2+0.3^2+0.1^2}}=0.44$	0.64	0.55	0.44·0.2 = 0.09	0.19	0.27
Traditional till	0.1	0.2	0.3	$\frac{0.1}{\sqrt{(0.6^2+0.3^2+0.1^2}}=0.15$	0.21	0.33	0.15·0.2 = 0.03	0.06	0.16

After computing the normalized weights, we compute the *dominance matrix* for ranking

	Dominance matrix		
	No till	Reduced till	Traditional till
No till	0	1	1
Reduced till	0	0	1
Traditional till	0	0	0

In the dominance matrix, we see the ranking based on aggregation of concordance and discordance matrices. In our case, the No Till alternative clearly outperforms the other two, while the Reduced Till alternative outperforms the Traditional Till one.

5. GISCAME is an object-oriented SDSS developed to simulate land use changes and their effect on water erosion and the delivery of ES by agroecosystems (Sect. 8.4.1). The five components of GISCAME are:

- Creating a spatial dataset with land use layers and *Ecosystem Services* (ES) relevant to the catchment;
- Selecting indicators to rank the land uses we chose in the first component, and normalizing them;
- Computing the evaluation matrix of the land uses and the ESs, and normalizing the scores;
- Creating planning and environmental rules which will guide us in the next step;
- Mapping the rules we created for our dataset, to see if the states are improved compared with the original state.

6. A *radar diagram* (Fig. 8.8) illustrates data values from several variables by means of a graph of several axes emanating from a single origin. We can see that Pest control went through the highest change and for that reason it can be considered the most efficient alternative. A large difference, as is explained in the text, means that the recommendations were successful in reducing the risks.
7. *Transition probabilities* (Sect. 8.4.1, Fig. 8.7) can be assigned based on statistical data or expert's knowledge. They help us understand the effect of one land use on others and also on the environmental factors. To formulate a transition probability, we need consider the current catchment conditions, and then assign, based on expert knowledge, for example, the predictions for the future state of land uses in the catchment.
8. A single-line definition of each of the steps in the *cellular automata* components of GIS-CAME can be found in Sect. 8.4.1:

- Collecting data of the current state of the catchment conditions;
- Applying transition probabilities based on the experts or statistical data;
- Assigning the conditions in the previous step to the neighborhood grid-cells;
- Execute the recommendation to create the new conditions of the catchment.

References

Akbari M, Memarian H, Neamatollahi E et al (2020) Prioritizing policies and strategies for desertification risk management using MCDM–DPSIR approach in northeastern Iran. Environ Dev Sustain 23(2):2503–2523

Barshad G (2019) Mitigating land degradation in semiarid catchments using a spatial decision support system. Ben-Gurion University of the Negev

Bauder ET (2005) The effects of an unpredictable precipitation regime on vernal pool hydrology. Freshw Biol 50(12):2129–2135. https://doi.org/10.1111/j.1365-2427.2005.01471.x

Bhattacharya RK, Chatterjee ND, Das K (2020) Sub-basin prioritization for assessment of soil erosion susceptibility in Kangsabati, a plateau basin: a comparison between MCDM and SWAT models. Sci Total Environ 734:139474

Bodini A, Giavelli G (1992) Multicriteria analysis as a tool to investigate compatibility between conservation and development on Salina Island, Aeolian Archipelago Italy. Environ Manag 16(5):633

Braden B (1986) The surveyor's area formula. Coll Math J 17(4):326–337

Cohen MJ, Shepherd KD, Walsh MG (2005) Empirical reformulation of the universal soil loss equation for erosion risk assessment in a tropical watershed. Geoderma 124(3–4):235–252

Costanza R, d'Arge R, de Groot R et al (1997) The value of the world's ecosystem services and natural capital. Nature 387(6630):253–260

Crossland MD, Wynne BE, Perkins WC (1995) Spatial decision support systems: an overview of technology and a test of efficacy. Decis Support Syst 14(3):219–235

de Azevedo RG, de Souza Filho FA, Nelson DR et al (2020) Development of a drought vulnerability index using MCDM and GIS: study case in São Paulo and Ceará Brazil. Nat Haz 104(2):1781–1799

De Groot RS, Alkemade R, Braat L et al (2010) Challenges in integrating the concept of ecosystem services and values in landscape planning, management and decision making. Ecol Complex 7(3):260–272

Dortaj A, Maghsoudy S, Ardejani FD et al (2020) Locating suitable sites for construction of subsurface dams in semiarid region of Iran: using modified ELECTRE III. Sustain Water Resourc Manag 6(1):1–13

Dragan M, Feoli E, Fernetti M et al (2003) Application of a spatial decision support system (SDSS) to reduce soil erosion in northern Ethiopia. Environ Model Softw 18(10):861–868

Feick RD, Hall GB (2001) Balancing consensus and conflict with a GIS-based multi-participant, multi-criteria decision support tool. GeoJournal 53(4):391–406

Ferretti V, Montibeller G (2016) Key challenges and meta-choices in designing and applying multi-criteria spatial decision support systems. Decis Support Syst 84:41–52

Frank S, Fürst C, Witt A et al (2014) Making use of the ecosystem services concept in regional planning—trade-offs from reducing water erosion. Landscape Ecol 29(8):1377–1391

Fürst C, Frank S, Pietzsch K, et al (2015) GISCAME 2.0-Manual 2.4

Fürst C, Volk M, Pietzsch K et al (2010) Pimp your landscape: a tool for qualitative evaluation of the effects of regional planning measures on ecosystem services. Environ Manage 46(6):953–968

Ghaleno MRD, Meshram SG, Alvandi E (2020) Pragmatic approach for prioritization of flood and sedimentation hazard potential of watersheds. Soft Comput 24:15701–15714

Grêt-Regamey A, Sirén E, Brunner SH et al (2017) Review of decision support tools to operationalize the ecosystem services concept. Ecosyst Serv 26:306–315

Hein L, Van Koppen K, De Groot RS et al (2006) Spatial scales, stakeholders and the valuation of ecosystem services. Ecol Econ 57(2):209–228

Holland JM (2004) The environmental consequences of adopting conservation tillage in Europe: reviewing the evidence. Agr Ecosyst Environ 103(1):1–25

Inkoom JN, Frank S, Greve K et al (2018) A framework to assess landscape structural capacity to provide regulating ecosystem services in West Africa. J Environ Manage 209:393–408

Jaiswal RK, Lohani AK, Tiwari HL (2015) Statistical analysis for change detection and trend assessment in climatological parameters. Environ Process 2(4):729–749

Jat RA, Sahrawat KL, Kassam AH, et al (2014) Conservation agriculture for sustainable and resilient agriculture: Global status, prospects and challenges. Conserv Agric Glob Prosp challenges 1–25

Jelokhani-Niaraki M, Sadeghi-Niaraki A, Choi SM (2018) Semantic interoperability of GIS and MCDA tools for environmental assessment and decision making. Environ Model Softw 100:104–122

Kahneman D, Slovic P, Tversky A (1974) Judgment under uncertainty: heuristics and biases. Cambridge University Press, New York

Kangas A, Kurttila M, Hujala T, et al (2015) Single-criterion problems. In: Anonymous decision support for forest management. Springer International Publishing, Switzerland

Kaur R, Srivastava R, Betne R et al (2004) Integration of linear programming and a watershed-scale hydrologic model for proposing an optimized land use plan and assessing its impact on soil conservation—a case study of the Nagwan watershed in the Hazaribagh district of Jharkhand, India. Int J Geogr Inf Sci 18 (1):73–98

Keenan BP (2003) Spatial decision support systems. In: Mora M, Forgionne G, Gupta JND (eds) Decision making support systems: achievements and challenges for the new decade IGI Global. USA, pp 29–39

Keenan BP, Jankowski P (2019) Spatial decision support systems: three decades on. Decis Support Syst 116:64–76

Keeney RL (1982) Decision analysis: an overview. Oper Res 30:803–838

Koschke L, Fürst C, Frank S et al (2012) A multi-criteria approach for an integrated land cover-based assessment of ecosystem services provision to support landscape planning. Ecol Ind 21:54–66

Kumar A, Sah B, Singh AR et al (2017) A review of multi criteria decision making (MCDM) towards sustainable renewable energy development. Renew Sustain Energy Rev 69:596–609

Kuok KK, Mah DY, Chiu PC (2013) Evaluation of C and P factors in universal soil loss equation on trapping sediment: case study of Santubong River. J Water Resource Protect

Lilburne L, Eger A, Mudge P, et al (2020) The land resource circle: Supporting land use decision making with an ecosystem-service-based framework of soil functions. Geoderma 363(114134)

Lorenz M, Fürst C, Thiel E (2013) A methodological approach for deriving regional crop rotations as basis for the assessment of the impact of agricultural strategies using soil erosion as example. J Environ Manage 127(Supplement):S37–S47

Macary F, Dias JA, Figueira JR et al (2014) A multiple criteria decision analysis model based on ELECTRE TRI-C for erosion risk assessment in agricultural areas. Environ Model Assess 19(3):221–242

Malczewski J (1999) GIS and multicriteria decision analysis. John Wiley & Sons Inc., Canada

Malczewski J (2006) GIS-based multicriteria decision analysis: a survey of the literature. Int J Geogr Inf Sci 20(7):703–726

Malczewski J, Rinner C (2015) Multicriteria decision analysis in geographic information science. Springer, New York

Massei G, Rocchi L, Paolotti L et al (2014) Decision support systems for environmental management: a case study on wastewater from agriculture. J Environ Manage 146:491–504

Meshram SG, Singh VP, Kahya E et al (2020) The feasibility of multi-criteria decision making approach for prioritization of sensitive area at risk of water erosion. Water Resour Manage 34(15):4665–4685

Montgomery DR (2007) Soil erosion and agricultural sustainability. Proc Natl. Acad Sci 104(33). https://doi.org/10.1073/pnas.0611508104

Morgan RPC (2009) Soil erosion and conservation. John Wiley & Sons, Hoboken

Morin E, Gabella G (2007) Radar-based quantitative precipitation estimation over Mediterranean and dry climate regimes. J Geophys Res Atmos 112(D20):1–13. https://doi.org/10.1029/2006JD008206

Newman JP, Maier HR, Riddell GA et al (2017) Review of literature on decision support systems for natural hazard risk reduction: current status and future research directions. Environ Model Softw 96:378–409

Patel DP, Srivastava PK, Gupta M et al (2015) Decision support system integrated with geographic information system to target restoration actions in watersheds of arid environment: a case study of Hathmati watershed, Sabarkantha district Gujarat. J Earth Syst Sci 124 (1):71–86

Perez J, Salazar RC, Stokes A (2017) An open access database of plant species useful for controlling soil erosion and substrate mass movement. Ecol Eng 99:530–534

Roy B (1968) Classement et choix en présence de points de vue multiples (la méthode ELECTRE). RIRO 8:57–75

Shahbazi F, Jafarzadeh AA (2010) Integrated assessment of rural lands for sustainable development using MicroLEIS DSS in West Azerbaijan Iran. Geoderma 157(3–4):175–184

Simon HA (1976) From Substantive to Procedural Rationality. In: Anonymous 25 years of economic theory. Springer, Boston, MA, pp 65–86

Simon HA (1991) Bounded rationality and organizational learning. Organ Sci 2(1):125–134

Souissi D, Zouhri L, Hammami S et al (2020) GIS-based MCDM–AHP modeling for flood susceptibility mapping of arid areas, southeastern Tunisia. Geocarto Int 35(9):991–1017

Srivastava AK, Gaiser T, Cornet D et al (2012) Estimation of effective fallow availability for the prediction of yam productivity at the regional scale using model-based multiple scenario analysis. Field Crop Res 131:32–39

Svoray T, Ben-Said S (2010) Soil loss, water ponding and sediment deposition variations as a consequence of rainfall intensity and land use: a multi-criteria analysis. Earth Surf Proc Land 35(2):202–216. https://doi.org/10.1002/esp.1901

Svoray T, Levi R, Zaidenberg R et al (2015) The effect of cultivation method on erosion in agricultural catchments: Integrating AHP in GIS environments. Earth Surf Proc Land 40(6):711–725. https://doi.org/10.1002/esp.3661

Tayyebi A, Meehan TD, Dischler J et al (2016) SmartScape™: a web-based decision support system for assessing the tradeoffs among multiple ecosystem services under crop-change scenarios. Comput Electron Agric 121:108–121

Teklewold H, Mekonnen A (2017) The tilling of land in a changing climate: empirical evidence from the Nile Basin of Ethiopia. Land Use Policy 67:449–459

Xiong M, Sun R, Chen L (2018) Effects of soil conservation techniques on water erosion control: a global analysis. Sci Total Environ 645:753–760

9 Final Thoughts

Abstract

This chapter summarizes the main findings of the book. It describes possible geoinformatics applications that may be used to extend water erosion studies beyond soil loss assessments, to comprehensive soil health scoring, and the provision of various ecosystem services. The chapter is founded on the knowledge framework famously described by the former US Secretary of State Donald Rumsfeld under the title "known knowns". This framework is interpreted in this book to include the following techniques: (1) the commonly available geoinformatics applications that each soil conservationalist can apply to the catchment she is responsible for; (2) the more complex methods that are available in the scientific literature, but less commonly used, and deserve to be made known more widely; (3) future directions of current research that should be further developed and applied, in the short term; and (4) the long-term research that represents our future breakthroughs in water erosion geoinformatics.

Keywords

Future directions · Geoinformatics applications · Known knowns

9.1 The General Framework

Identifying areas at risk of soil degradation and reducing the on-site and off-site consequences of accelerated water erosion to agroecosystem components are of interest to scientists, practitioners, and decision-makers. These adverse consequences include, for example, eutrophication of water bodies and contamination of neighboring fields downslope, diminished carbon sequestration, structural failures of dams and other structures, reduced water storage capacity in the soil, stagnant water and decreased aeration, and, of course, soil loss.

These and other effects of water erosion account for billions of dollars' worth of damage every year that has been around for a very long time, and it spans large swathes of the earth's surface in various climatic regions and lithologic environments. Soil degradation—one of the most pernicious outcomes of water erosion—can bring diminished food production, supply, and security, and decreased provision of many other important *Ecosystem Services* (ES) such as water purification, flood protection, and pest control.

With the emergence of the information age—and of geoinformatics in particular—new opportunities have arisen to conduct more sophisticated analyses of spatial patterns of water

T. Svoray, *A Geoinformatics Approach to Water Erosion*,
https://doi.org/10.1007/978-3-030-91536-0_9

erosion and soil degradation. Geoinformatics can shed new light on the processes and factors that control water erosion in time and space, on landscape formation and the destruction of various entities, and on the possible use of decision-making systems to improve planning and management of agricultural catchments.

The use of geoinformatics to improve catchment management is not entirely new. The roots of this book lie in the wide range of Remote Sensing (RS) studies, DEM-based geomorphometric computations, physically based and empirical models, geostatistics, and GIS-based spatial analyses used by previous authors to study primarily soil loss and deposition budgets. These works have successfully and accurately quantified soil and water flow processes over several time scales. They have also been successful in isolating *clorpt* factors and exploring their effect on the entire soil loss process, and have documented, over more than three decades, soil degradation processes in agricultural catchments. Studies of this sort are, and will continue to be, important because the loss of the soil itself is the most evident and destructive impact on the soil layer in the long run (see Chap. 1).

However, by adopting a more comprehensive approach, more can be done with quantitative geoinformatics tools to assess water erosion consequences in agricultural catchments, beyond soil loss. Soils in agroecosystems are complex and are affected by various physical, chemical, and biological processes giving rise to compound surface mosaics involving multiple interrelated components. Given the innate heterogenous structure of soils, the reduction in their function in providing ES is multifaceted, and quantifying the damage to soils and agroecosystems is more complex than apparent from merely modeling soil loss and deposit budgets. Natural soil processes are complex enough, but to these we must add the highly intensive human activities that dramatically and adversely affect the provision of ES by agricultural soils. These human activities include, for example, soil cultivation and fertilization, driving in unpaved roads, the impacts of heavy machinery due to their weight pressure, harvesting, crop rotation, and conservation practices.

The aim of this book is to survey the use of diverse and versatile geoinformatics tools in the study of spatio-temporal variation of water erosion in agricultural catchments. It begins with the early stage of problem definition and ends in the decision-making phase. While this book does not set out to provide the reader with all possible methodologies for this purpose—these are described in more detail in many journal papers—it does aim to propose a general framework for the study of water erosion in agroecosystems using geoinformatics.

This framework—divided into three levels of data analysis—includes the use of RS data, GIS-based data models, spatial modeling, machine-learning techniques, geostatistics, expert systems, and spatial decision-making systems (Fig. 9.1).

In the first level of data analysis—the level of Threat—the analyst acquires spatial and temporal information on the agroecosystem—including monitoring *extrinsic* and *intrinsic* conditions, and the on-site/off-site consequences. This is a rather explanatory level, in which RS data is used to collect updated information about the catchment conditions, by converting spectral reflectance data into a spatial pattern of environmental and human-induced conditions. Other sources of spatial data are, for example, existing databases of governmental bodies, soil samples geotagged with GPS, and electronic theodolite measurements. When we map vegetation growth, roads and tillage direction, field ownership, soil properties, and rainfall intensity, we are not only describing the catchment, but also laying the groundwork for quantifying water erosion risk. Next, information about channeling and gully heads, organic matter content, surface hardness, root health, and many other soil properties is used in the second and third levels, as indicators of soil health, to prioritize soil conservation and management actions.

In the second level—the level of Approach—the information acquired in level 1 is processed to identify areas at risk, to map soil health over the catchment, and to execute heuristic

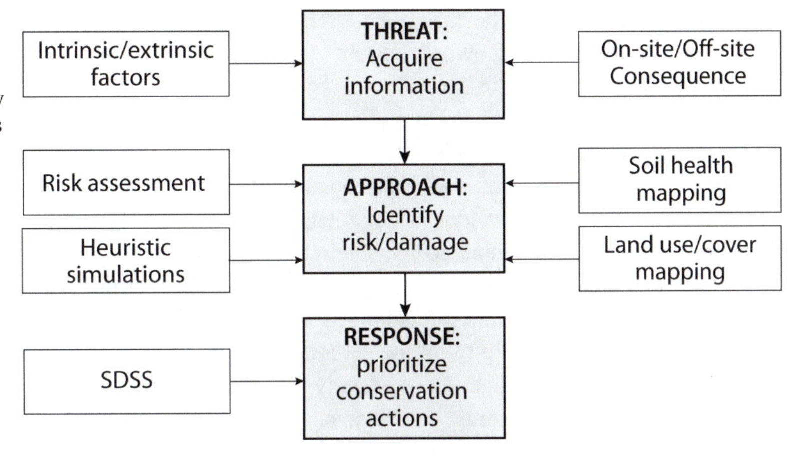

Fig. 9.1 Schematic representation of the three levels of the geoinformatics framework of this book (gray rectangles). The white blocks illustrate the components of each level

simulations of futuristic scenarios. Thus, geoinformatics paradigms, supported by expert systems, machine learning, and other physically based and geostatistical tools, can be used to bypass the need to model the exact amount of soil loss and/or soil deposition. Alternatively, the analyst can focus on classifying the level of risk, which is a simpler task and therefore reduces possible error in prediction. Also, in this level, the analyst is focused not only on soil and water budgets but on predicting soil health including all physical, biological, and chemical aspects.

The third level—the level of Response—concerns the use of spatial decision-support systems to prioritize actual soil conservation actions to reduce, and even prevent, water erosion processes. Thus, the analyst aims to suggest the most suitable conservation action, under the local conditions, to prevent future damage and, as a result, to improve soil function to provide various ES.

We suffer consequences of water erosion globally since the days of the hominids with a large boost since the agricultural revolution. Nowadays, the field of water erosion studies is experiencing a fascinating era in which the urgent need to save our soils, due to the modern and extensive use of intrusive cultivation, meets with the rise of rich geographical datasets and sophisticated data-driven and theory-driven methodologies in geoinformatics that can be exploited to mitigate soil degradation.

9.2 Usage of Geoinformatics to Study Water Erosion

Geoinformatics data models and analytical tools can be used to gather, store, manage, and analyze information about agricultural catchments from diverse aspects. This book covers many geoinformatics methodologies and there are several ways to summarize them. I have chosen to use the framework of knowledge acquisition that is based on the quadratic matrix of combinations between known and unknown information, denoted as the "*known knowns*" matrix. This knowledge framework has received a worldwide exposure by Donald Rumsfeld, then the United States Secretary of Defense, in a Pentagon news briefing, in February 2002, and it can be most clearly described in Rumsfeld's own words:

> Reports that say that something hasn't happened are always interesting to me, because as we know, there are known knowns; there are things we know we know. We also know there are known unknowns; that is to say we know there are some things we do not know. But there are also unknown unknowns—the ones we don't know we don't know. And if one looks throughout the history of our country and other free countries, it is the latter category that tends to be the difficult ones.

The "known knowns" framework was not originally coined by Rumsfeld, and had been in use before him within the intelligence community. Furthermore, the theoretical framework for the "known knowns" matrix was developed by two psychologists using a social sciences paradigm known as *the graphic model of awareness*, or the *Johari Window* (Luft and Ingham 1955). This matrix was subsequently also used to describe knowledge acquisition in the natural sciences (Little et al. 2011), and in the soil sciences, in particular, by McBratney (2006).

Here I offer my own interpretation of the "known knowns" matrix to present the summary of geoinformatics knowledge on the study of water erosion in agricultural catchments, as described in this book (Fig. 9.2).

According to the matrix in Fig. 9.2, the known known framework may represent the geoinformatics tools described in this book, through the following four aspects:

Known knowns (1) are the geoinformatics algorithms that many of the potential users/practitioners "know how" to apply and can easily execute, using data from their own agricultural sites, with readily available software. These methods are usually robust, and can be practiced in agricultural catchments, under various clorpt conditions, with relatively high certainty.

The Unknown knowns (2) are the geoinformatics methods that are already developed and published in the scientific literature, but are more complex and need more effort—such as validation over a wide parameter space, simplification, or exposure to users—before they can be applied under real-world conditions by practitioners. In other words, these methods are known to others (mainly scientists) and have the potential for good use by practitioners. These existing tools are still hidden in the literature, and must be promoted for more widespread use, in a bid to improve soil conservation and management.

The Known unknowns (3) is the group of geoinformatics methods that I believe need more research and further development to be used in real-world applications. In other words, these are the short-term research and development directions that need to be explored with more empirical evidence, or more established theoretical framework.

The Unknown unknowns group (4) is the last component of the matrix, and perhaps the hardest to predict. These are the methods and research directions that mainstream science in the discipline has not yet identified, and they may represent the next long-term breakthrough in geoinformatics water erosion.

9.2.1 Known Knowns

The methods in this category are relatively robust, have been used with success in many case studies, and can be applied in a variety of catchments, after model parameterization and field validation:

Vegetation Indices (VI)—especially the NDVI—are among the most common applications for vegetation mapping, in RS, in general, and in water erosion studies, in particular. However, because VI can reach saturation in fully grown fields, they are used mainly for estimating crop cover and soil fertility at the earlier crop growth stage, when canopy cover is

	Known	Unknown
Known	(1) Known knowns *Current practice*	(2) Unknown knowns *Current science*
Unknown	(3) Known unknowns *Needs further R&D*	(4) Unknown unknowns *Scientific breakthroughs*

Fig. 9.2 The knowledge discovery of geoinformatics tools for studying water erosion in agricultural catchments based on the "known knowns framework"

at a low level. The fields in this state are especially sensitive to water erosion processes because when vegetation cover is sparse overland flow can still be generated.

Hard classification techniques of surface covers offer high accuracy in labeling soil, rock, vegetation, and built objects, including roads and buildings. Supervised classification can provide higher (occasionally, much higher) accuracy than unsupervised classification so the effort to gather training data appears to be worthwhile. However, when applied to real-world problems, it is still important to test the classifiers against validation data, because despite little algorithmic differences between them, predictive accuracy can vary greatly.

Drainage structure modeling is a well-established field since the 1980s and offers important geomorphometric information. This includes the *Topographic Threshold* (TT), which has been used and verified in many catchments around the world. TT relies on two of the topographic factors that connect the dots between all erosion models surveyed in Chap. 3 and other studies: The *contributing area,* and the *hillslope gradient*. Although TT is a binary index, and data-mining and expert-based methods outperform it, it is recommended for use in reconnaissance studies. Contributing area and overland flow direction can be computed with high certainty using the D8 algorithm, or one of its successors. Moreover, the aforementioned classification tools can be used to map surface covers and to attribute non-stationary runoff coefficients from the literature to the mapped surface cover objects, using simple algebraic equations.

Extracting expert knowledge of practitioners or scientists is an attractive approach in geoinformatics, because it can replace a demanding data-driven methodology that usually requires large data gathering effort. While the *pairwise comparison matrix* may have some limitations, it is a robust and systematic method for extracting expert knowledge in the effort to predict soil degradation risk, for example.

The fuzzy rule-based approach can be used to simulate water erosion processes with flexible *membership functions*. Evidence from the past two decades shows that this approach has achieved reliable representation of reality on the ground. Perhaps more importantly, fuzzy logic is a technique with high generalization capabilities that can be applied to catchments of varying conditions.

Despite the advantages of RS-based soil indices, **soil health mapping** still relies mostly on point measurements of various soil properties. The stratified random technique represents surface, and particularly soil, heterogeneity, through GIS layers, more successfully than random and matrix sampling techniques.

Spatial interpolation can be applied based on *Tobler's Law* using *IDW*, with d^{-2} representing the effect of distance on the estimated variable. *Kriging interpolation* has also been widely used successfully, but it requires quantification of spatial autocorrelation. Despite the achievements of *ordinary kriging*, *kriging with covariates*—and *cokriging*, in particular—have been found to be more successful, and are highly recommended.

In the field of **decision-making**, the MCDM —both in the form of *weighted summation* and the *concordance–discordance analysis*—is a simple method that can be applied to produce recommended prioritization of soil conservation actions, based on several criteria. Alternatively, GISCAME is a more detailed SDSS approach that provides recommendations on land use changes and soil treatments to improve the provision of ecosystem services.

9.2.2 Known Unknowns

These methods represent short-term research directions. At the present time, they can be applied—with caution—but further development is recommended.

Soil Evolution Models (SEM) were evolved from landscape-evolution models, and are a promising physically based modeling approach that can be used to estimate soil loss and deposition dynamics. Of particular interest is further development of SEM to simulate the effect of short-term (seasonal scale) erosion processes, in general, and that of human intervention and cultivation methods, in particular, on channeled and sheet erosion in agricultural catchments.

Further study of **Spectral signatures** of soil properties—in particular, the analysis of spatial variation in *soil indices*—may be useful in producing reliable catchment scale mapping of soil variation. In particular, a meta-analysis of hyperspectral signatures of soil properties under heterogeneous conditions (clorpt variations and shadowing/atmospheric effects, etc.) may advance the mapping of soil health from RS data (Silvero et al. 2021).

Perhaps one of the leading tools for studying spatial patterns of soil loss-deposition budgets is the **change-detection analysis** of elevation data in agricultural fields from LiDAR data or high-resolution drone-derived DEM data. Although this line of research is relatively straightforward, and the technology and mathematical formulation is relatively simple, we do not yet have sufficient evidence on between-events spatial patterns of soil loss-deposition budgets in agricultural fields.

Sub surface processes have received less attention in the literature, despite their large impact on soil and water transport. A major challenge in piping studies is the monitoring of the pipes network dimensions and connectivity. One important direction to promote this research avenue is advanced use of Ground-Penetrating Radar (GPR) (Got et al. 2020).

More study can be useful in quantifying relative contribution of **water and wind** as agents of soil loss (Katra 2020). We do not have enough evidence or understanding of how much soil and other particles leave the field due to water or wind activity. A theoretical framework to quantify these dynamics does not yet exist, and can be very valuable.

Finally, **the farmers** are the most influential human agents governing water erosion in agricultural catchments. Through their use of cultivation methods, cropping systems, roads, and structures—and, of course, conservation actions and drainage engineering—farmers crucially affect on-site and off-site consequences. One way to enlist farmers to engage in pro-environmental behavior is outlined in the *Dragons of Inaction* (Gifford 2011, and Chap. 2). But the application of the Dragons to water erosion studies is still overlooked, and more empirical studies are needed to validate this framework, and to interpret what makes farmers support conservational actions.

9.2.3 Unknown Knowns

This section of the matrix summarizes the geoinformatics tools that are available in the scientific literature, and can be applied, but the practitioners community is not sufficiently aware of them—either because these tools are not embedded in the commercial GIS software, or because practitioners are insufficiently conversant with the scientific literature, or for any other reason.

Spatial data models are not always optimally used in water erosion studies, and often the developers use raster data—or more rarely, *Thiessen Polygons* with vector data. Park et al. (2001)'s *hillslope algorithm* can be used to divide hillslopes into theoretical catenary units, thereby enabling a more process-based analysis of soil and water movements. Similarly, the subcatchment as a spatial object for data aggregation can be used for catchment scale analysis. Another approach that may be meaningful in water erosion studies is parcel segmentation for object-based analysis. Although an accurate image segmentation into field boundaries is not an easy task, many users obtain data from existing GIS databases, and can use the vector field boundaries for *zonal statistics* or *object-based classification*. While the latter can be prone to error, it can smooth out the results, providing less noisy information and a division of the catchment into more administratively meaningful units (e.g., ownership, the associated village, etc.) that can be important in terms of the cultivation method being used.

Physically based models that quantify soil loss and deposition, such as the GeoWEPP, are available online and can be applied to provide information about the *on-site* and *off-site consequences* of water erosion in the catchment, under various environmental and human-induced conditions, in a spatially and temporally explicit manner.

The **spectral unmixing method** is widely described in the RS literature and has a considerable potential in quantifying sub-pixel data. This approach is essential, because the information needed for soil conservation implementation is often more detailed than the spatial resolution provided by most available RS data. Wider use of this method can help to stem soil degradation in heterogenous catchments.

Machine-learning and **data-mining** procedures are a hot topic in the scientific literature that provide a wide range of classifiers for accurate categorization of surface formations, consequences of water erosion, and identification of areas at water erosion risk. The data science paradigms also provide advanced tools for detailed accuracy assessment.

Soil health indices are used to study and monitor the health of agroecosystems after erosion events. These have been widely studied over the past three decades, and should be more commonly applied by farmers and practitioners. Although the literature provides indices and theoretical framework for monitoring the function of agroecosystems and their services (Lehmann et al. 2020), real-world applications are still few and far between.

Spatial autocorrelation of soil properties has also been widely studied in the soil and pedometrics literature, and the progress in this field can help improve soil health mapping applications. As we saw in Chaps. 4 and 7, three measures can be used for this purpose: The *nugget–sill ratio*, the *Moran's I*, and the *variogram envelope*.

9.2.4 Unknown Unknowns

The unknown unknowns are the research questions that the mainstream scientific community has not yet addressed—either because they appeared to be unachievable or less important, or because focus was directed at other trends.

Predicting the unknown unknowns is naturally difficult, especially in the extremely dynamic world of geoinformatics. Some might even say that it is an oxymoron—as how can one predict what one doesn't know? But I have chosen to take the risk although possibly some of the suggestions below will not be bold enough and others will not be materialized in the near future.

Real-time monitoring of water erosion processes can be used to identify thresholds in channeling initiation points, runoff development, and water ponding as they occur. Such ambitious monitoring capabilities require RS measurements through rain, fog, and at night. SAR-based data analysis may be one such method, or some other technology yet to be developed. Currently, however, documentation of water erosion processes from sensors in real time is scarce or nonexistent.

The very small scale (<1 cm) is overlooked, especially in the case of *in situ* geoinformatics studies. Sheet flow is affected by roughness at the scale of millimeters—including small disturbances that can lead to incision and channeled flow. These can result from small pits created by a tractor turning around, or due to bioturbation such as lair-digging activity by voles, and the like. Sheet flow is also affected by the state of a rock fraction—between loosely resting and embedded—which can cause large variation in runoff connectivity. Water flow at this scale has been studied in lab experiments (Danino et al. 2021), but not mapped over wide regions. Close-range remote sensing using drones or other platforms can be a step forward in this research gap.

The effect of **intrinsic soil factors** on soil variation is less well known than the effect of clorpt factors. We have effective statistical and geostatistical means of quantifying the effect of system complexity, but less developed theoretical frameworks of cause-and-effect scenarios. Development of such a framework could be very useful in studying water erosion risks at the field, hillslope, or catchment level.

Space–time analysis has also received less attention in the geoinformatics literature. Specifically, the temporal aspect of geostatistics has not been developed as much as it should be. This has led to a major research gap in the study of water erosion processes, in which time plays a crucial role.

Human behavior is a highly influential factor in water erosion processes and is overlooked in many studies. Issues in need of study include the awareness of the public to water erosion; the movements of agricultural machinery; and the cultural and restorative impact of agricultural areas on individuals. Geotagged social network data, sentiment analyses, and tracking technologies can be used for this purpose.

Soil production from scratch—i.e., the acceleration of soil production processes—is another possible line of research, but one that many would consider to be a long shot (Van Breemen and Buurman 1998). That is probably why science has usually disregarded it. Although the subject is not totally neglected, most of the effort in this area is on anthropogenic soils—especially in urban settings, archeological sites, and mines—and much less in the study of soil formation in large agricultural catchments (Howard 2017).

9.3 Summary

This book surveys the problem of water erosion in agricultural catchments and their consequences in terms of diminished soil health and ecosystem services. It offers a geoinformatics framework to provide spatially explicit information that may help to reduce this damage. The idea behind this collection of methods is to enlist scientific theory and empirical evidence to develop real-world applications in soil conservation and drainage engineering. Application of the tools suggested here may benefit from the following recommendations:

1. We must acknowledge that the problem is long-lived, expensive, irreversible, and will not go away if we do not take decisive action.
2. We must keep it simple. The world of data science offers many new approaches, but we must select the ones that are most robust, repeatable, and applicable. Otherwise, we may be swamped with prediction errors when applying a new and seemingly promising method to a new database.
3. Scientific knowledge on the controlling processes and spatial variation of soil has substantially improved the development of various geoinformatics tools but, in addition to developments in basic science, we must stress the crucial importance of testing theory against the notoriously variable real world, and the need for ever more replications, to enhance the accuracy of predictions.
4. We must acknowledge that soil erosion in agricultural catchments has a human dimension, in the form of the environmental behavior of farmers, the policy-making by authorities, and professional recommendations by practitioners.
5. Last but not least, we must remember that although soil loss is perhaps the most dire—and certainly the most documented—consequence of water erosion, it is not the only one. It is essential, therefore, to study the effect of water erosion on agricultural catchments using measures that quantify the effect of water erosion on physical, chemical, and biological traits of the soil.

References

Danino D, Svoray T, Thompson S et al (2021) Quantifying shallow overland flow patterns under laboratory simulations using thermal and LiDAR imagery. Water Resour Res 57:e2020WR028857

Gifford R (2011) The dragons of inaction. Amer Psychol 66(4):290–302. https://doi.org/10.1037/a0023566

Got JB, Bielders CL, Lambot S (2020) Characterizing soil piping networks in loess-derived soils using ground-penetrating radar. Vadose Zone J 19(1):e20006

Howard J (2017) Anthropogenic soils. Springer International Publishing, Switzerland

Katra I (2020) Soil erosion by wind and dust emission in semi-arid soils due to agricultural activities.

Agronomy (Basel) 10(1):89. https://doi.org/10.3390/agronomy10010089

Lehmann J, Bossio DA, Kögel- Knabner I et al (2020) The concept and future prospects of soil health. Nat Rev Earth Environ 1:554–563

Little JL, Cleven CD, Brown SD (2011) Identification of "Known Unknowns" utilizing accurate mass data and chemical abstracts service databases. J Ameri Soc Mass Spectrometry 22(2):348–359

Luft J, Ingham H (1955) The Johari window, a graphic model of interpersonal awareness. Los Angeles: University of California, Los Angeles

McBratney AM (2006) Musings on the future of soil science (in ∼1 k words). In: Hartemink A (ed) the future of soil science international union of soil sciences, pp 86–88

Park SJ, McSweeney K, Lowery B (2001) Identification of the spatial distribution of soils using a process-based terrain characterization. Geoderma 103(3):249–272. https://doi.org/10.1016/S0016-7061(01)00042-8

Silvero NEQ, Demattê JAM, Amorim MTA et al (2021) Soil variability and quantification based on Sentinel-2 and Landsat-8 bare soil images: a comparison. Remote Sens Environ 252(112117). https://doi.org/10.1016/j.rse.2020.112117

Van Breemen N, Buurman P (1998) Soil formation. Kluwer Academic Publishers, Netherlands

Index

A

Accuracy assessment, 166, 168, 175, 179, 211, 341
Agisoft, 192, 193
Akaike's Information Criterion (AIC), 134, 145, 147
Analytical Hierarchical Processes (AHP), 45, 46, 217, 309, 312, 314, 315, 324
Ancillary Information, 129, 141, 161, 284
Area Under Curve (AUC), 211, 235, 236, 242, 243, 245, 246
Artificial Neural Net (ANN), 234, 236, 238, 239, 242–245, 260
Autocorrelation, 107, 127, 129, 131–137, 140, 143–147, 267, 272, 276–283, 285, 289, 297, 298

B

Band ratio indices, 153
Block kriging, 141
Buffer zone, 48, 181, 188, 189, 241

C

CAESAR-Lisflood, 75, 76, 81, 83, 100–103
Catchment, 4, 6, 8–10, 12–14, 18, 19, 24, 26–33, 39–48, 50, 51, 53–55, 57, 60, 61, 64, 66–68, 76, 77, 79–82, 84–87, 89–94, 97–103, 107–109, 112, 113, 115–123, 125–127, 129–131, 138, 144–146, 151–155, 159–161, 166, 168, 171–175, 177, 184, 188, 190, 191, 193, 195, 197–199, 205–207, 209–217, 222, 223, 225, 226, 228–230, 232, 234, 240–243, 246, 250, 255, 257–260, 265–269, 272, 274, 276, 277, 279, 280, 282–284, 287, 289, 293, 295, 297, 298, 305–312, 315, 320–324, 326–328, 330, 331, 335–342
Cellular automata, 82, 100, 101, 313, 320, 323–325, 329, 331
Change detection, 86, 151, 172, 191, 195, 196, 198, 200
Classification, 119, 120, 151–153, 161–164, 166–173, 175, 177, 178, 186, 188, 191, 196–200, 213, 233–236, 239, 241, 242, 244, 245, 248, 259, 260, 312, 339, 340
Climate, 3–5, 7, 13, 17, 33, 58, 60, 62, 63, 75, 86–90, 108, 111, 126, 259, 266, 311, 327, 328
CLORPT, 5–8, 33, 39, 41, 42, 54, 108, 109, 111, 117, 118, 120–122, 125, 126, 145, 146, 155, 156, 199, 206, 210, 217–219, 222, 223, 225, 226, 228, 232, 233, 243, 244, 247, 269, 276, 289, 298, 300, 306, 308, 310–314, 328, 336, 338, 340, 341
Cokriging, 141–143, 265, 284–287, 289–291, 295, 298, 300, 339
Compromised Programming Technique (CPT), 222, 223, 225, 227, 232, 259, 260
Conceptual modeling, 76
Concordance-discordance Analysis (CA), 186, 315, 317, 318, 320, 329, 339
Confusion matrix, 166, 168, 198, 199, 225, 234, 259
Contour, 11, 26, 44, 46, 47, 113, 87, 130, 178, 184, 191, 210
Contributing area, 5, 41, 42, 47, 50, 51, 53, 54, 67–69, 75, 77, 80, 87, 91, 101, 110, 113–119, 145, 146, 177, 185–187, 198, 199, 206–212, 216–222, 225, 228, 233, 237, 241, 249, 250, 254, 255, 257, 259, 279, 284, 289, 291, 307, 309, 310, 315–317, 325, 339
Covariance, 131, 132, 167, 271, 293
Covariates, 141–143, 147, 278, 284, 289, 291, 296, 298, 300
Criteria, 114, 134, 217, 220–222, 225, 226, 228, 231, 305–311, 313–320, 328–330, 339
Cropping system, 12, 39, 42, 48, 67, 68, 121, 122, 144, 217, 311, 312, 315, 340
Crop rotation, 11, 26, 97, 161, 283, 310, 312, 336
Cultivation, 8, 11, 12, 17, 18, 20, 22–24, 26, 27, 29–31, 33, 39–46, 48, 53, 55, 57, 59, 62–64, 67–69, 75, 90, 93, 94, 101, 108, 109, 113, 121, 122, 125, 126, 144, 159, 171, 197, 206, 216, 227, 228, 233, 236, 237, 276, 280, 282, 297, 298, 308, 310, 311, 313, 322, 325–328, 336, 337, 340

D

Data mining, 205, 206, 217, 233, 236, 243, 244, 246, 259, 265, 339, 341
Data model, 77, 80–82, 85, 89, 90, 101, 107, 108, 110, 119, 121–123, 126, 128, 136–138, 143, 144, 146, 206, 238, 250, 336, 337, 340
Decision support, 305, 307, 308, 337
Decision tree, 167, 168, 236, 237, 242, 244, 246, 260
Degradation, 13, 18, 19, 24, 31, 65, 133, 160, 174, 175, 268, 311, 324
Dendrogram, 271, 272, 299

T. Svoray, *A Geoinformatics Approach to Water Erosion*,
https://doi.org/10.1007/978-3-030-91536-0

Deposition, 5, 10, 23, 25, 30, 32, 40, 56, 59, 60, 67, 75, 76, 78–83, 85–90, 93, 97, 100–103, 110, 111, 113, 122, 144, 151, 190–192, 195, 196, 198, 205, 223, 227, 228, 231, 232, 258, 267, 336, 337, 339, 340
Descriptive statistics, 125, 196, 242, 277
Detachment, 10, 11, 32, 41, 43, 48, 52, 53, 56, 75, 76, 86–89, 94–98, 100, 101, 107, 108, 111, 123, 258, 311
Digital Elevation Model (DEM), 79–82, 84, 87–89, 98–100, 110–116, 118–120, 123, 125, 142–147, 151, 153, 177–181, 183–185, 187, 190–195, 197, 198, 206, 207, 209–212, 255, 256, 259, 289, 336, 340
Dragons of Inaction, 63, 64, 66, 340
Drainage area, 47, 77
Drainage density, 117, 118, 240, 309, 314
Drainage structure, 12, 33, 110, 146, 153, 177, 178, 183, 188, 197, 199, 280, 339
Drone, 151–153, 166, 169, 178, 190–195, 197, 198, 207, 340, 341
Drought, 159, 196, 309, 314, 322, 324, 325, 327

E
Ecosystem Services (ES), 12, 39, 40, 57, 265–269, 276, 297, 298, 305, 308, 310–313, 320–325, 327, 328, 330, 335–337, 339, 342
Eight-flowdirection matrix, 177
ELECTRE, 309, 312, 318, 320
Empirical modeling, 76, 94, 205
Environmental factors, 3, 7, 26, 39, 48, 67, 68, 107, 144, 278, 314, 325, 331
Erosion, 1, 3, 5, 7, 9–17, 23–34, 39–48, 50, 51, 53–64, 67–69, 75–90, 93, 94, 97–103, 107, 110, 111, 113, 115, 116, 120–123, 125, 127, 130, 137, 144, 151–153, 157–163, 171, 173, 177, 190, 191, 194–198, 200, 205–207, 209–212, 216–218, 221–223, 225–236, 238–244, 246–250, 253, 255, 257–259, 265, 266, 276, 297, 298, 305–314, 320, 321, 324–328, 330, 335–342
Erosion processes, 1, 4, 9, 10, 12, 13, 15–17, 23, 25–27, 30–33, 39–41, 48, 51, 52, 56, 57, 59, 61, 67, 75–77, 79–81, 88, 93–95, 97, 100, 101, 107, 108, 112, 113, 116–119, 121, 127, 135, 144, 153, 159, 173, 175, 197, 215, 227, 228, 230, 247, 249, 250, 253, 255, 257–260, 265, 268, 297, 311, 312, 326, 337, 339–342
Error assessment, 188–190
Error matrices, 166
Expert knowledge, 7, 206, 217, 222, 227, 232, 252, 255, 265, 322, 323, 325, 331, 339
Expert system, 41, 217, 225, 226, 336

F
Farmer, 1, 15, 16, 18–21, 23–27, 31, 34, 39, 41–44, 46, 48, 53, 55, 56, 59, 61–70, 87, 100, 102, 103, 109, 121, 122, 144, 214, 225, 226, 232, 234, 265, 266, 268, 269, 275, 276, 298, 305, 308, 310, 312, 328, 340–342
Flow accumulation matrix, 113, 141, 285
Flowcharts, 114
Flow direction matrix, 110, 146
Fuzzy logic, 7, 206, 217, 246–250, 255–260, 339

G
Geographic Information Systems (GIS), 45, 46, 48, 86, 87, 89, 90, 93, 98, 103, 113, 119, 123, 129, 139, 146, 171, 181, 183, 196, 336, 339, 340
Geographic Information Systems Cellular Automaton Multicriteria Evaluation (GISCAME), 305, 306, 312, 313, 320–331, 339
Geology, 141
Geomorphometry, 138
Geostatistics, 108, 127, 131, 132, 135, 141, 277, 278, 295, 336, 342
Graphic model of awareness, 338
Gullies, 15, 25, 46, 47, 49, 51, 54, 55, 59, 67–69, 75, 81–86, 93, 100, 115, 207, 209–211, 214, 217, 233, 238, 240, 242, 259, 260, 336

H
Harod catchment, 13, 14, 43–45, 47, 53, 90, 91, 118–120, 122–125, 129, 130, 132–137, 139, 143, 159–162, 174–176, 183, 189, 190, 207, 211, 213, 214, 216, 222–224, 226, 228, 229, 247, 253, 255–259, 265, 270, 272, 273, 276, 277, 279–285, 287, 289, 291, 295, 313, 314, 324, 328
Hillslope catena, 32, 77, 103, 107–109, 111–113, 115, 119, 144, 257, 258, 267, 282
Hillslope curvature, 42, 110, 111, 116, 146
Hillslope gradient, 5, 11, 12, 30, 41, 42, 45–47, 50, 53–56, 67–69, 75–78, 80, 83, 84, 87, 89, 91, 92, 95, 98, 101, 102, 110–112, 115, 116, 118, 119, 125, 129, 145, 146, 177, 183, 184, 187, 191, 193, 194, 206–212, 216, 218–220, 222, 223, 225–228, 236–238, 240, 241, 247–250, 252, 254, 255, 257, 259, 279, 289, 300, 308, 309, 315, 322, 325, 326, 339
Horton–Strahler, 117–119
Hydrology, 108, 110, 119, 121, 177, 190, 254, 255, 309, 314
Hyperspectral, 155, 157, 160, 168, 173–176, 178, 191

I
Ideal point, 222, 223, 227, 232
Indicator kriging, 141
Intrinsic soil factors, 4, 8, 9, 31, 126, 276, 341
Inverse Distance Weighted (IDW), 140–143, 145–147, 283–285, 287, 293, 295–297, 339
Iterative Self Organizing Data Analysis Technique Algorithm (ISODATA), 163, 164

J
Johari Window, 338

K
Kappa coefficient, 166, 168, 198, 199, 211, 225, 227, 259
K-means, 163–166, 168–170
Knowledge Discovery in Databases (KDD), 233
Known knowns, 335, 337, 338
Kriging, 13, 107, 140–145, 147, 277, 283–285, 293, 295–298, 339
Kriging with an external drift, 141–143

L
Landsat, 9, 126, 129, 152, 160–162, 165, 174, 175, 183, 197, 198, 255, 256
Landscape Evolution Models (LEM), 79–81, 86, 101, 102, 177
Landslide, 57–61, 80, 234, 312
Land use, 19, 22–24, 26, 30, 75, 86, 89, 90, 93, 94, 98, 101, 121, 122, 126, 151, 152, 163, 171, 191, 198, 225, 226, 228, 231, 230–232, 247, 250, 308, 309, 312–314, 318, 320–323, 325, 328, 330, 331, 339
Light Detection and Ranging (LiDAR), 87, 184, 192, 340
Lithology, 309

M
Machine learning, 7, 153, 163, 167, 197, 337, 341
Membership functions, 221, 247, 249, 252, 253, 255, 259, 260, 339
Meteorological radar, 6, 50, 82, 109, 123–125, 146, 206, 225, 226, 325
Moran's I, 107, 131, 135, 144, 145, 265, 279, 282, 283, 298, 300, 341
Morgan-Morgan-Finney (MMF), 28, 75, 76, 93–102, 103
Multicriteria Analysis (MCA), 206, 217, 218, 232
Multicriteria Decision-Making (MCDM), 305–309, 311–318, 320, 323, 328–330, 339

N
Nitrogen, 4, 57, 59, 60, 63, 121, 126, 158, 159, 163, 236, 270, 282, 316
Normalized Difference Vegetation Index (NDVI), 112, 146, 153–155, 157, 159, 163, 197, 199, 251–257, 338
Nugget–sill ratio, 131, 133, 134, 279, 282, 283, 298, 300, 341

O
Object-based classification, 151, 171, 172, 197–199, 340
Off-site consequences, 39, 56–61, 67–69, 87, 90, 309, 311, 335, 336, 340, 341
On-site consequences, 39, 56–59, 61, 67, 68, 90, 309, 311, 335, 336, 340, 341
Optimization, 308–311, 313, 329
Ordinary Kriging, 123, 141, 143, 265, 284, 339
Organic matter, 2, 3, 8, 10, 11, 13, 16, 25, 31–33, 60, 68, 88, 121, 127, 135, 152, 156, 157, 159, 160, 171, 173, 236, 265, 268–270, 272, 278, 280, 282, 306–308, 310, 315, 316, 336
Overland flow, 11, 25, 29, 46–50, 53, 56, 69, 75, 77, 79, 80, 86–88, 95–98, 100–102, 108, 111, 117, 121, 171, 209, 217, 228–230, 241, 251–254, 257, 276, 311, 339

P
Pairwise comparison, 218–223, 225, 227–230, 232, 247, 250, 259, 260, 274, 300, 311, 317, 339
Parent material, 3–6, 8, 14, 26, 33, 39, 41, 51, 56, 67, 68, 77–80, 82, 83, 108, 111, 127, 129, 152, 160, 173, 175
Permutation, 5, 7, 8, 131, 136, 280, 282
Physically based models, 76, 246, 248, 250, 255
Piping, 39, 51, 55, 56, 58, 59, 67, 68, 197, 236, 241, 340

Q
QGIS, 119, 214

R
Rainfall, 4–8, 10–14, 21, 26, 28–30, 32–34, 39–41, 43, 44, 47, 48, 50, 52, 53, 56, 58, 61, 67, 68, 75, 76, 79, 80, 82–89, 91, 93–103, 108, 116, 118, 121–126, 138, 146, 151, 177, 185, 186, 188, 195, 198, 206, 208, 210, 215, 221, 222, 225–233, 241, 248, 250–252, 255–257, 259, 260, 284, 289, 300, 308, 309, 315, 316, 322, 325–328, 336
Random forest, 163, 167–171, 234, 237
Range, 13, 24, 25, 28, 44, 55, 77, 79, 80, 85, 108, 114, 129, 132–136, 140, 144, 152–157, 161, 166, 170, 191, 195, 197, 198, 206, 214, 219, 222, 228, 234–236, 238, 240, 242, 247, 255, 266, 269, 272–274, 276, 278–283, 297, 308, 309, 318, 319, 321–323, 329, 336, 341
Ranking, 16, 167, 218, 219, 221, 250, 259, 275, 305, 307, 309, 312, 314–318, 320, 321, 328–330
Raster, 9, 80–82, 90, 93, 98, 107, 109, 110, 119, 123, 125, 128, 137, 140, 141, 143, 146, 153, 181, 182, 184, 188, 206, 210–212, 214, 215, 223, 225, 226, 228, 236, 238, 239, 247, 250, 252, 255, 260, 340
Rating, 67, 68, 166, 218, 219, 259, 274, 275
Receiving Operator Curve (ROC), 211, 233, 235–237, 242, 246
Regolith, 77–80, 100
Regression, 33, 94, 95, 98, 134, 158, 159, 174, 234, 236, 238–240, 242–246, 260
Regression kriging, 141, 284
Residue, 2, 42, 43, 86, 87, 158, 159, 280, 328
Rill, 47, 50, 51, 53–55, 59, 67, 75, 87–89, 93, 100–102, 209

Roads, 14, 15, 46–49, 55, 57, 59, 69, 93, 113, 121, 161, 168, 183, 185, 186, 190, 208, 210, 228, 241, 311, 321, 322, 325, 336, 339, 340
Runoff, 10, 11, 25, 26, 40, 41, 44, 45, 47, 48, 50–54, 56, 58–60, 67, 69, 76, 77, 80, 82, 83, 88–90, 93–95, 97–103, 112, 113, 116, 117, 121–123, 127, 151, 159, 171, 177–179, 184–188, 190, 199, 210, 228, 232, 250–255, 257, 268, 276, 280, 309, 311, 312, 315, 316, 325, 339, 341

S
SAGA, 113
Sample size, 127–130, 144, 176, 283
Sampling, 22, 97, 107, 127–130, 132, 138, 141, 142, 144, 145, 147, 155, 199, 207, 214, 226, 234, 257, 278, 279, 284, 328, 339
SCORPAN, 7
Sheet erosion, 15, 39, 53, 54, 57, 67, 100, 160, 340
Simple kriging, 141, 142, 284
Soil degradation, 1, 9, 12, 13, 16–21, 24, 30–34, 39, 61–64, 67, 116, 123, 151, 159, 160, 200, 205, 223, 225–227, 229, 230, 232, 234, 246, 260, 297, 305, 315, 324, 335–337, 339, 341
Soil erosion, 3, 9, 10, 12–27, 29–34, 39, 40, 49, 57, 61, 63, 66, 67, 76, 77, 79–81, 86, 87, 90, 93, 94, 96–102, 108, 119, 121, 123, 151, 152, 161, 196, 211, 226, 228, 238, 240, 249, 250, 254, 255, 258, 259, 265–267, 321, 322, 324, 328, 342
Soil Evolution Model (SEM), 75, 80, 81, 85–87, 98, 102, 267, 323, 339
Soil fertility, 25, 265, 266, 314, 338
Soil health, 1, 3, 12, 13, 26, 31–33, 40, 42, 57, 58, 61, 63, 75, 132, 135–137, 157, 159, 161, 174, 265–270, 272–280, 282–284, 289, 293, 295, 297–299, 305, 308, 309, 311, 313, 315, 316, 328, 335–337, 339–342
Soil horizons, 1, 2, 42, 108
Soil-land inference model, 110
Soil loss, 1, 9–15, 18, 19, 23–34, 40, 43, 48, 51, 52, 56–59, 68, 75, 76, 85–94, 97–103, 107, 110, 116, 119, 146, 151, 152, 160, 171, 190–192, 195, 196, 205, 207, 211, 218, 221–223, 227–232, 234, 236, 247, 251, 252, 257–260, 266, 267, 305, 310–312, 314, 315, 321, 324, 325, 335–337, 339, 340, 342
Soil moisture, 8, 9, 82–84, 96, 98, 123, 126, 137, 141, 142, 151, 152, 160, 197, 198, 233, 250–252, 280
Soil processes, 3, 4, 7, 9, 32, 33, 86, 87, 120, 131, 157, 247, 268, 276, 336
Soil properties, 1, 4–8, 12, 13, 21–23, 33, 41, 48, 87, 88, 93, 94, 107–112, 116, 120–122, 126–132, 134–137, 140, 141, 143–147, 151, 154, 157, 159–162, 197, 216, 218, 234, 251, 252, 257, 265, 267–274, 276–285, 287, 289, 290, 292, 293, 295–300, 322, 336, 339–341
Soil quality, 9, 12, 13, 23, 31–33, 59, 68, 107, 197, 266, 268, 275, 280
Soil security, 266
Spatial autocorrelation, 107, 128, 131, 132, 135, 136, 145, 146, 265, 268, 276–278, 280–283, 289, 339, 341
Spatial Decision Support System (SDSS), 305–314, 316, 318, 320–322, 324, 328–330, 339
Species-distribution theory, 8
Spectral indices, 151, 153, 154, 157, 159–161, 197, 198
Spectral Mixture Modeling (SMN), 172–177
Splash erosion, 52, 53
Stratified random, 129, 130, 144, 145, 226, 265, 279, 284, 339
Sub-pixel analysis, 144
Support Vector Machine (SVM), 236, 240–246, 260
System complexity approach, 7

T
Texture, 12, 59, 75, 88, 98, 101, 111, 127, 137, 160, 161, 163, 197, 216, 218–220, 250, 270, 272, 273, 278
Thiessen polygon, 128, 129, 137–140, 145, 147, 283, 340
Tillage, 11, 19, 25–28, 41–46, 55, 63, 64, 67, 69, 88, 121, 122, 126, 159, 163, 190, 217, 230–232, 237, 276, 280, 283, 306, 307, 310, 314, 316, 318, 322, 323, 325, 326, 328–330
Tillage direction, 39, 44–47, 67, 68, 93, 208, 209, 225, 226, 228, 230–232, 241, 325, 336
Time series analysis, 131, 153, 192
Topaz, 89
TOPMODEL, 82, 83
Topographic threshold, 47, 50, 54, 191, 194, 205–207, 209, 259, 339
Topographic wetness index, 115, 116, 119, 193, 252, 284
Topography, 8, 11, 26, 39, 41, 42, 44, 50, 67, 68, 75, 77–79, 98, 100, 101, 108–112, 115–117, 120, 121, 123, 125, 126, 141, 143, 145, 153, 177–179, 209–211, 217, 223, 228, 229, 242, 252, 278, 280, 289, 296, 300
Triangulated Irregular Network (TIN), 81, 137–139, 145, 147

U
Uncertainty, 63, 309
Universal kriging, 141–143, 265, 284, 285
Universal Soil Loss Equation (USLE), 10–12, 33, 98, 121, 152, 161, 252, 310, 321, 322, 325
Unmanned Aerial Vehicle (UAV), 191, 192, 241
Unmixing model, 174
Unpaved roads, 39, 46, 47, 49, 67–69, 100, 206, 209, 336
Unsupervised classification, 163–165, 199

V
Variogram, 127, 131–134, 137, 141, 278, 279, 281, 282, 284, 285, 295, 297
Variogram envelope, 107, 131, 135–137, 144, 145, 265, 279, 280, 282, 283, 341
Vegetation, 2, 5, 7, 13, 17, 18, 21, 23–26, 29, 39–41, 44, 46, 51–53, 68, 77, 78, 87, 93, 96, 98, 100, 108,

109, 111–113, 118, 119, 126, 146, 151–159, 161, 162, 166–168, 170–178, 185–187, 190, 191, 197–199, 208, 210, 223, 226, 228, 230, 233, 234, 241, 251, 253, 300, 325, 336, 338, 339
Vegetation indices, 153–155, 157–160, 167, 191, 197, 253, 338
VenμS, 160, 171, 172, 178, 179, 181, 183, 184

W
Water Erosion Prediction Project (WEPP), 75, 76, 86–90, 93, 101–103
Water holding capacity, 268
Weighted Linear Combination (WLC), 186, 208, 221–223, 225, 227–230, 232, 238, 259, 260, 311, 315
Weighted Summation (WS), 221, 315, 316, 318, 320, 329, 339
Weights, 42, 53, 89, 110, 113, 134, 140–143, 147, 182, 185–188, 198, 199, 217–222, 225, 227–230, 232, 236, 238–240, 242, 247, 250, 252, 253, 255, 257–260, 274, 275, 283, 284, 293, 298–300, 306, 307, 311, 313–319, 328–330, 336
Wetness index, 119, 191, 194, 254, 255

Y
Yehezkel catchment, 5, 6, 90, 92, 115–117, 171, 172, 176, 199, 209, 211–213, 230, 241–244

Z
Zonal statistics, 102, 108, 123, 145, 159, 161, 191, 206, 214, 215, 326, 340